A HISTORY OF INVENTION

BENZ·SENDLING MOTORPFLÜGE GMBH BERLIN NW7

A History of INVENTION

FROM STONE AXES TO SILICON CHIPS

TREVOR I. WILLIAMS

UPDATED & REVISED BY

WILLIAM E. SCHAAF, JR.

WITH ARIANNE E. BURNETTE

Little, Brown & Company

LONDON · BOSTON · NEW YORK

A LITTLE, BROWN BOOK

First published in Great Britain in 1987 by Macdonald & Co
This updated and revised edition published in 1999 by
Little, Brown and Company (UK)

A CIP catalogue for this book is available from the British Library

Design by Andrew Barron & Collis Clements Associates

Printed and bound in Italy

Little, Brown and Company (UK)
Brettenham House
Lancaster Place
London WC2E 7EN

Contents

Picture Acknowledgments

The publishers would like to thank the following for their kind permission to reproduce the illustrations in this book:

Aberdeen City Libraries;
Aerofilms;
Aldus Archive, London;
Arcaid/Richard Einzig;
Archiv für Junst u. Geschichte;
Archivio IGDA, Milano;
IGDA/Bevilaqua; IGDA/Caprotti;
IGDA/Ciccione; IGDA/Costa;
IGDA/Dagli Orti; IGDA/Di Gregorio;
IGDA.Nimatallah; IGDA/Nelson Museum; IGDA/Novaresio;
IGDA/Perissinotto;
IGDA/Pubbliaerfoto; IGDA/Rizzi;
IGDA/Seemuller; IGDA/Wergani;
Arts AntiquaArt Ancien S.A.;
Aspect/Peter Carmichael;
BASF A.G. Ludwigshafen;
Ken Baynes, London;
BBC Hulton Picture Library;
Bernard Beaujard;
Bildarchiv Preussischer Kulturbesitz, Berlin;
Bilderdienst Süddeutscher Verlag, Munich;
Brenard Press;
Bridgeman Art Library;
Britain on View;
British Aerospace, Weybridge;
British Museum, London;
British Telecom, London;
Buffalo & Erie County Historical Society, Buffalo;
Bulloz;
CEGB, London;
Christie's Colour Library, London;
Ciba/Geigy, Fribourg;
Bruce Coleman/T O'Keefe;
Bruce Coleman/N Tomalin;
Bruce Coleman/N de Vore;
Bruce Coleman/J T W;
Daimler Benz, Stuttgart;
Darlington Museum;
Design Council, London; Design Council/Sylvia Katz; Design Council/R Newport; Design Council/BIP;
Deutsches Museum, Munich;
James Dixon & Sons, Sheffield;
Drake Well Museum Titusville;
ECPA;
Esto/Peter Aaron;
ET Archive;
Explorer/Boutin; Explorer/Delu;
Barry Finch/Courtaulds plc, London;
Werner Forman Archive;
Paul Forrester, London;
John Frost Historical Newspaper Service;
Glamorgan Archive Service, Cardiff;
Glaxo-Holdings, London;
Gutenberg-Museum, Mainz;
Sonia Haliday Photographs;
Robert Harding;
Harvard Business School/Baker Library, Boston;
J G Heatley, Oxford;
Helston Folk Museum;
Hermitage Museum, Leningrad;
Hoechst A G, Frankfurt;
Michael Holford; Michael Holford/British Museum, London;
Michael Holford/National Maritime Museum, London; Michael Holford/Science Museum, London;
Holt Studios;
Angelo Hornak; Angelo Hornak/National Maritime Museum;
Alan Hutchinson;
ICI, Billington;
Image Bank;
Impact/Martin Black;
Imperial War Museum, London;
Johnson Matthey, London;
David King, London;
Library of Congress, Washington DC.;
Light Fantastic, London;
Los Alamos National Laboratory;
Lufthansa Cologne;
MAN Archives, Augsburg;
MacQuitty International Photographic Collection;
Mansell Collection;
Marconi Co. Ltd., Chelmsford;
MARS; MARS/Aerospatiale;
McDonnell Douglas, St. Louis;
Milk Marketing Board of England & Wales, Thames Ditton;
Marion & Tony Morrison;
Münchner Stadtmuseum, Munich;
Museo Civico, Locarno;
NASA, Washington;
Nationalmuseet, Copenhagen;
National Library of Scotland, Edinburgh;
National Maritime Museum, London;
National Trust/C Waite, London;
Naval Photographic Centre, Washington DC;
Network/Barry Lewis;
Network/Mike Goldwater;
Peter Newark's Western Americana;
Nobel Foundation, Stockholm;
Nobel's Explosives Co., Stevenston;
Novosti;
Richard Nowitz, Jerusalem;
Oxford Scientific;
Photri;
Picturepoint;
Polaroid Corp., Cambridge, Mass;
Popperfoto;
Public Record Office, Kew;
Rare Breeds Survival Trust, Kenilworth;
Betty Rawlings;
Peter Roberts;
Roger-Viollet;
Ann Ronan;
Science Museum, London; Science Museum, London/T William;
Science Photo Library/Dr J Burgess;
Science Photo Library/Tim Davis;
Science Photo Library/Hank Martin;
Science Photo Library/Andrew McClenaghan; Science Photo Library/Hank Morgan; Science Photo Library/Dr B Richards; Science Photo Library/Alex Tsiaras;
Scottish Tourist Board, Edinburgh;
Seaphot/Planet Earth/Tim Francis;
Ronald Sheridan;
Siemens AG, Muncih;
The Slide Centre;
Sony, Tokyo;
Sperry Ltd., London;
Sutcliffe Gallery, Whitby;
Wim Swann Archive, London;
Telegraph Colour Library;
Union Pacific, New York;
United Technologies Sikorsky Aircraft, Stratford, Conn.;
Universitetets Oldsaksamlingen, Oslo;
VDN Picture Library;
Waterways Museum, Stoke Bruerne;
John Watney;
Wedgwood, Barlaston;
Henry Francis Du Pont Winterthur Museum, Winterthur, Del.

The publishers would also like to thank the following for their kind permission to reproduce additional illustrations in the revised version of this book:

Allsport, 274;
John Frost Historic Newspapers, 316;
Oxford Scientific, 273;
Science Photo Library, 280, 283, 285, 291, 299, 302, 306, 309, 311, 312, 320, 327, 337, 338, 341, 345;
Telegraph Colour Library, 276, 295, 325, 327, 335.

Preface

Trevor Williams wrote the original text with great skill and in a manner which stands well against the test of time. Yet time moves on, and in just over a decade since this book was first published, vast changes have swept the world, defining alternative future paths and altering our perception of the past. Nations have risen and fallen, global alliances have been forged, and new systems of communication and construction have arisen which have impacted invention in every facet of modern life.

One of the first questions to be asked of any revised edition is what changes have occurred. I therefore felt it salient to describe here my role as I saw it. In general, the intentions of this revised edition were to revise text referring to obsolete nations or systems, to edit dates and descriptions which new evidence refutes and to add new material on a few relevant inventions. In doing so, however, I endeavoured not to obfuscate or distort Mr. Williams' voice or personal views, even where they differed from my own. To maintain the continuity of his storyline, I attempted to match his writing style so that new material could be interwoven into the existing structure rather than reorganising the content, or simply adding new chapters.

Some tasks were clear, such as eliminating references to the Soviet Union and 'current' Cold War tactics. Others required slight modification to passages in order to reframe the current state of affairs – some military and computing technologies, in particular, have taken radically different strategic paths since 1987.

A few passages in the opening sections of the book were revised based on new research which re-dates existing finds. New archaeological discoveries continually change our beliefs on the dates and sequences of historical events. Even with a lack of new discoveries, modern reconstructions, built using original tools and methods, enable researchers to understand better what was possible with the original products. Many currently held beliefs on the intentions and capabilities of Viking shipping and early computers have been shaped by such reconstructive research projects.

Finally, I have added sections for some technologies which did not exist in 1987 or which had not developed prominence at that time – the explosion of the World Wide Web and genetic engineering markets, for instance. The greatest challenge facing me was not to write too much or on too many inventions. When a book covers the entirety of human tool-making, it is difficult not to place undue importance on our current age. I have not attempted to write on all new developments, but rather have selected a few which I feel are relevant and epitomise concurrent developments in many areas.

Encapsulated within every history is a personal viewpoint – an interpretation of events from a particular point in time. Although all writers strive for objectivity, we cannot declare, in all honesty, that personal viewpoints and beliefs do not come into play in historical accountings. The very act of editing itself – selecting what will be discussed with which emphases – defines a personal stance and places knowledge and beliefs in context. For the most part, this is Trevor Williams' history and I have done my best to ensure it remains so. It is only in the latter stages that I permitted some small evidence of my own beliefs to emerge in a discussion of today's 'state of the art' and possible future trajectories. This, I hope, was handled in a manner which is harmonious with Mr. Williams' text and sets the material in a context of which he would have approved.

William E. Schaaf, Jr.
with Arianne E. Burnette
August 1999

This remarkable Roman mosaic of the first century BC, from Palestrina (Praeneste), depicts many aspects of life in the Nile Delta – shipping, building, wildlife and agriculture, among others.

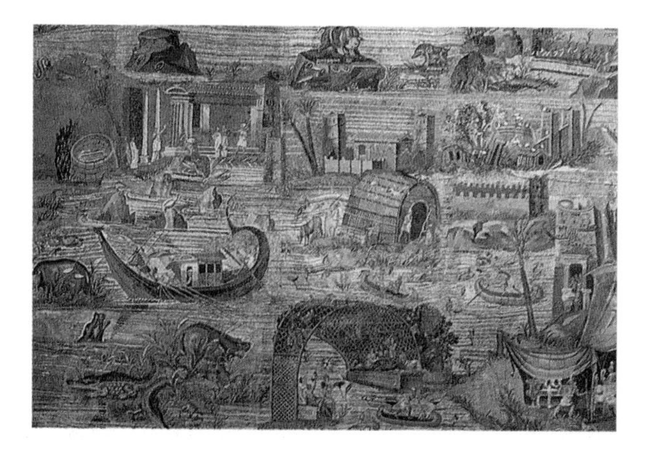

Introduction

Over the last thirty years I have devoted much time to the history of technology, and written and edited a number of books of a general nature, dealing with particular themes and historical periods. The fact that these have been well received has encouraged me to embark on this new work which is both more ambitious and simpler. More ambitious, in that it encompasses the whole history of material civilization from the working of flint and the dawn of agriculture to the launch of the Russian space-station MIR-1 and the commissioning of Superphénix, the first fast-breeder nuclear power station, at Creys Malville in France, in 1986. Simpler, in that footnotes, text references, and other trappings of scholarship have been abandoned in favour of an uninterrupted discursive text that resists the temptation to explore by-ways, interesting though these may be. In the modern vernacular – itself the reflection of recent technological advances – a strong reliance is placed on visual impact by far more extensive use of colour pictures and artwork than I have previously been accustomed to.

Civilization has many facets, but how man lives depends very much on what he can make. The object of this book is to arouse wider interest in the way in which technological factors have shaped – and continue to shape – human history; but if this is achieved, interest must be sustained. There is, therefore, an extensive bibliography at the end for those who wish to read further. For much of the period here reviewed inventors are anonymous and must always remain so: we cannot tell who made the first wheel or smelted the first copper. Indeed, as we shall see, it is likely that many basic inventions were made independently at different times and different places. In the words of a Russian proverb, the best of the new is often the long-forgotten past. As the story unfolds, however, the situation changes: increasingly, particular inventions can be attributed to particular individuals sensitive to social and economic opportunity. Even so, we shall note several instances within recent times where rival inventors have filed key patents within days, or even hours, of each other. The idea was, so to speak, in the air ready to be seized. This growing importance of the individual has been taken into account by the inclusion of a biographical dictionary. However, while the creative and imaginative genius of the individual is the mainspring of invention, we are perhaps to some extent now moving towards a new period of anonymity: much technological progress now stems from the work of teams, not publicly identified, within the laboratories of governments or large corporations. This reflects yet another significant development: the complexity of modern technology is now so great that often only very large institutions can afford to deploy the resources necessary to venture further into the unknown.

The traditional idea that necessity is the mother of invention – first postulated by classical writers such as Aristophanes – is at best no more than a half-truth. What I hope this book will show is not that technology serves only to meet perceived human needs but that society can profit by taking advantage of new technical advances Which factor is dominant in a particular instance is sometimes hard to distinguish, but of the reality of the powerful interreaction between technology and society there can be no doubt.

Trevor I. Williams, 1987

On land and sea the 19th century saw a revolution in transport due to steam propulsion. This picture shows Boulton and Watt's drawing for Robert Fulton's *Clermont* (1807). (Birmingham Public Libraries, Boulton and Watt Collection)

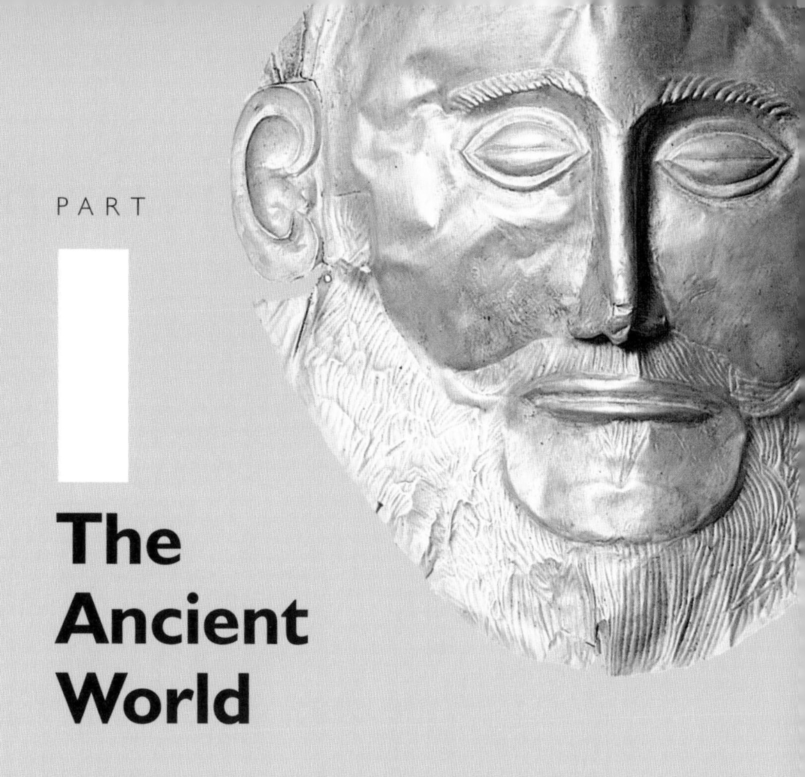

PART

I

The Ancient World

THE ANCIENT WORLD

This is a convenient but very general term, representing a span of some 4,000 years and stretching geographically from Western Europe to China – a broad belt within which a succession of great contiguous empires rose and fell.

THE ROMAN EMPIRE reached its greatest extent in the 3rd century AD. It then comprised and had assimilated the cultures of most of modern Europe except Germany, the north coast of Africa, the whole of Egypt, and much of the Middle East.

THE EGYPTIAN EMPIRE
THE ASSYRIAN EMPIRE

THE EGYPTIAN EMPIRE reached its peak geographically about 1450 BC, when it extended southward beyond Khartoum on the Nile and northward to the Euphrates.

THE ASSYRIAN EMPIRE grew from small beginnings on the middle Tigris. By the seventh century BC it ran from northern Egypt to what is now western Iran, but collapsed in 612 BC with the sack of its capital Nineveh by the Medes and Babylonians.

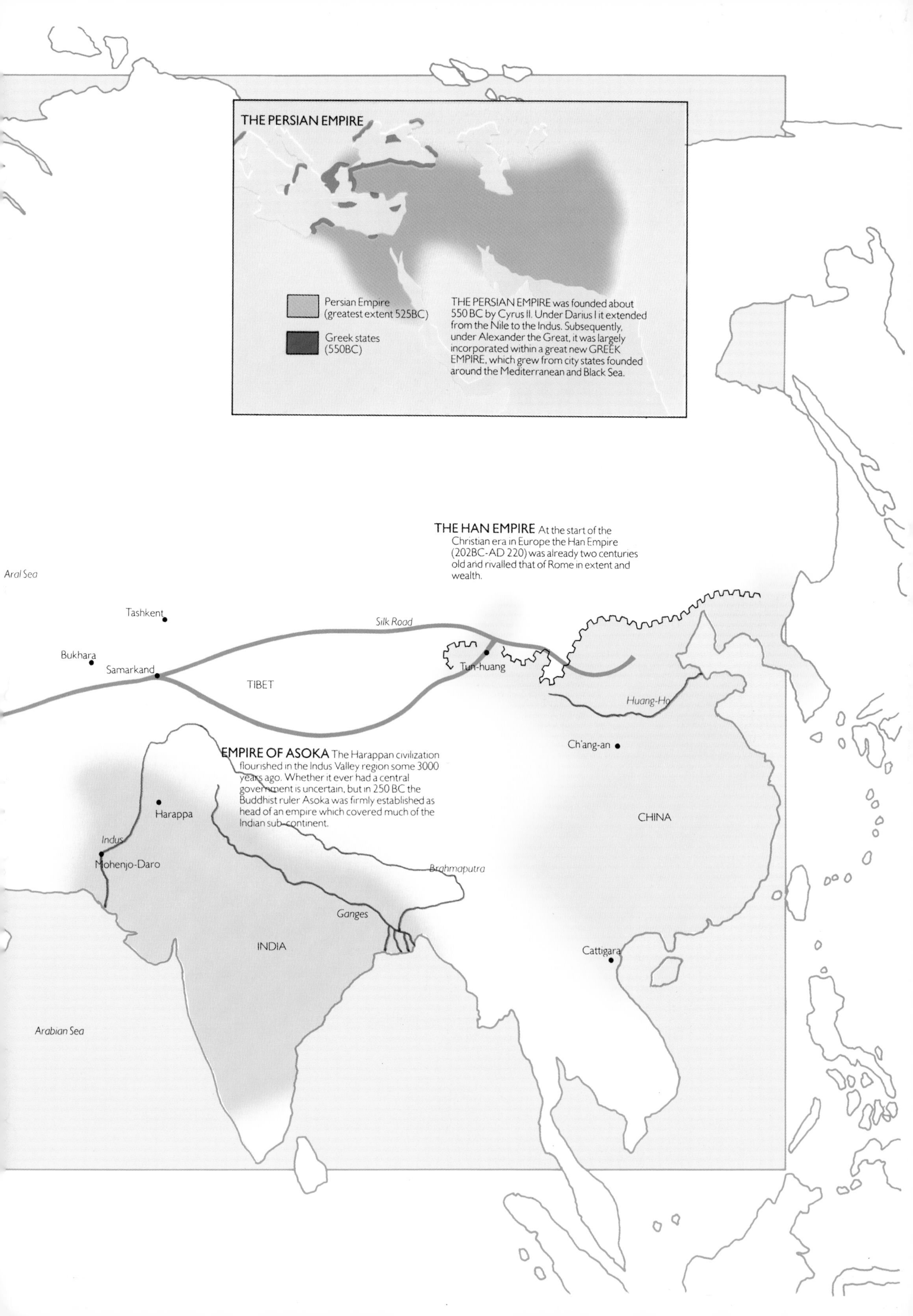
THE PERSIAN EMPIRE
Persian Empire (greatest extent 525BC)
Greek states (550BC)
THE PERSIAN EMPIRE was founded about 550 BC by Cyrus II. Under Darius I it extended from the Nile to the Indus. Subsequently, under Alexander the Great, it was largely incorporated within a great new GREEK EMPIRE, which grew from city states founded around the Mediterranean and Black Sea.
THE HAN EMPIRE At the start of the Christian era in Europe the Han Empire (202BC-AD 220) was already two centuries old and rivalled that of Rome in extent and wealth.
Aral Sea
Tashkent
Silk Road
Bukhara
Samarkand
TIBET
Tun-huang
Huang-Ho
Ch'ang-an
EMPIRE OF ASOKA The Harappan civilization flourished in the Indus Valley region some 3000 years ago. Whether it ever had a central government is uncertain, but in 250 BC the Buddhist ruler Asoka was firmly established as head of an empire which covered much of the Indian sub-continent.
Harappa
Indus
Mohenjo-Daro
CHINA
Brahmaputra
Ganges
INDIA
Cattigara
Arabian Sea

CHAPTER ONE

The Beginnings of Civilization

Civilization cannot be precisely defined, for not only is it an evolutionary process but over the ages it has manifested itself in very different forms. Thus the civilization of Rome, which itself varied very considerably within its huge extent, differed greatly from that of Polynesia, of which the West was not even aware until the great Pacific voyages in the 16th century, or that of the Incas of South America, extinguished by the Spaniards. There were manifold differences in religious belief, in social customs, in forms of government and in artistic creation. Yet there is one facet of civilization which is of fundamental importance to all. This is technology, which in the broadest sense can be taken to mean the application of knowledge to practical purposes.

Today, technology is virtually synonymous with applied science, but the basic technologies – such as agriculture, building, pottery and textiles – were originally empirical and handed down from one generation to another. Science, in the sense of a systematic investigation of the laws of the universe, is a comparatively recent phenomenon. Technology was fundamental in that it provided the wherewithal for organized societies, and organized societies made possible not only a division of labour – as between land workers, potters, sailors and the like – but also a milieu in which creative arts, not necessary for day-to-day existence, could flourish. Most of these arts were dependent on some kind of technological support: the sculptor required tools, the writer needed ink and papyrus (or later paper), the dramatist needed specially built theatres. Very often, the boundary between craft and art was indistinct. From a very early date the potter's vessels were often richly ornamented, demanding a practical skill in glazing; the weaver used dyes to colour his wool and manipulated his loom to produce intricate and colourful designs; the exquisite products of the jeweller needed knowledge of the working of metals and of specialist techniques such as enamelling and granulation.

This book sets out to recount the history of technology from the dawn of civilization to the present day when, for better or worse, it has become a dominant factor in the lives of the peoples of the Western world and, by its lack, in the Third World too. It is not by any means a simple history, unfolding steadily and logically. On the contrary, it is far from

	6000 BC	3500 BC	3000 BC	2500BC
AGRICULTURE	first agricultural stage; emmer wheat; flax; einkorn; millet; saddle quern; barley; flint sickles; domesticated animals	ploughs; early irrigation in Middle East	irrigation in Egypt; hemp grown in China	bronze sickles; seed drills in Babylonia
DOMESTIC LIFE	weaving; pottery; oil lamps	pottery fired in kilns	potter's wheel in Mesopotamia; glass made in Egypt; silk in China	animal and vegetable dy; writing c papyrus
TRANSPORT	dug-out boat in Holland	pack animal (ass); wheeled vehicles in Mesopotamia	reed boats on Nile; sail boats; wheeled vehicles in China; wooden ships in Mediterranean; Cheops funeral ship	skis used ir Scandinavi
BUILDING		kiln-fired bricks in Middle East	stone building in Egypt; step pyramid of Zoser (stone); arch appears in Egypt	masonry dam ove Wadi Gerrawi
POWER MACHINERY AND WARFARE	simple bows and arrows; spears		levers and ramps used to move heavy loads; adze and bow drill used in Egypt; chariots in Sumer	
METALS	alluvial gold	copper/bronze	soldering and welding; silver ornaments at Ur	cire-perdu casting; silver shee metal work
CITY CULTURE	Jericho; Catal Hüyük	Sumerian civilization writing; Egypt Nile culture calendar	early cuneiform writing; Egyptian hieroglyphs; Egyptian Old Kingdom; Babylonian calendar; hieratic script	Indus Valley civilization – Harappa an Mohenjo-D

uniform – while the first men were walking on the moon, others in the remotest parts of the world were still living virtually in the Stone Age. Nor do we have a simple process of new techniques being invented at a single centre, from which they spread out to distant lands. This is often true, but in addition we have many examples of independent invention and strong interactions where separate cultures meet. Turbulent though its history has been, China has enjoyed some 4000 years of unbroken civilization, and among its exports to the West were such important inventions as the magnetic compass, paper, printing and gunpowder. Later, particularly after the Jesuits were established there, Western technologies became absorbed into Chinese civilization. Tracing the routes by which both manufactured goods and new ideas were exchanged over the centuries is one of the more fascinating aspects of the history of technology.

Olduvai Gorge, in the Serengeti Plain, is one of the world's most famous archaeological sites, and gives its name to the earliest known human stone tool technology.

The earliest men

The proposition that technology is an essential ingredient of civilization at once poses the question of when man became a civilized being, distinguishable from his hominid ancestors. There can, of course, be no precise point of transition, but the criterion generally accepted is that man is distinguished from his primate ancestors by his ability to make tools.

Since Bishop James Ussher's confident assertion in the 17th century that the Creation occurred in 4004 BC, man has been granted a much greater antiquity. Remains discovered in the Olduvai Gorge in East Africa, and associated with stones crudely shaped as tools, are estimated to be 2.6 million years old. The earliest known human stone tool technology is therefore called Oldowan. Human remains from as far away as Beijing are between 350,000 and 500,000 years old and earlier *Homo erectus* fossils have been found in Java dating to between 600,000 and 1.8 million years ago. However, the study of such early remains falls largely within the province

2000 BC	1500 BC	1000 BC	500 BC	AD 1	AD 500
cotton grown in Indus Valley; shaduf	paddy field rice cultivation in China; maize grown in America	iron-tipped plough; chain of buckets	cotton in Middle East and China; Archimedean screw for irrigation	tea grown in China	
kaolin clays used by Chinese potters	Chinese li cooking pot	locks and keys in Egypt; looped knitting; candles in Asia Minor	Roman guild of dyers; salt trade established	silk imports to West; blown glass	horizontal kilns for pottery in Rome and China
horse-drawn vehicles; spoked wheels	reed boats in Peru; horses ridden	Greek penteconter; Phoenician bireme; trireme	Nile-Red Sea canal; magnetic compass world map	wheelbarrow in China; iron horseshoe; metal stirrup; padded saddle in China	clinker-built boats in Scandinavia; Chinese ships with transverse bulkheads
Stonehenge I	beehive tomb at Mycenae	Assyrian-Median wall	Parthenon; iron reinforcement of buildings	Great Wall of China; Stone arches in China; ogival arch	Hadrian's Wall; Pantheon, Rome; Pont Du Gard aqueduct
war horses used by Asian steppe peoples; siege towers	simple pulleys used by Assyrians; composite bows	incendiary mixtures; woodworking lathe	Sinjerli concentric fortress; cranes and complex pulleys; cross-bows	torsion artillery; chain mail; undershot waterwheels; Roman wood plane; crank in China	Barbegal watermill complex
copper sulphide ores smelted	iron working by Chalybes	use of iron spreads; two-piece mould casting	tin mining in Cornwall; Chinese bronze and copper-nickel alloys	Steel made by co-fusion; Chinese stack casting; Use of brass in China	Iron Pillar at Delhi
ptian um; earliest Chinese script; Phoenician cities dominate Mediterranean trade	alphabet; sundial clocks in Egypt	Carthage founded; Collapse of Hittite empire; demotic script	Greek city states; Rome founded	Latin shorthand; Julian calendar	paper manufacture in China; block printing with stone in China

ancient civilizations 10–11, 20–25 **gunpowder** 136–137, 183 **Inca Empire** 14, 90–91 **magnetic compass** 112–113 **paper and printing** 99–104

Rock paintings provide vivid pictures of the activities of early man. This example from the Tassili n'Ajjer Plateau, Algeria, dates from around 6000 BC. It depicts a hunting scene, including giraffes, now found only south of the Sahara.

of the anthropologist, and to find men with whom we ourselves could identify we cannot reasonably go back more than, say, 25,000 years, although the species *Homo sapiens sapiens* developed much earlier.

Even to go so far back brings us within a period of wide climatic variation marked by successive advances and retreats of the glaciers, which at times covered much of North America, Europe and Asia. The severity of the weather made survival possible only by technological responses – mastering fire, constructing shelters or improving natural ones, and making warm clothing. These climatic changes had another very important consequence, in that at its greatest extent the enlarged ice cap immobilized vast quantities of water. This changed the level of the oceans, dropping it to a level about 100 metres (328 feet) below that of today. Human – and animal – migration was favoured by the emergence of land bridges which had not existed before, opening the way, for example, across the Bering Strait from Asia to North America, and from Britain to Continental Europe. Conversely, as warmer periods intervened, the oceans rose and newly occupied areas became isolated.

Although these climatic changes were most severe in northern latitudes, the effects were global. They profoundly affected the regional fauna and flora, and man could survive only be adapting his techniques to new conditions. The Sahara is a case in point. Today it is a desert, but during the European Ice Age it was parkland, and there are abundant traces of early settlement: many rock paintings depict animals long since vanished. Moreover, it was a highroad for human migration, and not the barrier it is today.

Climatic considerations as well as the fossil evidence suggest Africa as the cradle of mankind – though the possibility of more than one centre cannot be ruled out. From there migration occurred to the accessible parts of the world where the climate was favourable. Fifty thousand years ago human colonies certainly existed in Europe and Asia, Japan and Australia. North America, however, seems to have been virtually unpopulated until about 20,000 years ago, when hunters from Mongolia crossed the land bridge into what is now Alaska and Canada in pursuit of rich herds of mammoth and reindeer. Later they spread southwards to Mexico, then along the Isthmus of Panama and down the west coast of South America. An Ice Age settlement far down in the south of Chile, at Monte Verde, dates back to about 11,000 BC. From such stock arose the short-lived civilizations of the Incas and the Aztecs.

For the moment, however, we must go back to consider what kinds of accumulated skills were bequeathed to modern man by his remote ancestors. If we accept the definition that man is a tool-making primate, we may look on a pointed stick hardened in the fire or a roughly chipped stone as evidence of technological achievement. But if we are to talk of civilized beings in the generally accepted sense, we must look for more than this. Apart from manual skills, we shall look to capacity for rational conceptual thought and ability to communicate through a spoken, and ultimately written, language. Only thus can the achievements of one generation be passed on to, and improved by, the next. We tend to regard information technology as a modern innovation, but it is in fact one of the oldest and most fundamental: a system of keeping public records and accounts was one of the first requirements of organized communities.

Man the hunter-gatherer

The Paleolithic or early Stone Age was one in which man lived off the land as a hunter and gatherer of food rather than as a deliberate producer. Success in this demanded the acquiring of a range of skills, many of them quite subtle and

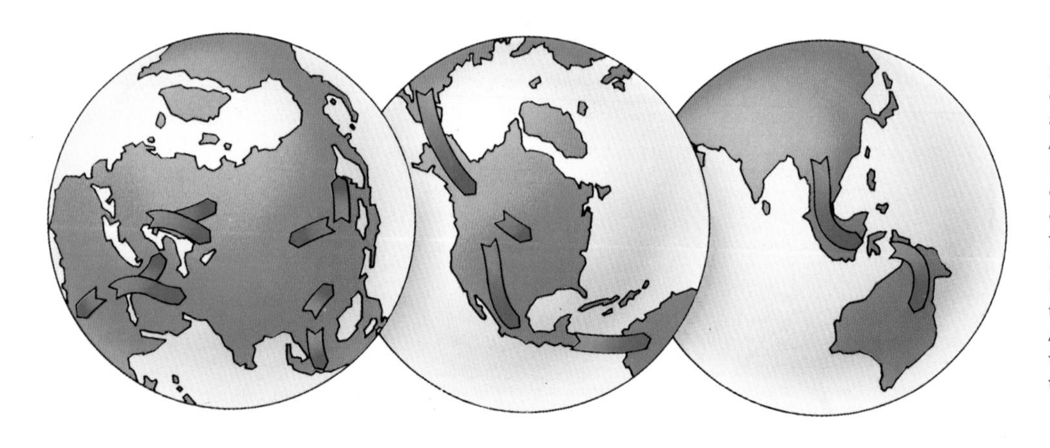

Sea levels fell during the last Ice Age, creating land bridges between Britain and Europe, Asia and North America, China and Indonesia, and New Guinea and Australia. As they developed means of surviving in colder climates, early humans began to spread out from Africa about 50,000 years ago. The new land bridges enabled them to colonize first Europe and Asia, and then Australia and finally America, reaching the tip of South America not less than 13,000 years ago.

complex: primitive man not only survived, but survived comfortably, in circumstances in which modern civilized man would quickly perish. It is not feasible to put the acquisition of these skills in any sort of chronological order, for practice varied from place to place and time to time depending on local circumstances. The remarkable skill of the flint knapper (or breaker), for example, would not be practised at all in areas where this kind of stone was not available. So we must be content with summarizing what may be called the basic technologies of Paleolithic man, by which we mean, essentially, man who had no knowledge of metals.

Initially, man did not make tools but depended on what came readily to hand – a sharp stone, a broken bone, a cleft stick. Then he began to shape wood, bone and stone into tools. Understandably, wooden artefacts rarely survive, though their form can often be deduced from the stone devices – such as axe-heads and spearheads – with which they must have been used. Initially, the pointed hand axe, still used by Australian aborigines, was the basic all-purpose tool, useful for cutting, scraping, crushing and splitting. But by a very early date there is clear evidence not only of the deliberate making of tools, but also of a surprising degree of specialization.

Many of the weapons devised for the essential fundamental activity of hunting remain in use to this day, suitably modified as new materials and techniques became available. The earliest spears were doubtless no more than wooden staves sharpened to a point and hardened in the fire. One of the earliest surviving pieces of woodwork is just such a spear, made of yew, recovered from water-logged soil – a good preservative of wood – in England. It is, perhaps, more than a quarter of a million years old. With such simple weapons large prey were stalked: a similar spear, also of yew, was found within the skeleton of an elephant in Germany. Later, heads made from flint or bone were used, and the power of the hunter's unaided arm was increased by the use of spear-throwers like those used in historical times by the Eskimos, the aborigines of Australia and the natives of New Guinea. The bow was the earliest device for suddenly releasing concentrated energy, and is clearly represented in rock paintings in North Africa that date from perhaps 20,000 BC or earlier: doubtless they were then already of great antiquity. For hunting small game, throwing sticks were used, and the sickle-shaped Australian returning boomerang is a highly specialized form of this. Boomerangs of different design evolved in Africa, India and elsewhere: a boomerang predating the earliest known Australian boomerangs was found in southern Poland in the Obła-Zowa Rock Cave and is thought to be 21,000 years old. To design such a device with the benefit of modern knowledge of aerodynamics would be a task for an expert; how it was evolved empirically by primitive man remains a mystery. The sling, too, was an early invention: shaped slingshots have been found at many Paleolithic sites.

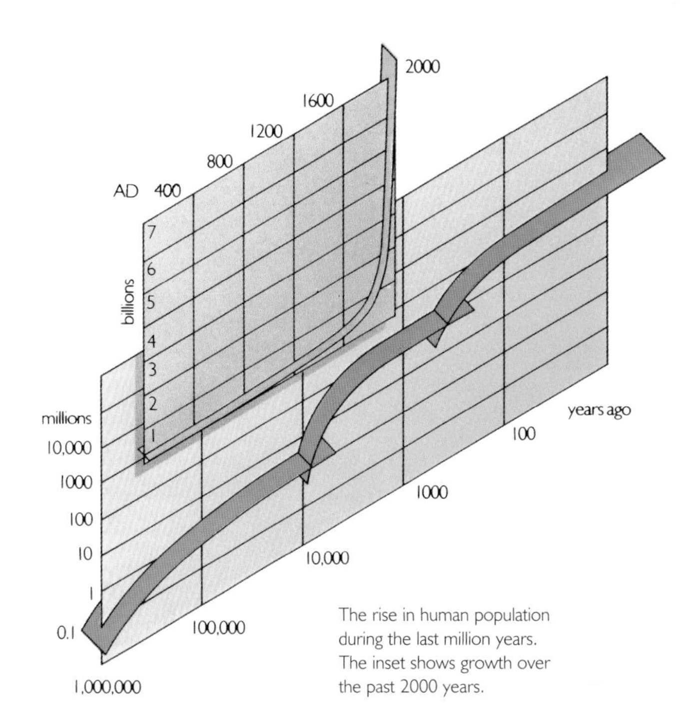

The rise in human population during the last million years. The inset shows growth over the past 2000 years.

For people living near the sea or rivers, fish formed an important source of food, and various kinds of fish-hooks have been recovered from Paleolithic sites. They range from the simple gorge hook – a double-pointed spike to the middle of which the line was attached – to true hooks made from flint, bone or shell.

Egyptian pictures dated earlier than 3000 BC show the use of the lasso and the bolas. The latter consisted of two weights joined by a short rope and is thrown with a whirling movement. It wraps itself round the legs of the quarry to bring it down. Both devices are probably very ancient weapons and may well have been invented independently in more than one place: lassos were also used by the Indians in pre-Columbian America, and the bolas was used in the South American region of Patagonia.

Stone was of unique importance because of its hardness and ability to provide a cutting edge, but wherever he was man made use of all local resources. While animals were hunted primarily for meat, they also provided bone, horn and ivory. These were worked up not only into useful articles but also other objects, such as figurines, which were either ornamental or had religious significance. Antlers served as picks and shoulder blades as shovels. Both are commonly found near the remains of ancient mines and quarries, such as those at Grime's Graves in England, which were opened to provide flint.

Animals also provided hides which could be scraped clean of fat and hair with flint scrapers and then made soft and supple, perhaps by chewing as the Eskimo has done in historical times. This was worked up into clothing, essential for protection in the cold northern winters.

Wood served many purposes. In northern climates, where it was plentiful, it was an essential fuel and source of light. Fire was also used in hunting to stampede animals into an ambush or towards pitfalls: excavations in Spain, for

antler tools 16 **flint tools** 16–17 **Grime's Graves** 17 **stone-working** 16–17

Early man made skilful use of all materials ready to hand. Bone and horn found many uses: shoulder blades, for example, made excellent shovels and antlers served as picks.

These harpoons made from antler horn came from a Mesolithic site (*circa* 6000 BC) at Stone Carr, Yorkshire, England.

example, have shown the skeletons of a herd of elephants surrounded by burnt brush. Wood, shaped by chopping and scraping, served many other practical purposes. It provided handles for tools, giving a mechanical advantage. And it must also have served in making the frames of shelters, like the leather-sheathed wigwams of American Indians, the portable tents of Lapps and the yurts of the Mongols. Although true boats cannot be identified much before the seventh millennium BC, simple rafts must have been used much earlier. Under even the most primitive conditions, simple containers and drinking vessels were desirable, and there were many natural sources of these before the advent of pottery. Sea shells, hollowed-out horns, gourds, coconut shells and the like could easily be utilized, as they still are in some parts of the world. People who could sew leather garments could no doubt sew leather bottles as well. And although there is no direct evidence, we may reasonably assume that simple basketry was known at an early date as osiers, reeds and other suitable materials were widely available. Indeed, as we get within sight of historical times, we may reasonably infer that man made the same sort of use of the resources lying close to hand as primitive tribes do today. In similar situations, men tend to react in the same way, if only because the number of options is limited.

The working of stone

Because of its perishable nature, our knowledge of how wood was used must be largely a matter of inference. With stone, however, we are on firmer ground as there is a vast number of surviving artefacts. Throughout the world flint was the favoured material because of its unusual physical properties. While a heavy blow will shatter it, reducing large boulders to manageable proportions, lighter blows produce what is called a conchoidal (shell-like) fracture. A flake of flint becomes detached, leaving a shallow depression. By controlling the force and direction of the blow, a skilled workman can regulate the size and shape of the flakes with remarkable precision.

Two basic techniques were practised. In one, pieces were chipped off a substantial piece of flint until the desired object, say a hand axe, was produced. The results are called core tools. Alternatively, the stone was worked to produce flakes of a size and shape suitable for use as scrapers or knife blades. These are called flake tools. Inevitably, the two techniques would sometimes overlap. The core-tool maker would naturally collect some of the resulting flakes to work up into flake tools. These flint-shaping techniques have not entirely vanished in modern society. At Brandon in Suffolk, flint knapping was practised from 1790 until recently to provide flints for flintlock firearms in Africa.

If properly delivered, quite a light blow will detach a flake. Generally, the flint to be shaped would be held in one hand and struck with a pebble held in the other, though experiment and observation of primitive people in recent times show that a piece of hard wood or bone could also be used. An alternative technique was to strike the stone to be shaped against the edge of a larger one used as an anvil. In pressure-flaking, a pointed piece of wood or bone was pressed hard against the flint edge to be shaped and small flakes became detached at the point of contact. In the so-called tortoise-core technique, a flint was first shaped until it resembled an inverted tortoise. If struck repeatedly around the edge, this yielded a succession of uniform sharp flakes which could be used without further treatment as skinning knives or for similar purposes.

Certain other siliceous stones, such as obsidian and chert, show the same useful property of conchoidal fracture and were used in regions where they occur. Glass and ceramics also fracture conchoidally, and in modern times the aborigines of Australia have readily adopted discarded bottles and insulators from telephone lines as valuable raw materials.

Other siliceous stones were put to use in regions where flintlike stone was not available, or when the readiness of flint to flake was a disadvantage, as in making heavy tools such as tree-felling axes. Fine-grained igneous rocks, such as basalt, were roughly trimmed with the aid of other stones, and then finally shaped by grinding. This might be done with the aid of a block of coarse rock such as sandstone. Alternatively, a mixture of sand and water would serve. This technique was also used by the Egyptians to prepare flat faces for building stones. It would give a satisfactory working finish, but some surviving axes have so high a polish that they had clearly been further burnished, perhaps with leather and a fine polishing powder such as clay or fuller's earth.

Questions of racial difference in physical skill and mental acuity are delicate and controversial. It is a matter of fact, however, that the attainments of Paleolithic man varied greatly, for surviving artefacts clearly demonstrate the emergence of distinct cultures. Thus in the Acheulian culture of northern France, which originated in Africa, the use of core tools predominated, while in the Clactonian-Mousterian culture, which may have arrived in Europe from Asia, flake tools were favoured. The Aurignacian culture, which included the Cro-Magnon, was characterized by the extensive use of bone and horn. To the expert, the precise way in which a flint is worked or an antler is carved can be as revealing as the brushwork on a painting to a modern art historian.

A

B

C

D

The most primitive stone tools were little more than chipped pebbles, like those recovered from Olduvai Gorge (A). These contrast with Neanderthal flake and core (B), scraper and flint (C), arrowheads (F), and the beautifully worked flint dagger from Denmark (E), which probably dates from the early Bronze Age. At an early stage, the making of flint tools became an organized industry and characteristic products are distributed along trade routes. One such centre was that at Grime's Graves, Norfolk, England (D). There, Neolithic miners sank shafts to a depth of 10 metres (33 feet) with radiating galleries. Similar flint mines have been discovered elsewhere in England and in many parts of Europe.

E

F

first boats 49–50 **Olduvai Gorge** 13 **pottery** 38–41 **rock engraving** 18 **stone figurines** 18

At an early stage in mankind's history, art already took many forms. The carving of figures was one. This early Stone Age ivory figurine (right), perhaps a fertility symbol, comes from France. The rock engraving (below) – a different art form from rock painting – comes from Algeria.

Decorative arts

As we approach historical times – say 25,000 BC – new cultural criteria emerge. Initially, man's tool-making efforts were strictly utilitarian, though pride in craftsmanship, evidenced by a finer finish than need demanded, soon became apparent. Then artistry was practised for its own sake, or at least for mystical rather than practical purposes. So-called Venus figurines, a few centimetres high and often carved in ivory, have been found on sites as far apart as Russian, Italy and France. A particularly interesting development was cave art. This vividly depicts many of the activities and tools – especially those associated with hunting – which would otherwise be a matter of inference. While the best known examples are from western Europe, as at Chauvet, Lascaux and Altamira, they have also been found – in a surprisingly similar style – thousands of kilometres away in the Urals. Many of the paintings are crude, but others are magnificently powerful compositions, showing acute observation and great artistic skill. While outlines are sometimes incised or drawn in black with soot or charcoal, a range of colours was used, showing a familiarity with natural pigments such as ochre and pyrolusite.

If we look back 10,000 years, to say 8000 BC, we see a rather paradoxical situation. On the one hand, over a great part of the globe, man had achieved a high degree of skill in working not only stone but also a wide range of other natural materials and he had developed a considerable artistic talent. Nevertheless, by modern standards, he was still a barbarian. His economy was strictly that of a food gatherer, hunting wild animals, fishing and enjoying such fruits of the earth as the seasons provided. Archaeological evidence suggests that population units were quite small, perhaps a couple of dozen families, though these might collaborate from time to time to hunt larger animals like the mastodon. This way of life had scarcely changed for tens of thousands of years, although it had gradually become much more sophisticated. The development of painting and decorative arts is proof that survival was no longer a full-time occupation. One commentator has described this as the original affluent society. Why, then, after such a leisurely progress to a comfortable and satisfying way of life did the tempo suddenly change? Why did vastly more happen in the next hundred centuries than had happened in the past 5000?

The reasons are by no means clear, but climatic change was undoubtedly one of the most important factors. Throughout man's previous history, the world had been subjected to a series of Ice Ages, but the last of these ended about 10,000 years ago. Thereafter the world gradually warmed up. While this favoured man, making life less harsh, it led to the decline and extinction of some of his principal sources of food, such as the mammoth. This was made worse by over-hunting by the growing human population. Under the pressure of events, man turned from merely keeping the environment at bay to seeking actively to control it.

The Agricultural Revolution

Thus occurred what has been called the Agricultural Revolution. In the new age (the Neolithic), stone remained of paramount importance for the making of tools, but agriculture – the controlled growing of crops and the herding of cattle – increasingly became the mainstay of life. Early agriculture and its attendant technologies form the subject of the next chapter and for the moment we need consider only the general nature of the transition and its immense implications for the future of mankind. The word transition is used advisedly, for the term 'revolution' suggests a rate of change which in reality never occurred. The first stage – which some peoples never embarked upon at all – must have been the

The great megalithic monuments of Europe are evidence of considerable skill in the dressing of stone, the handling of heavy loads and the organization of labour. This site at Carnac in north-west France contains more than 3000 massive stones arranged in parallel rows, some extending to more than 5 kilometres (3 miles); originally there were probably 10,000 stones. The many burial chambers of the period – often rich sources of artefacts – made similar use of heavy stones and the whole chamber was buried beneath a mound of earth.

supplementing of hunting and gathering by a little primitive agriculture. Indeed, food-gathering has never been wholly displaced. Although it now uses very sophisticated techniques, the world's fishing and whaling industry is still predominantly a food-gathering enterprise.

Bones from the middens of early settlements give a clear indication of the animals on which their occupants fed. While many were clearly fairly omnivorous, often there is evidence of a concentration of effort upon a single species. Thus the Magdalenians, who flourished in Europe around 10–15,000 BC, devoted their attention mainly to reindeer. In the Levant, at about the same period, the Natafians relied largely on gazelles. So far as it goes, this supports the theory that the domestication of animals arose from so close an association with wild herds that the animals became used to man's control.

At this early stage the first cities of the Eurasian world were still in the distant future, but we can see evidence of the beginnings of settled communities living in permanent dwellings grouped in villages. About 17,000 years ago, in Mezhirich in the Ukraine, mammoth bone houses provided a base camp for the hunter-gatherers, and sheltered an estimated 30 to 60 people for part of the year.

Modest though they appear, these developments laid the foundations of a new sort of society. The beginnings of a settled – rather than nomadic – way of life paved the way for the much more sophisticated city life of the early empires. We can detect, too, the beginning of a capitalist economy, for stores of grain and herds of cattle represented assets which could be used in a variety of enterprises.

The Neolithic Revolution was an important milestone in human history, but it marked a fork in the road rather than a staging-post on a continuous highway. As the new way of life spread slowly through Europe, it changed rather little. Village life in simple buildings and a mixed hunting/agrarian economy became the normal order of things. Society was still apparently organized on a tribal basis, though the great megalithic (large stone) monuments of the second millennium BC – such as those of Stonehenge and Carnac – demonstrate not only a surprising degree of engineering skill but also argue a higher degree of social organization than has been generally supposed. But except for the Mediterranean area, and even there much later, Europe could show nothing remotely comparable with the civilizations that flourished in the Middle East and Egypt from the fourth millennium BC.

The next few chapters are devoted to their achievements in such diverse fields as irrigation and metallurgy, pottery and transport, building and weaponry. As these span a period of some 4000 years, an area greater than that of Europe, and were subject to an ebb and flow of power still imperfectly understood, it is perhaps simplest to take a succession of historical snapshots rather than attempt a continuous view.

early agriculture 26–33 **rock painting** 14 **Stonehenge** 45, 57, 59 **tool-working** 15–17

The Ancient Empires

The Middle East and the Nile Valley

If we accept the invention of writing as marking the transition from barbarism to civilization, it is proper to focus on the Middle East as the cradle of human culture. Here the Sumerians, who perhaps came from central Asia via Persia, were firmly established by about 3500 BC. They lived in organized cities, they had knowledge of metal-working, their agriculture was sustained by irrigation systems, they had well-established trading relations, and – above all – they had a written language. If we look at the map again over a thousand years later, the Sumerians have been displaced by the Akkadians, who in turn were followed by the Amorites: both were Semitic-speaking people from the Syrian desert. Far to the south, agricultural communities were beginning to arise in the Nile Valley. There, the Old Kingdom of pyramid builders was established about 2700 BC and lasted for some 500 years. This too had written records, though they used hieroglyphic symbols instead of the Sumerian cuneiform. On the far eastern fringe of the area, the agrarian communities of the Indus Valley were establishing themselves.

If we move forward in time again, to the beginning of the second millennium BC, a new pattern appears. City states had given way to dynastic empires, with Assyria and Babylonia emerging as major powers. Sargon I, King of Assyria, and Hammurabi, King of Babylon – the latter famous for his comprehensive code of laws – were early examples of individuals who rose to power and could command and deploy great resources. Asia Minor had been invaded from the north by Hittites, who established a base from which they expanded south to control much of Syria and east to sack Babylon in 1595 BC. Their success there was short-lived, however, as they in turn were overthrown by Kassites from the Zagros Mountains who ruled for four centuries. But the Hittites did carry cuneiform writing with them when they left Babylon. In the Nile Valley, Egypt had extended her rule further up the river into Nubia but in about 1700 BC fell victim to the Hyksos, a horse-borne nomadic race from western Asia. Their rule lasted little more than a century. When they were expelled, the New Kingdom was founded.

Progress did not lie only with the big battalions, however. In the Mediterranean, a highly sophisticated culture had been developing on the island of Crete, smaller than Wales. This was the Minoan civilization, so called after the legendary King Minos. A magnificent brick palace was built at Knossos, which at one time seems to have had a population of 80,000. Crete flourished from 3400 to 1100 BC, and was the centre of an Aegean civilization with influence as far afield as Troy, Mycenae and Tiryns. It also had a regular trade with Egypt.

Although there had been internal power struggles, Egypt had not sought expansion by conquest, except into Nubia for its gold. But the Hyksos' invasion encouraged Egypt to safeguard herself by occupying neighbouring territories. The conquest of Palestine and Syria created an empire which stretched north to the Hittites at the Euphrates and south to the Fourth Cataract of the Nile, an overall distance of some 2500 kilometres (1500 miles).

In addition to the Hyksos and Kassites, other nomadic peoples, originating on the steppes of central Asia and the Ukraine, launched successful attacks on the centres of civilization from about 1700 BC to about 1400 BC. Whereas the institutions of Egyptian and Mesopotamian society were largely preserved by their conquerors, in outlying areas the conquests were destructive, though laying the foundations of greater splendours in the future. In the west, Minoan civilization fell to the Achaeans. In the east, Aryan tribesmen destroyed the Harappan culture of the Indus Valley.

The successes of the nomadic conquerors stemmed from their mastery of a new technique of warfare: chariotry. The art of constructing light two-wheeled chariots may have originated in Sumeria around 2500 BC, and rapidly spread to the horse-raising peoples of the steppe. Later chariots, wrought in bronze and dependent on the skills of woodworkers, wheelwrights and leather workers, were expensive artefacts that could be possessed only by a small aristocracy. In battle each chariot usually carried both a driver and an archer who showered the enemy with arrows while protected by the speed of his own movement. The chariot enabled relatively primitive societies to subdue more advanced ones, and an aristocracy of warriors to dominate its peasantry.

Around 1200 BC the Hittites still ruled in Asia Minor, with their capital at Bogazköy, a wealthy city far greater than the Babylon from which they had been expelled. But their fortunes were in decline and their empire was finally disrupted by invaders from the north. As a result, knowledge of ironworking was dispersed throughout Asia Minor and the Middle East. This was to be yet another technology with profound social, military and political effects. The Bronze Age aristocracies of the Middle East were soon unable to withstand the assaults of such iron-using peoples as the Medes, the Chalybes and the Persians. The turmoil consequent on the violent irruption of Iron Age peoples into the heartlands of older civilizations finally saw the establishment of a few stable and extensive empires. Foremost among these was the Assyrian Empire, the greatest the world had yet seen.

At its greatest extent, at about the middle of the seventh century BC, it spread on its northern side in a broad arc from the Persian Gulf, north of Mesopotamia, to Tarsus on the Mediterranean. In the south, its boundary lay well south of the Euphrates, through Damascus, and south to northern Egypt. Only Arabia, over 2½ million square kilometres (1 million square miles) of inhospitable arid country, bounded on three sides by the sea and in the north by the formidable barrier of the Great Nafud Desert, remained aloof from the ebb and flow of power in the Middle East. Its destiny lay far in the future, when it became the birthplace of Islam.

The Assyrian Empire survived into the seventh century BC, when it came under attack from the Cimmerians and Scythians. These were nomadic steppe-dwellers from the north who had learned to ride on horseback – a skill that appeared later than the technique of chariotry. They could even control their horses on the battlefield while using bows and other weapons, and thus became the first cavalry in history. They achieved victories over the chariot-borne Assyrians that severely weakened the empire, and its subject peoples revolted. In 612 BC Nineveh, the Assyrian capital, fell under the pressure of attacks from Scythians, Medes and Babylonians, and the empire was dismembered.

The rise of Greece and Rome

Significant changes were taking place along the northern shores of the Mediterranean. There, Dorians from the north had occupied what is now Greece and this led to the establishment of city states from about 750 BC. Of these, Athens was to emerge as the leader, yet it had a population of no more than a quarter of a million even at its greatest. Greece is a mountainous country, not fertile enough to support a rapidly growing population, and of necessity the Greeks became great colonizers. They were not empire-builders, though, and these colonies themselves became independent city states around the Mediterranean.

Further to the west, the Etruscans arrived in northern Italy, probably from Asia Minor. Rome, according to tradition, was founded in 753 BC and in 509 BC the Romans expelled the Etruscans and founded the Republic. This was the beginning of another enormous empire, which eventually comprised not only most of Europe, but the territories of many of the ancient Middle Eastern empires, and the north coast of Africa, including Egypt.

The Alexandrian Empire

While Greece and Rome were beginning their rise in the west, the Medes and Chaldeans were creating a new Iranian (Achaemenid) empire in the east, following their sack of Nineveh. Under Cyrus, and later Darius, they extended their rule from the Hindu Kush in the east to the Mediterranean in the west, from the Persian Gulf in the south to the Caspian Sea in the north. Ambitions to push further west were thwarted by defeat by the Scythians in the Ukraine and by the Greeks at Marathon in 490 BC. The Greeks in their turn were subjugated by the neighbouring Macedonians who, under Alexander the Great, created yet another vast new empire.

The Assyrians attached great importance to astronomical observations and mathematical computations, which were carefully recorded in cuneiform characters on clay tablets (opposite, top left). This example, from about 1600 BC, sets out a number of problems in practical mathematics with diagrams relevant to their solution.

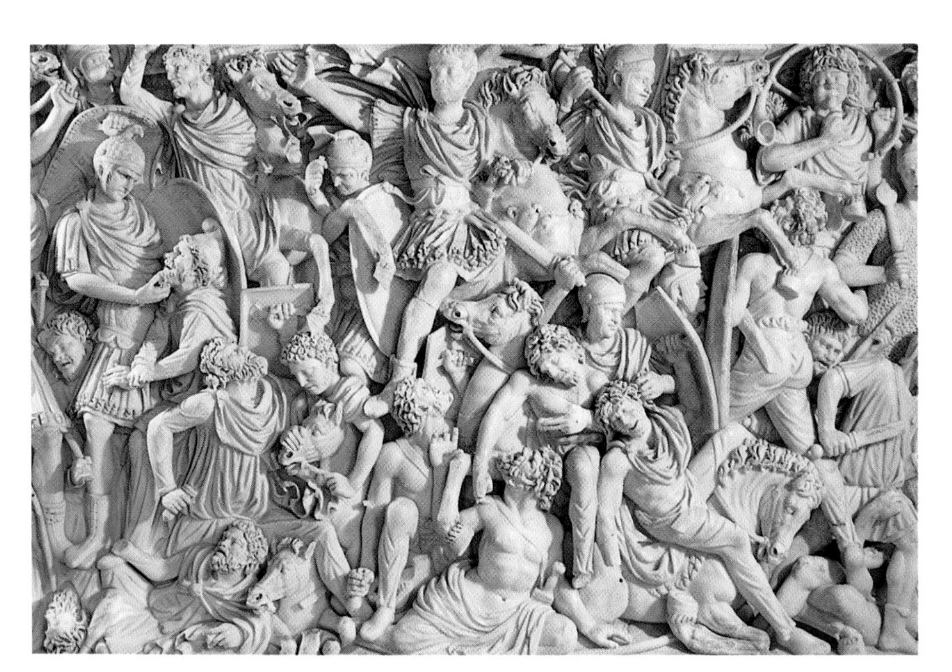

The Great Ludovisi sarcophagus, carved in marble (right), depicts a Roman battle scene of the mid-third century AD.

ancient civilizations 10–11, 20–25 **chariots** 46, 47, 49 **Roman Empire** 10–11, 24–25, 88 **writing** 31–32

Long after his death, Alexander the Great was a legendary figure in the ancient world: he is here powerfully portrayed in a mosaic of about 100 BC. This mosaic, from Pompeii, depicts the Battle of Issus. Mosaic is one of the oldest art forms, dating at least from the fourth millennium BC.

In 334 BC Alexander crossed the Hellespont with an army of 40,000 men into Asia and then passed south through Asia Minor and Syria into Egypt. Turning back, he advanced victoriously through Mesopotamia – where he defeated Darius at Issus and Gaugamela, capturing immense treasure – into Iran, past the south of the Caspian, on to Tashkent, Bukhara and Kashmir, stopping only when he had crossed the Indus. His return journey was less successful and his grandiose plans for conquest and civilization died with him at Babylon in 323 BC. Had he lived he might well have founded an empire that extended from the Pillars of Hercules to the Indus, spanning the whole of the ancient world. By any standards, it was an extraordinary expedition and it carried the ideas of Greek civilization to the eastern end of the known world.

For a century this Hellenized world survived, but the centre of gravity shifted from Athens to the newly established capital cities, particularly Alexandria, which became the cultural centre of the world. Nevertheless, it was the Roman star that was now in the ascendant, though this was not immediately apparent. Macedonia allied itself with Hannibal of Carthage, a city founded by Phoenicians which dominated much of North Africa, Spain, Sicily, the Balearics and Sardinia. During the Punic Wars of the third century BC, Rome all but fell to Hannibal, but the outcome was the destruction of Carthage in the second century BC; it was later rebuilt as one of the chief cities of the Roman Empire.

In 146 BC Greece became a Roman colony, the Seleucid Empire in Asia Minor was acquired in 188 BC and Egypt was taken after the Battle of Actium in 31 BC. In the second century AD, the Roman Emperor Trajan, and later Severus, advanced beyond Asia Minor into the Parthian Empire on the eastern Mediterranean and marched through Mesopotamia to the Persian Gulf. Although the Roman hold on Parthia was never very secure, the Macedonian Empire of Alexander thus effectively passed to Rome. Culturally, however, the influence of Greece remained enormous. The extent of the Roman Empire was, in fact, a great deal larger. It comprised also much of western Europe, including Britain. Estimates of its population must necessarily be approximate, but at its height it was probably around 70 million.

Conquest and technology

The rise and fall of the empires of the ancient and classical world reflect in various ways the concurrent development of technology. Superior military technology commonly made a direct contribution to victory, as when the Roman consul Crassus was routed by a numerically inferior Parthian force in 53 BC. The advent of composite bows, horse-drawn war chariots, cavalry, iron weapons, battering rams and other equipment for siege warfare all conferred advantages on those who could deploy them against less effective weapons. Occupation of new territory spread the technological skill of both conquerors and conquered in such fields as pottery and metal-working. While victory was sometimes followed by massacre, many of the vanquished were carried away into slavery, taking their native skills with them. The institution of slavery was of profound importance, for it was a major source of labour both for everyday work and for the grandiose projects, such as the pyramids of Egypt and the silver mines of Laurium, which are lasting memorials of their age. Slave markets were commonplace: contemporary accounts show that in both Athens and Rome as much as one-third of the population were slaves. Although many slaves were prisoners of war – when it was realized that these were more valuable alive than dead – others were criminals and debtors.

As new states developed, the improvement of transport by land and water became of paramount importance. By the time of Greece and Rome, there was a busy Mediterranean trade in grain, oil, wine, pottery, metal goods and glass. The needs of shipping involved the building of extensive harbour works – including the great port of Ostia to serve Rome and

Fortifications took many forms in the ancient world. One kind widely used was the defensive wall, strengthened by forts at intervals and usually with garrison towns for reinforcement in the rear. Very often these walls served a dual purpose: to keep hostile neighbours at bay and prevent the defection of dissatisfied subjects. The examples shown here are Hadrian's Wall in Britain (left) and the Great Wall of China (below).

its million inhabitants – and associated aids to navigation such as lighthouses. In Europe the natural routes of the great rivers were supplemented by a network of roads. Lead for the plumbing of elaborate water-supply systems in Rome was brought from as far afield as England.

China

Another great civilization had developed independently far beyond the limit of Alexander's penetration into Asia. In China thriving agrarian communities had been established, planting crops, weaving textiles and making pottery. Walled cities built in a grid-iron plan have existed since the 17th century BC. From these emerged states of modest geographical extent and often brief duration. The first of major consequence was the Shang, founded by invaders in the Yellow River area. Shang China flourished from about 1750 BC to 1100 BC. It is noteworthy for its extensive and superbly skilful use of bronze and the development of a pictographic writing. After an unsettled time known as the period of the Warring States, the Ch'i – expert in the working or iron – emerged as leaders. Their lasting memorial is the Great Wall of China, some 2400 kilometres (1500 miles) long: it surely ranks with the Seven Wonders of the Western World.

The Ch'i were short-lived and were succeeded in 202 BC by the Han Dynasty, which continued at least in name until AD 220. Under the Han, the population of China rose to about 60 million, of whom a quarter of a million lived in the governmental and cultural capital Ch'ang-an. Conquest extended Chinese influence to the north-west and to Korea and north Vietnam. In AD 90 the great general Pan Ch'ao reached the Caspian and established relations with the Parthians at the threshold of the Roman Empire. At the beginning of the Christian era the Han Empire thus rivalled that of Rome in size, wealth and culture.

This immense civilization – the longest-lasting the world has ever known – grew up virtually independently. Nevertheless, there was continuous and active contact with the West from a very early date. Although Rome and the Han Dynasty had no formal relationship, each was knowledgeable about the other and a vigorous trade was encouraged. In the first century AD, Ptolemy mapped the known world with surprising accuracy and specifically marked the port of Cattigara in south China. Overland, the Silk Road ran from Samarkand to Tun-huang on the Chinese border. From the east it carried silk, spices and precious stones; from the west came metals, pottery and glass. Trade was no longer solely on a barter basis. By that time coinage – probably invented by the Hittites – was in general use. Hoards of Roman coins have been found deep in Asia, including southern India. The existence of the Silk Road does not imply direct contact between East and West, however. Goods were conveyed in stages by successive carriers, few if any travelling the entire distance. In this way the intervening states got the benefit of the trade and taxes on goods in transit without the disadvantage of foreign infiltration. Pliny refers explicitly to a Stone Tower, probably on the northern edge of the Pamirs, where goods were exchanged; Ptolemy marked this on his world map.

Additionally, there was also an extensive trade by sea, following the discovery of the monsoon winds by Hippalus in the first century BC. Ships capable of carrying loads up to 500 tonnes plied from the Red Sea to Indian ports, where their cargoes were transferred to Indian or Chinese vessels.

Carthage 25 **Han Dynasty** 10–11, 27, 63, 136
Roman Empire 10–11, 24–25, 88 **silk** 36

The Roman Empire

If we take a look at the world situation as it was around the year AD 200, we see that the scattered civilizations with which we began our story had developed into a succession of contiguous ones running from the Atlantic to the Pacific, comprising western Europe, the Mediterranean basin, the Middle East, India and China. All had highly developed and often distinctive technologies and the products of these were exchanged by land and sea over well-established trade routes. At the time it must have seemed to many thoughtful observers that a settled world order had evolved which, despite internal strife, could continue to defend itself against the outside world. Perhaps nowhere would this feeling have appeared more justified than within the frontiers of the Roman Empire.

Although Edward Gibbon's great *Decline and Fall of the Roman Empire* contains much that is unacceptable to modern historians, most would accept his belief that the period from the death of the Emperor Domitian (AD 96) to the accession of Commodus (AD 180) was one of exceptional happiness and prosperity for those under Roman rule. Rome had made herself mistress of a vast land empire comprising most of Europe – until that time peopled by nations outside the frontiers of the high city civilizations – together with many of the Middle Eastern empires, Egypt and the coastal strip of North Africa.

Very consciously the heir of the Greek idea of civilization, as well as the conqueror of Greece itself and much of the Hellenistic Alexandrian Empire's successor states, Rome retained the trappings of a republican system along with a highly centralized bureaucratic administration. Her success derived not from any great technological originality but rather from the systematic and effective exploitation of existing technologies; from a highly disciplined army, officered by professionals; and from an unshakeable conviction of superiority. Under a succession of brilliant and hard-working emperors such as Trajan, Hadrian and Marcus Aurelius, the empire of the second century AD maintained its frontiers against the pressure of barbarian peoples beyond the Rhine and the Danube. The frontier itself was defended over large stretches by an elaborate system of walls, ditches and forts which together made up the *limes* (the boundary of the Roman Empire). The walls of Hadrian and Antoninus Pius in Britain formed the northernmost part of this system. Within the frontiers, the immense road network provided the most efficient system of communications the world had yet seen. In Gaul, Spain and Britain, cities were built which, if they barely emulated the splendours of Rome herself, brought stone-built architecture to a part of the world where it was virtually unknown. The public buildings of the imperial and town administrations, the temples and, above all, the great aqueducts brought to western Europe an entirely new concept of the possibilities of existing technologies.

Trajan's Column (right) is a dominant feature of the great Forum Traiani in Rome. Completed by the architect Apollodorus in AD 114, it is 38 metres (125 feet) high and part of the inscription records that this was the depth of rock that had to be excavated to level the forum. It is decorated with a series of spiral reliefs illustrating many aspects of Trajan's Dacian campaigns. This panel shows contemporary Roman shipping.

Carthage (far right) was founded by the Phoenicians in about 800 BC. At the height of its power (300 BC), the great Mediterranean port had 750,000 inhabitants. Destroyed by the Romans in 146 BC, it was restored as a major port by Augustus.

Like the *Pax Britannica* of the 19th century, the *Pax Romana* provided unusual opportunities for the development of a wide range of civilized activities. Yet in the technological field, for all its achievements, these opportunities were not developed to any remarkable extent: the scale of industrial activity certainly expanded enormously, but there was little innovation. Allowing for adjustment to very varying local conditions, crafts tended to be practised in traditional ways, but in more and bigger centres. This is not altogether surprising, for the urban communities which benefited most were comfortably sustained by a very strictly ruled peasant population. The very success of the Roman administration was a disincentive to change and the abundance of cheap labour, including slaves, gave no encouragement to the development of power-driven machinery. Only in the declining days of the empire, when conquests had ceased and the frontier defences were coming under increasing pressure, did shortage of manpower begin to change this conservative attitude, but by then it was too late.

The empire of the third century AD offered marked contrasts to that of Gibbon's 'golden age'. Barbarian incursions across the frontiers became increasingly frequent and threatening, while social divisions built up within. The state religion was weakened by the popularity of numerous Eastern cults. Among these Christianity was to emerge triumphant – not least because, from very early days, it organized itself on a centralized system modelled on the imperial bureaucracy itself. At the end of the third century the Emperor Diocletian inaugurated a series of thoroughgoing reforms: he divided the army into four principal commands and the administration between two emperors, each with an assistant 'Caesar',

and he increased taxes. Furthermore, in an attempt to unify society, he began to persecute the Christians.

The fourth century opened with civil war between the emperors Constantine I and Licinius from which Constantine the Great emerged victorious in AD 324. Attributing his victory to the intervention of the Christian god, Constantine promulgated the toleration of the religion throughout the empire under his personal protection, though he himself was baptized only on his deathbed. The long-term repercussions were enormous, not least upon technology. Christianity became the official imperial religion and eventually immensely powerful throughout Europe. Reluctant to change a social order that was favourable to it, and ruled by dogma, the Church chose to use its power to stifle the kind of scientific inquiry from which wholly new technologies were ultimately to be derived.

In Europe, the incursion of peoples of Germanic and Slav origin, relentlessly driven on by pressures from the Huns of the central Asian steppes, intensified. They were setting the stage for the appearance of a new kind of Europe, deriving its ideas from the north and east. It thus differed greatly from that created by the Romans, who were essentially Mediterranean in their outlook. When Constantine founded his new city of Constantinople on the site of Byzantium, a former Greek trading post on the Bosphorus (now Istanbul), the Mediterranean focus of the Roman Empire was reaffirmed. At first, the new Rome and the old Rome worked in harmony. There were often two emperors and in AD 395 Theodosius made the division permanent by appointing his two sons sovereigns of the eastern and western empires respectively.

The Byzantine Empire of Constantinople flourished as a bastion of Christianity. For over a thousand years it survived assaults by Persians, Arabs, Slavs and Turks. By contrast the Western Empire, torn by internal strife and barbarian attacks, disintegrated. The last of the shadowy emperors, a refugee in his 'capital' of Ravenna, was deposed in AD 476. Parts of Italy remained provinces of the Byzantine rulers for centuries but the Empire in the West was over.

There was no single reason for the collapse but technological factors have been plausibly adduced as contributing factors. These include exhaustion of the soil through over-intense cultivation; diminishing yields from the gold and silver mines; the insidious effects of malaria derived from undrained marshland; and even the debilitating effect of lead in drinking water, as a result of the widespread use of the metal in conduits and cisterns. Yet the fact remains that Rome, the last great power of the ancient world, fell victim to peoples at essentially a lower level of technological development than itself.

Creating an empire engenders in people a sense of pride and purpose and a readiness to accept hardship, but enjoyment of the resulting advantages does the reverse at all levels of society: rulers fall out among themselves in the quest for wealth and power, while the people at large reject the sacrifices necessary for their own security. Successful nations commonly fall ready victim to ambitious rivals ready to seize what they themselves are not ready to defend. Over-simple as this view may be in general, there is a great deal of truth in it as far as Rome is concerned. Like other empires before and since, the Roman West fell, in part, as a victim of its own success.

Alexandrian Empire 21–22 **Hadrian's Wall** 23, 87 **Roman Empire** 10–11, 88 **Roman technology** 58–63 **Trajan** 22

CHAPTER TWO

The Agricultural Revolution

Some time around 10,000 BC man began to lose his dependence on wild animals and plants and to domesticate them, with the result that his sources of food were more dependable and immediately available. This was when the next crucial step towards civilization – the so-called Agricultural Revolution – took place.

The term Agricultural Revolution needs qualification though, for it was not a rapid change but a gradual one spread over millennia rather than centuries. Even in the most progressive communities it began on a very modest scale: the archaeological evidence is clear that hunter-gathering activities continued side by side with the first tentative moves towards domestication. In many parts of the world domestication was never achieved at all.

It seems likely that the initial incentive to an agricultural way of life was not sheer necessity, but rather the attraction of security. Although plagues, droughts and floods were hazards of farming life, as they still are, it normally gave an assured surplus of food. Freed from the necessity to be constantly on the move, people found it feasible to build permanent dwellings and furnish them with a range of domestic appliances. How far the farming surplus created urban life is arguable, but it was certainly necessary to maintain it once it was established. Whatever the reasons, the transition was of crucial importance. Although those who continued solely as hunter-gatherers developed many practical skills and complex social customs, they were no longer in the main stream. Civilization as we now know it stemmed from the agricultural communities.

The role of cereals

Until the whole pattern of farming was disrupted by the advent of cheap synthetic fertilizers during the present century, a well-developed agricultural system involved a complex relationship between farm animals and crops. The crops, including grass, supplied food for both man and beast, and in addition to meat, dairy products, wool, leather and tractive power, the animals provided the manure essential to maintain soil fertility. Neolithic man, however, had no concept of such integrated systems. During a long transition phase, deliberate farming did no more than empirically supplement wild resources. Even within these limits, wide variation was obviously possible. Some pastoral communities might master the control and breeding of cattle yet depend on natural pasturage; others might grow crops on a limited scale to supplement wild resources.

Whatever form the transition took, there is no doubt of the crucial importance of cereals. These provide not only a nutritious basic food but also one which, after drying, could be stored for long periods – another necessity in a settled existence. Archaeological evidence based on seed remains

Although stylized rather than accurate, Egyptian wall paintings throw much light on everyday activities and techniques. These scenes from the tomb of Menena (16th century BC) are typical. Agricultural scenes predominate, though one of the subjects is a plank-built Nile boat of the day. An abundance of labour is evident, but oxen are used as draught animals – though here apparently treading the grain. The fans are being used to create a breeze for winnowing (blowing the chaff away from the grain).

shows that Neolithic man gathered wild cereals. This involved two technological innovations. One was the sickle, a necessary aid to harvesting. In its earliest form this consisted of sharp flints set in a wooden or bone frame: flints used for this purpose can be identified by a characteristic blunting of the cutting edge. Early sickles were straight; the now familiar curved sickle is of a later date. The harvested grain had then to be ground and this was commonly done by crushing and rubbing it between two stones, a smaller one held in the hand working against a larger base. These primitive hand-mills – called saddle querns – have been recovered in great numbers. The pestle and mortar was another way of crushing the grain.

In the harvesting and grinding of wild cereals, some grain would inevitably be scattered and in due course sprout. It is not difficult, therefore, to visualize a transition from harvesting wild cereals to reserving part of the grain for cultivation conveniently close at hand. It is a reasonable supposition that this would first have occurred not where wild cereals grew abundantly, but on the fringes of these areas, where some trouble was involved in collecting them. In any event, the first essential was a natural source of seed. Among such sources were the mountains of Anatolia, and the Zagros and Elburz Mountains lying south of the Caspian Sea. To the south of this region lay the area now known as the Fertile Crescent, where the soil and climate – and in particular the average rainfall – were suitable for the growing of corn. It stretched in a broad sweep of some 3000 kilometres (2000 miles) from Jericho, just north of the Dead Sea, to the Persian Gulf.

The wealth of archaeological evidence for the early establishment of farming in this area led to its general acceptance as the cradle of civilization. This was the centre from which agricultural and other techniques slowly diffused outward to the rest of the world, being adopted and modified according to local conditions. How far this assumption is justified we will consider later.

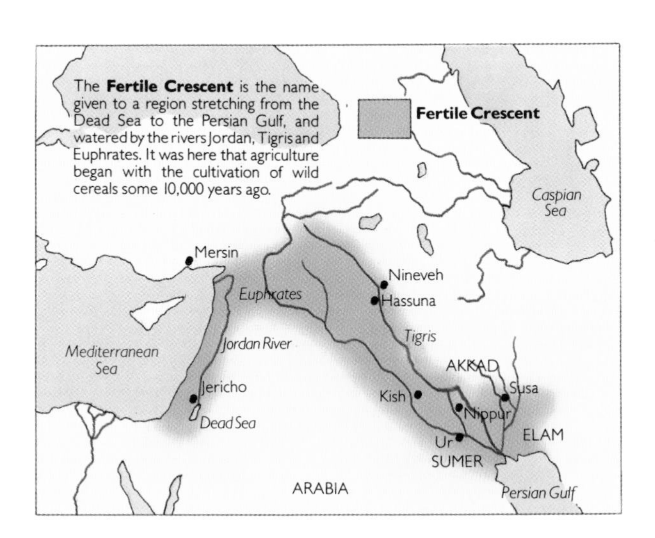

The **Fertile Crescent** is the name given to a region stretching from the Dead Sea to the Persian Gulf, and watered by the rivers Jordan, Tigris and Euphrates. It was here that agriculture began with the cultivation of wild cereals some 10,000 years ago.

The Chinese, too, left many pictorial records of their everyday life, as typified by this hunting and harvesting scene from a Han tomb at Cheng-tu, Szechuan.

The Agricultural Revolution predated not only the use of metals – first copper and later bronze – but even pottery. The earliest settlements were no more than villages, doubtless serving as a focus for a small area of surrounding countryside. The original walled city of Jericho, dating from about 8000 BC, spread over some 4 hectares (10 acres). Çatal Hüyük, in Anatolia, abandoned by 5000 BC, covered 13 hectares (32 acres). Houses were of stone or mud brick, and there is clear evidence of some sort of functional planning. As they fell into disrepair, they were flattened and new ones were built in their place. Stone was still dominant for containers and tools. Pits in the ground were used as granaries. Surviving frescoes showing hunting scenes and the bones of wild animals indicate that agriculture was not the sole means of subsistence.

The two important cereal plants in the Fertile Crescent were wheat and barley. The first kind of wheat to be domesticated was probably emmer, followed by einkorn perhaps a thousand years or more later. Barley seems to have been contemporary with emmer.

Early methods of cultivation probably consisted of no more than scratching the surface of the soil, scattering the seed and then, perhaps, trampling it in. When ripe it was harvested with flint sickles. The straw was used for thatching and other purposes. Successive crops inevitably impoverished the soil, but in those early days of low population density this could be overcome simply by moving to virgin land. Apart from plants grown for food, we know that flax was grown in the pre-pottery era as a source of both edible oil and textile fibre, an early example of an industrial crop.

Domestication of animals

The first animal to have been domesticated appears to have been the dog, derived from ancestral wolves. The Palegawra Cave, in north-east Iraq, has yielded bones dated back to 10,000 BC; European finds go back to 8000 BC. Probably the main incentive was the value of dogs as scavengers, helping

development of agriculture 152, 201–206, 291–301 **early tools** 16–17 **Fertile Crescent** 28 **hunter-gatherers** 14–16

Because of its crucial importance for year-to-year survival, harvesting figures prominently in early paintings. This scene is from a Han tomb in Szechuan, south-west China. As the labourers are evidently working in water, this is clearly a rice paddy field. Farther north, much dry rice was grown.

to keep campsites clean. But finds of skulls of young dogs with open braincases indicated that they were also used as food. Dogs were eaten in Europe until the end of the Bronze Age; in Mexico and Peru until the 16th century; and in China up to the present time. It is perhaps significant that one of the next animals to be domesticated appears to have been the pig, which also can flourish by scavenging waste human food. As they had no secondary uses in providing wool, milk or tractive power, they were killed young.

The small ruminants, sheep and goats (caprovines) were native to South East Asia and it is argued that, as with cattle, their instinct to follow a herd leader favoured their domestication. Though the size of cattle presented problems, their correspondingly large capacity to provide meat and milk made them attractive: later their strength led to their utilization as draught animals.

Other centres of agricultural development

The profound importance of the Agricultural Revolution has encouraged the belief that it was a unique event. The technological inventiveness which made it possible was assumed to be peculiar to the people of the Fertile Crescent, whence the new culture spread to the rest of the world. Those parts of the world not accessible to this cultural wave, such as Australasia, remained unenlightened.

There is a good deal of evidence in support of this. Much of European agriculture derived from that of the Middle East. Wild sheep and goats, for example, did not exist on the European mainland, so the domesticated variety must have been imported. Cattle, however, may have been independently domesticated in Europe. But European conditions were so different from the Middle East in both soil and climate that intelligent adaptation to local conditions was necessary for the agricultural impetus to be maintained. In the severe northern conditions, for example, oats were found to flourish where wheat failed.

Such demonstrable capacity for fundamental local innovation makes it plausible to suppose that the domestication of plants and animals might have originated independently elsewhere. A century ago the French botanist Alphonse de Candolle argued that agriculture had developed in three areas which at the time had no significant communication with each other: south-west Asia (including Egypt), China and intertropical America. Forty years later the geneticist Nikolai Ivanovich Vavilov, on the basis of worldwide surveys conducted by Russian botanists, concluded that there were at least eight independent centres of origin of the world's most important cultivated plants. Other evidence, genetic and otherwise, suggests that this may be an overestimate. But there is now wide agreement that at least South East Asia (including China) and Central America were centres of independent agricultural revolutions.

The earliest Chinese agricultural communities of which we have knowledge date from around 5000 BC, possibly earlier, and lived on river terraces in the north and north-west. As in the Middle East, farming originally did no more than supplement foraging. In the north, wheat and millet were the staple cereal, but in the south, later, it was wet-grown rice.

On present evidence, village communities in Central America, based on maize cultivation, date from around 2000 BC, but maize was probably domesticated at least 500 years earlier. And in the Andes, quinoa and canihua were grown perhaps as early as 2500 BC and are to this day unique to the region, although the present acreage is small.

Irrigation

All early agriculture, and much of it at the present day, depends on adequate rainfall to sustain the crop. Almost

inevitably, therefore, man's next step in controlling his environment was to try to make himself independent of the vagaries of the weather by developing systems of irrigation. Such systems were of two kinds: local irrigation by individual farmers or groups of farmers and more ambitious schemes aimed at satisfying the needs of considerable areas.

In the simplest form of irrigation, water was simply splashed on growing crops as their needs seemed to dictate, but systems of ditches soon appeared, supplied from rivers or wells. In southern China, wet rice was perhaps grown first in natural swamps, but by the middle of the second millennium BC raised banks were being used to prevent water draining away. This system of paddies necessitated very small units, rarely larger than 700 square metres (⅙ acre), owing to the difficulty of levelling larger areas: a very slight tilt would drain all the water to one side.

In favourable circumstances such simple systems can be highly effective, and they are commonplace today. But to control big, fast-flowing rivers with large and sudden fluctuations of water-level is a very different matter. This involves not only the building of massive dams and aqueducts but also a complex system of dykes and sluices. No less important, it demands an effective administrative system to distribute the water as needed over a large area and to levy taxes on individual users.

For early examples of such large-scale systems, we must turn to two areas which saw the early flourishing of urban life and the beginning of civilization as we now conceive it. These were the fertile areas of the Middle East drained by the Tigris and the Euphrates, and the Nile Valley. Both these great river systems drained huge areas and were subject to disastrous annual floods. But, once controlled, this abundance of water made possible a new kind of community – the city. With a population numbered in thousands, it not only dominated the surrounding countryside but often served as the administrative centre for an entire kingdom.

It was controlled by a ruler who exercised power through what were then innovations: an army, a judiciary, civil servants and a priesthood. There was an organized division of labour into crafts. Additionally, there was a merchant class who not only supplied local needs but also conducted import/export trade over routes extending for hundreds of kilometres. Power and stability were manifested in impressive public buildings and temples, and the prosperity of the community at large was demonstrated by housing and personal possessions far in excess of basic needs. This extraordinarily rapid transformation was made possible by the rise of a whole range of new technologies. But it was organized agriculture nourished by a controlled system of irrigation that was the foundation on which all else rested.

The surviving Code of Hammurabi, King of Babylon from 1792 to 1750 BC, testifies eloquently both to a highly organized social system and to the great importance of the

The Nile has always been the life-blood of Egypt and its regular annual flooding dictates the whole agricultural cycle. These rock inscriptions at Semna, high up the Nile in Nubia, record levels around 2000 BC.

irrigation system. Thus the tablet, 2½ metres (8 feet) high, on which the code was inscribed contains several clauses defining penalties for those who neglect their duty. For example: 'If a man has opened the waters, and the waters have carried away the field of his neighbour, he shall pay ten *gur* of corn per *gan*.' The systematic destruction of irrigation systems was an Assyrian method of ensuring that defeated enemies remained quiescent: that of Babylon was destroyed by Sennacherib, son of Sargon and King of Assyria.

Far to the south of Mesopotamia, civilization developed along rather similar lines in the Nile Valley, where the river floods annually in July with spectacular results and remarkably regularity, subsiding again in October. In Mesopotamia light rainfall in January was often enough for catch crops to be grown. But the Nile Valley had little significant annual rainfall, and in its absence the elaborate system of irrigation there was not merely desirable but vital. From the foundation of the Old Kingdom, in about 2700 BC, district governors were designated Diggers of Canals. Under the pharaohs, the whole valley was divided into a checkerboard of community dykes enclosing areas ranging up to several thousand hectares. Flood water was admitted through sluices to a depth of 1 metre (3 feet) or so and held for several weeks until the land was saturated and the water had deposited its fertilizing silt. The soft soil could then be cultivated and seed sown. Nice judgment was needed to exercise control even in normal seasons, and to assist the administrators an elaborate system of Nilometers (gauges for measuring the height of the Nile) was installed to give early warning of the start of the inundation upriver and provide detailed records for future years. Egyptian incursions into Nubia in search of gold were at least partially prompted also by a desire to establish observation points even higher up the Nile. When danger of an excessive flood was apparent, a state of emergency was declared and every available person was required to strengthen the dykes and attempt repairs when they were breached. The Nile has always been the life-blood of Egypt.

A feature of farming based on the flooding of these great rivers was that the deposited silt contained sufficient natural fertilizer to sustain the productivity of the land from year to

Ancient Egypt 10–11, 20 **aqueducts** 61–62 **dams** 60–61 **Nile** 10–11, 20

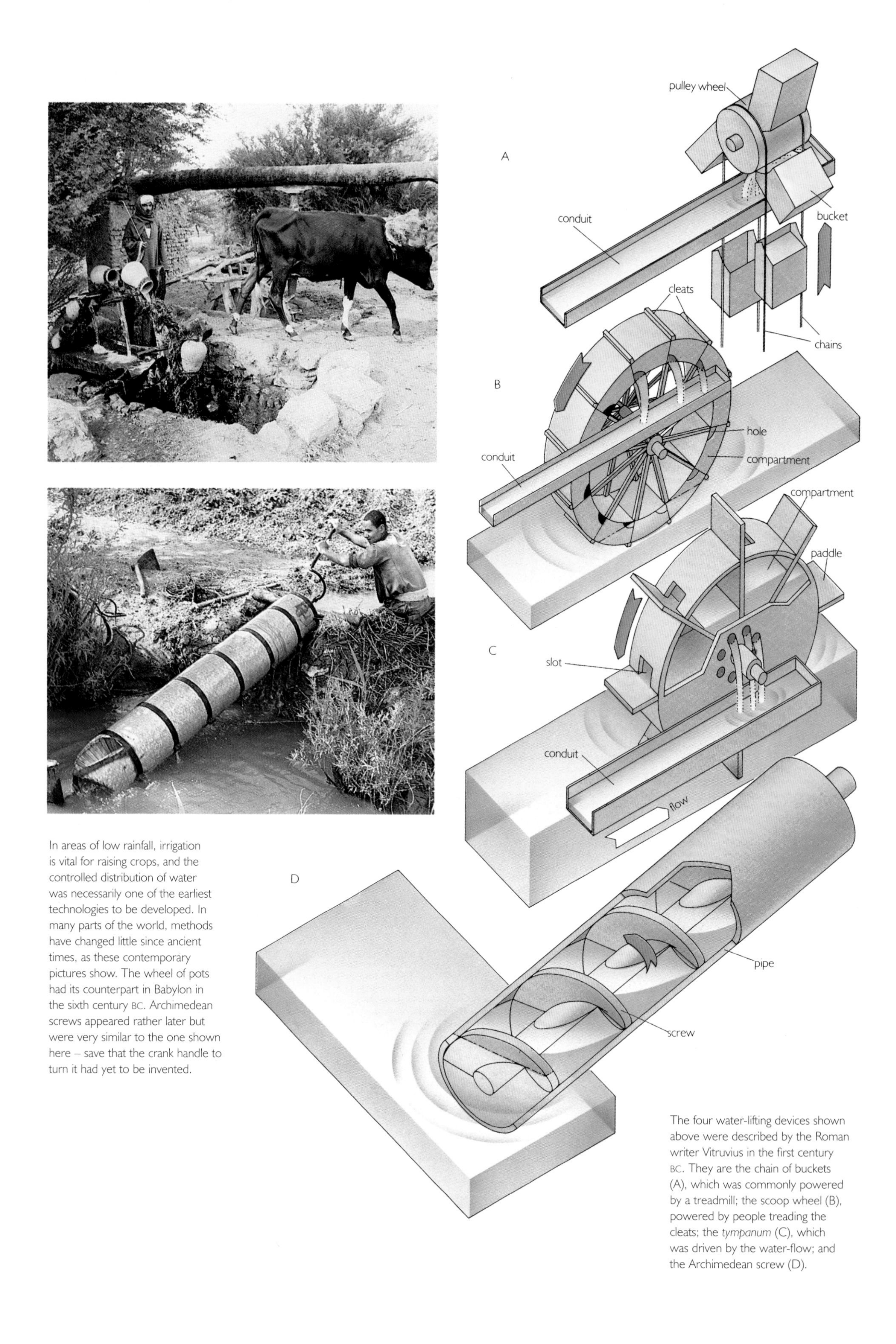

In areas of low rainfall, irrigation is vital for raising crops, and the controlled distribution of water was necessarily one of the earliest technologies to be developed. In many parts of the world, methods have changed little since ancient times, as these contemporary pictures show. The wheel of pots had its counterpart in Babylon in the sixth century BC. Archimedean screws appeared rather later but were very similar to the one shown here – save that the crank handle to turn it had yet to be invented.

The four water-lifting devices shown above were described by the Roman writer Vitruvius in the first century BC. They are the chain of buckets (A), which was commonly powered by a treadmill; the scoop wheel (B), powered by people treading the cleats; the *tympanum* (C), which was driven by the water-flow; and the Archimedean screw (D).

Writing is essential to organized societies. Egyptian hieroglyphs were in use by 3000 BC. The pictographs may represent whole words or syllables that make up words.

The cuneiform writing of Sumeria was another early form of writing. Here, words are expressed by forming different patterns of wedge-shaped impressions.

The alphabet ultimately triumphed – though in various forms. This inscription of the third century BC is written in the elegant Sabaean alphabet of 29 characters.

year. In the third century BC Theophrastus remarked on the extraordinary fertility of Babylonia: 'No country is more fruitful in grain.' A three-hundredfold return on grain sown was feasible and there could be two or even three harvests a year. An indication of the quantity of silt deposited in the Nile Valley is given by the fact that the level of the land has risen some 7 metres (23 feet) since pharaonic times.

While there were differences between the early civilizations of the Middle East and Egypt, the supreme importance of irrigation is a common factor. It is not surprising, therefore, that the same is true of the early civilization of the Valley of the Indus which rose in Tibet and flowed through the fertile plains of the Punjab before discharging itself into the Arabian Sea. This civilization has still been relatively little explored, however. It seems to have been established by about 2500 BC, and covered 1¼ million square kilometres (½ million square miles). The principal cities were Harappa and Mohenjo-Daro, where excavations have uncovered massive public buildings, a grid-iron plan of streets and extensive land drainage.

In China, the growing of wet rice in the south necessarily involved some system of irrigation to flood the paddy fields. But as there was normally no scarcity of water, this did not require any large-scale irrigation schemes. In the drier north, on the other hand, considerable irrigation systems were developed in the first millennium BC.

Writing and the calendar

The wall paintings of primitive man give us an occasional vivid glimpse of his daily life but they were not accompanied by any written record. With the rise of large organized agricultural societies, however, some form of record-keeping became essential. The administration of irrigation systems and other public expenditure had to be financed by the collection of taxes: the aphorism that the two certainties in life are death and taxes was doubtless as true among the Babylonians and Egyptians as it is today. Agreements had to be made about land tenure; records had to be kept of harvests; commercial transactions of all kinds had to be recorded; messages had to be sent to distant recipients.

The Code of Hammurabi (1751 BC) reflects a strictly ordered legalistic way of life. For example: 'If a man shall give silver, gold, or anything whatever to a man on deposit, all whatever he shall give he shall show to witnesses and fix bonds and shall give on deposit ... If an agent has forgotten and not taken a sealed memorandum for the money he has given to the merchant, money that is not sealed for, he shall not put in his accounts.'

Clearly, by the time of the Code written records were commonplace. Writing and numeration had indeed been invented some 2000 years earlier. The invention cannot be precisely dated, because it evolved from the use of simple tallies and tokens, and there is no distinct point of transition. Clay counters or tokens of many shapes and differing complexity were used to count goods in early agricultural communities from about 8000 BC. Such primitive systems are capable of very sophisticated development, as the Spaniards discovered when they invaded Peru. There the Incas, who had no written language, had devised a mnemonic system of *quipus*, based on skeins of many-coloured threads. Through this medium local *quipucamayus* furnished central government with a wealth of statistical information.

The advent of writing is generally regarded as marking the transition from barbarism to civilization. Its evolution from the earliest pictograms to modern alphabets is a complex story. For present purposes we need only note that the early writing of the Middle East was essentially a syllabary – a set of characters or symbols representing syllables. These were recorded in soft clay, subsequently hardened by heat. Stone was used only for monuments and major records. The script is called cuneiform – from the Latin *cuneus*, a wedge – because the pen was a piece of reed cut diagonally across, exposing a wedge-shaped section. In effect, different words were expressed by different patterns of wedges. As the material was virtually indestructible, a vast amount survives, ranging from laudatory records of the personal virtues and military successes of great monarchs to trivia which can be equated with today's household accounts. It is hardly surprising that a wealth of this writing is devoted to agriculture and irrigation.

Archimedean screw 66 **Incas** 14, 90–91 **irrigation techniques** 28–29 **water supply** 60–62 **writing materials** 35, 99–100

The Egyptians used a hieroglyphic system and, from about 2000 BC, began to write on leather in the form of parchment or vellum. Five centuries later they began to use the much cheaper papyrus, made by interweaving fibrous strips peeled from reeds which grew in profusion on the banks of the Nile. The thin sheets were sized with gum, providing a surface suitable for painting with an inked brush.

Agriculture is very much a seasonal occupation. Much depends on carrying out the various operations, such as sowing and harvest, at the proper time. In Egypt, it was very important to be fully prepared for the annual flood of the Nile. It is not surprising, therefore, that much attention was paid to making the astronomical observations necessary for devising an accurate calendar. For thousands of years astronomy was the only branch of science to be put on a precise mathematical basis.

Apart from the observations themselves, which we will consider later in another context, the basic problem is that a lunar month consists of 29½ days, and the sidereal year – the time taken by the earth to make a complete revolution around the sun – is 365¼ days. In consequence a year of 12 lunar months gets out of step by approximately 11 days a year. Over the centuries, various devices have been used to rectify this.

The Egyptians had an effective calendar as early as 3500 BC, but it failed to accommodate the six hours by which the sidereal year exceeds 365 days. Eventually, this was solved by the simple device of the leap year, instituted by Ptolemy III in 238 BC. The Babylonians had also established a calendar by about 3000 BC and the accuracy of the underlying astronomical observations is quite remarkable. They adopted the expedient of the intercalary month, inserting extra months into the 12-month lunar year over a period of 19 years.

The seven-day week is not related to any celestial motions, and seems to have been devised by the Assyrians, adopted by the Jews and taken over by the Christians in Europe. The Egyptians used a system of 10-day intervals, or decans.

In China, the situation was similar to that in the Middle East and the Nile Valley, though for different reasons. The basically agricultural economy depended on reliable prediction of the time of melting of the snows which fed the great river valleys and of the beginning of the rainy season. There, too, and quite independently, the device of the intercalary year was adopted. From an early date, the emperor himself was the keeper of the calendar, a measure of its national importance.

The crops

From the beginning of time, cereals – selected according to local climate and soil conditions – have been the basis of agriculture, but from a very early date a great variety of other plants have been grown to provide a more varied and attractive diet. Initially these, too, must have been derived from wild varieties, simply because there is no other source. Improved varieties would emerge later as a result of deliberate selection. The more palatable and productive types could then be widely established. Thus Tiglath-Pileser I, ruler of Assyria (1115–1102 BC), records: 'I brought cedar, boxwood and [?] oak trees from the countries which I have conquered, trees the like of which none of the kings my forefathers had ever planted, and I planted them in the gardens of my land. I took rare garden fruits, not found in my own land, and caused them to flourish in the gardens of Assyria.'

Various pulses – peas, beans and lentils – were grown as basic foods. Green and root vegetables were common. Tree fruits included figs, pomegranates, grapes, apples and pears.

The date palm was of particular importance, not only as an important source of food – traditionally 'seven dates make a meal' – but also for wood, coarse fibre for nets and ropes, and leaves for basketry. Like the sugar-rich grape, the date can be fermented to make an alcoholic liquor. In its wide utility it was matched in China by bamboo, from which an extraordinary variety of articles can be made. The date palm is suited to a hot, dry climate: it likes to grow 'with her head in hell-fire and her feet in the river.' The trees are unisexual, but the technique of artificial fertilization seems to have been mastered at a surprisingly early date.

The olive played an important part in the southern European economy early on, but was little grown in Egypt – except in the Fayum – or Mesopotamia. It is important as a source of oil, for lamps as well as food, and olive oil is known to have been imported into Egypt from the eastern Mediterranean from 2500 BC. In Mesopotamia, the sesame seed was the main source of oil.

While the primary use of crops was food, they also provided other useful materials. Among the most important of these were textile fibres. Flax was known in Babylonia and Egypt from around 3000 BC, but seems to have been cultivated in south-western Asia from about 7000 BC. Hemp, apparently originally from the Dnieper area of Russia, was the first fibre used by the Chinese. The botanical origin of cotton is obscure, but it seems to have been first used in the Indus Valley around 3000 BC. From there it spread west to Mesopotamia – the Assyrians looked on it as a kind of vegetable wool – and Egypt, and east to China. The working up of these vegetable fibres, together with wool and silk, involved the development of a whole range of new technologies.

Finally, we must mention plants used in medicine. Herbal remedies have been used since the beginning of time. Doubtless many were of little or no value, except to encourage the patient, but others have demonstrable therapeutic properties. The Ebers Papyrus, dating from the 16th century BC, lists no less than 700 herbal remedies, including poppy, castor oil, aloes and squill. The latter contains a cardiac stimulant similar to that in the drug derived from the foxglove,

Much of our knowledge of early Egyptian agriculture is gained from surviving wall paintings. This example from the tomb of Sinbin – chief artist to Rameses II – shows a typical field scene. Two dappled oxen – urged on with a scourge – draw a plough which clearly does no more than scratch the surface: there is nothing corresponding to the mouldboard of a modern plough. The seed is being sown behind it in the furrow scraped in the soil. The furrow would then have been closed with a rake (right). People also cultivated the soil by hand using a primitive hoe (above right).

digitalis. From this arose the first book on descriptive botany. In the first century AD Dioscorides listed over 500 plants in his *De Materia Medica*, which was still used in the 17th century.

The implements

No doubt bare hands and a pointed stick sufficed for the most primitive cultivation, but very soon a range of implements for specific purposes appeared. Many are not very different from their counterparts in use today or, at least, until very recently. Thus, while the earliest sickles consisted of flints set in a wooden or horn frame, the crescent-shaped copper and bronze sickles of the second millennium BC, clearly depicted in Egyptian and other frescoes, are virtually the same as the iron reaping hooks still sold in thousands by modern ironmongers.

In contrast to the wet, heavy soils of northern Europe, to the problems of which we shall come later, the light dry soils of the Middle East and Egypt needed little preparation. Indeed, in the area of the Nile floods, the custom was to sow the seed in the soft mud before it had time to dry. Elsewhere, the surface of the ground was prepared by picking at it with a hand hoe, larger lumps being broken up with a clod-hammer. From the hoe evolved the plough, the share of which cut a shallow furrow. For small areas a two-man team would often be used, one drawing the plough and the other guiding it. The breast-plough was operated by one man pushing against a crosspiece. From the fourth millenium BC in Europe and possibly earlier in Mesopotamia, draught oxen were being used to haul the plough, and there might be two handles for the ploughman to steer by. In its most primitive form the plough was made wholly of wood, using hard varieties such as holm-oak for the share. With the advent of copper and bronze, and later iron, the tip was protected with metal, making it more durable and more effective.

An interesting and very early accessory to the plough, rather than an essential part of it, was the seed-drill. This was simply a vertical tube, behind the share, down which seed could be fed to the newly opened furrow. The resulting crops grew in manageable rows, rather than haphazardly, as from broadcast seed. Such drills were favoured for vegetables and flax, rather than cereals. Seed-drills are depicted on Babylonian seals of the third millennium BC, and they were in use in India and in China, by the first century AD at latest. Surprisingly, they were unknown in Europe until the 18th century, when they were possibly introduced from China by Jesuit missionaries.

When rivers were in flood, water ran through the irrigation channels of its own accord, regulated by sluices. At other times, however, it had to be raised from rivers and wells, and for this two devices were – and still are – widely used. One was the shaduf, known from 1500 BC, though it is probably much older. A pole is pivoted at its mid-point between two pillars some 2 metres (6 feet) high. It carries a bucket at one end and a counterpoise at the other. The bucket is lowered into the water, raised and emptied into a channel or cistern. With even the simplest such device, a man could distribute some 2700 litres (600 gallons) of water a day. A later device, possibly used in the Hanging Gardens of Babylon in the sixth century BC, was the endless chain of buckets, dipping into the water.

The development of intensive agriculture, with farmers supplying consumers at a distance, posed related transport problems. Loads that were heavier than could conveniently be carried or dragged had to be moved considerable distances. How carts originated and evolved we will consider later. For the moment it suffices to say that a four-wheeled cart with solid wheels is depicted on a Sumerian tablet dating from around 3500 BC, and draught animals, usually oxen, were widely used.

calendar mechanisms 67 **carts** 47–49 **cereal cultivation** 26–27 **ploughs** 203–205 **seed-drill** 204–205

CHAPTER THREE

Domestic Life

When viewed today, the excavated cave dwellings of the distant past seem depressingly bleak and uncomfortable, especially as many were little more than shallow shelters exposed to the weather. But there is no reason to suppose that man was so indifferent to his creature comforts as to live like an animal in its lair. Although all the evidence has long since disappeared, we can visualize hides or brushwood stretched on wooden frames giving a fair degree of protection from wind and rain. Skins with the fur left on them – clearly depicted in very early rock paintings – would provide warm clothing and bedding. An area reserved for cooking – for which there is archaeological evidence from the ashes – was the forerunner of an organized kitchen. Such, indeed, was the way of life at La Grand Abri Troglodytique, in the Dordogne, for example, where local inhabitants were wont until recently to seek refuge in prehistoric cave dwellings during troubled times.

When man began to live in settled village communities – whether in the brick-built houses of the Middle East, or in the perishable thatched wooden huts of Europe – a new and more comfortable phase of domestic life began, supported by the improvement of existing crafts and the development of new ones. For clothing, skins were supplemented by cloth made from natural fibres; pottery, glass and metals were the raw materials for domestic utensils; simple furniture gave increased comfort; the art of cooking was developed to provide a more varied and attractive diet; and ornamentation was pursued for its own sake. By the end of our present period, a highly sophisticated urban life had developed, though – as always – progress was uneven and in many parts of the world virtually non-existent. Even within civilized nations there was, inevitably, a great difference between the lifestyles of the rich and most of the workers, including many slaves. Apart from the wealth of material that has survived by chance, our knowledge of these developments owes much to the custom of burying the dead with their most valuable possessions.

Since the earliest times, natural caves have been enlarged and modified for many purposes. These caves of the Dogon tribe of the Niger have been used both as tombs and for the storage of grain.

Leather

Leather is one of the most ubiquitous materials, widely and diversely used from the dawn of civilization until modern times. Although tanneries were mostly small and local, total production worldwide was enormous even before the start of the Christian era. In hot dry climates, such as that of Egypt, leather survives very well and a great deal of the material has been recovered. Additionally, the importance of leather resulted in much written and pictorial description of how it was made and used. In cold wet climates, however, it quickly perishes totally except – rather surprisingly – under permanently damp conditions where oxygen is excluded, such as the bottoms of old wells and peat bogs: these are profitable sources for archaeologists. For example, a large cargo of reindeer hides was salvaged in excellent condition from the sand-buried wreck of a Danish ship that sank in Plymouth Sound 200 years ago.

Apart from being intrinsically useful, leather was abundantly available. It literally came with the rations, for every slaughtered beast yielded a hide. Nevertheless, the preparation of leather and the making of useful articles involves a great deal more than skinning a beast: a range of preparative processes had to be evolved.

A basic problem is that raw hide is very perishable: unless quickly treated it will putrefy and go rancid. Drying away from direct sunlight will arrest the process of decay but leaves a skin too stiff to use. A first step in preparation was to scrape off the decomposable tissue on one side and – unless a fur rather than leather was required – the hair on the other. The hair was first loosened by steeping in fermenting solutions of flour, grape skins and so on. There were then three alternative preparative treatments. In tawing, the process which gave a white leather, the prepared skin was treated with alum. Buckskinning was favoured by many of the native populations of North America because all the materials were ready to hand after a hunt. In this process, the brains of the slain animal are heated and rubbed into the skin, separating and coating the fibres so that air could pass through but moisture

Leather, normally perishable, is preserved in damp conditions when air is excluded. These boots – unearthed at Oseberg, Norway – date from about AD 800.

would not. Tanning, the better known process, involves long soaking in pits containing tannin-rich bark of various kinds, or oak galls. After this, the treated skin might be softened and waterproofed by kneading with oil or fat; the Eskimos do this by prolonged chewing. At a surprisingly early date, quite precise directions for leather-making were published in cuneiform tablets of the Middle East. A tablet from the Assyrian city of Carchemish, of about 800 BC, states: 'Six shekels of alum, [?] *minae* of oak galls, ⅓ *ka* of fat, ⅓ *mina* and 2 shekels of myrrh for the tanning have been delivered.' The Code of Hammurabi even specifies the wages to be paid to tanners.

Other preparative processes were more unpleasant and involved treating the leather with highly offensive mixtures of excrement and urine – human and animal, including that of dogs and pigeons. As a result, both tanners and tanneries – found in most towns of any size and attached to every Roman garrison – were socially unacceptable. As a matter of social convenience and often of legal requirement, tanneries and associated leather works were sited in accordance with the prevailing winds. Among the Jews, whose attitude to unclean trades was religious as well as practical, practising the trade of a tanner could be grounds for divorce: 'And these are they who are compelled to put away their wives. He that is afflicted with boils or has a polypus, or that collects dog's dung, or is a coppersmith or a tanner, whether these defects be in them before they were married or whether they arose after they were married ... woe unto him who is a tanner by trade.'

The processing techniques were as varied as the purposes for which the leather was required. For most clothing a soft, supple leather was required, but for the soles of sandals hard-wearing toughness was needed. The fact that raw leather becomes stiff on drying was turned to useful account. Thus buckets for the shadufs and a great variety of other containers were made by moulding wet leather. In Neolithic settlements in Norway, and no doubt elsewhere, water was boiled by dropping red-hot stones into filled leather buckets. If carefully made, stitched joints would quickly swell and become watertight.

While most leather went into strictly utilitarian articles – such as tyres for chariot wheels, ropes for shipping, tents, sheaths for weapons, coracle-like boats, sails and harnesses for oxen and horses – it was also used for highly decorative work, such as caskets. It could be stained with a variety of vegetable dyes; ornamented with blind punches or excised patterns; and given a high polish by boning. Finally, leather in the form of vellum or parchment was a very durable alternative to papyrus, and later paper, as a writing material.

Textiles

A basic requirement for making clothing is the availability of fair-sized sheets of supple material, and for many thousands of years leather alone adequately fulfilled this need. But some of the earliest known dwellings show evidence of fabrics woven from natural fibres, both vegetable and animal. The archaeological record is patchy, however, because textiles also tend to decay. Under dry, warm conditions fabric survives remarkably well though, and there is a profusion of material from early Egypt and, of later date, from Peru. Under cold, wet conditions textiles, like leather, survive only in the rather rare circumstances that air is permanently excluded.

In the sense that weaving and the making of baskets and mats generally involve the interlacing of stranded material, both can conveniently be considered together: probably the latter is the older skill, though both were well established by 5000 BC. For making basketry and mats, reeds, grasses, flax stems and cereal straw were all common materials in the Middle East and Egypt; split bamboo was used in China and osiers in Europe. Variations on the simple criss-cross weaving made elaborate patterns possible. A very interesting variant was coiled basketry, in which the fibres were bound into a rope and then coiled and stitched into the desired shape. It is interesting, and surely significant, that the earliest pottery was made in just this way, using long ropes of soft clay. Making rope by twisting fibres was also a very ancient craft.

The fibres

As with cereals, the only original source of vegetable fibre was wild plants and these, too, would have been harvested in the wild before they began to be cultivated. Those in use at an early date are well identified from artefacts, but it is not always clear from contemporary writing which variety is referred to. Probably the first fibre plant to be cultivated was flax, which was well established in both Mesopotamia and the Nile Valley by 3000 BC. It seems likely, however, that it was originally prized for its oil-bearing seeds. It is suited to the climate of Europe, where it is still grown.

Flax stems were used after drying in basketry, but to extract a textile fibre they had to be subjected to a process known as retting, which involves long soaking in water

cloth 37 **cooking** 43–44 **glass** 41–43 **pottery** 38–41 **shaduf** 33

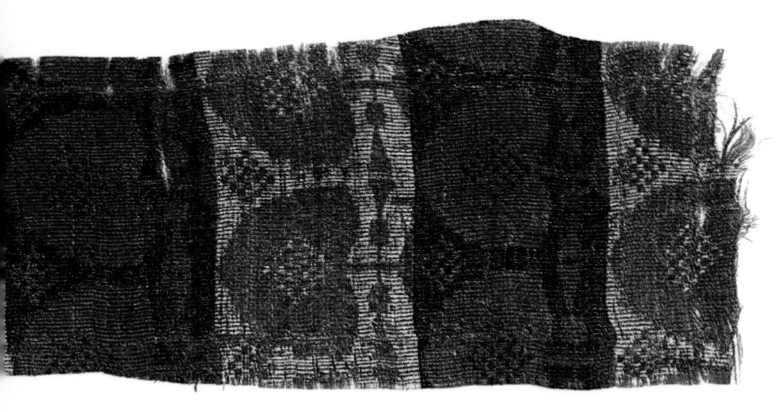

followed by beating. Hemp is another important oil-seed crop – and the source of a widely used narcotic – from which fibres can be obtained also by retting the stems. It was the source of the first Chinese fibre, and hempen ropes were produced there about 2800 BC. Early accounts are obscure because several different fibre plants are referred to as hemp. Whatever the source, it is a coarse fibre, largely used today for making rope and canvas. Another fibre plant processed by retting was jute, still widely grown in the valleys of the Ganges and Brahmaputra. It is possibly identical with the mallow mentioned in the Book of Job. Yet another plant in the same category is the nettle, but this grows so freely as a weed that it is difficult to know if it was cultivated. Nettle fabric was made in Europe up to the First World War.

These were all strong fibres, especially flax, but relatively coarse. A finer fibre is cotton, so fine indeed that many early writers confuse it with silk. It seems to have been of Indian origin and finds in Mohenjo-Daro date from no later than 2000 BC. It is found in both Egypt and China by 500 BC. The Romans established a large cotton manufactory in Malta, but the Indian product was still preferred.

All these were vegetable fibres. Sometimes they were combined as in fustian, a cloth made from linen and cotton. In addition, two fibres of animal origin were important, namely wool and silk. The former means generally the fleece of the sheep, which was widely domesticated throughout Asia and Europe, but that of some other animals – such as goats and lamas – was also used. Originally the wool was plucked from the beast – at certain seasons it naturally loosens – but shears were used by the Romans.

Silk is almost certainly Chinese in origin, and legend attributes the art of unwinding the continuous filament from the cocoon to the wife of the Emperor Huang-ti, about 2700 BC. The Chinese carefully guarded the secret of its production, and classical writers were inclined to the Assyrian view that it grew on trees, perhaps like cotton. This was, in a sense, not far from the truth, for avenues of mulberry trees were the normal habitat of the silkworms. The Chinese were, however, not averse to exporting made-up fabrics, which from around 200 BC found their way to the West down the Silk Road. They were costly, and could be afforded only by the very rich. Often they were unravelled to obtain the yarn, which was then rewoven at many eastern Mediterranean centres, often with flax: by skilful weaving it was possible to cover the flax completely so that the new fabric appeared to be pure silk. Sometimes, too, the silk threads were carefully

Silk has been used in China for nearly 5000 years, but it did not reach the West until about 200 BC, and silk production was not established there until the 10th century AD. The piece of silk damask (top), dyed in five colours, was found in a grave at Turfan in China and dates from the middle of the sixth century AD. The three lower pictures, which are all painted on silk, show the traditional processes of winding silk yarn from the cocoon (top); dyeing the resulting skeins of yarn (centre); and weaving the finished silk product (bottom).

split and worked up into gossamer-thin fabric. In the fourth century AD, the Chinese silk-producing industry spread to Japan and 200 years later an attempt was made to introduce it to Europe. It is said that in AD 552 silkworm eggs were smuggled to Constantinople inside a hollow stick. Not until the 10th century, however, was silk production firmly established in Europe, first in Spain and later in Italy. India had her own 'wild' silk industry, in which the moth was allowed to eat its way out from the cocoon before spinning. This gives short fibres, as opposed to the long ones produced in China. These were worked up into a coarser, uneven thread which for a long time was not recognized as being essentially the same as fine Chinese silk.

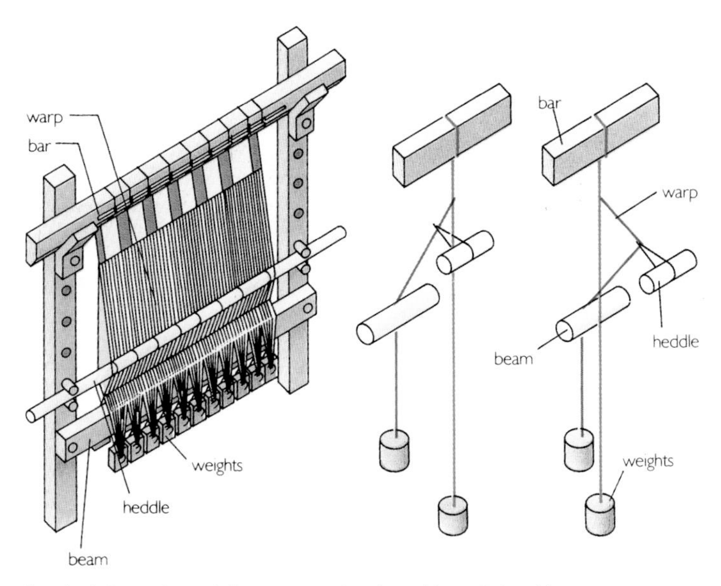

In a simple loom, the vertical warp threads hang from a bar and pass alternately before and behind a fixed beam at the bottom. Above this, the movable heddle separates the warp threads, and the weft thread is passed between them. The heddle is then drawn forward so that the alternation of the warp threads is reversed.

The preparation of fibres

Except for silk – consisting in its natural state of very fine continuous filaments which can immediately be twisted together to make a thread strong enough to weave – wool, cotton and other textile fibres come initially in tangled skeins, from which the fibres must be drawn out in alignment and twisted into thread. If necessary, this spinning process can be done without any equipment at all, simply by rolling raw fibre between the hands or between hand and thigh: some primitive tribes still do just this. In Greece, a half-pipe of clay, often richly ornamented and known as an epinestron, was placed over both thighs and the rubbing was done against this. But more usually some sort of suspended spindle was used, consisting of a short stick loaded with a heavy whorl (or disc) to give it enough momentum to keep it spinning and wind up the spun thread. The distaff was a similar device, but rather larger, often about a metre (3 feet) long. Spinning was almost universally an occupation for women: to this day the female side of a family is described as the distaff side. Often a favourite spindle was buried with the spinner's body and many have survived in this way.

Weaving

To convert the spun fibre into cloth some kind of loom was generally used. In principle – and very often in practice – this could be a very simple device. Its essential feature was some means of stretching the warp threads (in a horizontal loom, those stretched out lengthwise), while the weft was interwoven with them at right angles. In its most primitive form, this would be done with a needle, as in darning, but at a very early date alternative warp threads were attached to a heddle. By moving this to and fro, a 'shed' of warp threads was formed through which the weft was passed from side to side by means of a shuttle. After each passage of the weft, the cloth was pressed firm by means of a comb or sley.

Various technical changes were rung on this simple principle. One of the earliest looms – still encountered among primitive people – was the horizontal ground loom, which was simply a wooden framework pegged out on the ground. Alternatively, the frame was mounted vertically and so contrived that the weaver could sit while working. In both types the warp was kept taut by securing it to the frame. In the warp-weighted loom, the warp was fixed to a beam at the top but held under tension by being tied to loom weights – a common archaeological find – at the bottom.

The simplest form of weaving, in which warp and weft criss-cross alternately, is known as tabby weave. Variations gave attractive textured effects. The familiar canvas weave is made by passing two weft threads over two warps. Pleasing patterns could also be produced by using threads coloured with a considerable range of natural dyes – red from an insect parasitic on the kermes oak, violet orchil from lichens, yellow from saffron, blue from woad and indigo, and – prized above all – imperial Tyrian purple extracted at huge expense from a kind of Mediterranean shellfish. Apart from this, much more ambitious variations were attempted. As early as the second millennium BC double weaving was practised. In this, a pattern was obtained on both sides by using duplicate warp threads: the most outstanding surviving example is the girdle of Rameses III, dating from about 1170 BC.

Not all early fabric was woven, however: knitted fabrics created from a series of loops seem to have been made as early as 1000 BC. The advantage was that it conferred a degree of elasticity on the fabric, making it easier to adapt clothes to the shape of the human body. Much of it was done on frames, transferring the loops from peg to peg with a hooked needle. From this evolved the modern form of hand-knitting, in which all the loops are put on a single needle and then transferred to another.

Finishing processes

Even now the cloth was not finally ready for use. It was next subjected to fulling, in which gaps between the fibres were obliterated by trampling or beating the cloth in baths of

flax 35 **Mohenjo-Daro** 10–11, 31 **Silk Road** 10–11, 23 **synthetic fibres** 268 **textile machines** 147–150

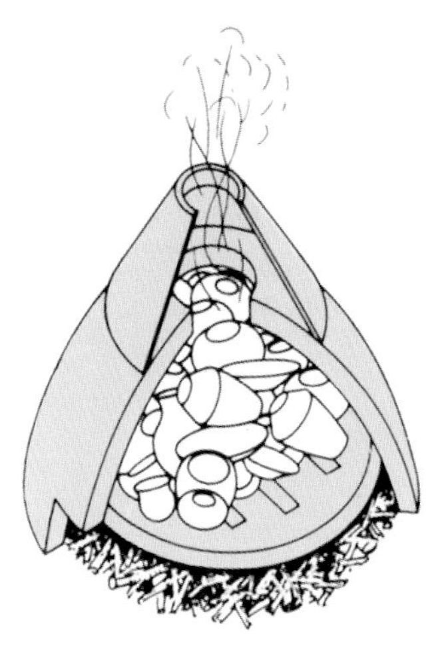

Early pottery was fired in a wood-burning vertical kiln (above). From a very early date, pottery vessels were decorated either with some incised pattern or a coloured glaze, as on this Chinese bowl with everted rim of the late fourth millennium BC, excavated in 1963 in Kiangsu (right).

The potter's wheel is first known in Mesopotamia from about 3000 BC and was not very different from its modern counterpart, with kickwheel (above). However, large vessels such as these Minoan storage jars from Knossos (left) were built up by hand.

Ancient pottery exemplifies the very uneven rate of progress of technology in different parts of the world. This simply decorated Neolithic vessel (below left) from Mildenhall, England, contrasts sharply with the elegant white pottery tripod jug (below) from Shantung, dating from about 2000 BC. The applied pellets of clay and the ridges of the shoulder possibly mimic similar vessels made from riveted sheet metal. The Greek amphora (left), dating from about 560 BC, illustrates the extraordinary skill of the best early potters in pictorial representation – from which, incidentally, we may learn much about life at that time.

detergent substances, such as soda, soapwort or fuller's earth. To give woollens a softer finish, the surface fibres were raised by brushing with teasels, or even hedgehog skins, and then shearing the nap.

Linen comes from the loom fawn in colour but, like cotton, it was prized for brilliant whiteness. This was achieved by long exposure to the sun in bleach fields. For the finest fabrics, the surface was then smoothed by putting it in a press or rubbing it with stones or blocks of wood. Woollen cloth was bleached with the fumes of burning sulphur.

Pottery

Because ceramic materials are virtually indestructible under most natural conditions, they provide our strongest single link with the lives of our distant forebears. They provide not only evidence of their needs and their skill in fulfilling them, but also much inferential evidence about their way of life. Different ethnic cultures made very characteristic wares, and the distribution of artefacts indicates not only the geographical limits of their makers but the principal routes along which goods were traded. Rarely, but sometimes very significantly, pottery bears the permanent imprint of some object long since perished – a grain of rice or a piece of fabric. Fluctuations in the pottery record are also informative. Because it is fragile, pottery is particularly suitable for settled communities, though it was certainly used in Neolithic times before urban life was established. So a decline in pottery may signify a decline in urban life and a return to less settled rural conditions: nomadic people used minimal domestic equipment and preferred it to be made of the less easily breakable wood, leather or, as it became available, metal.

Firing

Clay occurs widely on the earth's surface and it will set hard whenever a fairly strong fire is lit on it. It is very likely, therefore, that the firing of clay was not a single discovery that diffused outwards but one made independently at various sites. The oldest fragments of fired pottery in Japan are thought to be 13,000 years old.

The craft of the potter is threefold. First, the clay has to be kneaded with water until it is soft enough to work, but still stiff enough not to collapse as a vessel is shaped. Secondly, the desired vessel must be shaped. Thirdly, the clay must be fired to convert it from the plastic to a rigid condition.

Just as yarn can, if necessary, be spun with no equipment whatsoever, so clay can be shaped – and pleasingly shaped – by the hands alone. For this two techniques were used. First, a lump of clay was roughly shaped and then finished by repeatedly going round it to thin and shape the walls between finger and thumb. Alternatively, the vessel could be built up from long ropes of clay made by rolling between the hands: either a continuous coil was used or lengths cut to form a single course.

For all these methods, the potter would need to rotate the unfinished vessel as he worked, and for this a simple turntable would be helpful, even if it was no more than a flat stone with a convex base. From this it would be a relatively small step to the potter's wheel on which vessels could be 'thrown', the necessary energy deriving from the momentum of a flywheel instead of from the potter's hands. Generally, the flywheel was kept going by occasional kicks with the feet or jabs with a stick. For convenience, the clay was placed on a smaller disc which was on the same axle as the flywheel. Such wheels appear in Mesopotamia from about 3000 BC. If the wheel was true, elegantly symmetrical designs could be quickly executed. Such vessels as pitchers, with necks too small to admit the potter's hands, were made in two parts luted together. Some vessels were mass produced, as it were, by moulding on a wooden or pottery former. To give a fine impermeable surface, the vessel would finally be coated with a thin layer of fine clay or slip. Slip was also used to attach accessories such as handles.

If dried in the sun, or perhaps at the domestic hearth, such vessels would set hard enough to be usable, but to make them really strong and rigid they had to be fired at temperatures in the range of 450°–700°C (850°–1300°F), followed by very slow cooling to prevent cracking. While this can be, and in places still is, done simply by stacking the pots together and lighting a fire over them, more controlled results

Richly coloured glazes are achieved by applying various minerals before firing: much care is necessary to achieve the desired effect. This splendid Roman example, depicting the head of a god, dates from about 30 BC.

and higher temperatures are obtained with a kiln. In the vertical kiln, which probably dates from the fourth millennium BC, the fire roars up the middle and heating tends to be uneven. A better device is the horizontal kiln, in which the hot gases are reflected downwards from the roof on to the pots. Such kilns were used by the Romans, and in China by AD 200. The simplicity of the open-fire method, using almost any kind of fuel, would appeal to local village potters. Kilns were more suitable for the bigger outputs of urban potteries, though more expensive because they required prepared charcoal or billets of wood. With such kilns temperatures of a little over 1000°C (1850°F) could be attained: at this point some vitrification (converting into glass) occurs and even stronger and less porous products are obtained. Most clays will not completely vitrify below about 1600°C (2900°F), but in the middle of the second millennium BC Chinese potters discovered different kinds of clay, known as kaolins, which vitrify at much lower temperatures, yielding hard impervious stoneware.

Whatever method was used, large quantities of fuel were required, and where practicable potteries were sited where both clay and wood were available. Sometimes, however, it seems that – as is still done in the hill country south of Tibet – the purchasers took away unfired pots and fired them themselves at the first opportunity. Again making a reasonable inference from modern practice, some communities would be supplied by itinerant potters who made the best use they could of local materials.

Shapes and decorations

It is commonplace in technology for new products to mimic, initially, those they displace, before being accepted in their own right. It is not surprising, therefore, that the designs of some early pottery vessels clearly show the influence of contemporary ones in wood, basketry, leather or metal. Thus some Minoan vessels of around 2500 BC are decorated with functionless rivets; others were painted to resemble wood.

Decoration was of two kinds, not necessarily mutually exclusive. On the one hand, there was what might be called mechanical design – markings made with a stick, a fingernail or a piece of shell. However, what at first sight appears to be ornamentation may sometimes be no more than incidental to the making of the piece. Thus the string marking on some early Chinese and European ware may be no more than the trace of string used to keep the vessel in shape before firing: some pots certainly show attempts at deliberate erasure of the marks.

Apart from indentation and provision of a fine finish by applying slip before firing, pottery could be colourfully decorated by applying various minerals before firing. Most clays contain iron compounds and when fired with free access of air these impart a characteristic red colour to it. This can be enhanced by local application of iron oxide, a common

This magnificent pendant and necklace – made from gold, lapis lazuli, green feldspar, cornelian and turquoise – was among the treasures found in the tomb of Tutankhamun (*circa* 1350 BC). The theme of the pendant is a boat, with a scarab at its centre.

mineral. If air is largely excluded during the firing, the product is greyish in colour. If carbonaceous material, such as charcoal or oil, is applied, black ware is obtained, provided not all the carbon is burnt away. Other colours can be obtained by applying minerals such as sienna, umber and ochre. With these simple aids early potters achieved remarkable effects; great skill was necessary to judge the colour that would finally emerge after firing. Sometimes, as in the T'ang pottery of China, the emphasis was on uniform fine glazes and elegant design. Greek potters, on the other hand, favoured pictorial representation – a practice valuable in showing us scenes from the everyday life of the period.

Glass

Glass has never been manufactured on the same scale as pottery and as it is fragile and friable early examples are correspondingly rare. All ancient glasses – except those of China – are of similar composition, though the minor variations are sometimes helpful in identifying the source of particular items. In the Middle East and Europe, glass was made by fusing sand and limestone with natural soda (natron) or plant-ash. Traces of certain metals impart a colour to the glass. Iron is a common impurity and gives a greenish colour, but other metals were deliberately added, on an empirical basis: cobalt compounds, for example, impart a beautiful deep blue colour.

The earliest glass objects known from Egypt date from about 2500 BC, but glass vessels did not appear for another thousand years, when a number of small factories were

early pottery 38–39 **glazes** 122 **modern pottery and glass** 303 **porcelain** 150–152

Until about the beginning of the Christian era, glass was cast, as it was to make the small articles (top) from Carthage (sixth century BC). The technique of glass-blowing then developed, initially probably in Syria. Glassmaking was well established in the Near East by the middle of the second millennium BC, and a high standard of craftsmanship was developed, as evidenced by the Roman glass vessel (top left), probably of the fourth century BC, the first century AD Roman carafe (top right) and the fifth century AD conical beaker (far right) excavated at Kempton, England – though not necessarily of local provenance. The first blown glass was blown into moulds – as in the first century AD Roman beaker (above), which depicts circus scenes. Later, however, freestyle blowing also was adopted.

established. By that time they were also being made in Mesopotamia. Chinese glass appears much later, dating from about the sixth century BC, but is remarkable in containing large amounts of lead, the effect of which is to give a much more brilliant product: how far this was deliberate is, however, debatable.

Small solid objects such as beads and figurines were shaped from molten glass in clay moulds, but vessels were made by the sand-core technique. In this, sand tied up in a cloth bag was dipped into molten glass and the mass was shaped while still plastic by rolling it on a flat stone. When cold and set, the sand was extracted. Some large objects – such as a head-rest recovered from the tomb of Tutankhamun (*circa* 1350 BC) – were carved and ground from solid blocks of glass, no doubt precast to roughly the right shape, much as sculptors worked in stone.

By the middle of the second millennium BC, glassmaking was quite a flourishing industry in the Near East, but for some reason not understood it declined sharply between 1100 and 750 BC, after which there was a great revival when the Phoenicians became the greatest glass manufacturers and exporters of the ancient world. Just before the start of the Christian era there was a major development in the making of vessels, marked by the blowing of molten glass with a pipe, originally into clay moulds but later freestyle. Glassmaking extended to Italy, especially to the head of the Adriatic in the area of modern Venice – where to this day the craft is associated with the Murano district – and thence throughout the Roman Empire. Wealthy Romans had windows made of glass in their homes.

Preservation and cooking of food

The cultivation of crops and the domestication of animals were fundamental factors in assuring a regular and varied supply of food. As a great deal of pottery was used in one way or another in the context of food – either in the kitchen or to transport in bulk such comestibles as oil and wine – this is an opportune moment to say something about food and cooking, a very basic technology.

To maintain a surplus of food immediately raises the question of preservation. Cereals generally presented few problems: provided they could be harvested when ripe and the grain stored dry, they would keep for long periods. Nevertheless, in the wet and uncertain climate of northern Europe this was not always easy and, as now, vermin could cause big losses. Fruit and vegetables were mostly eaten fresh in season, though some with a high sugar content – such as dates, grapes and figs – could be preserved by sun-drying. The sugar-rich fruits were also fermented to make wine, and cereals to make beer. The Greeks were the first people among whom wine was drunk by all classes, and some of their vineyards extended to 40 hectares (100 acres).

Animal products, including fish, were a very different matter, for they are highly perishable, especially in hot climates. Several methods of preservation were practised from an early date. Meat and fish were dried on frames exposed to the air until the water content was too low to support bacterial growth. Such foods are represented in modern times by the pemmican of North America, the biltong of Africa and the dried cod of Europe. Strong sun was not necessary: the Eskimo managed to dry fish and meat very well in the strong cold northern winds. Alternatively, it could be frozen in cold climates or pickled in brine. Both meat and fish were preserved by liberal treatment with salt, which was subsequently soaked out. For this reason salt was one of the most widely traded commodities in the ancient world, as testified by finds of salt-cakes along the old trade routes. Salt was obtained – as it still is – partly by mining and partly by the evaporation of sea-water or the water of saline springs. In was, of course, much valued as a seasoning. In addition, there was a very large trade in spices, especially pepper, from the East.

Animals have never featured very prominently in Chinese agriculture, but elsewhere dairy products were of major importance. The life of raw milk is very short, especially in the hot climates of the Middle East and Mediterranean, but two derived products – butter and cheese – are much less perishable. A bas-relief from Babylon shows all the operations of a dairy. The farmer milks his cow – from behind, not from the side – and an assistant agitates the milk in a pottery churn. The globules of butter fat are then separated in a funnel. The product must have been similar to Indian *ghee* (clarified butter), in the general absence of any sort of cold storage – though there is a Chinese reference to ice-cellars before the first millennium BC, and wealthy Romans maintained ice-houses supplied from permanent mountain snow. The Egyptians were familiar with the cooling effect of evaporation and stored drinks in porous earthenware vessels kept constantly moist; exactly the same principle is used in modern unglazed butter and milk coolers.

When milk sours naturally, a lactic acid fermentation occurs, but the curd deposited is not a true cheese and will not ripen on storage: it is therefore eaten fresh. True cheese is made by coagulating the casein (the principal albuminous constituent of milk) with rennet, an enzyme present in the stomachs of young animals. This process, too, is of great antiquity. How it was discovered is a matter for surmise: a plausible explanation is that is arose as a result of storing milk in containers made from calves' stomachs.

The simplest form of cooking is by roasting in front of an open fire, and this is suitable both for meat and a dough of pounded grain. Ovens of clay or stone give better control of the cooking process and have been found among the earliest strata of Jericho. The general practice was to heat the oven by lighting a fire in it, and to rake this out when hot enough. Although much cooking was done at home, communal bakeries were used in Sumeria, Egypt and, much later, the Roman Empire.

With the advent of pottery, and later metal, vessels that could be placed directly on a fire or suspended above it, various forms of stewing were then possible. Meat, in particular, could thus be made much more tender. Some time before 1500 BC the Chinese developed an unusual form of cooking pot called a *li*. This had three hollow legs, the effect of which – whether deliberate or not – was to expose a much larger cooking area to the fire, an important feature in a land where fuel was scarce. Later, in the Shang period, these vessels were made in bronze and many were of very handsome design. The traditional Chinese way of making bread is by steaming, rather than baking, and this was done by fitting a perforated platform in the *li*, above the boiling water. These steamers, developed some 3500 years ago, are virtually the same as the familiar steamers of modern Western kitchens.

As we shall see later, the chemical arts grew – at least to some extent – from culinary ones. The familiar *bain-marie* (water-bath), designed not to steam food but to avoid overheating by keeping it at the temperature of boiling water, is reputed to have been invented about 1500 BC by a shadowy

early agriculture 26–33 **food preservation** 152, 299–300 **salt** 119–120

woman alchemist known as Mary the Jewess, who has been identified, on rather dubious evidence, with Moses' sister Miriam.

Furniture

With furniture, we have to make the same sort of reservations in respect of this early period as we have already done in other contexts: it varied from virtual non-existence in the homes of the very poor to an elegance and craftsmanship not seen again until the Renaissance. The question of how furniture was made we will defer until we look at woodworking tools, and now briefly consider only the kind of furniture that was made. As furniture supplies the same basic human needs the world over – to give comfort in sleeping, resting and eating, and to store and display personal possessions – it is not surprising that certain common features prevail widely. We must, nevertheless, allow for different regional customs. In some countries people prefer to sit cross-legged on cushions rather than in chairs; some eat upright at table, others in a reclining position. Many surviving artefacts are not complete – chairs and beds would have been upholstered with rush seats or leather thongs, and a marble couch would be comfortless without some form of mattress. We must remember, too, that virtually all everyday furniture has totally disappeared – though we catch glimpses of it depicted in vase paintings and frescoes – and what survives usually does so for special reasons: it may be part of the furniture of a nobleman's tomb, or made of bronze or some other particularly durable material.

A widespread feature of early furniture was the frequent use of animal motifs. The legs of tables and chairs, for example, were often modelled on those of animals, often with a distinction between front and hind legs. This continued in the ball-and-claw feet of tables in recent times. Such items had to be carved, but from the seventh century BC the Greeks made much use of lathe-turned woodwork. Wood has always been by far the most widely employed material and many different varieties were available. The Egyptians introduced the practice of covering cheaper wood with a thin veneer of a better one: for this purpose the Romans particularly favoured citron wood from Mauretania.

In primitive communities the only form of artificial light was that derived from the domestic fire. From this evolved torches, which were wooden staves, tipped with resin or bitumen. From these, in turn, were derived lamps and candles; the latter seem to have been used first in Asia Minor in the sixth century BC. Under the Romans, lamps became elaborate products of the metalworker's skill. At night pedestrians used torches to guide them, though there is evidence of street-lighting in the fourth century AD in a few cities such as Antioch, the Syrian capital.

Locks and keys became more widely used. Originally they consisted merely of a peg pushed through a hole to lift the latch on the inside, but the Greeks and Romans used keys with wards to fit a particular lock. As the security of personal possessions became a more pressing consideration, the complexity of locks increased. The locksmith's art developed in step with the growing prosperity of the town-dwellers of the ancient world.

At their best, the craftsmen of the ancient world produced articles that could scarcely be surpassed today with all the modern aids available. This ecclesiastical throne from Tutankhamun's tomb (top) is made of wood inlaid with semi-precious stones, ebony and ivory. The Roman lamp (right) indicates the care devoted to making everyday articles.

CHAPTER FOUR

Transport

Although the archaeological evidence is in the nature of things very patchy, it is clear that the popular conception of early agricultural settlements as self-contained communities growing their crops and tending their herds is not generally true. Even in pre-pottery Jericho, dating from around the seventh millennium BC, ample evidence for long-distance trade is provided by the discovery of cowrie shells from the Red Sea, turquoise from Sinai and obsidian from Anatolia. When Stonehenge was built between around 3000 BC and around 1000 BC, the massive bluestones were brought from the Preseli Mountains 240 kilometres (150 miles) away in Wales. This indicates not only remarkable engineering skill but also the existence even then of a social system in Britain capable of co-ordinating transport of very heavy objects over a long distance. More significant still, the design of a carving of a bronze dagger on one of the stones indicates trade contact – supported by other evidence – between Britain and Mycenean Greece as early as 1500 BC. Such examples can be multiplied many times over and it is clear that at a very early date transport was important not only for the local carriage of goods but also to support regular import/export systems.

By the beginning of the Christian era a highly sophisticated trade network had been established that stretched eastward as far as China, served all the countries bordering the Mediterranean and extended in northern Europe as far as Germany and Britain. From the south, it was fed by routes from southern Africa. Much of the trade was in luxury goods – silk and precious stones, glass and wine, gold and silver, oil and spices, wild beasts for the Roman arenas – but already transport in bulk had become a necessity. With a population of around one million, Rome was dependent on basic imports of many kinds, especially of grain and oil. This entailed elaborate dock systems – including huge granaries and lighthouses – of which the most important were those at Puteoli (Pozzuoli), two weeks' sailing from Alexandria, and later Ostia, commanding the mouth of the Tiber. Equally, although the provincial cities of the Roman Empire gradually became reasonably self-sufficient, they needed military supplies and all the variety of goods that combined to make up the Roman way of life. An important social consequence of the development of these trade routes was the rise of a distinct merchant class which, although held in no great social esteem, was of considerable influence.

For these reasons transport, and its systematic organization, early became of paramount importance. Water was favoured for transporting goods in bulk. By comparison with the Mediterranean, the Arabian Sea and the Indian Ocean, and such great natural waterways as the Nile, the Tigris and Euphrates, or the Rhine, Danube and Rhone, land transport was slow and expensive, but it was naturally the first to be developed. While the Romans established a comprehensive network of made roads, most of the other great arteries – such as the Silk Road – were routes rather than roads, punctuated by facilities for travellers at appropriate intervals.

Land transport

Up to a certain weight and size, burdens can be moved considerable distances with no equipment at all, but from the earliest times various simple aids were invoked. A Mesopotamian bas-relief shows men carrying loads on their backs with the aid of straps across the forehead, as labourers do in Africa today. Various forms of yoke have been used from time immemorial. The single yoke, with a load at each end, is carried across one shoulder, the double yoke over both. Loads too heavy for one man could be carried on a pole supported by the shoulders of two. This principle was extended to various forms of palanquin designed to carry people of importance. This survived up to the sedan chair of recent times, and indeed up to the present in the conveyance of high dignitaries on ceremonial occasions.

Loads too heavy to carry could be dragged, utilizing the strength of several men, but the mere act of dragging could damage the burden. It was, therefore, a logical step to use some form of sledge fitted with smooth runners. These were, of course, especially effective in the icy conditions of the far

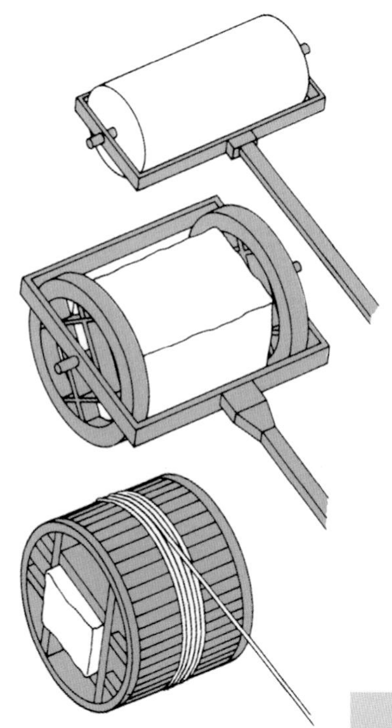

Stonehenge, apparently built mostly in the period 2100–1500 BC, illustrates the ability of ancient peoples to move very heavy loads: some of the stones weigh as much as 40 tonnes. The kind of devices shown right were used to move heavy pieces of masonry. At Stonehenge, some faint carvings of a dagger and axe-heads (above) indicate a Mycenean connection.

ancient trade routes 10–11 **early water transport** 49–53 **Mycenae** 55, 85, 86 **Silk Road** 10–11, 23 **wheeled vehicles** 47–49

north, but they were widely used elsewhere before the wheel became common: a Mesopotamian pictograph from Uruk, dating from about 3500 BC, clearly depicts such a sledge, similar to the travois of native Americans. It was only in the north, however, that skates, skis and snowshoes were developed to give individual mobility on snow and ice: Mesolithic rock carvings from Scandinavia very clearly depict the use of skis; they were known in China and central Asia from the seventh century AD.

An early four-wheeled chariot depicted in a Sumerian inscription of about 2500 BC. As was common with early vehicles, two draught animals (here horses) were employed, harnessed in parallel.

Beasts of burden

For countless generations man was his own beast of burden, though by communal efforts – particularly by the use of slaves – large quantities of goods could be moved if divided into manageable portions. With the domestication of animals, however, new possibilities arose: with their greater strength they could manage heavier loads and they could be trained to carry a rider. The earliest pack animal – possibly as early as 3500 BC – was the ass, which probably originated in what is now the Sudan. It is a hardy animal, resistant to harsh treatment, and can carry loads up to about 60 kilograms (132 pounds). The ass was soon joined by the onager, a closely related species from central Asia: it was domesticated in Mesopotamia by 3000 BC. The camel, well equipped for sandy deserts with its large feet and ability to go for long periods without water, was probably a comparatively late arrival, also from Asia. Although one is depicted on an Egyptian fresco of 3000 BC, this could well have been no more than a curiosity from a menagerie. The use of camels as pack animals cannot be firmly dated before the first millennium BC, in Mesopotamia. Two distinct breeds of camel were developed: the fast riding camel and the much heavier pack animal, which can carry loads of 300–500 kilograms (660–1100 pounds). These were totally different beasts, as different as racehorses and cart-horses.

Among beasts of burden the horse came to occupy a unique place because of its speed and versatility. It, too, originated in Asia and there is some evidence of its domestication in the Ukraine by the middle of the fourth millennium BC. However, finds of horse bones in association with human habitation may mean no more than that horses were eaten, as they still are. Horses were ridden by the nomads of the Asiatic steppes, including the Hyksos who conquered Egypt in 1700 BC, and they had a powerful influence on the conduct of warfare, first through the use of chariots and later as cavalry. Horse-drawn vehicles are depicted from about 2000 BC, but horses were not of widespread importance until the first millennium BC.

One very significant use of horses was in communication. After Cyrus, the Persian Empire embodied a road system stretching from Susa to Sardis, 2600 kilometres (1600 miles) away. Along the stages of this route sped regular messengers on horseback, completing the journey in nine days, an average of nearly 300 kilometres (190 miles) a day, a speed not bettered until Napoleonic times. It is interesting to note, however, that at the time of the Incas in Peru – where all travel was on foot – messages were conveyed along the main highway by relays of runners at the rate of 240 kilometres (150 miles) a day: it was even claimed that the royal table in the capital, 480 kilometres (300 miles) from the sea, could be served with fish only 24 hours after they were caught.

In hot dry climates horses suffer from the heat, and asses are too small for heavy work. The crossbred mule – tough-skinned, resistant to both heat and cold and requiring relatively little water – was the answer to this difficulty. When the cross was introduced is not clear, but Herodotus, writing in the fifth century BC, refers to Persian mules disturbing Scythian cavalry. Pack mules capable of covering up to 80 kilometres (50 miles) a day, were of far greater importance than horses for the transport of goods in classical times.

Horses were doubtless ridden bareback for a long period after their domestication, but the ease with which they can be controlled can be much improved by various forms of harness. This was important for the individual rider, but even more so in the control of chariots and cavalry. The padded saddle was known in China in the first century AD, but not used by the Romans until 300 years later. Apparently it evolved from the simple horsecloth used by the Scythians from the sixth century BC.

The stirrup first appears as a leather loop, also among the Scythians – and is depicted on a south Russian vase of *circa* 380 BC – but may well have begun as an aid to mounting rather than a means of control. The rigid metal stirrup, from which the foot can be quickly withdrawn in case of mishaps, may well have originated in China – possibly as early as the first century AD but certainly by the fifth – and found its way to Europe through the horse-riding tribes of central Asia. In mounted combat the stirrup provided both secure posture and greater manoeuvrability. There is no clear evidence of such stirrups in Europe until after Roman times.

The Romans had relied on horned saddles which permitted some manoeuvring by rotating their hips and pressing against the horns with their thighs. The reverse is true of spurs, which imply the use of riding boots. They were used in Europe from the fourth century BC but are a comparatively recent innovation in the Far East.

A

B

C

D

The earliest wheels were solid, assembled from planks and secured with crosspieces (A, B). These were clumsy and heavy, and the use of spokes made possible lighter construction without losing strength. The simplest form (C) involved a single, heavy transverse spoke, but by 2000 BC multi-spoked wheels had appeared in Mesopotamia (D).

Wheels could either rotate at the end of fixed axles (top right) or be fixed to the axle and rotate with it, the axle turning in some sort of trunnion (side projection) beneath the vehicle (bottom right). Wheels that turn freely have the advantage that they can rotate at different speeds, as is necessary if wheel-skid is to be avoided on bends.

The Celtic chariot (below), used in Britain, France and Germany against the Romans, had two wheels and semicircular sides. Drawn by two horses, it was a swift, narrow vehicle in which the warrior may have stood behind the driver.

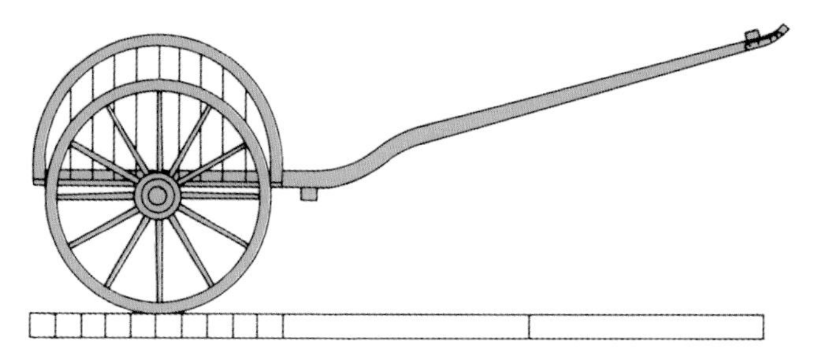

Horses run naturally with their heads forward: the bit, controlled through reins, was designed to enable the rider to force the animal's head up and so bring it to a halt. Simple snaffle bits were widely used in the ancient world, but the earliest known use of the fiercer curb-bit dates from a Celtic grave of the third century BC.

In the wild, or when confined to easy ground, horses' hooves grow quickly enough to compensate for wear and this is not so when they are used on made roads or other rough surfaces. The shoe was apparently a Roman innovation and took two forms. The long outmoded hippo-sandal was simply an iron plate tied over the hoof with thongs. Nailed-on iron shoes, similar to those today, date from the first century.

Wheeled vehicles

It is tempting to suppose that the first wheels derived from logs used as rollers, but there is no convincing evidence for this. The potter's wheel, turning on a central axle, is a more plausible source. All the earliest wheels were cut as a disc from three planks laid parallel: the three pieces were then held together by two crosspieces. If we accept a Middle Eastern origin for the wheel, as is likely, its mode of construction may have been dictated by the dearth of trees large enough to provide planks from which the wheel could be cut in a single piece. It is perhaps significant that in much later wheels from Denmark, where trees of the necessary size were available, wheels were cut from single planks and not – as one would expect if they derived from rollers – by cutting successive thin cross-sections. Experience shows that wood so cut tends to split radially and fall to pieces.

A pictograph from Uruk of about 3500 BC shows side by side a sledge and a vehicle virtually identical but fitted with wheels, but there is no general evidence that this was the main line of evolution. From about 3000 BC there is an astonishing range of artefacts which give us a very clear picture of early wheeled vehicles. These include actual vehicles, as well as models of them, from royal tombs; carvings on stone; frescoes; and vase paintings. From Europe, some have been recovered from bogs, perhaps cast in as votive offerings. From China, wheeled vehicles are known from 2600 BC; a little later carriage-building was sufficiently specialized to be particularly identified with the Xue clan.

From these sources it is not clear whether the wheel turned on a fixed axle – advantageous on bends – or, as in some bullock carts of traditional design in modern India, the wheels and axle turned together. We see the same things in more modern times: in old-fashioned railway wagons axles and wheels turn together, but the non-driven wheels of automobiles rotate on the axle. As a circular hub is normally depicted, however, and there are sometimes indications of a

ancient civilizations 10–11 **development of road transport** 176–177, 246–251 **potter's wheel** 38, 40 **wheels** 48–49

linchpin to secure the wheel, the free-turning wheel was probably normal practice, though there must have been plenty of local variation. In either event, we can reasonably suppose that the moving parts would be lubricated with water or animal or vegetable fats. A Chinese work of the fifth century AD refers to lubrication with mineral oil.

The sections of these tripartite wheels were held together by wooden crosspieces and for common use this no doubt sufficed to finish construction. From as early as 3000 BC, however, we find the wheel strengthened by a rim (felloe): this might be of wood studded with nails or – from 2000 BC – a copper strip. Surviving felloes indicate that the wheels were thin, little wider than 2½ centimetres (1 inch) in some cases, but they must nevertheless have been heavy and clumsy. This disadvantage was mitigated by the introduction of wheels with spokes, originally four in number, which appeared in Mesopotamia around 2000 BC. They were in use in Egypt some 400 years later, in China by 1300 BC and in Scandinavia by 1000 BC. Wheels with six or eight spokes were in use in the West by the beginning of the second millennium BC, and in China vehicles of the same period having 18-spoked wheels have been excavated.

The widespread distribution of the tripartite wheel suggests dispersal of a single invention, perhaps made in the Mesopotamian region somewhat before 3000 BC, but the possibility of invention at more than one point cannot be altogether ruled out. Certainly the vehicles to which wheels were attached were very diverse, though they had one point in common – they had a central pole and were designed for pairs of draught animals. These were exclusively oxen until about 1000 BC. Where more than two animals were used, they normally worked in line abreast rather than in file. Broad-shouldered oxen pushed against a heavy wooden yoke secured to collars. When horses were introduced, this was anatomically impracticable and they worked against a breast-band, but this, too, was unsatisfactory because it constricted breathing and tended to restrict the main blood vessels supplying the brain. Although this defect was recognized, it was – rather surprisingly – not overcome until the advent of modern harness in medieval times. A plausible explanation is that oxen were used for all heavy work, and horses normally worked so much within the limit of their capacity that the inefficiency of their harness was of no great consequence.

Vehicles with two shafts, as opposed to a single pole, were common in China around the beginning of the Christian era, but the rare mentions of them in Roman sources suggest that they were then not popular in the West.

Except for ceremonial purposes almost all early vehicles were two-wheeled carts rather than four-wheeled wagons, which were common by Roman times for heavier loads. The wheels were rather small, rarely exceeding 1 metre (3 feet) in diameter. Roman wagons must have been clumsy, even though the wheelbase was short, for it seems that both axles were firmly fixed to the body of the vehicle. The swivelling front bogie was probably a Celtic invention made in the first century BC. It seems to have been adopted or reinvented by the Romans, for a Diocletian edict of AD 301 concerns the prices to be charged for wagon parts refers to a *columella* (little pillar). The same word was then used for vertical spigots of various kinds, such as those on which the rotors of oil-presses turned and probably refers here to the spigot of a bogie. The Romans were much impressed by the barbarians' skill at vehicle building and indeed adopted many Celtic terms for their own use.

A minor but interestingly and important variation was the wheelbarrow, essentially a one-wheeled cart. This originated, and was extensively used, in China from the first century AD but did not appear in Europe until medieval times, when it is frequently depicted in illustrations of mining and building

Top: We can reasonably suppose that racing is as old as civilization. This picture on a Greek vase of *circa* 550 BC is very reminiscent of modern pony-cart racing. Note the light body and wheels; the two animals are mules.

Left: A wagon (*circa* AD 800) excavated at Oseberg, Norway. In overall design, it is strikingly similar to farm carts used in northern Europe more than 1000 years later.

This spirited picture, from a Han tomb rubbing, shows a light vehicle and harness as used in China about the beginning of the Christian era. Note the single horse and paired shaft: in the West, a single shaft and a pair of animals was then usual. The wheel shown has six spokes – though this may be artistic licence – but by that time many vehicles in both China and the West might have had as many as 18 spokes.

activities. It is a convenient and economical means of negotiating both narrow urban alleyways and field paths.

Although wheels and harness showed little variation, there was a great diversity of vehicles: sometimes different bodies were mounted on the same chassis, according to need. For quick personal journeys, and for local use, light two-wheeled vehicles were preferred. For several fare-paying passengers, Roman firms favoured four-wheeled covered vehicles in which up to 150 kilometres (100 miles) could be covered in a day. Liquids such as oil and wine were carried in barrels mounted on a wheeled chassis. War chariots were a special case and will be considered in a later chapter.

The performance of these early vehicles can be judged by that of their recent counterparts, but the Theodosian Code gives some interesting figures. According to this, the load of the lightest category of vehicle, the *birota*, should not exceed about 70 kilograms (154 pounds) and that of the heaviest, the *angaria*, 500 kilograms (1100 pounds). Clearly, special vehicles must have been used for really heavy loads such as quarried stone and timber. Both Assyrian and Egyptian wall paintings of around 2000 BC show enormous blocks of stone – some already carved as monuments – being laboriously moved on runners, possibly placed over rollers; they are being dragged by teams of men from the front, assisted by others using massive levers at the back. While we must not take the artists too literally, it is clear that they were trying to represent teams consisting of upward of a hundred men.

Two interesting variants of this have been described. In the first century BC the Roman engineer Vitruvius fixed short axles to the ends of massive stone columns and then harnessed them like rollers to teams of oxen. His son Metagenes extended the idea to squared stone, which could not roll, by fitting strong wheels directly to the stone. By such means individual loads exceeding 50 tonnes were moved. Where possible, water transport was used. An Egyptian rock carving of about 1500 BC shows two obelisks, each weighing some 350 tonnes, being floated down the Nile on a specially constructed lighter some 60 metres (200 feet) in length: there remained, however, the problem of loading and moving to the final site.

Wheeled vehicles were, of course, normally driven over fairly level ground, though not necessarily prepared roads, which were rare outside the confines of the Roman Empire. There is, however, some evidence for the use of tramways by the Greeks. These consisted of grooves in stone paving, in which the wheels ran, so that the vehicle was self-directed. Those that survive, as in Malta today, may have been no more than ruts resulting from long use – a familiar sight on farm tracks today – but the constancy of the gauge and the fact that the grooves almost always pass across the centres of the stone slabs both suggest deliberate intent. Such tramways were common in medieval mines, but there is little evidence of intermediate use in Europe. In China, Shi Huang – who achieved the unification of the country in the latter part of the third century BC – included standardization of axle ruts in a huge building programme.

Water transport

The history of water transport presents a now familiar picture, the most primitive means surviving alongside the most advanced: the dugout canoe and the Welsh coracle are contemporaries of submarines and hovercraft. Nevertheless, a clear line of evolution is apparent as technical advances were made to meet growing needs.

While the sea presented a major obstacle to primitive man, as is apparent from the pattern of spread of early cultures, rivers did not daunt him. From the chance use of a floating log, it is a relatively small step to the dugout canoe, shaped by fire or stone axes. The earliest known example is the Pesse boat from Holland, dated at 6400 BC, but there is no reason to suppose that it was then an innovation: it merely happens to have survived and been discovered. This boat was some 4 metres (13 feet) long, and a rather later example from Britain extends to 12 metres (39 feet).

In northern Europe trees of the necessary size for boatbuilding were abundant, but in the Nile Valley and

boats and ships 50–53, 105–111, 172–176, 254–255 **chariots** 21, 46 **development of road transport** 176–177, 247–251

The earliest boats were dugout canoes, like this Bronze Age vessel (top). Later, light boats consisting of hide stretched over a wooden frame appeared in many regions; the curagh of Ireland (above left) is made in this way. The earliest Nile boats were made from reeds. Modern reconstructions (above right) have been made of these boats and they have been used to duplicate voyages possibly taken by early peoples. The boat under construction is Thor Heyerdahl's *Tigris*.

Mesopotamia this was not so and for common use more readily available materials had to be used. On the Nile, the buoyant papyrus reed grew in abundance, and bundles of reeds were used to build rafts and boats. So too, at least from 1200 BC, did the Chimu Indians of Peru, where similar *caballitos* are still used for inshore fishing. On the Tigris and Euphrates there were no reeds and different kinds of boat were developed. For heavier loads, rafts were made by lashing inflated skins to a wooden framework. Such rafts, known as *keleks*, are clearly depicted in a relief from Nineveh of the seventh century BC. For lighter work *quffas* were, and still are, used. These were circular and made of leather stretched over a wooden framework. Commonly, they were some 4 metres (13 feet) in diameter and they have counterparts in the kayaks of the Eskimo, the Welsh coracle, the Irish curaghs and the earliest Scandinavian ships.

Such vessels could be manoeuvred with paddles but moved upstream only against the lightest currents. In the case of the *keleks*, it was possible that they were dismantled on completion of a downstream voyage, when the timber was sold and the floats taken back upstream by pack-horses. Even today, timber is transported in Scandinavia and elsewhere in the form of rafts that are floated downriver and dismantled at their destination. The much smaller and lighter coracles could be carried back upstream on a man's back, and straps were specially fitted for that purpose.

Conditions on the Nile were very different. Apart from the great loop below Luxor, the river flows almost due north, the direction from which the prevailing wind blows. It was, therefore, generally possible to drift downstream, assisted by paddles if need be, and then hoist a simple square sail to make the return journey. These factors influenced the design of reed boats. By 3100 BC, as a picture on pottery attests, sails were in use to take advantage of the wind. To sail before the wind it is an advantage to have the mast well forward. In the absence of a keel there was no central point at which to anchor the mast, and accordingly this was of bipod construction with two legs – sometimes splayed at the base to distribute the load – attached to the sides of the boat. The sail was hoisted and lowered by halyards leading aft, and the boat was steered by twin broad-bladed paddles at the stern. The hull may have been waterproofed with pitch. Such boats were in use in the third millennium BC, and ply today on Lake Titicaca in Bolivia, but they were not confined to inland waters. The prophet Isaiah refers explicitly (*circa* 740 BC) to Egyptian ambassadors sent 'by the sea, even in vessels of bulrushes', and there is no reason to suppose that this was then a new practice. Reed boats were cheap and easy to construct, but in the longer term the future lay with wooden ships, which were not supplanted until the advent of iron vessels, effectively after the beginning of the 19th century.

Wooden ships

When the first wooden ships were built in Egypt is not known, but they were certainly trading in the Mediterranean around 3000 BC, and at an early date two main lines of development can be seen. On the one hand were the broad-beamed merchantmen, with a length:beam ratio of around

4:1, which for economic reasons necessarily relied mainly on sails: a full team of oarsmen would have been prohibitively expensive. On the other was the long narrow galley with a length:beam ratio nearer 10:1. It was used mainly for battle, propelled solely by oarsmen, and fitted with a powerful ram. It was much favoured by the Mediterranean nations up to the Battle of Lepanto in 1571, and even then was not wholly discarded for another two centuries.

Apart from these differences in propulsion, there were important differences in construction. The Egyptians, and later the Greeks, favoured what is known as carvel construction, in which the horizontal timbers are laid edge to edge. Originally, they were literally stitched together with ropes, but these were soon replaced by mortise and tenon joints and later nails. In northern Europe, and particularly in Scandinavia, clinker-built ships predominated. In these, successive rows of planks, running continuously from end to end of the vessel, overlapped slightly. The Hjortspring boat, from south Jutland, dates from the third century AD.

As commonly occurs when new methods arise, the earliest Egyptian wooden ships were closely modelled on their reed-built predecessors. Only gradually did designers learn to take advantage of the different properties of the new material. Initially, there was no keel and no ribs: such a hull can be likened to a saucer. External pressure of the water tended to press the timbers together, but the weight of any internal cargo would tend to open up the seams. For this reason, a load-carrying deck was built on transverse beams joining the top planking: this conferred some lateral stability. There remained the problem of longitudinal stability: as the hull rode through the waves it tended to 'hog' – buoy up at the middle and sag at the ends. To counter this, a rope known as a hogging truss was run from bow to stern. This was tightened by twisting it with what is now called a Spanish windlass – a wooden bar passed through a bight in the rope. Later, the Greeks introduced underbelts running along the outside of the hull and similarly tightened as the need arose. Acts 27.15–17 vividly describes their use during a tempestuous voyage in which St Paul was being carried off to Rome in AD 62: 'when the ship was caught and could not bear up into the wind we let her drive ... they used helps undergirding the ship; and fearing lest they should fall into the quicksands, strake sail, and so were driven.'

We can get a very clear notion of the construction of the earliest Egyptian wooden ships from the surviving funeral vessel of Cheops, builder of the Great Pyramid at Giza, who died about 2530 BC. It was built from some 1224 separate pieces of wood, mostly cedar, held together by tenons and thongs, and including both butt (end-to-end) and scarfed (overlapping) joints. There is no true keel, but tall stem- and stern-pieces are attached to the central longitudinal timber. The deck carries a finely worked cabin. It was propelled by five spear-shaped oars on each side, working against wooden pegs or tholes, to which they are attached by leather thongs: U-shaped rowlocks were a much later invention. The oarsmen appear to have stood. At the stern were two steering oars. Cheops' funeral ship was 43 metres (141 feet) in length, but as it was built as a tribute to his supreme power it may perhaps have been rather larger than seagoing contemporaries. Nevertheless, within a few centuries we have records, though not pictures, of far larger vessels – 60 metres (200 feet) long, 20 metres (65 feet) wide (a 'fineness' or length:beam ratio of 3:1), and carrying a crew of 120 men.

For detailed pictorial representation we have to go forward to rock carvings at Thebes showing a five-ship trading expedition dispatched by Queen Hatshepsut to Punt (probably modern Ethiopia or Somalia) about 1500 BC. Some noteworthy changes have taken place and we see most of the basic features of wooden ships for the next 3000 years. Very significantly, the mast is dead amidships, necessary for sailing at sea with the wind on the beam rather than almost dead astern as on Nile voyages. However, sailing more than very slightly into the wind would be impossible, and with the wind abeam a good deal of leeway would be unavoidable. Not surprisingly, rowers were retained, 15 on each side. The mast is single, not a bipod, indicative of a keel and some sort of timber framing, though it probably did not have the ribs to which planking was nailed in later ships. The single sail is now much broader and is attached to a spar at both top and bottom. Rigidity is conferred by deck beams passing through the planking: the hogging truss is retained but is tightened not by a Spanish windlass but by pulling it down towards the deck like a bowstring.

The Egyptians were not great empire-builders but they traded in the eastern Mediterranean and captured ports in the Levant. They had warships to protect their trade, but up to about 1450 BC – when they were conquered by the Myceneans – it was the Cretans who effectively policed this area: the absence of defensive walls in ancient Cretan cities is interpreted as a measure of their confidence in their naval power. But for many centuries, from roughly 2000 BC to 350 BC, the great trading nation in that part of the world was Phoenicia, a collection of city states headed by Tyre, Byblos and Sidon. The Phoenicians eventually traded not only throughout the Mediterranean, but also ventured beyond the Pillars of Hercules in search of tin from Cornwall and amber from the Baltic and as far out into the Atlantic as the Azores. According to Herodotus, an expedition of Phoenician-manned ships was dispatched from Egypt about 600 BC with instructions to circumnavigate Africa from the Arabian Sea, returning via the Pillars of Hercules and the north coast of Africa. The 25,000-kilometre (15,500-mile) voyage is said to have taken three years. Herodotus himself was sceptical of this story, which he heard more than a century after the alleged event, but there is no doubt that it was within the ability of the Phoenician ships of the day.

Battle of Lepanto 108 **development of boats and ships** 105–111, 172–176, 254–255 **Mediterranean civilizations** 10–11 **Queen Hatshepsut's ship** 53

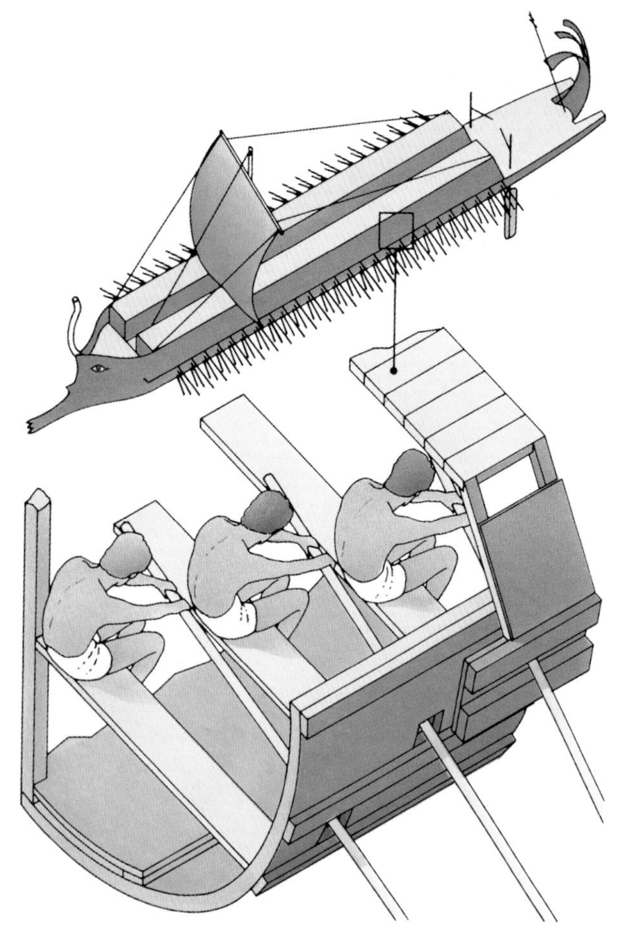

A Greek trireme of about 500 BC. There were three banks of oars pulled by 170 rowers, probably one to an oar. Sails supplemented the oars, except during battle. Note the powerful ram.

For their long voyages they relied on two kinds of vessel – the so-called hippos, or the Ships of Tarshish, frequently mentioned in the Old Testament. The latter name seems to have derived from the important metal trade with Tartessus (now lost, but somewhere in south-west Spain). The smaller hippos were very rounded vessels – with a length:beam ratio of about 2.5:1 – fitted with a central mast carrying a single square sail, though later ones show important innovations. As depicted on a sarcophagus found at Sidon in 1914, it had become a two-masted vessel, the second mast raked and resembling the later bowsprit. Additionally, there was a low aftercastle, such as we see very much later on Spanish carracks, and a raised cargo hatch. There were still the two steering oars at the stern but, if the artist's representation can be trusted, there was no longer any provision for rowers.

Tyre fell to Alexander in 332 BC after a long siege, and for a time much of its trade passed to its ancient colony, and later rival, Carthage. This was totally destroyed by the Romans in the Third Punic War. The true heirs to the Phoenicians, however, were the Greeks. They seem never to have ventured beyond the Mediterranean, except to their trading settlements ringing the Euxine (Black) Sea. Their somewhat piratical style of trading is reflected in the design of the penteconter, designed for both cargo-carrying and attack. Typically, this was some 25 metres (82 feet) in length and 3 metres (10 feet) in the beam: in addition to a square sail on a central mast, there was provision for 50 rowers, which is how it derived its name. Such dual-purpose vessels would be fast and manoeuvrable, more than a match for small broad-beamed merchantmen, such as those on which the Greeks themselves relied in protected waters.

Greek galleys

The penteconter was akin to a privateer, but for serious naval warfare a more powerful vessel was required. For this purpose the Greeks adopted the Phoenician bireme but made it lighter and faster. This had two banks of oars on either side and a single mast carrying a square sail, but this was discarded when action was imminent. The main weapon was a massive ram extending beyond the stem. As the strength of the opposition increased, still more powerful vessels were called for, which led to the trireme during the sixth century BC. Typically, this was a long sleek vessel some 43 metres (141 feet) in length and 6 metres (20 feet) in the beam: the bronze-covered ram was 3 metres (10 feet) long. As the name implies, there were three banks of oars, with one man per oar. Additionally, there were two masts to carry sails. It carried a crew of 200, of whom 170 were rowers. The superiority of the Greek trireme was convincingly demonstrated in the overwhelming defeat of a larger Persian fleet at Salamis in 480 BC.

Roman ships

Unlike the Greeks, the Romans were not a seafaring race, being soldiers rather than sailors. Nevertheless, they had to import vast quantities of goods by sea and needed a battle fleet to protect their trade: under Pompey this numbered some 500 vessels. The Roman triremes, though modelled on those of Greece, were of heavier construction, designed not to ram but to board the enemy.

Roman merchant ships differed little from their Mediterranean predecessors. Commonly they were two-masted, but with two triangular topsails set above the square sail; and a third mast, which later defined itself as the mizzen, seems to have been introduced in the first century AD. They were still tubby ships, generally some 55 metres (180 feet) long and 14 metres (46 feet) wide.

Chinese shipbuilding

Ships from the Red Sea trading ultimately with China would generally have transhipped their goods into the Chinese multi-masted 'sand ships', which can be traced back at least to the period of the Warring States in the fifth century BC. They were square in both bow and stern, with a rather weak keel, and had a shallow draught. They were thus very suitable for use on a long coastline abounding in sandbanks, hence their name. By the third century AD, well constructed ocean-going ships up to 70 metres (230 feet) in length were being built. Probably around the fifth century AD – more than a thousand years ahead of Europe – the Chinese made the very important innovation of solid transverse bulkheads, possibly inspired by the familiar internal division of bamboo, so that if the boat was holed the leak could be contained. As early as the eighth century BC there are references to square junks, or *fangzhou*, which have been interpreted as double-hulled vessels akin to catamarans; however, the earliest, at least, may have been no more than large rafts with two floats carrying a load-bearing superstructure, not too different from the *keleks* of Mesopotamia. On the slender evidence of the dimensions of a dry dock at Alexandria, it is possible that the Romans, too, had twin-hulled vessels.

The performance of early ships

The capacity and speed of these early vessels is necessarily largely a matter of surmise, the more so as it would depend considerably on their maintenance. Thus an encrustation of barnacles would slow them considerably, as would water-logged timber. Despite coatings of pitch, the larch and fir commonly used for the hull – because of their lightness and ease of working – absorbed a lot of water, and galleys, in particular, spent much of their time beached, simply to dry out.

The writings of Pliny and several recovered wrecks indicate that thin lead plates were used to protect the hull from the wood-boring Teredo worm, rife in the warm Mediterranean waters. This was doubtless effective, but would add seriously to the weight. Not until the 18th century was the practice of metal-sheathing revived, when British naval ships in the West Indies were coated below the water-line with copper.

The basic cargo ship of those days was probably a vessel about 20 metres (65 feet) long and 8 metres (26 feet) in the beam, with a carrying capacity of, say, 150 tonnes. The ship in which St Paul was wrecked in AD 62 is said to have had 276 people aboard. Much larger vessels were not uncommon, however, and Lucian, in the second century AD, refers specifically to the *Isis*, a ship with dimensions more than twice as great and a cargo capacity of around 1200 tonnes. With a following wind such ships would average perhaps 5 knots, rising to 6 knots under very favourable conditions of wind and tide: this would be rather more than half the speed of a penteconter.

However, the particular conditions of the weather in the Mediterranean ensured that favourable conditions were relatively rare. Long open-sea voyages were normally embarked upon only between April and September. During this season the prevailing wind was north-west and this had very important consequences for the annual importation of corn to Rome from Egypt, following the Battle of Actium in 31 BC. The annual import was around 150,000 tonnes, one-third of Rome's total need. From Ostia, the voyage to Alexandria might well be accomplished in three weeks but the return journey – beating into the wind, for which the ships' rigs were still ill-adapted – might take 10 weeks. Allowing for loading and unloading, and occasional storms and other delays, few ships achieved two round trips a year. Thus a very large fleet had to be maintained: even supposing one-tenth of them could carry 1000 tonnes of grain each, we have to think in terms of around 200 ships for the plying of this trade alone.

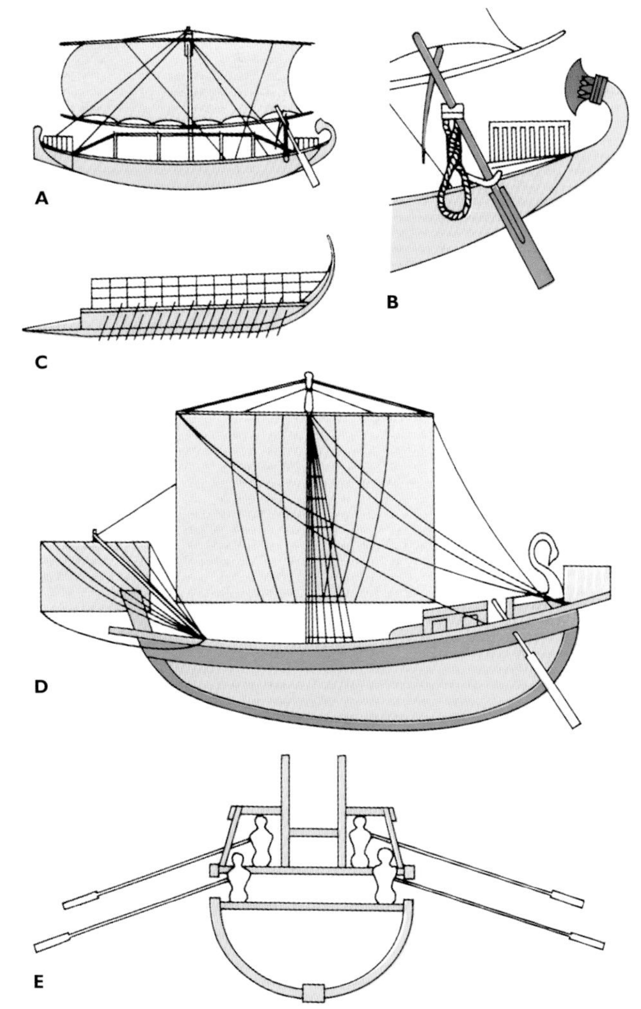

A trading ship (A) of the type that the Egyptian Queen Hatshepsut dispatched to Punt about 1500 BC. It had a single broad sail set amidships and 30 rowers. The steering system (B) consisted of a large paddle fixed to the side of the boat so that it could be turned like a rudder.
An Assyrian bireme (C) of about 700 BC had a ram extending in front of the short hull to attack enemy ships. The vessel was powered by two banks of rowers (E) placed low inside the hull, giving it stability in rough seas and in action, and armed men were carried on deck to overcome the enemy crew.
A Roman merchant ship (D) of the second century AD possessed three sails – a triangular topsail mounted above the broad mainsail and a foresail at the bow. The ship was steered by steering oars near the stern.

Classical literature contains some references to very large vessels indeed. The poet Moschion describes a monster ship, the *Syracusia*, built by Hiero II of Syracuse about 250 BC, allegedly under the supervision of Archimedes. It was heavily armed against pirates, carrying some 200 soldiers, and could take aboard nearly 2000 tonnes of cargo. It appears, however, that after its maiden voyage to Alexandria the *Syracusia* never sailed again – perhaps it could be called a sort of pre-Christian *Great Eastern*.

Archimedes 66 **Carthage** 24 **development of boats and ships** 105–111, 172–176, 254–255 ***Great Eastern*** 147, 173, 175

CHAPTER FIVE

Building Construction

Because of their durability buildings are the most evident and abundant memorials to the technological achievement of past civilizations. Many, such as the pyramids of Egypt, the megalithic circles of northern Europe, the Pantheon in Rome and the Great Wall of China survive virtually intact. Even where they have fallen into ruins or been demolished to ground level so that only the footings of the walls remain, much can still be deduced about their size and construction. Even where thatched wooden building predominated, as in northern Europe, the replacement soil in the original post-holes can give a valuable indication of size and layout when all else has long since perished. In addition, for the really outstanding buildings, we have contemporary descriptions and illustrations. Overall, therefore, we have a pretty clear idea of what ancient buildings looked like and the materials used in their construction. But we shall find ourselves on less certain ground when we come later to consider how they were erected.

Building materials

While certain scarce materials can be imported from great distances if needed, buildings – even the most prestigious ones – have necessarily to be constructed largely with locally available materials and skills. This, in turn, considerably influences architectural style – the constraints imposed by building in brick, for example, are very different from those faced by the stonemason. Climate, too, is a limiting factor: sun-dried mud bricks are durable in hot dry climates but useless where there is heavy rainfall.

Much of the Middle East, and particularly that area now represented by Iraq, is a flat alluvial plain where no stone is quarried and wood is scarce. From Sumerian times the popular building material was, almost inevitably, sun-dried bricks, which were cheap and easy to make. In Egypt, the situation was different: good building stone was abundantly available and from the earliest times was widely used for major public buildings. It was, however, far too expensive for common use, and for general purposes the normal material was again sun-dried bricks. The Chinese, too, used bricks at an early date but only for ancillary buildings such as terraces and boundary walls. They used tiles for roofing, but for the rest of their houses they regarded wood as the only fitting material.

Although ancient bricks survive in abundance and many frescoes depict the making of bricks, we have little direct evidence of the techniques used. However, it was most certainly not a fast-moving technology and the detailed account given by Vitruvius in his *De Architectura*, of the first century BC, doubtless represents long-established practice. Possibly the earliest bricks were simply shaped by hand, but the uniform size of bricks in even the lowest levels of Jericho indicates the use of open-frame wooden moulds, like those still employed today. In early Egypt the mould was slightly overfilled, and the protruding clay was smoothed with the hands to give a slightly curved top surface to the finished brick. Vitruvius recommends soils rich in clay, though local builders would have to do the best they could. To prevent the brick crumbling as it dried, the clay soil was mixed with chaff and chopped straw. He advises that manufacture should be restricted to spring and autumn, as in the intense heat of summer the outside will dry hard and leave the interior still pasty: indeed, he goes on to quote a local by-law in Utica requiring bricks to be dried for several years before use. This seems excessive, especially in the case of later Roman bricks, which were normally no more than 4 centimetres (1½ inches) thick, though 45 centimetres (18 inches) long and 30 centimetres (12 inches) wide. Bricks of this thickness made today in Iran are laid within a few days. However, one can imagine that Babylonian and Etruscan bricks, both up to nearly 15 centimetres (6 inches) thick, would have required a much longer drying time. Indus Valley bricks, 28 centimetres (11 inches) long, show a nice regularity, the ratio thickness: width:length being 1:2:3.

The kiln-fired brick was much more durable but required a good clay rather than a soil rich in clay. The early Egyptians

In dry climates, satisfactory bricks can be made from sun-dried mud reinforced with chopped straw.

This very ancient method is still used virtually unchanged, as this Egyptian scene illustrates.

Until the advent of the arch, pillar-and-beam construction was universal. It could take very different forms, as (above left) in the Lion Gate, Mycenae (*circa* 1350 BC) and the ruined temple (above right) at Selinunte, Sicily (sixth century BC). The construction of the temple pillars as a series of drums is very evident.

did not use them at all, but in the Middle East they were made as early as the fourth millennium BC. Burnt bricks were in use in Greece in the fourth century BC, but rather rarely. Although they came to be widely used in the Roman Empire, including Egypt, there is no record of them before the time of the Emperor Augustus (63 BC–AD 14): Vitruvius makes no allusion to them.

Cobwork is in a sense the mud-brick mould writ large. In this widely used technique, earth is rammed between parallel boards, which are then removed. The wall is allowed to dry and then plastered to throw off water. Although most suitable to dry climates, it is still sometimes encountered in northern Europe. An early example of its massive use was in the Great Wall of China. The original section, constructed in the third century BC, was built entirely of rammed earth. Stretching for some 2400 kilometres (1500 miles), its building occupied 300,000 men for 10 years.

For bonding bricks, the Egyptians used a mortar made from gypsum (plaster of Paris): hard-setting mortars that were made from slaked lime and sand did not appear until Greek and Roman times. Rather surprisingly, gypsum mortars are also found in Egyptian masonry though the massive stone blocks could be preshaped with sufficient precision for dry-jointing to have been feasible. It has been suggested that the 'mortar' was used not to bond the blocks together but to serve as a lubricant when sliding them into place. In describing how mortar should be made, Vitruvius is very particular about the kind of sand used. Ideally it should make a grating sound when rubbed in the hands: in modern terminology, it should be sharp. The grains of sand from river-beds, he says, are too rounded by the actions of the water, while sand from the sea-shore contains too much salt, which will gradually make its way to the surface of the mortar after it has set.

The Romans also made much use of concrete, particularly a type made by mixing pozzolana, a volcanic earth from the Alban Hills, with slaked lime and sand. It not only set very hard, but would even do so under water. In building work, this kind of concrete was extended by adding coarse aggregate, but the Romans seem not to have followed the modern practice of mixing the two together before pouring them into the shuttering. Instead, they filled the space with aggregate, broken stone or rubble and then poured the freshly mixed concrete on top. The disadvantage of this is that it is difficult to get a homogeneous mix, even with shaking and stirring.

For building in stone, locally available materials must generally, though not invariably, be used. To produce florid effects the Romans imported varieties of coloured marble and other building stones from many sources. Very often, however, there may be a local choice and various factors will decide which is used. These will include ease of working, mechanical strength and durability. The Egyptians began to use stone for permanent buildings from the middle of the third millennium BC. Initially limestone was preferred because it was easily shaped with picks or, possibly, chisels: the much harder and more durable granite, mostly obtained from Aswan, had to be laboriously shaped by pounding with boulders, around 5 kilograms (11 pounds) in weight, held in both hands.

Although much of Iraq is an alluvial plain, stone was available in the north, especially in the region north of Nineveh, and it began to be used about the eighth century BC. For general building purposes limestone was used, but the softness of gypsum favoured its uses for sculptures, despite its bad weathering properties. For the Greeks, the situation was different again. Their country consists almost entirely of limestone, with many sources of fine marble. The finest of all was that which came from Mount Pentelicus, near

Great Wall of China 23, 87, 125 **megalithic monuments** 19, 45, 57, 59 **Pantheon** 127 **Parthenon** 56, 58 **pyramids** 65

The Parthenon, the most famous of all the many elegant Greek temples, was built under Pericles, 447–432 BC. Built entirely of Pentelican marble, it originally had great doors at either end and metal grilles between the columns.

Athens. Pentelican marble was used in such outstanding public buildings as the Parthenon and the Erechtheum: the latter was decorated with black marble from Eleusis.

Like stone, timber tended to be used according to its local availability. There was, nevertheless, a substantial trade in wood from a very early date, both to relieve shortages and to provide woods with prized properties, particularly for cabinet-making. For building purposes, cedar was valued in the south, but is rarely encountered north of the Alps, where oak was commonly used: in Roman times beams up to 15 metres (50 feet) long were hewn from a single trunk. Just as stone was carefully selected, and often left to weather for a couple of years in the open to detect defective blocks, so too was timber. Early in the third century BC Theophrastus gave careful directions as to how and when trees should be felled, and listed the best seasons for felling different kinds: conifers in the spring and oak at the beginning of winter.

Wood, stone and bricks were the basic building materials, but from a surprisingly early date extensive use was made of metals, especially iron. A temple at Agrigentum, built about 470 BC, contains a massive iron beam no less than 5 metres (16 feet) long and others not much smaller form part of the roof of the Propylaea, built in 435 BC. Extensive use was also made of iron and bronze clamps and dowels to hold masonry together securely. Lead clamps were used even in Mycenaean times and the Romans used much lead for roofing and for conduits in water-supply systems.

Building design

Although masonry buildings vary enormously in their size and appearance, and in the uses to which they are put, they involve only a very small number of basic components: in essence, walls, pillars, beams and arches. The nature of these components depends very much on the materials and labour available and the technological ingenuity of the builders. When we come to consider the appearance of a building, as distinct from its function, we stray into the closely allied field of architecture. Thus the dearth of long timber in Mesopotamia imposed architectural constraints unknown to northern Europeans with ready access to tall forest trees: until the construction of the arch was mastered, individual spans were relatively short whatever material was used.

Walls

Although lighter and more perishable materials might be used for common dwellings, the walls of important buildings were made of either bricks or masonry. Until we come to the elegant buildings of classical Greece, the most striking feature of most early architecture is its massiveness. The Mesopotamian ziggurats – found in all the principal cities and reminiscent of the much later *teocallis* of the Aztecs – were immense brick-built structures. That at Ur, dated at 2000 BC, covered a ground area of 75 × 54 metres (246 × 177 feet) and towered up to a height of 26 metres (85 feet): the exterior walls, surrounding a solid core, were 2½ metres (8 feet) thick. In public buildings even dividing walls might be several metres thick. The Egyptians commonly used immense stone blocks, some weighing several hundred tonnes. On their outward, visible faces these were dressed with great accuracy but often this extended inwards only a few inches: the gaps between stones were filled with whatever rubble came to hand. The Incas adopted a similar building style. The walls, of porphyry or granite, were of great thickness and made from blocks carefully preshaped so that no mortar was necessary. Their height was not impressive, though, rarely exceeding 4–5 metres (13–16 feet).

While large construction works such as canals and aqueducts served important practical purposes, large public buildings were to a great extent demonstrations of national power. It is, therefore, surprising – though perhaps symbolic of the transience of power – that for a long time little attention was paid to foundations. The walls of Nineveh, more than 20 metres (65 feet) thick, rested on only a layer of rubble, and the great walls of the Temple of Karnak had for foundations no more than half a metre (1½ feet) of sand and the columns rested on a layer of rubble. The perils of building on sand are referred to in St Matthew's gospel: perhaps the writer had in mind this widespread defect.

Beams and columns

Even in hot countries with low rainfall, where much of life is lived in the open air, buildings must be roofed both to give protection during wet seasons and coolness and shade at other times. Supporting the roof of a large building was a major problem for early builders, for large spans were impossible. Even where it is locally available, or can be imported, wood has the double disadvantage of being perishable and inflammable. Even so, the Greeks made much use of timber roof members and they fireproofed vulnerable timbers with alum. Although stone is very strong, in the sense that it will stand great compression, it is weak under bending forces. In practice, a limestone block used as a beam was limited in length to about 3 metres (10 feet) and even the sandstone which the Egyptians got from Silsila, far up the Nile near the First Cataract, was not safe beyond 10 metres (33 feet).

The solution to the problem was to use rows of columns separated by the maximum safe length of the beams available. The disadvantage of this was that the columns so cluttered the interior that no single large place of assembly could be constructed, apart from open-air courtyards.

Such early buildings had flat, or nearly flat, roofs. The ends of the beams rested on either the outer wall or the interior columns. Some central columns might have to support the ends of as many as four beams – two running crosswise and two lengthwise – and the ends had to be mitred so that they interlocked. A consequence of this was that the columns had to be of large cross-section to give the necessary bearing area for each beam. Within a building the interlocking of the beams provided stability, but in more isolated structures some additional security was needed.

At Stonehenge, for example, the top surfaces of the upright stones have bosses which fit into corresponding depressions in the lintels. It is plausibly argued, with other supporting evidence, that as this is an ancient woodworking device, Stonehenge replaced, or was modelled on, some earlier wooden structure. The builders of Stonehenge were certainly heirs to a well-established local tradition of building in wood on a massive scale. Thus the huge defensive works built about 3400 BC on Hambledon Hill, some 50 kilometres (30 miles) away, contain no less that 10,000 heavy oak posts embedded in the ramparts.

Although the weight of the stone beams and of the stone or earthenware roofing slabs was together enormous, the flat-roof construction had one major advantage. The load on the columns was entirely vertical, and the beams and roofing slabs were thus not liable to move, provided the foundations held. With a pitched roof, and more particularly with the advent of the arch, a new problem had to be faced. If the pitch is high, a lateral as well as a vertical thrust is exerted on the supporting walls and pillars, and some means has to be found to offset this. Although the Greeks were not wedded to perfectly flat roofs, they never exceeded a pitch of 30°.

Although, as we have seen, the civil engineers of the ancient world were not daunted by the purely physical problems of shaping and moving massive stones up to as much as 500 tonnes in weight, columns were commonly built in sections. Each drum-shaped unit was prefabricated, making allowances if necessary for some tapering of the final structure, and then dropping on a vertical spigot set in the one below it. Even with this subdivision, the sections were quite formidable loads and had to be delicately handled. When the

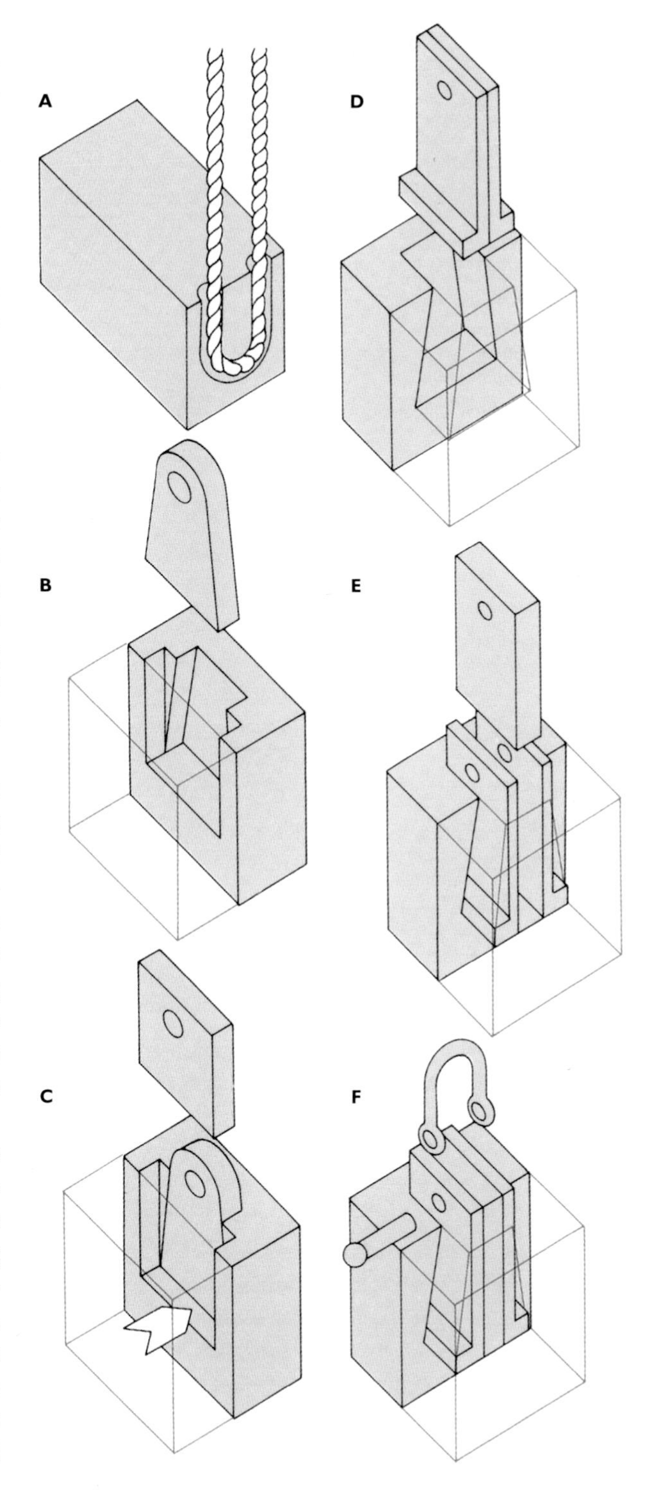

An early method of lifting stone building blocks was to cut a groove in the end of the block (A) so that a lifting rope could be fixed inside it. Hero of Alexandria described a method of using an iron hanger (B) that could be fitted into a hole in the block with a piece of wood to keep it in place (C). A rope was then attached to the hanger. A similar method used a hanger made of two L-shaped pieces (D) and a central piece (E). A bolt was slipped through the hanger to attach it to a handle (F).

ancient building 64–65 **arches** 58–60 **development of building** 124–132, 258–262 **Stonehenge** 45, 59

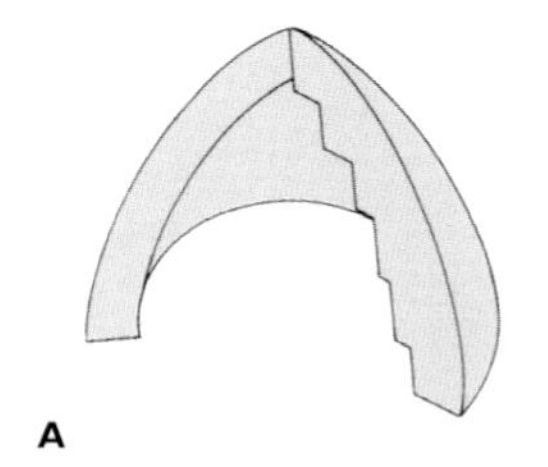

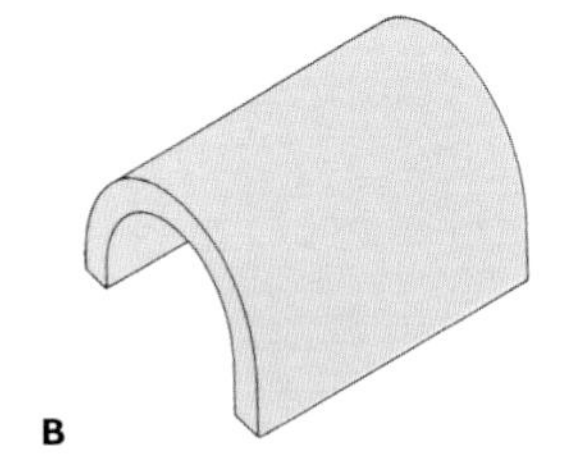

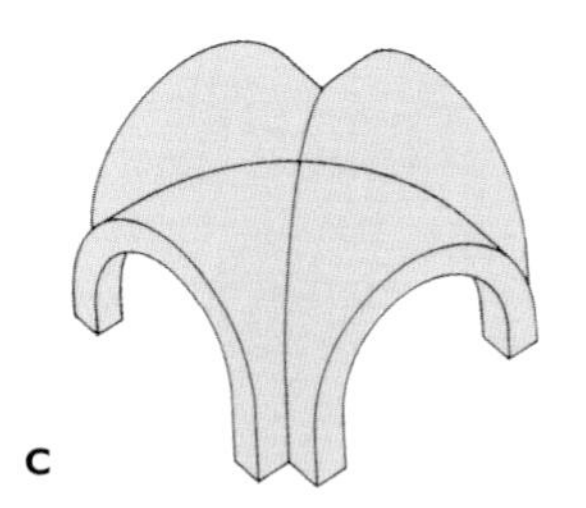

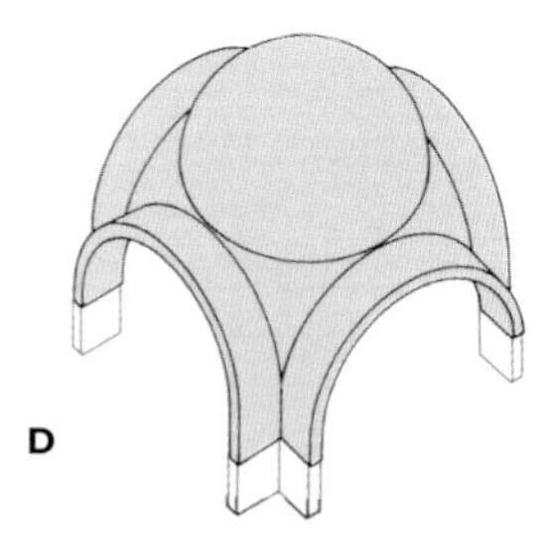

Above: A corbelled vault (A) is constructed with courses of masonry overhanging those beneath. The projecting corners are then chipped away to produce a smooth curved surface. The barrel vault (B) is an extended arch, and a cross-vault (C) consists of two barrel vaults joined at right angles. The Byzantine dome (D), built by corbelling, rests on pendentives that raise it above the walls.

Right: The advent of the arch opened up new architectural possibilities by increasing the length of single spans and enabling the spans to be built from small manageable units instead of single massive ones. The Roman arch – here exemplified by a fifth century AD temple at Qalaat, Syria – was limited by the fact that it was semicircular, so that its span necessarily equalled twice its height.

Parthenon was built in the fifth century BC, each of the columns, some 10 metres (33 feet) high, was built of 11 drums each weighing 8 tonnes.

The arch

The introduction of the arch was in two senses a major architectural advance. It ultimately much increased the length of a single span, and it enabled such spans to be built from relatively small, manageable units instead of single massive beams. Essentially, arches are of three kinds. First, there is the simple arch; secondly, there is the barrel vault, which is a linearly extended arch; and thirdly, there is the dome, which is a three-dimensional arch.

Important though it was, and commonplace as it now is, the arch evolved only slowly. Excavations at Ur have revealed vaulted roofs to sewers, whereas ones of similar date at Mohenjo-Daro were covered by long flat bricks. At Ur there were also small domed burial chambers, similar to the beehive-shaped 'Tomb of Agamemnon' at Mycenae, *circa* 1450 BC. Such domes were made by corbelling – that is, by allowing each successive course of masonry to extend inwards slightly.

The Egyptians, like the Greeks, favoured pillar-and-beam construction, but they were familiar with the arch from a very early date. A granary at Thebes, built about the middle of the second millennium BC, includes an arch spanning 4 metres (13 feet). Far away in China, stone arches can be dated back to the third century BC. Later it is possible to see definite routes for the dissemination of different kinds of arch, such as the Gothic. But early on it is difficult to know whether we are dealing with a single invention that was disseminated or a multiple one.

The first great exponents of the arch were the Romans, who introduced various new devices. They used temporary timber centring to support the arch during construction and wedge-shaped stones or voussoirs, culminating at the top in a keystone locking the whole structure firmly in position. A limitation to their architectural design was their use of the semicircular arch, restricting its height to half its span. The ogival arch, which is pointed and has S-shaped sides, was used in Buddhist India in the second century AD, but it was nearly a thousand years before it was adopted in Europe.

Although the Romans made extensive use of concrete, which is lighter than stone, the lateral thrust on the supports

The main characteristics of ancient architecture was its sheer massiveness. The Egyptians, among others, were accustomed to using enormous blocks of stone, as indicated by this picture of a quarry at Aswan (top right). The hypostyle hall at Thebes (above left), begun by Seti I in about 1293 BC, is characterized by immense stone pillars, some 22 metres (72 feet) high. In northern Europe, too, early builders were not daunted by massive masonry, as exemplified by Stonehenge (bottom left): note the boss on the top of the left trilithon which would have engaged with the socket in a (fallen) lintel. The true arch, presaged by the corbelled arch – here illustrated by a relatively late example from Chichen Itza, Mexico (centre right) – made lighter construction feasible.

aqueducts 61, 62 **arch bridges** 62, 63, 125, 262 **dome construction** 127 **Gothic arch** 130–131 **Mohenjo-Daro** 10–11, 31

In irrigation systems in hot climates, much precious water is lost by evaporation from open channels. The Nabataeans were well aware of this and built an extensive system of underground aqueducts. This is their ancient city of Avdat.

of their largest arches and domes was enormous. In the main, this was relieved by massive cross-vaults serving as buttresses, which gave some impressive results. Thus the vaulted roof of the Emperor Diocletian's palace had a span of 35 metres (115 feet), while the great hemispherical dome of the Pantheon, commenced in AD 120 and completed in only four years, spans nearly 50 metres (164 feet).

Those responsible for domestic and public buildings in the Middle Eastern and Mediterranean areas had mainly to provide for hot, dry climates. As their empire extended northward, however, the Romans had to give increasing attention to keeping warm rather than cool. For heating they often used braziers, but in northern Europe – Britain, Gaul and Germany – much use was made of hypocausts, which channelled warm air through flues beneath the floors and in the walls. As damp is a normal accompaniment of cold, it is not surprising to find Vitruvius giving instructions for building a ventilated cavity wall.

Domestic architecture

For most people, home was no more than what would today be designated a hovel, a rough one-roomed building constructed from whatever materials lay readily to hand. For a privileged few, however, homes were of a vastly higher standard. Reconstruction of residential blocks of the second century AD at Ostia, the port of Rome, revealed buildings four or five storeys high, in which one family might occupy five or six rooms, each served by its own wooden staircase. The buildings were of brick, with a core of concrete to the walls, and one or more floors might have a balcony. The windows were unglazed – although the use of glass was not uncommon in public buildings of the time – and there was no individual water supply; water was obtained from a cistern in the street. Additionally, communal baths were often provided, as well as gardens for recreation.

Civil engineering projects

So far, we have considered buildings mainly in the context of those occupied by people, either as public buildings or private dwellings. At a very early date, however, building spilled over into a variety of other fields relevant to the increasingly sophisticated life of urban societies. Although the materials and techniques used were much the same, their purposes were so different that they need separate consideration. Not surprisingly, in view of the paramount need for assured water supplies for both domestic and agricultural needs, ambitious water engineering projects loomed large in the ancient world.

Water supply

The blocking of watercourses with accumulated debris is so common that the notion of the artificial dam readily suggests itself; what is surprising, though, is the scale of damming operations at a very early date. Because of their massive nature, and because they are often remote from the populations they serve and thus have been protected from being used as quarries, many ancient dams survive – indeed, a few are still in use. Perhaps the oldest is the dam built across the Wadi Gerrawi, in the eastern desert of Egypt, about 2500 BC. Faced with masonry, it is 90 metres (295 feet) thick and 125 metres (410 feet) long. A far larger one, originally built about a thousand years later by Egyptian engineers, blocks the valley of the Orontes in Syria to form the famous Lake of Homs, 50 square kilometres (19 square miles) in area.

Sites suitable for dams are not easily found and water engineers had to be prepared to conduct the water collected over long distances. This involved not only problems of construction but of grading the canal as it passed over irregular country: often both tunnels and raised aqueducts had to be built. A case in point is the aqueduct built by Sennacherib in 691 BC to carry water to Nineveh from Bavian, fully 80 kilometres (50 miles) away in the hills. In places, it is 20 metres (65 feet) wide and at one point it crosses a valley on a five-arched aqueduct 300 metres (1000 feet) long. This alone required two million blocks of masonry, and the whole canal is lined with stone and sealed with bitumen. The stone came from a quarry at Bavian – none was available at Nineveh – and it seems that as the canal grew the engineers used its bed as a road to transport the stone needed further along. These were outstanding works but literally thousands of lesser dams were built. The Nabataeans, who flourished in the Negev and southern Jordan from the first century BC to the first century AD, with Petra as their capital, were extraordinarily diligent in conserving the flash floods of occasional showers. Within 130 square kilometres (50 square miles) around the old city of Ovdat nearly 20,000 low dams have been identified, averaging 50 metres (164 feet) in length.

The Romans were familiar with the Nabataean irrigation systems and with earlier Greek ones, and they embarked on great projects of their own throughout their empire. They were good engineers and organizers, and rich: at Nero's instigation two dams were built on the Aniene simply to provide lakes for his pleasure. One of these was 45 metres (150 feet) high, a height not exceeded until the building of the Alicante Dam in Spain in 1594. Most early dams were straight barriers, in contrast to modern ones built in the form of an arch laid on its side so that the pressure of the water compacts the masonry and the thrust is absorbed by the riverbanks. Possibly a Roman dam built in the second century BC at Glanum, in Provence, was of this type. Procopius, writing in the sixth century AD, refers to a masonry dam, arched upstream, built by Chryses at Dara on the Persian frontier.

In hot climates much water is lost by evaporation when it is led through open channels, which is why the Nabataeans extensively used underground conduits. Surface ducts were often covered over, which also excluded dirt. Frequently, however, tunnelling was necessary when high ground stood in the way, and with an abundance of cheap labour this presented no great problem. A tunnel built by Eupalinos at Samos in the sixth century BC passes through a hill nearly 300 metres (1000 feet) high: it is 1100 metres (3600 feet) long and 1¾ metres (6 feet) square. It was excavated from both ends, and the precision with which the two teams met in the middle is a tribute to the accuracy of the surveying methods used. Common practice in tunnelling was to mark the line on the surface with posts and then sink a succession of shafts, easily made vertical by the use of plumb-lines. The

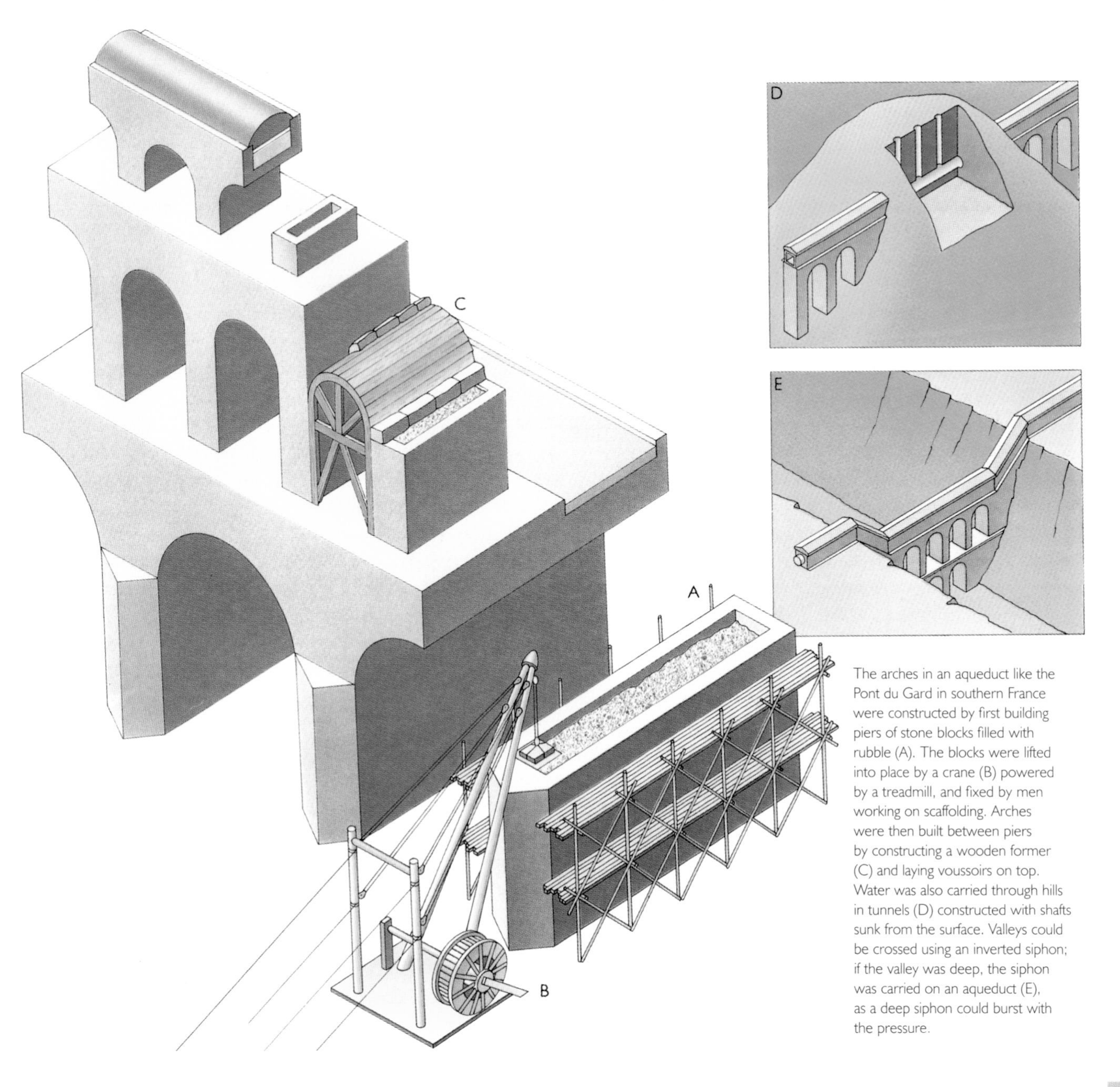

The arches in an aqueduct like the Pont du Gard in southern France were constructed by first building piers of stone blocks filled with rubble (A). The blocks were lifted into place by a crane (B) powered by a treadmill, and fixed by men working on scaffolding. Arches were then built between piers by constructing a wooden former (C) and laying voussoirs on top. Water was also carried through hills in tunnels (D) constructed with shafts sunk from the surface. Valleys could be crossed using an inverted siphon; if the valley was deep, the siphon was carried on an aqueduct (E), as a deep siphon could burst with the pressure.

brick and stone 54–56 **classical Japanese building** 126 **irrigation** 28–31 **rock building** 124 **timber-framed house** 132

The Romans made extensive use of the arch in many different kinds of buildings. These Roman structures are the Pont du Gard aqueduct built by Agrippa in 19 BC to supply Nîmes, and a three-arched bridge across the Afrin in Syria (below).

bottom of these shafts, which later provided access for maintenance, were then joined.

Where shallow valleys had to be crossed, the water channel was carried on a masonry-faced embankment, but for heights above about 2 metres (6½ feet) arched aqueducts were used. For heights up to about 20 metres (65 feet), stable structures could be built provided the pillars were massive and the arches narrow. Above this, however, such structures were very vulnerable to gales or subsidence; the solution was to build a second tier of arches on the first and, in extreme cases, a third tier on that. The most famous surviving example is the Pont du Gard near Nîmes, built in the first century AD: still functional, it towers 55 metres (180 feet) above the floor of the valley.

Today, aqueducts in the classical style are rare: the water is carried in a closed pipe down one side of a valley, across its bottom and up the other side. Such a system is now known as an inverted siphon: the Greeks and Romans called it a 'stomach' (*koilia* or *venter*). At first sight, it seems strange that neither the Greeks nor the Romans made much use of this, for they were well aware of the underlying principle, which is simply that water finds its own level. The reason lies in the engineering problems: very considerable pressures are generated at the lowest points in such systems and it was difficult to contrive leak-proof joints. Moreover, there is additional strain at any sharp bend, which is difficult to avoid. Despite this, a few inverted siphons were built, the best known being that at Pergamon (Bergama) constructed about 200 BC. It was evidently unsatisfactory, however, for within a century it was dismantled and replaced by a conventional arched aqueduct. Its failure is not surprising, for simple calculation shows that pressures up to 18½ kilograms per square centimetre (260 pounds per square inch) – roughly 10 times that within an automobile tyre – had to be sustained.

Water distribution

In AD 97 Frontinus – a former military governor of Britain – was put in charge of the water supply of Rome. From his book *De Aquis Urbis Romae* we have a remarkably detailed picture of this very complex system, complex not only in the technical sense but also from the administrative point of view. We know that he commanded a work-force of 700 men and, by implication, was responsible for the distribution of around 900 million litres (200 million gallons) of water a day passing through 400 kilometres (250 miles) of piping of one sort or another. Water was supplied to both public cisterns and private households, the latter paying a water rate proportional to the size of the spout (*calix*) that supplied it: there were 24 different sizes, and from what he says there was evidently a good deal of malpractice by enlarging or exchanging them.

Frontinus says little about the pipes themselves, but Vitruvius describes both earthenware and lead ones. Interestingly he recommends the former, remarking that the grey pallor of plumbers is indicative of the toxicity of the metal with which they work – an early example of an occupational health hazard. Earthenware pipes had the additional advantage that they could be laid and joined – using a cement consisting of quicklime mixed with oil – by ordinary bricklayers. Lead pipes, by contrast, had to be soldered, a much more skilled operation.

Frontinus, like Vitruvius, epitomizes the businesslike approach to practical problems that was the basis of the Romans' success. At one point in *De Aquis* he enthuses: 'With such an array of indispensable structures … compare the idle Pyramids or the useless, though famous, works of the Greeks.' This is, in stark contrast to the Greek philosophy, succinctly stated by Xenophon in the fourth century BC, in *Oeconomicus*: 'What are called the mechanical arts carry a social stigma and are rightly dishonoured in our cities. For these arts damage the bodies of those who work at them and who act as overseers, by compelling them to a sedentary life … this physical degeneration results also in deterioration of the soul.'

China was in this respect closer to Romans than the Greeks. The ninth-century AD Islamic philosopher Al-Jahzi, stationed geographically between them, wrote: 'The curious thing is that the Greeks are interested in theory but do not bother about practice, whereas the Chinese are very interested in practice and do not bother very much about the theory.'

Bridges

Stone-arch bridges were as suitably for carrying roads over rivers as for carrying aqueducts. The Romans built many in Europe and the Chinese also built many at much the same time: a famous example is the Traveller's Bridge near the Luoyang palaces about 200 BC. The Roman semicircular arch made for inconveniently steep access, especially for wheeled vehicles. Not until much later did the elliptically arched bridge appear: a fine early example is the Arigi bridge in China, built about AD 600.

The arched bridge was, of course, a relatively late and sophisticated development. Early bridges, and many later ones up to the present day, were simply built on the pillar-and-beam principle. As in buildings, however, there was the difficulty that spans were necessarily short and for long crossings many piers were needed. Deep wide rivers defeated the most determined bridge-builders. For such crossings, regular ferries plied on important routes, but an alternative was the floating pontoon bridge. Such a bridge was used by Darius to carry his troops across the Bosphorus in 516 BC.

Roads

At the height of their power the Romans had a system of paved roads covering some 80,000 kilometres (50,000 miles), extending from Hadrian's Wall in the north to the Euphrates in the east and the Sahara in the south. By the fourth century AD it was falling into disrepair and nothing comparable took its place in Europe until the 18th century. Outside the Roman Empire no such road system existed, though this is not to say that there were not well developed and widely used lines of overland communication – but these were routes rather than roads, beaten tracks changing course when over-worn and requiring little formal maintenance.

Roman roads were not laid down originally for military purposes, as evidenced by the fact that their building generally followed, rather than preceded, occupation. Their main purpose was to serve the needs of administration and trade, though this naturally involved the ability to move garrison troops quickly if need arose. Ordinary roads were 5–6 metres (16–20 feet) wide, but main highways could be as wide as 10 metres (33 feet). The surface was normally paved with stone slabs, laid on successive layers of rubble, crushed stone

Roman roads, like their buildings, are noteworthy for their massiveness: they are commonly much thicker than modern roads, carrying far heavier traffic. For this reason they have a high survival rate: this picture shows a section of the Appian Way, built in 312 BC.

and sand. The most remarkable feature of Roman roads is their great thickness, often well over a metre (3 feet): by contrast, modern roads designed for very heavy motor traffic are rarely more than half this thickness, even though they are subject to greater stresses.

Transport canals

So far, we have considered canals mainly in the context of water supply, but they were used also for transport. As early as 510 BC the Emperor Darius had a canal cut to link the Nile and the Red Sea. The Romans made extensive use of the great rivers of Europe and linked some of them by canals, partly in the interest of transport but often for flood control.

It was the Chinese, however, who were the great canal builders of the ancient world. Among many important early examples was the 150-kilometre (90-mile) canal built in 133 BC to link the Han capital of Ch'ang-an with the Yellow River. A much greater project was the Grand Canal, along which tribute grain was transported from the Huai and the Lower Yangtze to Luoyang. The first sections were completed in AD 610, allegedly with a labour force of five million men, and its total length was nearly 1000 kilometres (600 miles); by the eighth century it was carrying no less than two million tonnes of goods annually.

A major problem in canal construction is to allow for changes of level. The simplest device, used in China in the fourth century, is to connect successive level sections by slipways up and down which boats can be dragged. At other points, levels were controlled by sluice gates such as were commonly used on irrigation canals. The chamber lock is a medieval development.

development of roads 153, 176, 261–262 **expansion of canals** 152, 153, 261 **modern bridges** 262

CHAPTER SIX

Power and Machinery

The peoples of the ancient civilizations successfully embarked on projects which would not be undertaken lightly by modern engineering contractors. Long before the Christian era, we have noted such impressive achievements as the Pyramids, the Great Wall of China, the huge merchant ship *Syracusia*, Eupalinos' tunnel in Samos, the megalithic circles of northern Europe and the Pantheon in Rome. It is self-evident that such works could be accomplished only by using some kinds of machinery and expending a great deal of energy, but we have so far said very little about the nature of these resources.

In general, the only power sources available to ancient engineers were those of men and animals. Leaving aside the special case of sailing-ships, wind and water power made, overall, very small contributions. Steam power, which sustained, but did not initiate, the Industrial Revolution, was known, but only in the context of mechanical toys.

Manpower

The abundance of cheap labour, including much slave labour, was certainly a disincentive to the development of power-driven machinery. Nevertheless, it is easy to ignore the fact that manpower has to be properly deployed if it is to be effective. Although a man can carry a load of no more than about 40 kilograms (88 pounds) for long distances, he can briefly lift at least half as much again, say 60 kilograms (132 pounds). In theory, therefore, 20 men could lift into position a load weighing just over 1 tonne: this corresponds to a cube of stone less than a metre (about 3 feet) on a side. But, however much they jostled together, 20 men simply could not collectively get a grip on a block of stone this small; even if they could, they could not lift it more than a metre at a single heave. Some sort of mechanical aid was essential.

A sudden concerted lift is one thing, but a sustained expenditure of energy – as in a rowing galley – is quite another. Experiment shows that a man can generate about one-third of a horsepower for a few minutes, but cannot work steadily at more than one-tenth of a horsepower: for the purposes of comparison, we can equate 1 horsepower (746 watts) with the output of one small petrol or electric motor, such as those used to drive lawn-mowers and other similar appliances.

Translating this into the power problems of the ancient world, we find, for example, that the combined oarsmen of a Greek penteconter could sustain at a dozen horsepower. Making reasonable assumptions about such factors as water resistance, it has been calculated that this corresponds to a maximum sustained speed of less than 10 knots (18½ kilometres/11 miles per hour). This agrees very well with one famous voyage detailed by Thucydides, who lived in the fifth century BC. The Athenians had decreed that the male citizens of Mytilene, then in revolt, should be massacred and they despatched a trireme to give orders to that effect. They then had a dramatic change of heart and dispatched a second trireme to rescind the order. In the circumstances, these men – spurred on by the promise of extra pay as well as by compassion – must have rowed like fury, for it overcame the first vessel's 24-hour lead. The 345-kilometre (207-mile) voyage was completed at a probable average speed, under calm conditions, of a little over 8 knots.

Effectively, the oar is no more than a long lever working on the thole-pin as a fulcrum. It appears in many other guises, all designed to give some sort of mechanical advantage. Thus Assyrian bas-reliefs show huge loads being hauled on runners by teams of men pulling on ropes, assisted by others edging them forward by inserting massive levers behind them, each worked by as many as a dozen men. Given sufficient manpower – more easily marshalled in Egypt than in Greece or Rome – very large forces could be generated in this way. The mathematical theory of the lever was understood by the fourth century BC. In the third century BC Pappus boasted: 'Give me a place to stand on, and I will move the earth.'

As is powerfully expressed by this relief from Sennacherib's Palace at Nineveh, built early in the seventh century BC, the ancient world relied heavily on slave labour to execute massive building projects.

The three pyramids at Giza (top left), and the sphinx, are memorials to the kings of the Fourth Dynasty, and their court, who flourished around 2600 BC. Like later pyramids, they embody enormous blocks of masonry, some weighing upwards of 100 tonnes. They are reminders of the skill of early engineers in handling what even today would be regarded as formidable loads.

The Step Pyramid of Djoser (bottom left), dating from the Third Dynasty (*circa* 2650 BC), is conspicuously different from those of Giza. It represents a short-lived phase during which relatively small building stones were used.

The wedge is another device of great antiquity – the stone axe is really no more than a wedge mounted on a haft. Mechanically the wedge is akin to the lever: a long forward movement effected by blows with a hammer is translated into a short vertical lift. It could be used, for example, to lift a heavy weight or to split stones in the quarry. For civil engineering works, much use was made of the inclined plane or ramp which is, again, no more than a giant wedge. The difference is simply that while the wedge moves and the load is stationary, the inclined plane is stationary and the load is dragged and levered up it. The use of earthen ramps has been plausibly postulated, for example, as the means by which the massive lintels were put in position on top of the tall uprights of Stonehenge and other megalithic circles. How far they were used in building the pyramids is another matter.

The original Step Pyramid of Djoser at Saggara, built about 2650 BC, is unusual in that it is built of worked stones of manageable size, rising to a height of 60 metres (200 feet). But the Great Pyramid of Cheops, built about a century later, is very different; it rises to 146 metres (480 feet) and is built of enormous blocks, some weighing hundreds of tonnes. Given the incentive, and vast resources of human labour, the overland transport of these on sledges was relatively straightforward: Herodotus gives a vivid description of the great stone causeway, a kilometre (1100 yards) long, built for the purpose of carrying them from the quarry to the site. To raise stones for the lower courses, ramps may well have been built, but these would have had to be immensely long to reach the upper levels, given a workable slope. Herodotus states confidently that a succession of levers was used to raise the stones from level to level, but we must remember that he was describing events that took place 2000 years before his time – as distant as the birth of Christ from our own. The great causeway Herodotus saw with his own eyes, but there

Great Wall of China 23, 87, 125 **megalithic monuments** 19, 45, 57 **oar-powered boats** 50–51 **Stonehenge** 45, 57, 59

For lifting heavy loads, large cranes powered by human treadmills were widely used in the ancient world. This detailed illustration of current building methods is taken from the Atterii Monument, which was erected in Rome in the first century BC.

can by then have been little direct evidence of the building methods used.

In the same category as these early mechanical aids is the screw, by which a powerful linear thrust can be generated by a modest sustained torque (turning power). It certainly appeared far later than the lever or the wedge, but it is difficult to date precisely: Hero of Alexandria clearly described a double-screw press in the first century AD. In a mechanical sense, of course, the thread of a screw is no more than an inclined plane coiled up into a helix.

For raising objects to a height, the simplest device is the rope and pulley, and this was certainly used by the Assyrians. If one wheel only is used, there is no mechanical advantage – though several men can tail on to a single rope – but the Greeks and Romans used much more sophisticated systems. In the *pentapaston*, for example, the upper block had three pulleys and the lower two; by winding the rope through these in succession a 5:1 mechanical advantage could be gained. Greek dramatists – particularly Euripides in the fifth century BC – made much use of the god descending from heaven, the *theos ek mechanes*, and it is clear that some system of pulleys was used to produce this effect. According to Plutarch, Archimedes drew a three-masted ship along single-handed with the aid of multiple pulleys. This is only just plausible, but with growing complexity friction losses are a major problem in such pulley systems.

Pulley systems were normally incorporated in some sort of crane. The simplest of these was a bipod shaped like an inverted V, kept steady by rope stays: such shear-legs are still widely used today. In a more elaborate version, the pulleys are attached to a jib, hinged at ground level, which could be raised and lowered.

In simple cranes, the excess rope might be left lying on the ground, but generally it was wound on to some sort of drum. For this, and for many other purposes, rotary motion was necessary and this was achieved in various ways. The simplest was the capstan or windlass in which handles – effectively short levers – were thrust at right angles through the shaft to be turned. A problem here is that if the operator relaxed for a moment the weight of the load would start to unwind the rope and turn the shaft backwards, perhaps uncontrollably. The modern answer to this is the pawl and ratchet and there is evidence that this device was known as early as the fifth century BC: Hero of Alexandria describes it fully as a method of preventing a military catapult discharging prematurely.

There is some dispute about the origin of the crank, today a very familiar means of turning a shaft. In China, it is quite clearly depicted in a tomb-model of a winnowing machine which is certainly not later than AD 200, but some historians believe that in the West it cannot be certainly dated before a picture in the Utrecht Psalter of AD 830. This may be true if we regard a crank as specifically a shaft doubly bent at right angles, but it is quite permissible to look on the origin of the crank as a handle mounted at right angles to the circumference of a wheel, as in old-fashioned mangles. On this basis, the rotary quern, widely used to mill corn in Europe before the Christian era, was certainly crank operated: the upper circular millstone had a hole near its rim into which a wooden handle was inserted.

For intermittent use – as for loading and unloading ships' cargo – the windlass was simple and convenient. For other purposes, however, such as grinding corn or raising water or in major building operations, a steady source of power was required, and for this the Romans commonly used a treadmill. This consisted of two vertical wheels connected by rungs, like those of a ladder. As the operator climbed this endless ladder, his weight caused the wheel to revolve: the more he exerted himself, the faster it turned. Some massive wheels, of which pictures survive, were worked by teams of men.

The torque exerted by the weight of the operator depends on his horizontal displacement from the shaft: it is greatest at the 3 o'clock position but obviously this is impossible for a man inside such a giant squirrel-wheel. However, this does not apply if – as in some wheels used to raise water for irrigation – the operator is outside. Working at the 3 o'clock position, he could then get maximum torque all the time. The principle of the treadmill was applied in other ways. For example, large Archimedean screws used for raising water for irrigation often had wooden slats on the outside so that they could be turned by treading them with the feet.

The axle of a treadmill had of necessity to be horizontal, but for certain purposes – such as turning millstones to grind corn – a vertical shaft was necessary and so some sort of gearing was needed. That commonly used in Roman times was what we could now call a pair of crown wheels, but by the first century AD all the simple kinds of gears were well known, and were described at length in Hero's *Mechanica*. Of particular interest – for it is another means of transmitting power from one shaft to another at right angles – was the worm gear, which is in essence a variant of the screw.

Gearing makes it possible not only to transmit rotary motion from one shaft to another but also to obtain a predetermined mechanical advantage. Thus if one gear wheel has 10 times as many teeth as the wheel that drives it, it will make one-tenth of a turn for every full turn of the latter. At the same time it will exert 10 times the torque applied to it by the driving wheel. Thus a weak force applied to the driving wheel is converted to a strong force delivered by the driven wheel. In theory this can be continued to any extent, and there are fanciful medieval descriptions of a man's breath turning heavy machines through trains of gears. In practice, of course, power losses in transmission severely limited the degree of gearing that was practicable.

A very exceptional example of early gear trains – though for a small mechanism rather than a heavy machine – is the so-called Antikythera mechanism, recovered from a sunken Mediterranean wreck in 1900. It was probably made in Rhodes about 87 BC and is an extraordinarily elaborate form of mechanical calendar, containing at least 25 gears cut in bronze. With it, the positions of the sun and moon could be predicted, and the risings and settings of certain stars. In heavy machinery, such as mills, the exact gear ratio was not important. In such calendrical devices as the Antikythera mechanism, however, the requirement was quite different and the gear wheels had to be cut with great precision in accordance with a mathematical formula, requiring a high degree of mechanical skill.

Animal power

As we have seen, draught animals were used from a very early date for transport and agricultural purposes and for economic reasons the ox rather than the horse was favoured. They were, however, little used to drive machinery. Donkeys were sometimes used in treadmills, but more commonly draught animals were harnessed to the ends of long beams attached to a vertical shaft: as the animal plodded round, so the shaft turned steadily.

Such an arrangement was ideal for driving millstones, when a vertical drive shaft is necessary in any case, and for oil-presses in which olives were crushed by a heavy edge-wheel running through a circular stone trough. We can see early examples of this in the great corn-mills which were found in the remains of a public bakery in Pompeii.

The Antikythera mechanism (top), a calendrical device made in Rhodes about 87 BC, is a unique example of early precision gearing. Generally, much heavier machinery was employed, much of it similar to that in current use, like an olive press in Cyprus (centre), or to these large corn-mills employed in a public bakery in Pompeii at the time of its destruction in AD 79 (bottom).

Archimedean screw 30 **catapults** 88 **cranks** 116 **gears** 69, 116, 166, 247, 248 **treadmills** 30, 61, 81

The constituent parts of an early Byzantine sundial calendar. Dating from the fourth to seventh century AD, it is a beautifully crafted example of early precision gearing.

By the fourth century AD labour, even slave labour, was becoming scarce in the Roman Empire and the unknown author of *De rebus bellicis* (*circa* AD 370) remarks on the neglected inventiveness of barbarian peoples. The author's task, in modern terminology, was to mechanize the army to save manpower. Among his most remarkable proposals was a paddle-propelled warship powered by oxen harnessed to a form of capstan. It is doubtful if such a vessel was built – like Leonardo da Vinci, the unknown author's imagination may have outstripped available resources – but it is an interesting early example of mechanical inventiveness.

Water and wind power

One advantage of the treadmill as a source of power was that it could be set up wherever required, then dismantled and carried – or even rolled, if the distance was not great – to a new site. By contrast, watermills could only work at fixed sites, and the wind was fickle. Nevertheless, water and wind power became very important until after the advent of steam. But it was water that was utilized by the Greeks and Romans.

Four basic water-wheels – the Greek or Norse mill (A), undershot wheel (B), overshot wheel (C) and breast wheel (D). The vertical shaft of a Greek mill can drive a millstone directly. For this purpose the other wheels need gearing.

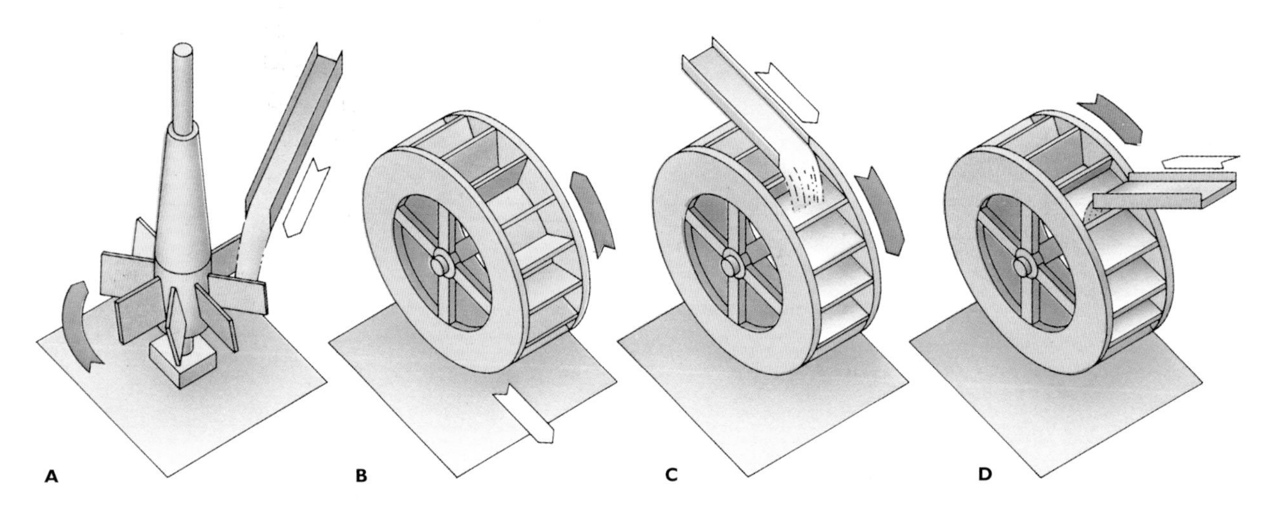

The history of the water-wheel is obscure because of the dearth of archaeological evidence and of literary and pictorial descriptions. It is perhaps easier to understand, however, if we first consider the three main types and their mechanical characteristics. First, there is the wheel attached to a vertical shaft, with a number of blades attached to it at an angle of about 60° to the horizontal: water impinges on the blades from a sluice and turns the wheel. The second type is the undershot wheel in which, as its name implies, a wheel carrying paddles dips into a running stream. And thirdly, there is the overshot type in which the circumference of the wheel is divided into separate segments, or buckets, which are successively filled with water from a spout at the top: the weight of the water causes the wheel to turn and each bucket empties itself as it reaches the bottom.

The vertical-wheel mill is often regarded as the most primitive and, by inference, earliest type, but there is no real evidence for this. Certainly one modern type of high-speed water turbine, a far from primitive device, has a vertical shaft. In the so-called Greek or Norse mill, used in northern Europe until recent times, the construction was entirely in wood, and the fact that none survive from antiquity is not surprising. More significant perhaps is that there are no early references to them, suggesting a late invention. Mills of this kind require fast-running streams, such as occur in mountainous areas. They cannot be identified in Europe before the Christian era, nor in China before the third century AD. For grinding corn their advantage was that the vertical shaft could drive the millstone directly.

The undershot wheel shares with the Norse wheel the advantage of needing little in the way of ancillary equipment. The power it generates is proportional to the difference between the speed of the current and the speed of the blades. Consequently, as the wheel slows down when the workload is increased, so the torque increases. As this type of wheel was described by Vitruvius, and he would not have done so had it been a rarity, we can safely date the undershot wheel at least as early as the second century BC. Similarly the fact

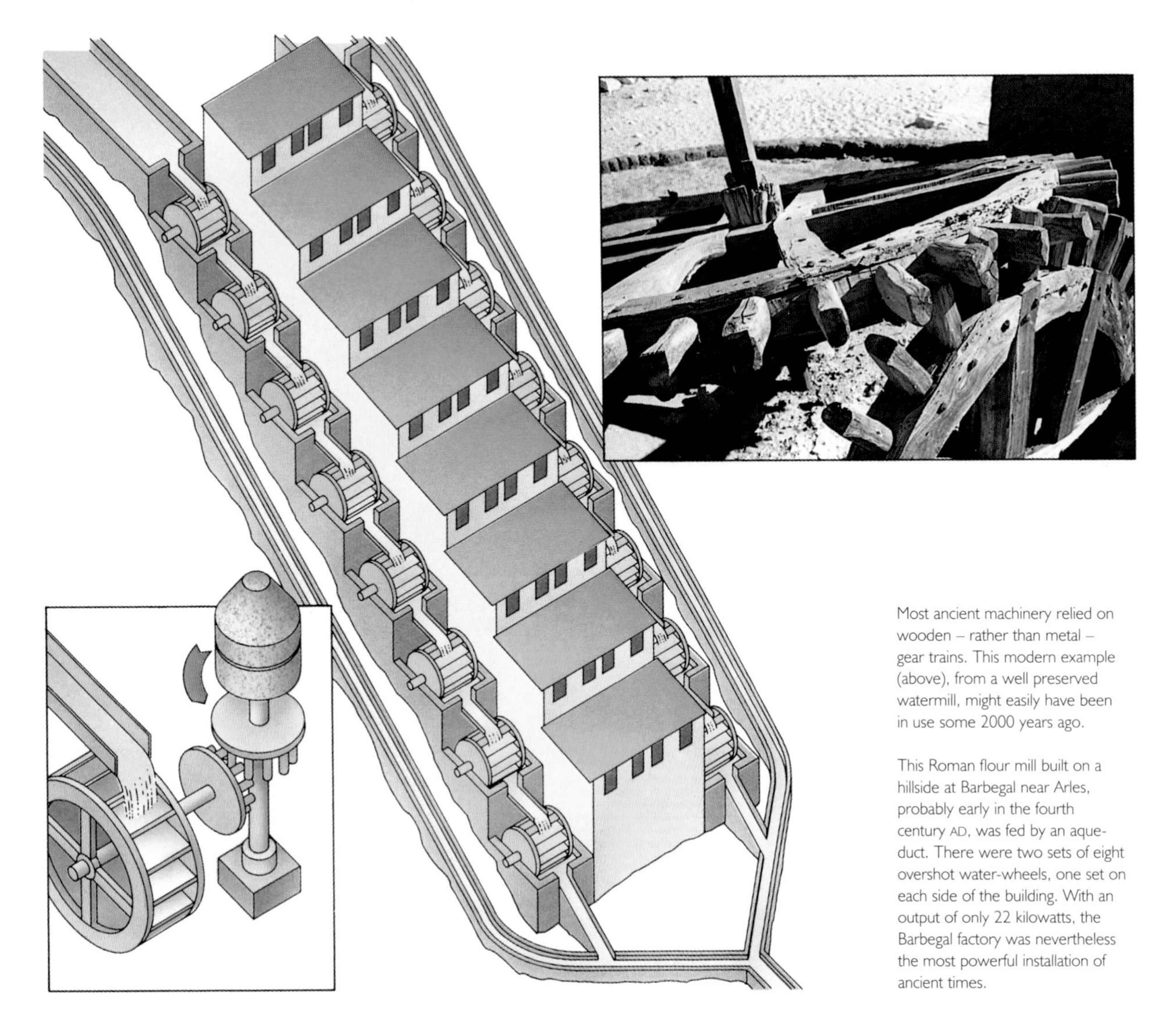

Most ancient machinery relied on wooden – rather than metal – gear trains. This modern example (above), from a well preserved watermill, might easily have been in use some 2000 years ago.

This Roman flour mill built on a hillside at Barbegal near Arles, probably early in the fourth century AD, was fed by an aqueduct. There were two sets of eight overshot water-wheels, one set on each side of the building. With an output of only 22 kilowatts, the Barbegal factory was nevertheless the most powerful installation of ancient times.

that he does not describe the overshot wheel suggests either that he was not aware of it or that it was too uncommon to justify description for a general readership.

In the overshot wheel the water has to be supplied to the top of the wheel. It must, therefore, be extracted from the river some distance upstream – 100 metres (330 feet) or so, according to the fall – and conducted through a permanent channel. Additionally, a millpond is required to give a steady head and, for maximum efficiency, the flow must be controlled by a sluice so that each bucket on the wheel is just filled as it passes the intake: if the bucket overfills water is wasted and if it does not fill completely the torque is less than it could be. Because of the extra work needed, the overshot wheel must have been far more expensive to install than simpler versions.

The origin of the overshot wheel is not clear, but it is tempting to suppose that it derived from the bucket-wheels used to raise water for irrigation. In effect, it is of course merely a bucket-wheel working in reverse. On thin evidence, a Greek poem of the first century BC has been interpreted as referring to an overshot wheel: it refers to 'nymphs [water] leaping down on top of the wheel ... to turn the heavy Nisyrian mill-stones'.

In Rome, the aqueducts supplying the city's corn-mills were strategically important, and a problem arose when they were cut by besieging Goths in AD 537. The military commander Belisarius solved it by mounting water-wheels between pairs of moored boats; such mills later became widely used in both Europe and the Middle East.

The power generated by water-wheels naturally depends on their size, the rate of flow of water, mechanical efficiency, and so on. However, we may suppose the output of an average undershot wheel to have been around one-tenth of a horsepower, compared with perhaps 2 horsepower for a large overshot wheel, which is basically more efficient.

One of the largest watermills in antiquity was that built at Barbegal, probably in the fourth century AD, to supply all the flour needed by the population of Arles, then totalling about 10,000, including a Roman garrison. Much of the masonry work survives and has been excavated to reveal a large installation consisting of eight pairs of overshot wheels. Estimates of its capacity vary, but it was probably capable of

floating mills 115 **hydroelectric power** 165, 231 **tidal power** 116, 277 **water turbines** 165–166

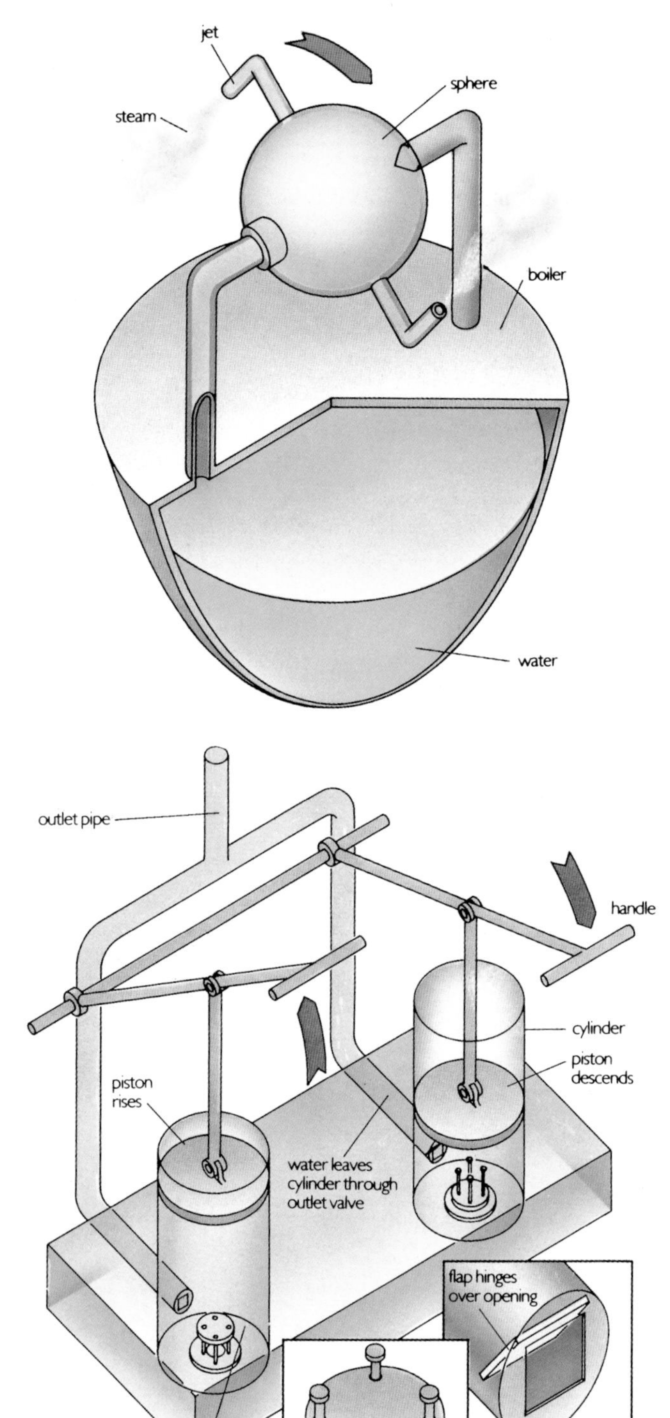

Hero's steam engine (top), built in the first century AD, was the first heat engine. Steam passed from the boiler into the sphere and escaped through the jets, causing the sphere to rotate. The water pump (bottom) built by Ctesibius in the second century BC had simple inlet and outlet valves that controlled the flow of water into and out of the cylinders as the pistons rose and fell.

grinding around 10 tonnes of grain a day. With all the wheels working, output was probably no more than 30 horsepower, but nevertheless this concentration of power was not surpassed in Europe for several centuries.

The siting of mills deserves comment. While properly dried grain can safely be stored for long period, this is not true of meal and flour. Corn-mills, therefore, were usually located in or near the communities they supplied.

As the grinding of flour was a daily necessity, it is scarcely surprising that most identifiable watermills were used for this purpose. Nevertheless, we have some evidence of other uses. Thus the Latin poet Ausonius, writing around the middle of the fourth century AD, refers to a stone-cutting mill on the Moselle, an area where a soft soapstone was quarried.

Wind and steam power

In view of the early appearance of sails to propel ships, it is surprising that the windmill was for all practical purposes unknown in the ancient world. It seems first to have appeared in Persia in the seventh century AD, and then consisted of sails mounted on a vertical mast. The only exception was a tiny windmill built by Hero to pump air for an organ, but it was scarcely more than a toy.

Steam power was very much akin to wind power in the ancient world: its possibility was recognized, but no effective use was made of it. Again, our only source is Hero, who describes what in modern terms we would call a reaction turbine, an advanced form of steam engine. In this, steam from a boiler was fed into a metal sphere from which extended, at diametrically opposite points, two narrow tubes bent at right angles. As steam was ejected from these nozzles, the sphere rotated by jet propulsion, like a catherine wheel. His description is sufficiently detailed for a modern model to have been constructed, and with this speeds of up to 1500 revolutions per minute have been achieved.

This engine was, however, no more than a toy, and there are various good reasons why a scaled-up version would have been of no practical value. One is that this kind of engine works effectively only at high speeds and for the needs of the day it would have had to be geared down enormously, through worm gearing, with great loss of power. More seriously, the constructional methods of the time would have been inadequate to contain the high-pressure steam. Apart from the risk of explosion, there would have been enormous loss of power through leakage and the fuel consumption would have been correspondingly high.

Nevertheless, the question has been asked: why did the Greeks and Romans not develop a simple low-pressure steam engine, such as the early beam-engines of the 18th century? They were, after all, familiar, through various pneumatic devices, with the basic components: the piston working in a cylinder as in suction pumps; the simple flap valve; and the steam boiler. That they failed to assemble these into a simple working engine was probably due to the fact that there was at that time no particular need for it: the tempo of life was slow, labour was still relatively abundant and cheap, and where power was needed existing devices were adequate.

CHAPTER SEVEN

Metals and Metal-working

The importance of metals has long been acknowledged in the major conventional divisions of early history according to the materials in common use – a Stone Age followed by a Bronze Age, itself succeeded by an Iron (and later Steel) Age. Such a division is convenient and certainly represents significant steps in the progress of civilization, but it must be regarded with considerable caution. We have already emphasized that in some remote parts of the world man has still not progressed beyond the Stone Age. Equally, while we can say with some confidence when bronze and iron were first used, their general adoption was very uneven. Many people had no means of making either bronze or iron, but might have acquired the latter, say, by way of trade. In that case, their Iron Age could have preceded their Bronze Age: this is certainly true of the Germanic races.

More important, but less generally appreciated, is that neither bronze nor iron is a sharply defined material. The properties of both vary greatly according to their composition, which depends not only on the mode of manufacture but also on the nature of the local ores. Today, metallurgy is a precise science and the relationship between the composition and fine structure of metals and their physical properties is well understood: as a result, metals conforming to a desired specification can be made consistently. In the distant past, the craftsman had no such advantage and had to rely on empirical knowledge laboriously acquired and passed down by word of mouth. Successful techniques did not necessarily travel well: a new source of ore, or even a change of fuel, might lead to a different product. That so much was achieved on a purely empirical basis is a reminder that an advanced technology was in no sense synonymous with the application of science.

It is commonly supposed that iron is generally superior to bronze. This is, in fact, not true: a work-hardened bronze, for example, could be given a far sharper and more durable cutting edge than, say, wrought iron.

Only seven metals were directly familiar to the ancient world: gold, silver, copper, tin, iron, lead and mercury. They did, however, unconsciously make use of others which were not specifically identified until much later. For example, iron from meteorites contains a good deal of nickel and many iron ores contain manganese. Archaeologically, the presence of these metals is often very helpful in pinpointing the origin of a particular object. Very occasionally maverick artefacts occur. For example, a few objects of pure antimony have been found in parts of Mesopotamia where copper-antimony ores are not uncommon.

Some of these metals, such as copper, are widespread in nature: others, such as gold and tin, are relatively rare – in any event, they must be sought. Some occur in the native (uncombined) state. Iron, for example, is found in fallen meteors and these were important early sources of the metal. Significantly, the early Greeks referred to iron as *sideros*, meaning 'from the stars'; in English, the metallurgy of iron and steel is still known as siderurgy. The ancient Egyptian name was *baaenepe*, meaning 'gift of heaven'. Copper, too, is sometimes found in the metallic state: deposits of the pure metal weighing several hundred tonnes have been found in Canada, for example. Gold, referred to as a noble metal because of its great reluctance to enter into any kind of chemical combination, occurs only as the native metal – either as nuggets of alluvial gold in riverbeds or as veins of ore, often dispersed in quartz.

Far more commonly, however, metals occur in chemically combined forms as minerals with properties far different from their own. Mercury, for example, is a mobile silvery liquid whereas cinnabar, its commonest mineral form, is a bright orange solid. The elements most frequently associated with metals are sulphur, carbon and oxygen. The story of how man learnt, at a remarkably early date, to drive these off and work the residual metals into a great variety of useful and ornamental objects is a fascinating story.

Pre-Christian metal-workers achieved standards not surpassed by modern craftsmen, but this is perhaps not surprising as the techniques have changed little. This Etruscan gold bowl, with two sphinxes on each handle, dates from the fifth century BC.

development of metallurgy 183–191, 213, 302 **iron working** 20–21, 75 **steam engines** 158–165, 174–180 **windmills** 114–115, 166, 231, 277

Because it is soft and malleable, gold is one of the easiest metals to work. Silver is harder and correspondingly harder to work; nevertheless, good silverwork is known from the third millennium BC. These elegant silver vessels and figurines of the third century AD were found at Chaource, France.

Gold and silver

Hesiod, a shadowy Greek poet of about the eighth century BC, spoke of five progressive stages of the human race, beginning with those of gold and silver, and ending with those of copper, bronze and iron. He was speaking speculatively, of unrecorded events already far distant in time, but he may well have described, fortuitously, the order of use of these metals, among some people at least. However, when the Egyptians first appeared on the historical scene, they were already familiar with copper, lead and iron, as well as gold and silver.

Gold is very distinctive in colour, and in some parts of the world it is not uncommon, though in very small quantities, in the gravelly beds of streams. It is relatively soft and easily beaten into shape. However, its very softness precluded its use for practical purposes and it was almost entirely employed for ornamental and ceremonial purposes.

Nuggets of alluvial gold were collected by washing away the lighter sand and fine gravel, just as modern prospectors win it in the field by panning. Much gold, however, is found finely dispersed in veins of quartz and to extract this was a very different matter. The Egyptians, who worked many state mines in Nubia – literally the 'land of gold' (*nub*) – used slave labour to pound the ore to a fine powder, from which most of the dross was washed away with water, while the gold was collected on sponges. Strabo – a Greek historian and geographer who flourished around the beginning of the Christian era – describes a similar method used somewhere in the Caucasus, in which the fleeces of sheep were used in place of sponges. So arose, it is supposed, the classical legend of Jason and his search for the Golden Fleece, which seems to have had its basis in a raiding expedition made around 1350 BC.

Natural gold usually contains some impurities, such as silver or copper, and these affect its properties, especially its colour. Ancient goldsmiths made use of these differing tints to produce attractive designs. When pure gold was desired, a technique was used which – in the absence of any knowledge of the basic principles – was remarkably sophisticated. Known as cupellation, it depended on the chemical inertness of gold. The impure metal was fired in a clay crucible with a little lead and salt, together with some carbonaceous material, such as barley husks. After several days, pure gold separated below a crust of slag in which the impurities were concentrated. Agartharcides, in his general account of Egyptian gold-mining projects in the second century BC, includes such a process, but it was already perhaps a thousand years old.

The Romans, who mined gold in many parts of their empire, but especially in Spain, introduced quite a different process. In this, the gold was dissolved from the crushed ore with mercury. From the resulting amalgam, the volatile mercury can be distilled and recovered, leaving the gold behind. A little of the necessary mercury was obtained as native metal, but most was obtained by roasting its orange sulphide ore, cinnabar – of which there are large deposits in Spain – in a current of air.

Although most silver is found in mineral form – especially in association with lead – it too is encountered in the native form, occasionally in deposits weighing as much as half a tonne. Such silver occurs particularly widely in Asia Minor, and the Hittites who controlled the mining area seem to have been the original source of supply for the ancient world. Traders from other nations came to seek the precious metal and it is significant that the ideogram or symbol used to represent the Hittite capital was identical with that for silver. The first use of silver cannot be precisely dated, but silver ornaments were found in the tomb of Queen Shub-ad of Ur, of the third millennium BC. The near-monopoly that the Hittites enjoyed from the third millennium BC did not last, and by the first millennium BC silver – much of it obtained as a by-product of smelting lead – was in common use throughout the Near East, and relatively cheap. An exception was Egypt, where imported silver was priced at twice the value of native gold. Not until Hellenistic times was this valuation reversed – silver then cost only one-thirteenth as much as gold. Silver and gold are rare metals and at an early date their scarcity value led to their widespread use in coinage. Probably the first to strike coins, in silver, were the Hittites.

As today, the scale of mining operations depended on the price obtainable for the metal. One of the largest mines in the Greek world was that at Laurium in Attica, a major source of silver and lead. It paid rich dividends in the fifth century BC, when the price of silver was high, but declined during the third, when the price was low. The nature of the smelting process is not clear but it must have involved roasting the lead ore (galena) to oxidize it, followed by reduction. The silver, present to the extent of about 2 kilograms (over 4 pounds) per tonne, was separated by crystallization from the molten lead. Whatever its precise nature, the process was not very efficient, for the slag-heaps contained at least one-third of the silver – sufficient to make it worthwhile to rework them several times at later dates. This reworking of slag, taking advantage of better technology and higher prices, has always been a feature of the metal mining industry; another well-known example is the reworking, up to modern times, of the slag-heaps of the Roman lead mines of Mendip, in the south-west of England.

It is something of a paradox that the Greeks, admired as founders of democracy, derived much of their wealth from the exploitation of slave labour in the mines under conditions of cruelty and degradation such as few barbarian states have ever equalled and could scarcely have surpassed: small wonder that they revolted in AD 103. The Roman record was no better until the Flavian reforms of the first century AD: slave miners were known as the *damnati in metallum*, a term that might have been applied to most miners in the ancient world.

Copper

The first metal to become available in sufficient quantity to be used for everyday purposes was copper. As we have noted, huge ingots of copper occur naturally, but most have been found in North America, where they afforded a valuable source of supply to the Eskimo and played an important part in the potlatch ceremonies of the native American people of the north-west American coast. The ancient world knew no such massive hoards, but small nuggets might be found and there is evidence for these having been used as early as 5000 BC: for all practical purposes, however, the metal had to be obtained by smelting the ore.

How this process was invented, some time in the fourth millennium BC, must be a matter of speculation, but it is plausible to suppose that it was a consequence of lumps of oxide or carbonate ores finding their way into a fierce wood fire. Under such conditions the ore would be reduced to crude metal, from which a better product could be obtained by hammering out the slag. It could then be further refined by a second smelting, and so on, until nearly pure metal was obtained. One of the best known of such ores, widely dispersed, was the beautiful green malachite; a famous source, from which small quantities are still obtained for jewellery, is one on the Red Sea, exploited by Solomon around 1000 BC. From the casual smelting of small quantities of ore in a wood fire, it would be a comparatively small step to the building of simple furnaces in which a mixture of ore and charcoal was fired.

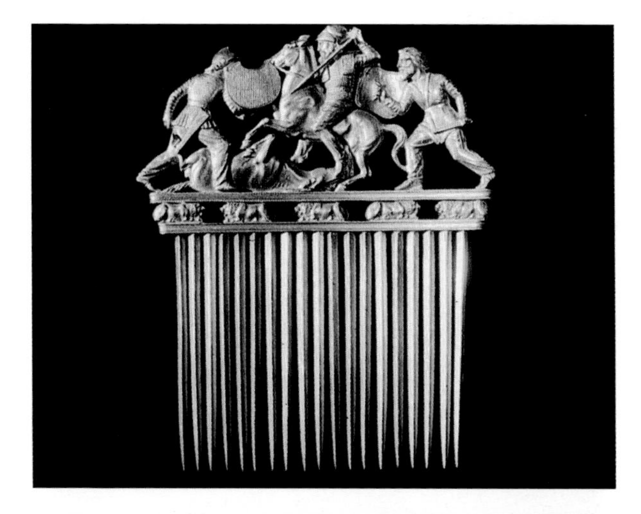

Early metalwork often throws light on contemporary life. This Scythian gold comb (top) shows in detail the dress and weapons of warriors of the day. The bronze stand (below) from Episkopi (12th century BC) shows a labourer carrying a standard bronze ingot.

Many copper ores, however, are sulphurous and the treatment of these is more difficult, especially as they usually contain many impurities, such as iron and antimony. We do not know the precise nature of the extraction process – and there must, indeed, have been many local variations – but it was inevitably laborious and repetitive. An initial roasting process would convert some of the copper sulphide to copper oxide and the latter would combine with unchanged sulphide to give copper and sulphur dioxide gas: at the same

copper alloys 74, 190 **copper mining** 81, 183 **early gold and silver artefacts** 76–77 **gold mining** 81, 190

Coins often present designs derived from daily life, though often in very stylized form. This Greek coin of Demetrius Poliorcetes shows a typical ship with ram (fourth century BC).

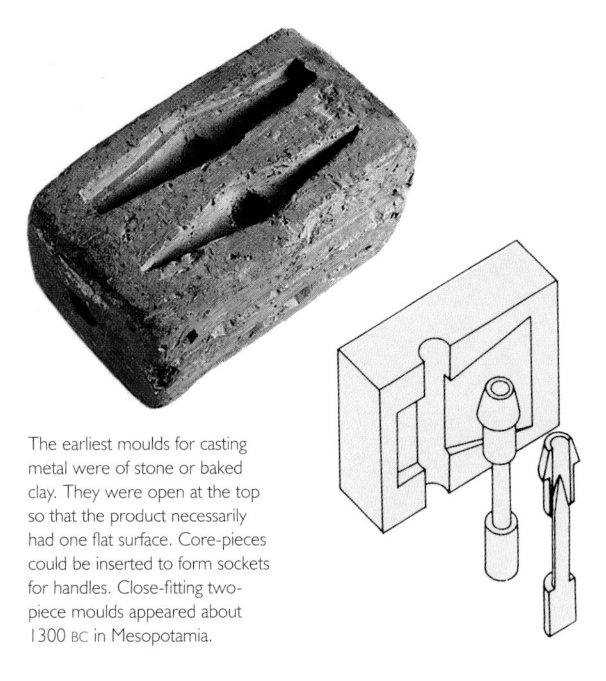

The earliest moulds for casting metal were of stone or baked clay. They were open at the top so that the product necessarily had one flat surface. Core-pieces could be inserted to form sockets for handles. Close-fitting two-piece moulds appeared about 1300 BC in Mesopotamia.

time, some of the impurities would be removed by volatilization. In a final stage of purification, leather bellows were used to blow air through the molten metal. Most impurities were oxidized and rose to the surface, whence they could be removed. Later, the Romans introduced the process known as 'poling', in which the molten copper is stirred with poles of green wood. The heat releases steam from the wood; this bubbles to the surface, serving much the same purpose as an air blast. Indicative of the large-scale use of the metal, it was finally cast into ingots, originally shaped as round cakes but later in a traditional hide-shaped form weighing around 70 kilograms (154 pounds), about as much as a man could lift. The utilization of sulphide ores began about the middle of the second millennium BC. One of the earliest known centres, dating from 1700 BC, was in the eastern Alps.

It is possible that Egypt passed briefly through what may properly be called a Copper Age, ending about 2000 BC, but from a very early date copper was almost entirely used not as such but in the form of bronze, its alloy with tin.

How this valuable alloy, much harder and more useful than pure copper, was discovered is again a matter for conjecture, but it has been plausibly argued that it was a consequence of accidentally working with copper ores that contained an admixture of tin ore. Geological evidence shows that the coexistence of such ores in nature is rather rare, but two particular sites are of much interest in this context. One is in the Taurus Mountains of southern Turkey: in Mesopotamia, bronze items are not uncommon before 3000 BC but rather rare for several centuries thereafter, suggesting the exhaustion of a mine producing the all-important mixed ore. The other site is the Erzgebirge in Bohemia, a famous mining area until recent times, where the Aunjetitz culture became famous for its bronze manufactures.

However it occurred, it was soon divined that the deliberate addition of tin – around 10 per cent – greatly improved the properties of copper. Although tin is fairly easily extracted from its main ore cassiterite (stannous oxide) by smelting with charcoal, it does not seem that bronze was made by melting the two metals together much before 1500 BC. Before that, and in many places until much later, the tin was introduced either by co-smelting the two ores, or by smelting cassiterite with crude copper. It was well recognized that the proportion of tin had to be carefully controlled according to the proposed use of the bronze. Thus in China the *Kao Gong Ji – The Artificers' Record* of the fifth century BC – lists six alloys varying from a 6:1 alloy for making bells (today usually 4:1) and large cooking vessels, through a 5:2 alloy for arrowheads, to a 2:1 mixture for mirrors.

Although tin is the most important metal that can usefully be alloyed with copper, it is not the only one. Analyses of archaeological specimens indicate the use of a wide range of alloys, though it is impossible to say how far these were deliberately prepared and how far the alloyed metals crept in adventitiously as normal impurities in the ores used. Thus some early bronzes contain substantial quantities of antimony and arsenic – excellent hardening agents – and the high content of lead in some Assyrian bronzes (as much as 7 per cent) indicates that it was a deliberate addition. Perhaps as early as the second century BC the Chinese were making and exporting 'China silver' by alloying copper with nickel. Overall, it is clear that, however empirical their methods, early copper-makers made conscious efforts to modify the properties of pure copper, which is very soft, by alloying it with other metals.

It is an inherent hazard of the mining industry that the life of mines is limited and sooner or later they become exhausted. In the distant past, as today, new and often far distant sources had constantly to be found. At an early date, perhaps by 600 BC, the Greeks were getting tin ore from Cassiterides – supposedly the Scilly Islands – which apparently served as a depot for the collection of Cornish ore. Apart from the problems of prospecting, development, sources of fuel and transport, new ores often behaved very differently from the old ones: the lot of the smelter was never easy. Even within the same mine, new veins might have very different properties.

Iron

The advent of ore-extracted iron was one of the great events of history, though where and when it occurred is unclear. Homer never referred to iron as anything but *sideros*, but later Greek writers often called it *chalybs*: the Greek tradition, borne out by other evidence, was that the first people to use iron – perhaps around 1400 BC – were the Chalybes, a tribe subject to the Hittites who seem to have monopolized its manufacture for some two centuries. By 1200 BC, however, following the collapse of the Hittite Empire, the working of iron was established in the Middle East, India and Europe – in southern Italy, at least. This date may be regarded as the beginning of the Iron Age, meaning not the time at which the metal was first used but when it became fairly generally available, either by local manufacture or through trade.

The early history of iron is most easily understood if we avail ourselves of modern chemical knowledge. Like copper, the properties of iron can be varied considerably by alloying it with other metals, but it is also particularly sensitive to the presence of carbon – which alters its fine structure – during the smelting process. If the iron contains little or no carbon – as in wrought iron – the metal is soft and malleable. If its carbon content is relatively high, in the range of 1.5 to 5.0 per cent, the resulting metal – cast iron – is hard and brittle. If the carbon content lies in the range 0.15 to 1.5 per cent, the product is steel, capable of taking a keen cutting edge. Today, control of the carbon content presents no great problem, but the iron-workers of antiquity had to work as best they could by empirical methods. It is remarkable what success they had firing in the dark at so small a target. In particular, they devised the two basic modern methods of steel-making: increasing the carbon content of wrought iron and decarburizing cast iron.

Initially, crude iron was made in very simple furnaces. The typical bloomery furnace was no more than a clay-lined pit fed with ore and charcoal: often the draught was increased, and the temperature raised, by the use of leather bellows. Although iron ore is reduced to metal at about 750°C (1380°F) – as compared with about 1100°C (2000°F) for copper ores, about the highest temperatures then attainable – such furnaces could not reach temperatures high enough to melt the metal so that it could be run off: this development had to await the first blast furnaces. Consequently we have virtually no early examples of iron 'pigs' analogous to the early ingots of bronze (except possibly from China in the Warring States period). Instead the crude bloomery iron, in lumps weighing up to about 25 kilograms (55 pounds), was repeatedly hammered to get rid of the slag. Sometimes – and especially in India, according to Aristotle – this iron was resmelted.

Inevitably, such simple processes yielded a very variable product, depending on the relative proportions of ore and charcoal and the temperature and duration of smelting. Sometimes steel was the natural product. When this was not so, malleable iron was carburized by heating it in charcoal for long periods, when the carbon gradually diffused inwards. This process, now known as case hardening, gave a hard coating of steel over a core of softer iron, and often this was a highly desirable outcome: in armour, for example, the hard outer skin would resist arrowheads and sword cuts, but the softer underlying metal would prevent it from shattering. Where a uniform product was desired, the metal was repeatedly forged to distribute the carbon evenly: a Chinese knife of AD 112 bears the inscription 'this knife has been forged 30 times'. The finest Chinese work was forged 100 times.

Multi-piece moulds for casting large vessels appeared at an early date in China: this *ting* or bronze four-legged ritual food vessel (above), dates from the Shang Dynasty (14th–11th century BC). The largest bronze vessels, highly decorated, might weigh as much as 700 kilograms (315 pounds).

Bronze Age 20–21, 71 **bronze artefacts** 67, 77, 78, 82 **Iron Age** 20–21, 71 **iron smelting** 155, 184, 187–189 **steel** 94, 155, 187–189, 213

Early metal-workers usually worked a single metal, as in the elegant silver cauldron (opposite, top right) of the Celtic Iron Age. Sometimes metals were used in combination, as in the Celtic helmet (opposite, centre left) of copper, iron and gold.

Although the techniques were similar, early metal-work showed a great deal of diversity and regional variation. Interest in the extraction and working of metals is indicated by the frequency with which it turns up as an artistic theme, as on a Greek vase (below centre).

Some of the metal craftsman's finest work was devoted to items of a ritual nature. Among such may be included articles as diverse as the gold death mask of an Achaean king, the so-called 'Mask of Agamemnon' (above); the Chinese bronze ritual *li* (opposite, top left) of the Shang Dynasty; and the sacred double-headed Minoan gold axe (below right).

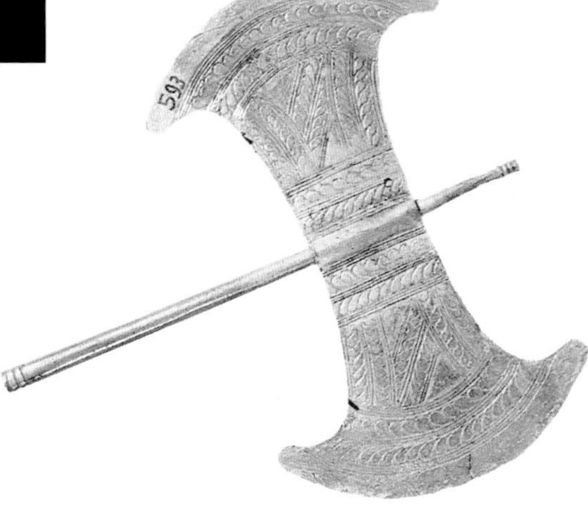

Metal objects – especially weapons and armour – were often richly decorated, as shown by a Chinese axe-head of the Shang Dynasty (opposite, bottom left).
The ceremonial copper head-dress (left), decorated with abalone shell, was made by native Americans.

A collection of bronze articles (left) dating from the seventh to eighth century BC. They comprise a curved knife blade decorated with a hound's-tooth pattern; a small axe-head; and a ladle fabricated from sheet metal.

A stack of Chinese moulds (below) dating from the Han Dynasty (202 BC to AD 220). Molten metal was poured into the stack to produce a column of identical castings, which were then separated into single units.

An alternative way of obtaining a desired carbon content was to fuse an iron too rich in carbon with one lacking it. This was often called the Chinese process and seems to have originated in China about AD 125.

The working of metals

The creation of useful metal objects involves two separate processes: first, shaping the metal to the desired form and, secondly, altering its properties by various treatments, according to the purpose for which the product is to be used. The properties of bronze and iron are so different that they require quite different treatments to achieve similar results. The speed with which the working of metals was mastered is astonishing: by 2500 BC almost all the techniques used today – except centrifugal casting and modern electrochemical methods – had been developed. Not only this, but they were also applied – to both massive objects and delicate ornamental trifles – with a skill and artistry never exceeded. The sole qualification necessary for this generalization is that such heights were not reached everywhere – nor are they today – but only in centres of excellence where wealth and patronage gave the craftsman a market for his skills.

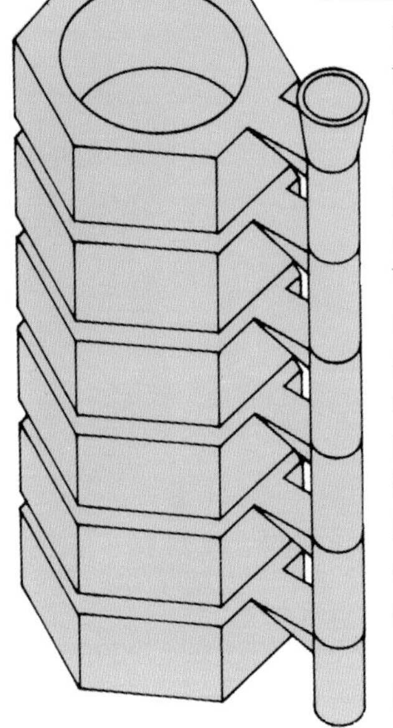

It is likely that metals found in their native state were originally shaped by hammering: to primitive man they must have seemed no more than special kinds of stone. But the discovery that they could be melted and would then solidify in the shape they had assumed in the liquid state at once opened up the possibility of moulding. The first moulds – in stone or baked clay – were open ones, so that one surface of the cast product was necessarily flat, but close-fitting two-piece moulds were in use Ugarit, and doubtless elsewhere in the Middle East, by 1300 BC. When axe-heads and similar tools were cast, core-pieces were inserted to provide sockets for wooden handles. For large castings, which would otherwise require excessive amounts of metal, sandy cores in bags were inserted, and removed when the metal had solidified, to give a hollow product. Clay moulds are relatively short-lived, and for repetitive work bronze or iron moulds were used: Chinese moulds made of iron, used particularly for making agricultural implements, date back to the fifth century BC.

The lost-wax, or *cire perdue*, process – still used today – made it possible to use a one-piece mould to make a rounded object. The shape was first modelled in wax – often beeswax, though a variety of vegetable waxes were available – and then enclosed in soft clay. When the clay was fired, the wax melted and ran out through pre-arranged drainage holes. The resulting cavity was then filled with molten metal. When this had cooled and solidified, the clay mould was smashed and the metal casting removed. The product would, of course, need some further work to remove the lugs corresponding to the drainage holes and to polish away any roughness from the surface. If a hollow moulding was desired, the wax model was built round a core of some infusible material.

For small mouldings, a single orifice and vent would suffice, but for large objects a much more elaborate system was required. Thus a temple painting at Thebes, dating from 1500 BC, shows the casting of a bronze door for a temple: molten metal is being poured into the mould through at least 20 funnels or sprues.

Sheet metal

In the ancient world much use was made of sheet metal and remarkable skill was used in working it. Very delicate and elaborate designs were achieved by hammering the reverse side of the metal, yielding *repoussé* work, and from an early date repetitive designs were made by use of a punch. Much use was also made of sheet metal to mimic large castings: for example, statues were often made by attaching sheet metal,

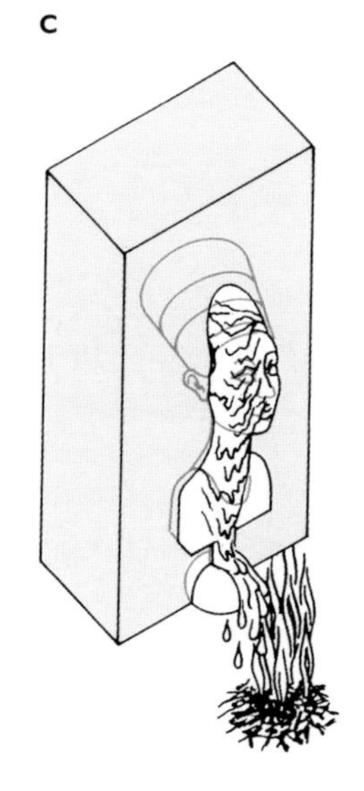

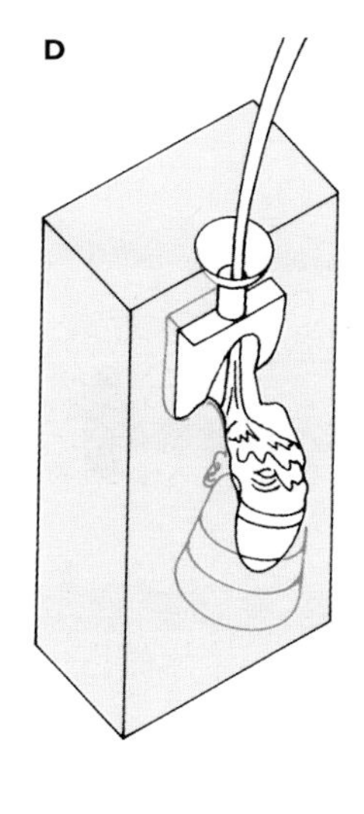

The lost wax or *cire perdue* process of metal casting begins with a wax replica of the object to be cast (A). The replica is enclosed in soft clay (B) and the clay is fired. The wax melts and runs out through a drainage hole (C), forming an earthenware mould that is then filled with molten metal through the drainage hole (D). When cool, the mould is smashed to obtain the casting.

suitably worked, to a wooden framework. An early example of such a statue is that of Pepi I of Egypt, *circa* 2300 BC. Gold, being exceptionally soft and malleable, could be beaten into leaf as thin as paper. A minor, but important, use of sheet metal was to give it a very highly polished surface for use as a mirror: copper was commonly used. (True 'looking glasses', made of glass 'silvered' with an amalgam of tin and mercury, appeared in Venice in the 14th century.)

The neat and strong joining of sheets presented a problem. In some very early work, the edges of adjacent sheets were folded over or sewn together with wire, as were the timbers of early boats, but with gold and bronze soldering and welding were quickly mastered – certainly by 2500 BC, as archaeological finds at Ur testify.

Surface metallurgy

Electrochemical processes that deliberately use an electric current to decompose metallic minerals or deposit thin films by electroplating are strictly developments of the 19th century. Nevertheless, ancient metallurgists were familiar with processes involving internal electrochemical changes, even though they had no inkling of their nature. The Romans, for example, were aware that iron could be coated with a thin film of copper by immersing it in a solution of a copper salt.

Rather more interesting are two techniques developed by the Moche Indians of the Andes, who flourished in the period 200 BC to AD 700. They succeeded in plating copper with gold, apparently by immersing it in a solution of gold in a mixture of corrosive salts: the exceedingly thin film was then bonded to the base by strong heating. They also produced some novel results with the copper-silver-gold alloy known as tumbaga. If this is repeatedly hammered and annealed – heated and cooled – the surface layer becomes disproportionately rich in silver, so much so that eventually the object has the appearance of pure silver. It has been plausibly suggested that such techniques are the reason why the Spanish conquistadors were often disappointed when they came to melt down plundered objects supposedly made of pure silver.

Forging

Copper and bronze could be moulded to shape, but for many purposes further treatment was necessary. If a hard cutting edge was required, this was achieved by hammering the metal when cold. For the craftsman engaged in repoussé work, however, this effect was a disadvantage: as he hammered the work, it became unmanageable. Fortunately for him, the metal can be restored to its original malleable state by annealing it.

As we have noted, early furnaces – even when augmented with bellows – could not reach a temperature high enough to melt iron. Pure iron melts at 1536°C (2797°F) and a 1 per cent steel at about 1475°C (2690°F); this compares with 1083°C (1981°F) for pure copper and a good deal less – 750–900°C (1380–1650°F) – for bronze, depending on its composition. Nevertheless, a simple forge could heat iron

The burial of the dead was often an occasion for the lavish sacrifice of treasure. This sumptuous burial dress of a Han princess was made from jade platelets sewn together with gold wire.

Today, iron nails are made mechanically from drawn wire, but in classical times they were forged by hand. These nails were taken from a Roman ship recovered at Nemi.

until it was soft enough to be beaten into shape on an anvil, the regular task of the village blacksmith up to modern times.

The final treatment of the metal depended on the purpose for which it was to be used. Unlike copper, iron and steel harden when they cool and the degree of hardening depends on the degree and rate of cooling. The biggest, but not always the most desirable, effect is produced by rapid quenching from red heat in very cold water. To produce an intermediate effect, the hot metal was plunged into oil or, if only slight hardening was desired, allowed to cool in a draught of air.

The scale of ancient metal-working

Today we are accustomed to the regular issue of detailed production statistics for all the major industries, but for the distant past we have no such information. Nevertheless, the extent of the ancient metal industries is not purely a matter for speculation, as we have some reliable information about particular enterprises and can make reasonable inferences about others.

It is instructive to look first at the kinds of objects made. In the main, we have been talking about relatively small items: axe-heads and trinkets, chisels and bowls, nails and ploughshares. Gold has never been abundantly available: 30 grams (1 ounce) per tonne of ore is quite a high content. In their heyday the Nubian mines probably yielded no more than 30 kilograms (66 pounds) a year from some 9000 tonnes of ore. There are, however, precise records of massive objects in other metals. Thus the huge bronze bowl made for Solomon's temple – described in the First Book of Kings – must have weighed overall around 200 tonnes. Another remarkable example is the famous iron pillar, probably made early in the fourth century AD, which now stands near Delhi. Made of exceptionally pure metal – which probably accounts for the total absence of corrosion over the centuries – it is 8 metres (26 feet) long and weighs 7 tonnes. It certainly could not then have been cast in one piece and must, therefore, have been welded from smaller pieces. Excavation at the palace of Sargon II (722–705 BC) yielded no less than 150 tonnes of iron bars: no doubt some of this was royal treasure rather than working materials for blacksmiths, but even so it is a very substantial amount of metal.

Study of the remains and slag-heaps of early copper mines in the eastern Alps suggests that this area alone produced some 20,000 tonnes of copper between 1300 and 800 BC. Examination of the slag-heaps at the copper mines of Tarshish indicate that Roman and pre-Roman slag amounts to around 20 million tonnes. Making reasonable assumptions about the richness of the ore and the efficiency of the extraction process, we can infer a copper output of around 2 million tonnes between about 1000 BC and AD 400.

Archaeological excavations provide plenty of evidence for what, in the context of the times, amounted to mass production of metalware, notably multiple moulds designed to produce large numbers of the same item in a single casting. From the third century BC the Chinese, for example, practised a system of 'stack casting' in which symmetrical and interchangeable moulds were piled high in layers. One find comprised no less than 16 stacks, consisting of over 500 sets in 36 specifications: this again argues a substantial and reliable supply of metal and a steady demand.

The 8-metre (26-foot) iron pillar at Delhi is one of the curiosities of the world. Although it has stood in the open for some 1600 years, it shows no sign of corrosion, supposedly in consequence of its purity.

Mining techniques

The technology of mining is one that progressed very little until the introduction of blasting and mechanization in comparatively recent times. As much as anything, this reflects the unchanging nature of the problems: roof support, transport, ventilation and drainage. Picks and shovels, usually wielded under cramped and unpleasant conditions, were the only tools except bare hands.

Where ore lay close to the surface, open-cast working was possible, but commonly veins of ore had to be located, or followed, underground through narrow tunnels. Timber supports began to be used about 2000 BC. Different levels were connected by vertical shafts up which men and ore might be hauled in buckets by ropes: more often, however, the only access was by crude ladders. Lighting was by primitive oil lamps. In metal mines these rarely posed the risk of explosion from inflammable gas that they would do in coal mines, but where sulphide ores were being worked quantities of noxious sulphur dioxide gas might be released. Foul air inevitably accumulated, and fires were often lit in attempts to produce through currents. Fire was also used, above ground as well as below, to shatter rock: it was first strongly heated and then water was dashed against it, a process known as fire-setting.

A sudden inrush of water in congested conditions could be fatal, and water had to be controlled to enable work to proceed. If the levels were right, it could be removed through special drainage passages (adits), which began to be used about 1000 BC. These were often on a considerable scale: one driven through solid rock by Roman engineers at La Zarza, in Spain, was 1½ kilometres (nearly a mile) long. Mechanical drainage, using a system of water-wheels and Archimedean screws, was introduced by the Romans. Piston pumps, some made of wood and others of bronze, and particularly identified with fire-fighting equipment, were also well known to the Romans. But despite all the ingenuity expended by engineers, water remained a constant hazard to the lives of the mineworkers.

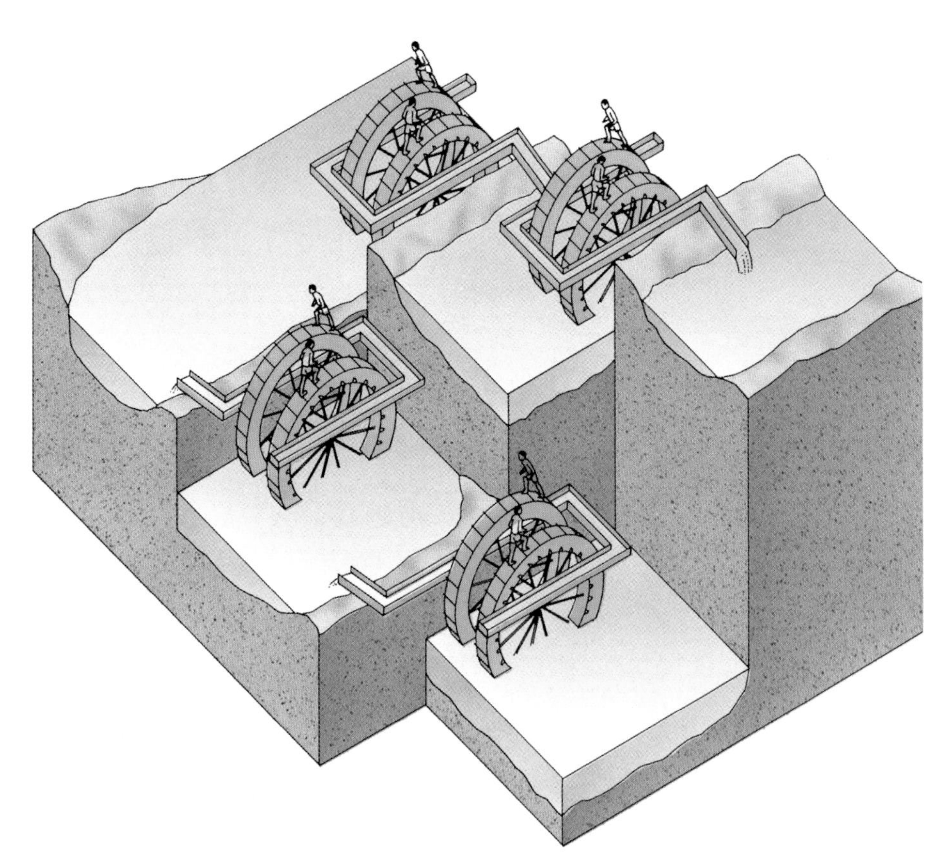

Above: When the Portuguese took possession of Brazil in 1500, they quickly removed large quantities of gold, possibly as much as 2500 tonnes. Today, Brazil is no longer regarded as a major gold-producing country, though primitive working on very small claims has been restarted at Serra Pelada under the aegis of the government.

Left: The Romans drained the copper mines of Rio Tinto in Spain with a series of wheels bearing containers in the rim that carried the water from one wheel to the next. A total of eight wheels raised the water 30 metres (98 feet) in height. The wheels were powered by people walking on treads fixed to the wheels.

Archimedean screw 30 **iron-working** 20–21, 75 **mining** 185–186 **water-lifting devices** 30

CHAPTER EIGHT

The Technology of War

The relationship between technology and war, which has achieved such great significance in recent years, has always been a close one. There is, of course, the obvious relationship at the level of weaponry and fortification, where superior skills conferred an advantage – though one that could be offset by other factors such as poor leadership and over-extended lines of communication. Additionally, as we have seen, the progress of technology on a broad front depended on the existence of substantial and well organized communities, with established trading relationships with the outside world. Such communities were commonly established by conquest or, at least, a convincing threat of conquest. Once established, they had to have a means of defence against ambitious neighbours or extending their own territory by aggression. Above all, the civilized world as a whole had in varying degrees to be on guard against the aggressive barbaric hordes forever lurking beyond the northern frontier.

Small wonder, therefore, that from the earliest times military technology received close attention. Necessarily, it evolved in different ways according to local needs and conditions. Thus the armoured war-chariot or cavalry could usefully be deployed only in comparatively open country; elsewhere, the advantage might lie with lightly armed mobile foot-soldiers. Catapults and other massive engines of war could be used to good effect by troops engaged in siege warfare, but were too cumbersome to be generally deployed in the field. Again, nations – such as the Romans – dependent on imported supplies had to ensure naval supremacy, a matter of less direct concern to the land-locked powers of the Middle East.

Hand weapons

Specialized weapons are a comparatively late development. For a very long time man was content to destroy his fellows with the same weapons as he used to kill animals: spears, axes, daggers, bows and arrows were all pressed into service. Much evidence concerning their nature comes from the widespread custom of burying warriors with their weapons, and the condition of skeletons often bears witness to their effectiveness.

The advent of copper and bronze brought new possibilities. Flint daggers were displaced by bronze ones. A longitudinal rib gave greater rigidity, and sockets and tangs made it easier – where the dagger was not cast in a single piece – to attach a hilt. The dagger was effective only at extremely close quarters and from it evolved two weapons for use where the opponents were a little more distant from each other. One was the halberd, developed in western Europe, which was originally a dagger fixed at right angles to a long shaft: the addition of an axe-head was a very much later development. The other was the sword, the manufacture of which severely taxed the coppersmith's skill. Copper itself was far too soft, and the ideal was bronze with a tin content of around 10 per cent; the blade had then to be carefully work-hardened by hammering on an anvil.

The spear was a weapon that could be used either as a thrusting weapon at close quarters or hurled from a distance: in either case balance was all-important. Copper spearheads were made with tangs to attach them to the shaft, but when the technique of casting bronze was mastered, socketed types were

As befitted great leaders, weapons figured prominently in their tomb furniture. The richly ornamented daggers (above) were recovered from the tomb of Tutankhamun. Most ancient weapons are, however, recovered in a state of decay. Those shown left were found at Marathon: they include a bronze dagger with an ivory handle, bronze arrowheads, and a lead slingshot.

preferred. Rather surprisingly, in the Bronze Age arrowheads continued for a long time to be made in bone or flint: barbed and socketed metal arrows appeared comparatively late.

When iron became available, it was the sword – a weapon that struck with a particularly powerful impact – that was most improved. By a combination of carburizing to form steel and quenching the hot metal, the requisite combination of toughness and hardness could be achieved. Early iron weapons are particularly identified with the Hallstatt culture in what is now Austria. This developed about 1000 BC, but iron weapons of much earlier date are known. A particularly fine example is an iron dagger of about 350 BC found in Tutankhamun's tomb: from its design and workmanship it was clearly imported, but its origin is uncertain.

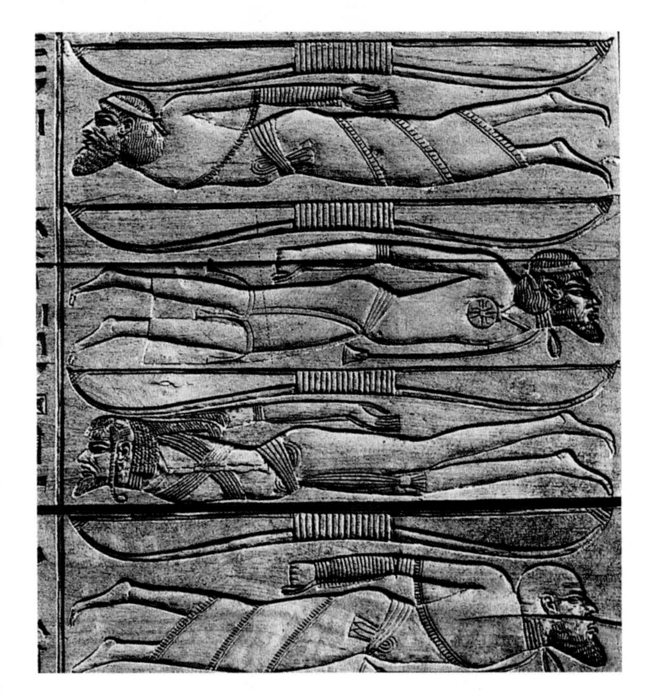

A stool from Tutankhamun's tomb shows Syrian captives with longbows.

The bow

The bow is represented in rock paintings dating back at least 30,000 years and is doubtless of much greater antiquity. It is of special interest as the first example of a device in which energy is stored slowly and released quickly, giving the human frame a power which it could not generate directly.

The size and construction of a simple bow is determined by several factors. When fully drawn, the distance between the centre of the bow and the centre of the string cannot exceed the distance from a man's ear to the end of his fully extended arm – in practice, a little under 1 metre (3 feet). The arrow must be a little longer. If the bow's lower end is to be clear of the ground when it is extended, the length cannot much exceed 2 metres (6½ feet). Most men cannot exceed a draw-force of around 45 kilograms (100 pounds), which determines the stiffness of the bow.

These limits were approached in the medieval English longbow, but most early bows were a good deal shorter. For men on horseback or in chariots, a longbow would in any event be difficult to manage. The simplest bows were made of springy wood: yew if it was available, but alternatives such as palm and split bamboo were also satisfactory. They were strung with gut. They were not kept permanently taut, but strung when required. From a very early date, however – certainly by the 13th century BC – powerful composite bows incorporating wood, horn and sinew were used. Bows such as these were, of necessity, used by people, such as the Eskimo, who lived in virtually treeless areas.

In the middle of the second millennium BC a military revolution occurred in Asia. Horses began to be employed in light chariots, to be followed later by cavalry. The short, powerful composite bow suited this kind of warfare admirably and it long remained a favourite Asiatic weapon. The Greeks, by contrast, relied primarily on heavily armoured infantry (hoplites) supported by more mobile peltasts, foot-soldiers whose main weapon was the hurled javelin: they defended themselves with a light leather shield (*pelta*). Their forces also included archers and slingers, but in comparatively small numbers. In the right terrain, such forces could hold their own against cavalry: for example, at Plataea in 479 BC Persian cavalry failed to shake the Greek hoplites, who were protected on one flank by a river and on the other by mountains. But at the Battle of Cunaxa, north of Babylon, in 401 BC, the Greek army under Cyrus the Younger was defeated by the Persians under his brother, Artaxerxes. The Greek hoplites overcame the more numerous Persian infantry, but the remainder of Cyrus's forces, thrown into disorder by archers, were routed by the Persian cavalry, manoeuvring freely on open ground. Cyrus himself was killed, and his infantry extricated themselves only with difficulty in a long and famous retreat. The lesson was clear: no Western army could hope to defeat the Persians on their own ground unless they could contain their cavalry. It was learned by the great Macedonian general Alexander, who crossed the Hellespont in 334 BC at the head of an army of some 40,000 men, of whom about 8000 were cavalry. His 25,000 infantry, deployed in the famous phalanx, were more mobile than their predecessors and supported by skirmishers armed with bows and slings. Backed by superb generalship, it was a formula that proved unbeatable in the terrain in which he was operating.

The Romans are an illustration *par excellence* of a nation that built a vast empire by military conquest. Originally the army, organized in legions of some 6000 men, was armed very much on Greek lines: the main weapons were thrusting spear (*hasta*) and sword, supported by body armour and shield. From about 400 BC, however, two of the three main lines within each legion's fighting formation were armed with a throwing spear (*pilum*) weighing around 2 kilograms (4½ pounds). At that time, according to Polybius, the armed forces of Rome and her Italian allies totalled some 800,000 men. By 100 BC, the iron tip of the spear was not fixed

chariots 46–47 **development of warfare** 133–140, 215–222, 316–327 **galleys** 50–52 **seige weapons** 87–88

firmly but was secured with two light pins, so that it became detached on impact and the enemy could not throw it back. A rather later development achieving the same result indicates the skill already attained by the blacksmith: the shaft behind the tip was treated to make it soft, so that it would bend easily if it hit the ground. Each legion had a regular quota of 300 cavalrymen, but this was abolished under the reforms of Marius in the first century BC, although some light cavalry could still be deployed.

This organization served well enough under European conditions, which were then their particular concern, but in 53 BC it proved as disastrous to Crassus, who had embarked on a campaign in the much more open country of Parthia, as it had to Cyrus. Beyond the Euphrates he was faced near Carrhae by a smaller Parthian force comprising 1000 cavalry and 10,000 mounted archers. The cavalry were different in kind from those that had confronted Cyrus, however. The new troops were called cataphracts and both riders and horses were heavily mailed. They carried a lance as well as a bow: in all essentials, they were the forerunners of the medieval knights, formidable when mounted but too weighed down by armour to fight on the ground. The archers themselves were only lightly armoured, but the power of their composite bows drove their arrows through Roman armour. Their skilful horsemanship is remembered in the Parthian shot, a parting arrow shot to the rear as they wheeled and withdrew. Despite their humiliating defeat, the Romans seem not to have raised squadrons of cataphracts themselves until some two centuries later.

For the mounted warrior a longbow was an embarrassment and a shorter bow, as depicted here on an Etruscan bronze statuette (*circa* 500 BC), was favoured. Note the heavily built cavalry horse.

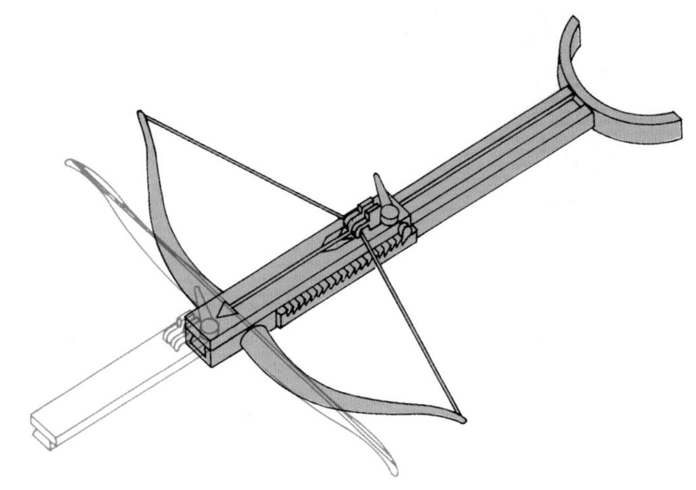

The *gastraphetes* was a form of crossbow first used in ancient Greece about 400 BC. It enabled an archer to use both arms to bend the bow. It fired a bolt at considerable speed, but it took some time to prime the weapon.

Engines of war

As with a bullet, the penetrating power of an arrow depends on its velocity, which in turn depends on the power of the archer to draw the bow. Composite bows could be made far stronger than one unaided human arm could bend fully. The archer could bend a much more powerful bow only if he could exert the full strength of both his arms, but then he could not hold and direct the arrow. The first answer to this difficulty was the Greek *gastraphetes*, which means literally a belly-shooter. This was essentially a crossbow and was so called because it was primed by resting one end against the *diostra* (slider) on the ground and pressing against the other with the stomach: with both hands the bow was then bent until the bowstring was caught by a hook. The bolt to be fired was laid in a trough and, after aiming, the weapon was discharged by a trigger which released the bowstring. We do not know when the *gastraphetes* was first used, but it was probably shortly after the Peloponnesian War (431–404 BC), for its use in that conflict is not recorded by Thucydides, who took a keen interest in such devices.

Using both hands, a bowman could exert a force of about 90 kilograms (200 pounds) to prime his weapon. Although bigger and stronger bows could be made, a man could not bend them in this way even with his two hands. The next development, therefore, was to draw back the bowstring with a small winch controlled by a pawl-and-ratchet mechanism. A weapon so designed and constructed could discharge a heavy metal bolt – contemporary accounts describe one up to 2 metres (6½ feet) long and about 4 centimetres (1½ inches) in diameter. Such a bolt would not only penetrate any kind of body armour with ease but could also pierce heavier defences such as wooden stockades.

By this stage the *gastraphetes* could no longer be used as a hand weapon – even if rested on a stand or wall – and was becoming a piece of artillery. For its future development we have to rely largely on some information given by Biton, a Greek writer on warfare who flourished about 240 BC. He

refers to weapons with a bow 5 metres (16 feet) long capable of firing stone balls weighing nearly 20 kilograms (44 pounds): credence is lent to this seemingly improbable assertion by the discovery at Pergamon of scores of stone shot of just this size. And it is related that Hiero II's vast merchant ship *Syracusia* included in its armament a huge catapult – supposedly designed by Archimedes himself – capable of throwing an 80-kilogram (176-pound) shot up to 200 metres (656 feet). Such huge bows could have been bent only by some kind of rope and pulley mechanism capable of giving a very big mechanical advantage, for we are now talking of draw-forces of several tonnes. How the bows were constructed we can only speculate, but some combination of horn and sinew must still have figured prominently: these were the best energy-storing materials then available, and far superior to wood, whose capacity diminishes rapidly above 25°C (77°F), a not unusual Mediterranean temperature.

Although the catapult was a more powerful weapon than the simple bow, it fired much more slowly. Among the disadvantages of this was that an alert enemy could move safely in the predictable minimum time between shots. It is interesting that Philo of Byzantium, an important if fragmentary source of information about early weapons, describes in the second century BC what he called a *polybolos* ('multi-thrower'). This was in effect an automatic catapult rapidly reloaded from a magazine.

The weapons so far described derive their power from the bending of a bow, though sometimes one divided into two separate halves. But the Romans made extensive use of an alternative device in which the energy was stored before release by twisting a skein of elastic materials: on release, these actuated two rigid horizontal arms, corresponding to the ends of a bow. Ancient writers indicate that the favoured materials were sinew and women's hair, but are vague on the details of how these were worked up. By the latter years of the Roman Empire, mobile horse-drawn catapults, known as *carroballistae*, were systematically allotted to the army at the rate of one for every hundred men: each was served by a crew of 11 men.

Finally, brief mention must be made of the onager, a device also based on torsion-stored energy but actuating a rigid arm mounted on a horizontal pivot. In effect, it was a mechanical sling and derived its name from the violent recoil of the back legs of its frame when it was fired.

Fortifications

In the broadest sense a fortification is any material barrier separating opponents, but it can take many forms according to the exact function it is intended to perform. It may enable a small force to resist a much stronger one until reinforcements can arrive, or a largely civilian community to survive until such time as the besiegers lose heart and retire. It may be a purely local construction, or it may be a continuous line of defence running for long distances across country. It may be a largely political device rather than one of immediate military use, an omnipresent demonstration of government power over the surrounding countryside. Whatever their purpose, their builders lavished much care on their construction and, as they were built for strength, ancient fortifications are among the most impressive monuments to the technological achievements of past ages, whether they be the hilltop fortress of Mycenae, the Great Wall of China or the simple stone brochs of Scotland.

As man first used hunting weapons to attack his fellows so, for defence, he used the methods he had already adopted to protect his settlements from marauding animals. Throughout the world the earliest forts of which we have knowledge consisted of simple earthworks. Earth from a

Above: All warriors relied on some kind of body armour. These examples show a suit of bronze armour from Dendra in southern Greece (left); a fifth-century BC Greek helmet worn by a Chalcidean soldier (centre); and a fragment of scale armour worn in Roman Britain (right).

Alexander the Great 21–22 **body armour** 140 **crossbow** 139 **iron working** 20–21, 75 **Roman Empire** 10–11, 24–25, 88

ditch was thrown up to form a rampart behind it: usually a space (berm) was left between the two so that earth trickling down from the rampart would not fall into the ditch and make crossing it easier. A narrow, easily defensible gap would lead into the interior. It is not to be supposed, however, that all such works – especially the larger ones – were forts in the true sense. In some, the ramparts were clearly too long to have been defended by the numbers of men we can reasonably expect to have been available and some were probably no more than large pounds in which cattle could be mustered: thus they belong to agricultural rather than military history. Simple forts of similar construction were built as a matter of duty by Roman soldiers while campaigning in the field and are very different from the frontier fortresses established later.

Where sustained defensive fortification was the main purpose, more sophisticated methods were soon adopted. The ring of the ditch was completed and it was crossed only by a timber bridge which could be easily removed in the face of attack. The outer face of the rampart could be made stronger and nearly vertical by timber revetment, and the rampart itself made more durable with a core of stone. A succession of ramparts made it possible for the defenders to withdraw if need be, leaving the enemy with new obstacles to overcome.

Such forts could, of course, be built only where the ground was relatively easily excavated. In rocky situations stone or brick buildings were normal. In Scotland, they took the form of circular brochs, dating from about 500 BC, of which many survive. Designed to protect small communities for short periods from small marauding bands, they are correspondingly modest affairs. They were generally about 12–15 metres (40–50 feet) in height and of roughly the same diameter at the base. Above the solid foundations, the wall was built hollow and included protected stairways and galleries. It was roofed at the top. Such forts were no more than temporary refuges, never intended for prolonged occupation.

While such simple fortifications served the needs of small communities on the fringes of civilization, the situation was very different in the case of the new and densely populated civilizations of the Middle East and the Mediterranean. By the middle of the second millennium BC, massive fortress-palaces had been built incorporating well-conceived design features. One such was the citadel at Tiryns, which about 1400 BC began to assume the form now exposed in ruins. It was designed, taking advantage of the terrain, so that potential attackers were obliged to take a devious route that gave the advantage to the defenders. In particular, the advancing attackers were obliged to turn their right sides, unprotected by shields, to the defenders: this became a regular feature of later designs and was part of a general policy of convoluting the access routes to delay and expose the enemy as much as possible. Roughly contemporary was another hilltop fort at Mycenae, with its famous Lion Gate guarded by a square

Fortresses took many forms, depending on local circumstances and building materials. The Scottish brochs were designed to provide temporary refuge from marauding bands.

tower. Protected by a single curtain wall, it incorporated an unobtrusive postern gate for escape to the mountain side.

A more sophisticated fortress was that built by Rameses II at Medinet Habu about 1200 BC. This had a double wall and access was by the customary gatehouse. The outer wall is of particular interest, for it embodies one of the earliest examples of a crenellated parapet. The full-height sections (merlons) gave protection to archers while they fired through the spaces between (embrasures). An even more striking early fortress was that built at Sinjerli, in Syria, about 640 BC, which was divided into wards by no less than three concentric walls, entry to each being through a strong gatehouse. In the centre, and therefore heavily protected, was the main strongpoint. It thus showed many of the features that were to appear in the great castles of medieval Europe.

Such fortress-palaces were designed for permanent occupation by powerful monarchs and their retinues, including a strong permanent garrison. Their accommodation was limited, however, and for the populace at large city walls became the main protection, with retreat to the citadel a last resort. Apart from military use, such walls were valuable for controlling traffic into and out of cities so that trade could be regulated and taxes levied. Like Mycenae and Tiryns, Troy was originally a fortified royal residence, but by 1500 BC, after five earlier rebuildings, it had become a prosperous and populous city enclosed in massive walls dominated by several square towers. These walls ultimately fell not to an enemy but apparently as the result of an earthquake. Rome was first protected by such walls in the fifth century BC, but the earthen ramparts were breached by a Celtic attack in 386 BC. They were replaced in 378 BC by a massive stone wall – the so-called Servian Wall – with towers and a ditch some 10 metres (33 feet) deep in front of it. They seemingly deterred Hannibal in 211 BC, for he withdrew without testing them. Gradually they fell into disrepair, as the Romans increasingly relied for defence on holding potential enemies at bay on their distant frontiers.

Continuous lines of frontier defence, as opposed to isolated strongpoints, were then no novelty. As early as 800 BC

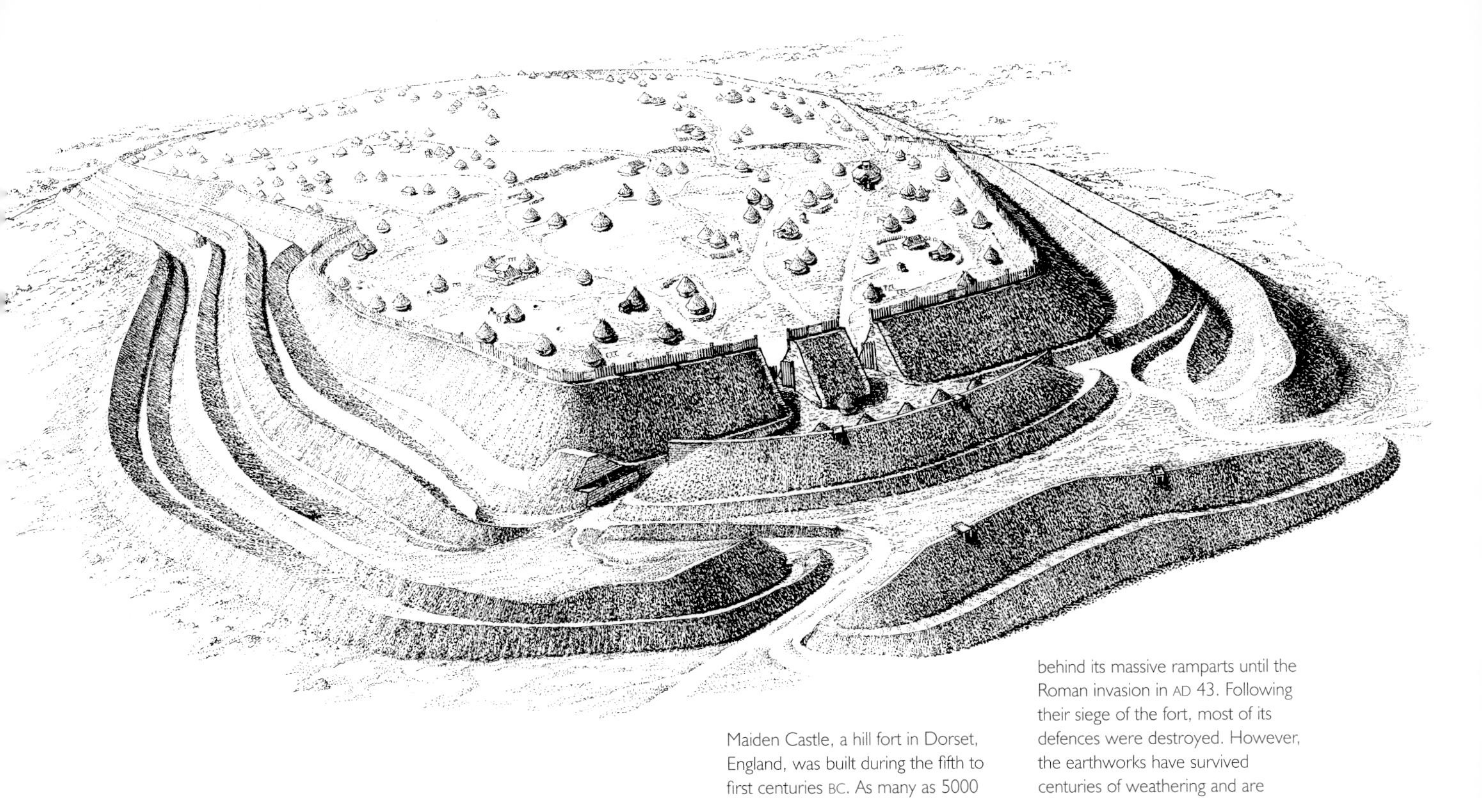

Maiden Castle, a hill fort in Dorset, England, was built during the fifth to first centuries BC. As many as 5000 occupants lived in wooden huts behind its massive ramparts until the Roman invasion in AD 43. Following their siege of the fort, most of its defences were destroyed. However, the earthworks have survived centuries of weathering and are clearly visible today.

the Assyrians built a continuous wall – the Medean Wall – some 80 kilometres (50 miles) long between the Tigris and the Euphrates to protect Nineveh and exclude the Medes. This it failed to do when the latter joined forces with the Chaldeans. The Great Wall of China eventually extended some 2400 kilometres (1440 miles) from the Gulf of Chihli almost to the Tibetan frontier. Like many such fortifications, it served a dual purpose: to keep the Huns at bay and the Chinese contained – a Berlin Wall of ancient times. The Romans, too, undertook some massive works of this kind. Perhaps the best known is Hadrian's Wall in Britain, running 80 Roman miles (120 kilometres or 112 miles) from Wallsend-on-Tyne to Bowness-on-Solway, and originally built in AD 122–8. The eastern part was built of stone, continuing westward as a turf rampart, which was eventually wholly rebuilt in stone. There was a continuous parapet, with small forts at every mile and 17 larger ones at greater intervals: reinforcements for threatened points were available from other garrison forts to the rear. It was a major undertaking, and in its heyday sustained a garrison of 15,000 men. This was run down under Albinus to AD 196, with the result that it was overwhelmed by Picts and Scots from the north. Though restored 10 years later under Severus, it was breached again a century later and finally abandoned in AD 383. This was not, however, the Romans' first venture in wall building. From the time of Tiberius, the Germanic tribes had posed a constant threat to their northern frontier and a continuous fortified line, nearly 500 kilometres (310 miles) long, was established between the Rhine and the Danube. In Britain the Mercians, under Offa, built a 200-kilometre (120-mile) Dyke between the Dee and the Severn in the eighth century AD to contain the Welsh.

Siege weapons

Assault on fortifications presented technological problems quite different from those of campaigns in the field. At first sight, the simplest option for the attacker was simply to invest the fortress and wait until starvation or thirst drove the occupants to surrender. But the advantage did not necessarily lie with the attacker, for resolute and well-provisioned defenders could hold out for surprisingly long periods, especially if they had absolute control of their water supply. It was often the attackers who had to give up through lack of supplies, or because they were under pressure elsewhere. Ancient history records many instances of long and often

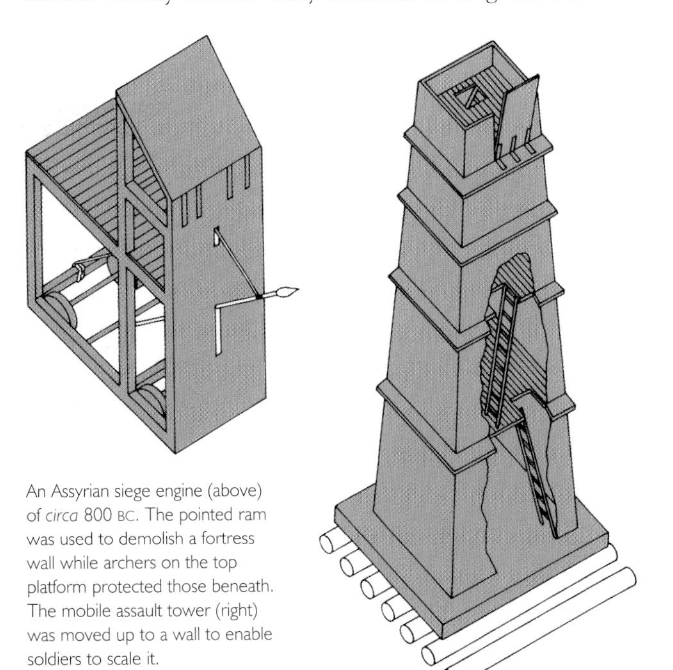

An Assyrian siege engine (above) of *circa* 800 BC. The pointed ram was used to demolish a fortress wall while archers on the top platform protected those beneath. The mobile assault tower (right) was moved up to a wall to enable soldiers to scale it.

crossbow 139 **Great Wall of China** 23, 125 **fortresses** 133–135, 138-139 **Mycenae** 55

This frieze from the palace of Ashurnazirpal II (883–859 BC) at Nimrud (Iraq) shows his army assaulting a city by breaching, tunnelling and battery, while bowmen oblige the defenders to keep their heads down.

unsuccessful sieges. Thus by retiring to its island fortress and destroying the causeway connecting it with the mainland, Tyre survived a 13-year siege by Nebuchadnezzar in the sixth century BC. Much later, in 332 BC, it fell to Alexander in less than a year after he had built a huge mole across the sea to enable his troops to advance.

To impatient attackers determined to force the issue there were basically three options: they could get over, through, or under the walls. The simplest, but perhaps most hazardous way, of gaining access to the ramparts was by the use of scaling ladders, deployed at several points under covering fire to divide and pin down the defenders. Gaining the ramparts was no guarantee of success, however, for the surviving defenders could then withdraw to the towers and harass the exposed attackers from there.

Success depended very much on being able to get a considerable number of men in position very quickly, and for this purpose wheeled siege towers were constructed. These were built to the height of the wall to be scaled and pushed into position: if necessary, a section of any defensive ditch had to be filled in beforehand – a clear indication of where the attack was to be mounted. A connecting ramp was then extended and the attackers swarmed across. Such towers usually contained several stages connected by ladders, so that the men who left the top deck could be quickly reinforced.

Such towers were constructed of timber on the spot and then abandoned: not surprisingly, therefore, none have survived. Reliable contemporary descriptions show that they might have been very large indeed. In his account of those built by Alexander for his assault on Rhodes, Vitruvius tells us that the largest were 10 metres (33 feet) square at the base and 60 metres (200 feet) high. Allowing for wheels, armour and so on, such a structure would have weighed upwards of 100 tonnes: but there is no inherent improbability about this, for, as we have seen, engineers of the period were accustomed to moving single stone blocks that were far heavier.

To breach a wall or gate, use was made of the heavier kind of catapult already described, but the heavy iron-tipped battering ram was probably more effective. The earliest types were simply swung rhythmically by a gang of men, but later the ram was suspended by ropes within a cage and swung like a pendulum. To protect the men while they worked the ram was enclosed in a *testudo* (tortoise), a wooden frame covered with wood or green hide. A ram might also be carried on the lower deck of a siege tower. A simpler kind of *testudo*, depicted on Trajan's Column *circa* AD 110, was no more than the interlocked shields of the soldiers, carried over their heads.

If the walls could not be scaled or breached, recourse was made to mining. This involved tunnelling beneath them and shoring up the exposed foundations with timber. Unless this was interrupted by counter-mining by the defenders, the supporting timber was then fired and – with luck – the wall above collapsed.

This was by no means the only use of fire in siege warfare. Incendiary mixtures based on some form of tar were used by the defenders in attempts to burn siege towers and *testudos* and by attackers to set fire to buildings, often thatched, within the walls. Assyrian bas-reliefs show this technique to have been used as early as the ninth century BC. Xerxes used it in the capture of Athens in 480 BC, and during the siege of Tyre in 332 BC the Phoenicians launched a fireship in an attempt to destroy the mole being built by the Macedonians.

The sack of Rome

By the third century AD the Roman Empire was weakened by internal strife and external pressure from the Franks and Saxons; the latter were sea raiders. As a measure of protection, the Saxon Shore – which ran roughly from the Wash to Portsmouth in Britain and on the Continent from Ostend to Brittany – was protected by a chain of coastal forts. Some of these – for example that at Reculver – still survive. The policy of relying for the defence of Rome itself on the integrity of distant frontiers became increasingly hazardous. When Aurelius became emperor in AD 270, one of his first tasks, other than strengthening the border defences, was to erect the first of a succession of new city walls in Rome. These were some 20 kilometres (12 miles) in length, 7 metres (23 feet) high, and 4 metres (13 feet) thick: the 13 gateways were dominated by nearly 400 towers. The work took 10 years and Aurelius did not live to see it completed. Later, under the Emperor Diocletian, the walls were further strengthened and raised. By that time, however, the decision had been taken to divide the empire, and Rome itself became a symbol rather than the reality of imperial power.

The sack of Rome by Alaric and his Visigoths in AD 410 was not a decisive military victory – for Alaric quickly withdrew – but it clearly marked the decline of Roman power and the end of the *Pax Romana* that had for so long unified much of Europe and the Mediterranean region. But as the West entered a long period of darkness, new light was appearing in the East.

Alexander the Great 21–22 **fortresses** 133–135, 138–139 **Great Wall of China** 23, 125 **Hadrian's Wall** 23 **Trajan's Column** 24

PART

2 Discovery of a New World

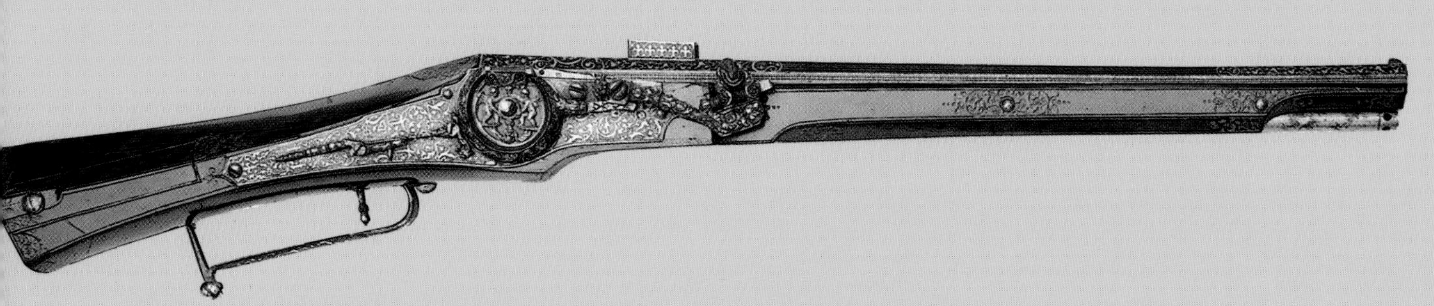

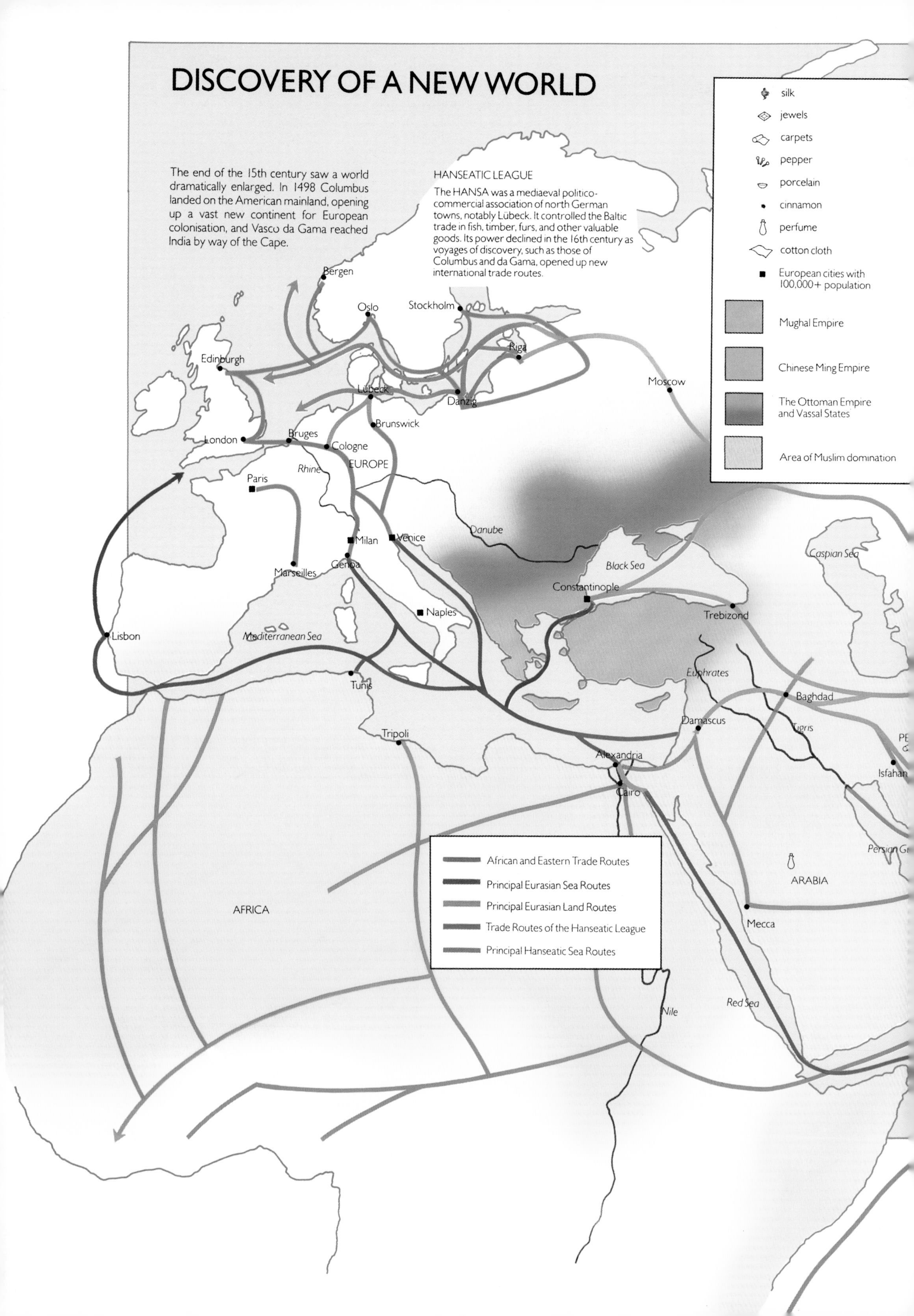
DISCOVERY OF A NEW WORLD
The end of the 15th century saw a world dramatically enlarged. In 1498 Columbus landed on the American mainland, opening up a vast new continent for European colonisation, and Vasco da Gama reached India by way of the Cape.
HANSEATIC LEAGUE
The HANSA was a mediaeval politico-commercial association of north German towns, notably Lübeck. It controlled the Baltic trade in fish, timber, furs, and other valuable goods. Its power declined in the 16th century as voyages of discovery, such as those of Columbus and da Gama, opened up new international trade routes.
silk
jewels
carpets
pepper
porcelain
cinnamon
perfume
cotton cloth
European cities with 100,000+ population
Mughal Empire
Chinese Ming Empire
The Ottoman Empire and Vassal States
Area of Muslim domination
Bergen
Oslo
Stockholm
Riga
Edinburgh
Moscow
Lübeck
Danzig
Brunswick
London
Bruges
Cologne
EUROPE
Rhine
Paris
Danube
Milan
Venice
Genoa
Marseilles
Black Sea
Caspian Sea
Constantinople
Naples
Trebizond
Lisbon
Mediterranean Sea
Euphrates
Tunis
Baghdad
Damascus
Tigris
Tripoli
Alexandria
Isfahan
Cairo
African and Eastern Trade Routes
Principal Eurasian Sea Routes
Principal Eurasian Land Routes
Trade Routes of the Hanseatic League
Principal Hanseatic Sea Routes
ARABIA
AFRICA
Mecca
Nile
Red Sea

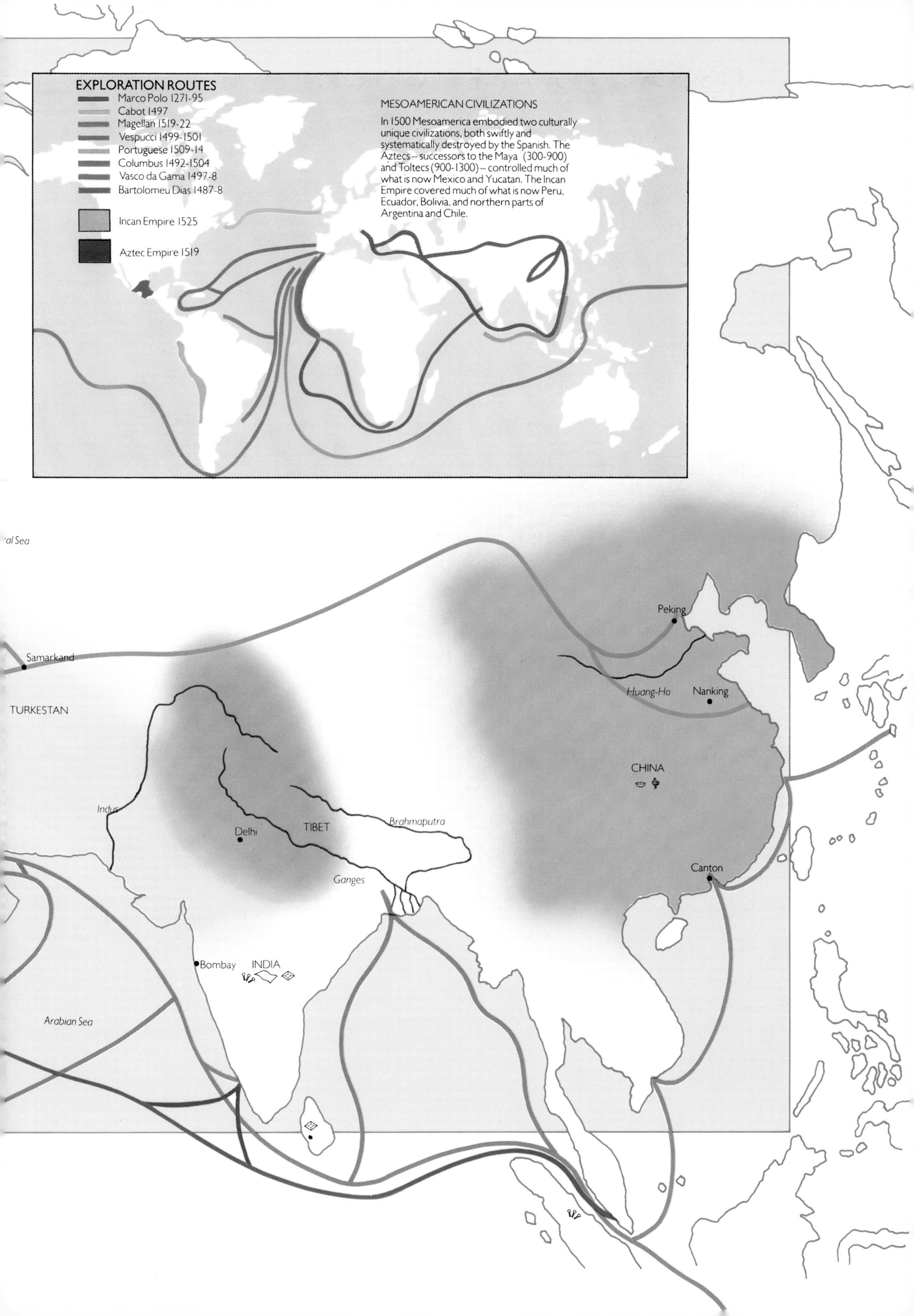

EXPLORATION ROUTES
Marco Polo 1271-95
Cabot 1497
Magellan 1519-22
Vespucci 1499-1501
Portuguese 1509-14
Columbus 1492-1504
Vasco da Gama 1497-8
Bartolomeu Dias 1487-8
Incan Empire 1525
Aztec Empire 1519
MESOAMERICAN CIVILIZATIONS
In 1500 Mesoamerica embodied two culturally unique civilizations, both swiftly and systematically destroyed by the Spanish. The Aztecs – successors to the Maya (300-900) and Toltecs (900-1300) – controlled much of what is now Mexico and Yucatan. The Incan Empire covered much of what is now Peru, Ecuador, Bolivia, and northern parts of Argentina and Chile.
Samarkand
TURKESTAN
Indus
Delhi
TIBET
Brahmaputra
Ganges
Peking
Huang-Ho
Nanking
CHINA
Canton
Bombay
INDIA
Arabian Sea

CHAPTER NINE

From the Rise of Islam to the Renaissance

In AD 642, the armies of the Caliph Omar entered Alexandria to complete the Arab conquest of Egypt. 'I have taken a city,' reported the victorious general, 'which I shall not attempt to describe. Suffice it to say that I discovered 4000 villas, 4000 baths, 40,000 tax-paying Jews and 400 pleasure parks worthy of a king.' For three centuries, with only a brief interlude of Persian occupation, Egypt had been a province of the Christian Roman Empire ruled from Constantinople. Henceforward it would be governed by the laws of Islam. The Christian provinces in Syria had already fallen to the Islamic conquerors, and by the end of the decade so, too, had the Persian Empire itself. And this was only the beginning of a whirlwind conquest which would transform the political geography of half the ancient world.

For centuries that world had been dominated by the empires of Rome and Persia: for centuries they had contested a frontier running north–south through what is now Iraq – the Arab lands to the south caused neither of them much trouble. Then, in 607, the Persian armies under their Emperor Chosroes II suddenly began a victorious progress. In 10 years they overran the Byzantine fortresses of northern Mesopotamia to conquer Syria and Egypt. In 622, the Eastern Roman Emperor Heraclius began to mobilize his forces for a counter-attack. The territories under his control were a rump of the Roman Empire, while Europe was divided among barbarian kingdoms established in the wake of the invasions of the fifth century – the Franks, the Lombards in northern Italy and the Visigoths in Spain. In political terms the Christian world was fragmented.

Even in the eastern Mediterranean the Roman emperor's lands had shrunk dramatically since the glories of Justinian, the builder of Hagia Sophia in Constantinople and San Vitale in Ravenna, who had died in 565. With the recent Persian triumphs, Constantinople was embattled. However, in six years of brilliant campaigning Heraclius recovered the lost eastern provinces, while Chosroes was deposed and murdered by his own nobles. But the Christian triumph, and the restoration of the ancient balance of power, was shortly to be upturned from an unexpected quarter.

The emergence of a new faith

The vast region of Arabia, one-third the size of Europe, was *terra incognita* to the ancient world, except for its coastal regions, where the so-called Incense Route wound its way from Sabaea on the Red Sea through Medina and Petra to Syria. Much of it was true desert, waterless and devoid of vegetation and had little to attract a conqueror. The Arabs were

	AD 500	600	700	800	900
PAPER AND PRINTING	paper making reaches Korea and Japan from China		paper in central Asia	Chinese print with wood blocks	Version of *Diamond Sutra* (earliest dated printed work 86
SHIPPING AND NAVIGATION		Viking ships adopt keel and mast for sail		Chinese paddle boat; lateen sail; Oseberg ship	Gokst ship
MECHANISATION	floating water mills at Rome	windmills in Persia	mechanical clocks in China	crank depicted in Europe	
CHEMISTRY			Alchemical works of Jabir ibn Heyyan (Geber)		
BUILDING CONSTRUCTION	Santa Sophia built	Grand Canal extended (China); glazed windows in churches	pointed arch in Islamic architecture		caustic alkalis discovered
WARFARE			Greek fire used in defence of Constantinople		

Although Mohammed's life as a religious leader was brief, his impact was enormous and enduring for the world at large. This painting from a 16th-century Turkish manuscript is typical of thousands showing him preaching to his first followers.

a pastoral people consisting of many tribes, each laying claim to one of the large oases where water and vegetation were to be found. Basically, they were idolators, but their religion contained elements of Persian fire-worship, Christianity and Judaism. Their religious capital was Mecca, where was lodged the sacred black stone known as the Kaaba, reputedly a gift to Abraham from Heaven. There, about 570, was born Mohammed. Although he came from the important tribe of the Koreishites, guardians of the Kaaba, he spent many years as a shepherd. He claimed that the angel Gabriel came to him in the desert and ordered him to preach a new faith, the principal tenet of which was: 'There is but one God and Mohammed is his Prophet.' He initially attracted few followers, but his heresy angered the Koreishites and in 622 he fled to the neighbouring town of Medina: it is from this *hegira* (Arabic for 'flight') that Moslems reckon their dates. In Medina, Mohammed's preaching gained much support; he gradually emerged as a warrior and he and his followers systematically raided the rich caravans bound for Mecca. There was inevitable retaliation, but in 630 Mohammed re-entered his native city as a conqueror. Within two years he was dead, but not before he had inspired his followers to spread his doctrine through the world by the sword.

The spread of Islam

From Persia the Arab armies carried the new faith northwards among the Turkic and other tribesmen of central Asia. From Egypt they spread along the whole coast of North Africa,

1000	1100	1200	1300	1400	1500
movable clay type in China	papermills in Europe		wooden movable type in China; first European block prints	block-printed books in Europe; Gutenberg uses movable type	
rsemen onize eenland	magnetic compass first used at sea in China	sternpost rudder used in West	mariners' charts (Rutters)	caravels; 3-masted carracks in Mediterranean	quadrant and astrolabe used for navigation; Portuguese sail round Africa
watermills widely used in Europe	post windmills in Europe	cog appears in Europe	verge-and-foliot clocks; tower windmills; Dondi's astronomical clock; treadle lathes		spring-driven clocks; crank and connecting rod depicted
	alcohol prepared by distillation	Roger Bacon's *Opus Tertius*; Albertus Magnus's *Libellus de alchimia*	sulphuric acid discovered		Thomas Norton's *Ordinall of alchimy*
	stained glass windows; Abbey St Denis heralds Gothic				
motte-and-bailey castles	Krak des Chevaliers fortress	Chinese explosive devices based on gunpowder	Chinese cannon; cannon in Europe	handheld firearms	

At its peak, the geographical extent of the Islamic Empire was enormous and, through voyages of exploration and trade, its influence extended even farther afield. This picture, dating from the 13th century, shows an Arab ship trading in the Persian Gulf.

overrunning the Byzantine provinces (Carthage fell in 698) and the Berber tribes in what is now Morocco. Only the narrow Straits of Gibraltar now separated them from Europe and these were crossed in 711. The following year the Visigothic Christian capital of Toledo fell and within a generation the whole Iberian peninsula, except for a narrow coastal strip in the north, was in Islamic hands. Next, the armies crossed the Pyrenees into the Frankish kingdom, and it seemed as though the whole of Europe must fall to them. Then, in 732, exactly a century after the Prophet's death, they suffered a crushing defeat at the hands of Charles Martel (Charles the Hammer), grandfather of Charlemagne. At the other end of the Mediterranean, the Arabs reached the Bosphorus by 688 but Constantinople was saved by a spirited defence under Emperor Leo III, aided by the military pyrotechnic known as Greek fire.

It was one of the most extraordinary and significant centuries in the history of mankind: at its beginning the Arabs were virtually unknown nomads; at its end they were the most powerful rulers in the world. In the process, they had defeated the largest and technologically most advanced armies of their day without themselves having the benefit of any revolutionary new military technology. How this came about we cannot pursue here, but it was very much a triumph of the spirit over the material advantages of their enemies. The Arabs had for generations lived hard, expecting little in the way of comfort. Moreover their belief was absolute and fanatical: to die in battle against the infidel was a certain passport to paradise.

Within their vast new domains, Arabic was the official language. Acceptance of the Islamic faith was not obligatory outside Arabia, though those who rejected it were obliged to pay tribute. A significant result of this was that Jewish communities continued to flourish, and Jewish scholars and translators were allowed to play a leading role in the great Moslem centres of learning.

In 750, the Abbasid Dynasty established itself in Baghdad on the Tigris, where it ruled for 500 years. There, especially in the late eighth and ninth century the court – notably under the legendary Harun-al-Rashid – became a great cultural centre that contrasted markedly with conditions in most of contemporary Europe, outshining even the court of Charlemagne. Stretching from the Atlantic to the frontier of India, the Islamic domain incorporated the ancient empire of Alexander and exceeded in sheer territorial extent that of Rome. Inevitably, the centre could not maintain control, and rival caliphates were established in Cairo and Cordova.

Expansion did not stop with the heroic age. In the late ninth century, Sicily became a province of Islam and, even after its reconquest 200 years later, Arab influence lingered under the rule of the Norman kings: in the 12th century the great Arab geographer Idrisi held an appointment at the court of Roger II. At the end of that century, the Afghan ruler Mohammed of Gor established the sultanate of Delhi, which brought the science and mathematics of India within the sphere of Islam. The geographer Ibn Batuta, born about 1305 in Tangier, spent eight years at the court of Delhi in mid-century and from there went on an embassy to the court of the Yüan Dynasty in China. His account of his travels, which took him as far as Moslem trading posts on the coast of Ceylon (now Sri Lanka) and the Sultanate of Achin in Sumatra, provides a survey of the greatest international community the world had yet seen.

The culture of Islam

The Arabs did little directly to advance technology; but, if not innovative, their craftsmen produced some of the finest copper-work, glass, perfumes, textiles, leather goods, ceramics and dyes of the Middle Ages. Not surprisingly, perhaps, in view of their dedication to spreading the faith by the sword, they developed great skill in iron- and steelworking. The swords of Damascus were prized throughout the world for the keenness of their blades, and were noted for their characteristic wavy pattern – poetically likened to the tracks of small black ants – made by inducing crystallization effects in the metal. The Damascus swordsmiths' raw material was 'wootz' steel, imported from southern India.

In the long term, the Arabs had a profound effect on world history. Their scholars accumulated and transmitted the learning of the ancient classical world, and made influential philosophical contributions of their own. Their translations into Arabic were drawn not only from the Hellenistic tradition but also from the classical texts of China, India and Persia, among others. In consequence of their rise to the status of a wealthy trading people – well placed geographically to link East and West – they encouraged the mathematics necessary for accounting and surveying; the standardization of weights and measures; astronomical observations relevant not only to navigation and the construction of a calendar but also to such theological needs as the determination of the precise direction of Mecca and the commencement of Ramadan. From India they imported (and transmitted to the West) the system of decimal numeration – as well as the numerals themselves – known in the West as 'Arabic', which were an enormous improvement on the clumsy Roman notation. They were not content to gain knowledge by passive observation, but resorted to carefully contrived experiments to advance chemistry, physics and medicine. No field of knowledge was unexplored and the results were enshrined in a wealth of specialized texts and several immense encyclopaedias. Great new libraries were established in centres such as Baghdad, Cairo, Basra, Toledo and Córdoba – the latter alone holding half a million volumes by AD 1000. These became centres of teaching and disputation, so that the scholarship of Greece and Rome was augmented by the writing of Islamic and Jewish scholars. All the learning in the world was gathered within Islam.

Though dedicated to extending its dominion by force of arms, Islam was essentially a religious and cultural movement. Inevitably, places of worship were of exceptional importance – like cathedrals in the Western world – and correspondingly lavish attention was devoted to them. This representation of a mosque (left) is typically stylized. The Arab influence on architecture extended widely, as exemplified by the 12th-century cathedral at Palermo, Sicily (above). Note the ogival arches.

China

In the Far East the early seventh century witnessed an event which, if not as sensational as the founding of Islam, was nevertheless of considerable importance. In 618, the first ruler of the T'ang Dynasty brought renewed unity to the ancient culture of China, which for the next three centuries flourished with renewed vigour. Although halted in central Asia by Islamic armies in the battle of the Talas river in 751, Chinese influence extended in the south in trade with India. Within their dominions the T'ang Dynasty established central authority through a professional civil service selected by competitive examination and transformed internal communications and trade with immense canal-building programmes.

Under them and the succeeding Sung Dynasty (960–1279), cultivation of tea and cotton became major industries, and improvements in the art of printing were developed, while the Chinese invention of paper began to spread outwards: 751 saw a paper factory established at Islamic Samarkand. The Sung armies were equipped with pyrotechnic weapons using explosives, and their merchant ships were navigating with the magnetic compass. But Sung power came to an end when the nomadic Mongols – successors to the central Asian empire of Genghis Khan – established their power at Cambuluc (modern Beijing). The Mongol Empire had already pacified the turbulent peoples of the steppe and it was during this *Pax Mongolica* that Marco Polo and other European travellers made journeys to China itself. With the overthrow of the Mongol Yüan Dynasty in the 1360s by the native Ming, China began to close in on itself. But there had been a fertile period in which some important technological developments had diffused westwards to feed the burgeoning civilization of Europe.

The T'ang Dynasty in China (618–907) was a period of great cultural activity backed by technological skill, as exemplified by this great winged horse from a Chienling tomb.

The Middle Ages in Europe

In 476 the last emperor at Rome was deposed, but the fabric of the empire had crumbled long before. The next three centuries, the Dark Ages of Europe, were a period of endemic warfare and intellectual stagnation as the descendants of the invaders struggled to establish orderly if often despotic governments. As the only institution which preserved something of the tradition of the defunct imperial administration, the papacy at Rome slowly achieved increasing influence, buttressed by the fact that, whatever their political rivalries, the rulers were Christian. Under Charlemagne, the Franks established an empire that stretched from the Elbe to the Ebro and southwards into Italy. At Rome on Christmas Day 800, Charlemagne was crowned by the Pope and proclaimed Emperor of the Romans. His reign witnessed a revival of learning and art, symbolized by the royal chapel at Aachen, which some later scholars have dubbed a renaissance. The ninth-century geographer and natural historian Al-Jahiz pithily summarized the relative strengths of the West, the Middle East and the Far East in those times: 'Wisdom hath alighted on three things, the brain of the Franks, the hands of the Chinese, and the tongue of the Arabs.'

During the ninth century this Frankish Carolingian Empire broke up, while Europe found itself under new threats from the longships of the Norse Vikings raiding up her rivers. These sleek, shallow-draught vessels, the culmination of a long tradition of shipbuilding, were well suited to their purpose. Deploying what almost amounted to a new technology of naval warfare, adventurous war-band leaders founded the duchy of Normandy and the state of Kiev in Russia. In 982, Erik the Red discovered and colonized Greenland.

On Europe's eastern land frontier, the rulers of the newly emerging German empire repulsed the nomadic Magyars at the historic battle of Lechfeld in 955. By this time European warfare had been transformed by the evolution of the mounted man-at-arms, capable – thanks in large part to the developments in harness, notably the stirrup – of delivering a mass shock charge. European arms were to win their first triumph outside Christendom with the First Crusade of the 1090s. An expensive military establishment was supported, in a society barely on the edge of a true money economy, by a social structure known as the feudal system, which was based on man-to-man allegiance in exchange for military and labour services.

The international institution of the Church was becoming integrated into this feudal society. While the Church stifled original thought, and those who practised it were condemned as heretics, it helped at least to preserve the old crafts and classical learning and, as a major landowner, to promote agriculture. From about 1100, the Abbey of Cîteaux and its sister houses made a further contribution to the development of Europe. Seeking the life of devotion away from the

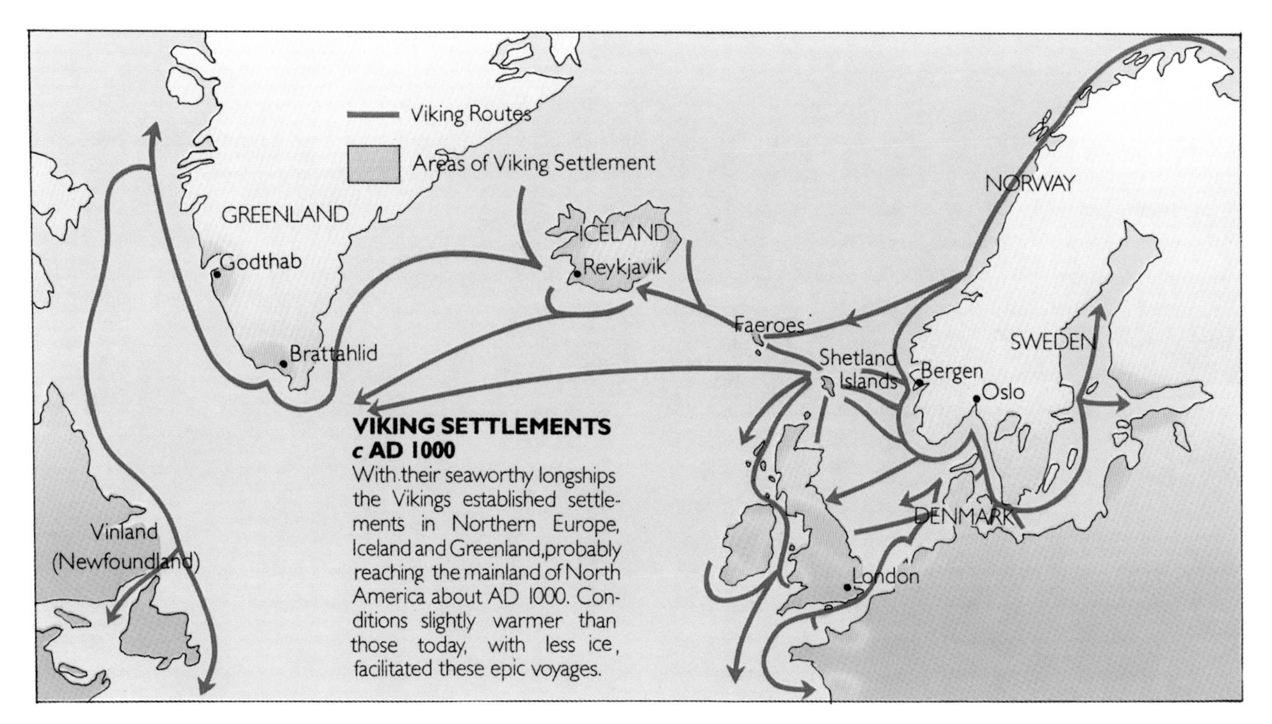

VIKING SETTLEMENTS c AD 1000
With their seaworthy longships the Vikings established settlements in Northern Europe, Iceland and Greenland, probably reaching the mainland of North America about AD 1000. Conditions slightly warmer than those today, with less ice, facilitated these epic voyages.

distractions of the world, the Cistercians established themselves in remote and wild areas, where they played a leading role in the clearing of forests and fostering the wool industry, maintaining large flocks of fine sheep.

In northern Italy the rapidly expanding trade of the towns was laying the foundations of a prosperity which would lead to their transformation into independent states. Perhaps still more important, intellectual life was being liberated as scholars began to desert the monasteries and the cathedral schools for venues of their own choosing, setting the pattern for the future universities.

The intellectual ferment in Islam was by then beginning to stagnate, but the impetus was not lost. Long-forgotten classical texts, enriched by such Arabic commentators as Avicenna (Ibn Sina) began to come back to Europe, as translations from the Arabic into Latin. Classical Latin had no equivalent for many Arabic technical terms, so the words were simply transliterated into the Roman alphabet. Such words as 'alembic' (a type of retort), 'alcohol', and 'alkali' derive from this source. European scholars made their way to the great Arab libraries, especially those of Moorish Spain. Gerald of Cremona (1114–1187), who worked at Toledo, epitomized their reaction: 'beholding the abundance of books in every field in Arabic, and the poverty of the Latins in this respect, he devoted his life to the labour of translation.' Another indefatigable and versatile 12th-century translator was Adelard of Bath. He sharply contrasted European subservience to clerical authority with the intellectual freedom of the Islamic schools.

By this time, the intellectual life of Europe was focused at the *studia generale* in Paris and elsewhere. There scholars organized themselves as corporations, analogous to the guild corporations of craftsmen and merchants emerging in the towns. In the 13th century, St Thomas Aquinas endeavoured to integrate the new learning into the Christian theological tradition, but his thought, too, was heavily coloured by writers such as the Jewish philosopher Maimonides and, above all, Averroes (Ibn Rushd), whose commentaries on Aristotle remained influential even into the 15th and 16th centuries. Pre-eminent among the early universities were Paris, Bologna and Oxford, but by the year 1400 some 80 were scattered over Europe. Their chief role was to pass on the wisdom of the past rather than to extend the frontiers of knowledge, which was still a distinctly hazardous pursuit.

This garnet and gold brooch from Linon dates from the second half of the seventh century AD. It is a fine example of barbaric work.

Chinese printing 100–103 **invention of paper** 99–100 **magnetic compass** 112–113 **Marco Polo** 90–91 **Viking longships** 106–107

CHAPTER TEN

Paper and Printing

In earlier chapters, the development of a written language was singled out as the first criterion of a civilized society. It is, of course, important for the communication of ideas, and for literary purposes, but initially its most important use was to record everyday transactions of state and commerce, without which the early urban communities could not have existed.

From the days of classical antiquity, important documents were written on parchment – the untanned skin of sheep or goats. But the high cost of this – a single copy of a work corresponding to 200 modern quarto pages would require the skins of a dozen sheep – led to the much cheaper, but less durable, papyrus being favoured for everyday purposes. This, in turn, was replaced by paper of Arab manufacture. This change caused the production and reading of books to increase enormously and paved the way for the revolution that was to be brought about by printing.

Paper

Although the word paper derives from papyrus, it is technically different, consisting of felted, as distinct from regularly interwoven, vegetable fibres. This was undoubtedly a purely Chinese invention, later exported to the West. Tradition asserts that it was first made in AD 105 by Cai Lun, a eunuch at the imperial court, who told the emperor: 'Bamboo tablets are so heavy and silk so expensive that I sought for a way of mixing together fragments of bark, bamboo and fish net [hemp] and I have made a very thin material which is suitable for writing on.'

However, it appears that Cai Lun must have improved an existing product, rather than made a new invention, for paper made of hemp and ramie (another Asian fibre plant) has been found in Shangsi province which can be firmly dated at 49 BC. Paper was not invented specifically for writing: the Chinese used it for many other purposes, such as clothing, making small domestic articles, for wrapping and for the decorative arts. In any event, paper-making was firmly established in China by the second century AD, and by the fourth century it had largely superseded silk and bamboo and wooden tablets for writing.

Rattan and mulberry were adopted as additional raw materials, the latter possibly investigated because it was the main diet of silkworms. The raw material was finely chopped,

Paper-making was a purely Chinese invention, probably dating from the first century BC: it was not known in the West until nearly a thousand years later. The process changed little over the ages, as these 19th-century prints clearly show. The raw material – hemp, bamboo or other vegetable material – was chopped and pulped; the pulp was then spread thinly on porous screens, drained and dried in sheets.

Aztec and Inca civilizations 90–91 **development of writing** 31–32 **gunpowder** 136–137, 186 **printing** 100–104

Right: The Diamond Sutra is one of the most important of all Buddhist sacred works. This block-printed page is from an edition dated 'Fifteenth day of the Fourth Moon in the Ninth Year of Xiantong [AD 868]': it is the earliest printed book that can be precisely dated.

Below: The seal, capable of stamping the same symbol or inscription many times over – like today's rubber stamp – may fairly be regarded as the earliest form of printing. This Akkadian seal of 2300 BC shows (centre) the sun god Shamash rising between two mountains: on the right is Ea, the god of water, and on the left Ishtar, equivalent of the Greek Aphrodite.

凡欲讀經先念淨口業真言遍
脩唎脩唎 摩訶脩唎 脩脩唎 娑婆
奉請除災金剛 奉請辟毒金剛 奉請黄隨求金
奉請白淨水金剛 奉請赤聲金剛 奉請定除厄

boiled with wood ash (alkali) and then thoroughly washed to remove extraneous material. Thin layers of liquid pulp were then drained on porous screens and allowed to dry. Much later – from the sixth century onward – a new type of paper made from bamboo appeared: the fibres of this plant were much tougher and demanded much harsher treatment, involving repeated boiling in alkali.

The Chinese art of paper-making found its way through Korea to Japan in the seventh century, and a little later to Arabia. By the eighth century paper-making workshops were established with the help of Chinese craftsmen in Baghdad, Damascus and Samarkand. Paper from this source was exported to Europe, but papermills were not set up there until the 12th century – initially in Moorish Spain, then in Christian Spain and France, and later in Germany and Italy.

Paper was originally called cloth parchment in Europe. It was made mainly from linen rags – though other vegetable fibres such as cotton or straw were often included – which were pounded in water. Stamp mills, driven by wind or water power, were in use in Spain by 1150 to carry out the pulping process. The paper was made as it was in China, by lifting a quantity of pulp from the liquid slurry with a rectangular mould with a wire-mesh base. The limp mat was removed and placed between sheets of cloth and then pressed to expel surplus moisture. The sheets were then hung up to dry thoroughly. Such paper is very absorbent – like modern blotting paper – and from the 16th century it was sized to prevent the ink smudging. To improve its writing quality further, it was often treated also with some opaque mineral powder, such as the lime contained in tanners' waste.

Block printing

For the first few millennia of civilization all writing was by hand and thus necessarily slow, especially when, as with important documents, each character had to be drawn with great care. To some extent the Egyptians contrived to increase the speed of writing by simplifying their hieroglyphs into a cursive form, starting with hieratic ('priestly') script as early as 2800 BC and progressing to demotic about 800 BC. About 63 BC, Marcus Tullius Tiro, a friend of Cicero, had even devised a form of shorthand. This, however, did nothing to overcome the basic weakness that only a single copy could be made at a time. A small, but significant, exception was the seal, widely used as early as Sumerian times. With this, several signs could be impressed simultaneously and repeatedly, just as we now rubber-stamp documents with stock phrases such as 'cash on delivery' or 'received' with the date of receipt. This was the germ of printing in the modern sense, though it was a very long time before the transition was made. Again, it was certainly a Chinese invention that was only tardily transmitted to the West. It took place in two distinct steps: first, printing from full-page blocks, followed, some three centuries later, by the far more important technique of printing from movable type that could be broken up and used again.

Both these techniques were predated by a very early system of replication used in China about the fourth century which involved incising the characters on a smooth stone slab. Paper softened by moistening was then spread on the slab and gently worked into the incisions. When the paper was dry, it was wiped over with an inked pad, which failed to penetrate the depressed incisions. The result was essentially a negative print – white characters on a black background. It was a slow process, but it did allow many copies to be made from a single original.

The obtaining of positive prints required the characters to be in relief, rather than incised, and from about the sixth century this was done with the aid of wood blocks. The text to

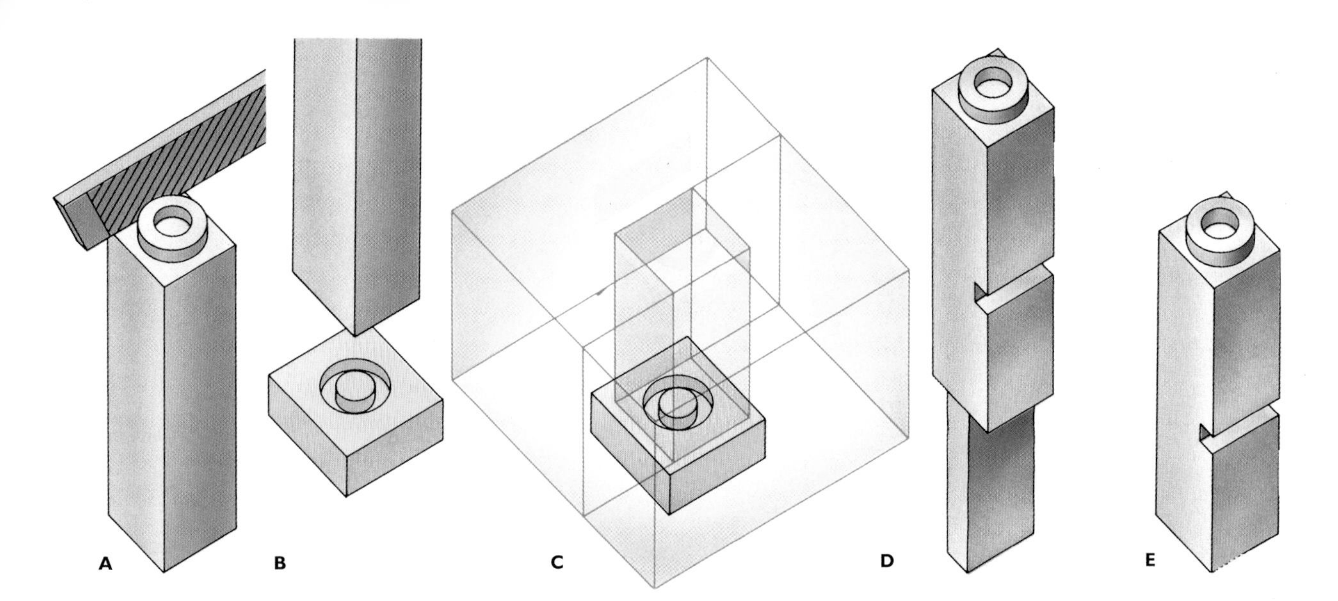

Movable type has to be of uniform size and shape to give straight and even lines. It was made in metal by filing a steel punch for each character (A) that was then used to impress a lead mould (B). This mould was then placed in a standard type mould (C) and each piece of type cast in tin-based alloy. The lug on the casting (D) was then removed to produce the type (E).

be reproduced was first handwritten on thin translucent paper and then glued face-downward on a block of fine-grained wood: that of the pear tree was particularly valued. The characters were then carved out with a sharp knife, leaving reversed letters standing up in relief. The residual paper was stripped off and the block was ready for printing. The surface was inked and a sheet of paper pressed gently and evenly on it.

A great diversity of printed matter was soon produced in this way, ranging from tax receipts and farming almanacs to poetical works. A particularly interesting work is a fine copy of the Diamond Sutra, a Buddhist work, made up as a scroll 5 metres (16 feet) long prefaced by a picture of Sakyamuni preaching. It bears the date 'Fifteenth day of the Fourth Moon in the Ninth Year of Xiantong [AD 868]'. It is thus the earliest printed book that can be precisely dated.

The volume of block-printed matter was enormous and much was of high quality. Many very ambitious publishing projects were undertaken in China before AD 1000. One of the most ambitious was an edition of the *Tripitaka*, the great canon of Buddhist doctrine. This consisted of 5048 chapters in 1076 volumes: the 130,000 plates required took 12 years to complete. By the 14th century at the latest, the technique of printing in different colours on a single sheet, using two or more blocks, had been evolved. The earliest example, printed in black and red, was an edition of the Buddhist work *Notes on the Diamond Sutra*.

By means of block printing, hundreds or even thousands of copies of an original could be produced at a time, the limit being the amount of wear the wooden block could stand before the print became too blurred. Nevertheless, carving the blocks was a time-consuming process and, of course, each was useful only for a single page of a single book. The next major step forward, made in the middle of the 11th century, was to use movable type, each unit being equivalent to a particular character.

Movable type

This seminal invention is specifically attributed to an experienced Chinese block printer named Bi Sheng. According to contemporary accounts, about 1045 he conceived the idea of making each character in clay, which was then fired to make it hard. A mixture of wax, resin and ash was put in the bottom of a shallow iron tray, which was then warmed until the mixture was soft. The clay characters were then pressed into it and levelled with a flat board: when the wax had set solid, the type was ready for printing.

About the beginning of the 14th century, movable wooden type began to be used, each piece of type being of uniform size and height. These units were mounted in wooden trays and held securely in place by wooden wedges. This development is attributed to Wang Zhen, an agronomist, who also devised a rotating frame to hold the type so that the typesetter could quickly pick out the character he wanted with bamboo tweezers. Because there are so many Chinese characters, these frames were very large, some 2½ metres (8 feet) in diameter. The compartments were indexed, and a rhyming mnemonic was used to enable the operator to find quickly the character he wanted. The number of duplicate types in each compartment was related to its average frequency of use. Wang Zhen described his invention in terms which printers the world over would have echoed until very recent times: 'Looking for the characters is very strenuous work for the typesetter, but it is much easier when they are made mobile. The rotating frames enable the typesetter to reach the character sitting down, and when the printing is through he can return the types as easily to their original places.'

Because of their potential hardness and durability, the Chinese also experimented with metal types. Even before Wang Zhen introduced his wooden type, trials were made with tin but failed because this metal would not hold the watery ink. Around 1500, however, success was achieved with copper type.

information technology 309–315 **modern printing methods** 232–235 **photocopying** 313

By the 14th century the Chinese method of printing had spread to Korea (A), but in Europe the use of movable type dates only from about 1450: Gutenberg's original press, now in the Gutenberg Museum at Mainz, is shown below (B). Many early works like the Gutenberg Bible (C) were attractive merely for their crisp black typography, often with ornate initial letters. Quite soon, however, as in a later Gutenberg Bible (D), of the late 15th century, elaborate abstract ornamentation was used, as well as coloured woodcuts such as that (E) depicting – in the vernacular of the day – the shipbuilding methods used for the construction of Noah's Ark.

A

B

C

Omani sūt partis ytalie. Hij pueuti sunt a falsis apostolis: & sub nomine dñi nostri ihesu cristi in legem & pphetas erant inducti. Hos reuocat aplus ad verā & euāgelicā fidē scribēs eis a corintho.

Paulus seruus ihesu cristi·vocatꝰ aplus segregatus in euangeliū dei·qđ āte promiserat per pphetas suos i scripturis sanctis de filio suo·qui factus ē ei ex semine dauid scđm carnē: qui pdestinatꝰ est filius dei in virtute scđm spiritum

D

E

Although the art of printing is particularly identified with the production of books, many other kinds of printed matter – such as charts and tracts – were produced in very large quantities. This illumination shows a large 17th-century Italian printing works devoted to the making of playing cards. The latter were well known in China in the 10th century, but did not appear in Europe until the middle of the 14th. As all the earliest extant cards come from Italy, that seems likely to have been their country of origin in Europe.

Printing in Europe

The Chinese method of block printing was already in use in Japan and Korea by the 12th century, but how it was introduced into Europe is not clear: the Silk Road is an obvious route, with Persia as an important staging post. Italian merchants had established a complex and sophisticated trade relationship with the Far East on the one hand and the whole of Europe on the other. The remarkably complete business records – 150,000 letters and other documents – Francesco di Marco Datini, a merchant of Prato, show him engaging, without any feeling that he was doing anything remarkable, in a varied trade with France, Spain and England. Through their no less extensive eastern connections, the Italian commercial centres were quite familiar with printed paper money and saleable block-printed items such as playing cards.

By the end of the 13th century, such block prints were being produced locally in southern Germany and northern Italy and books printed page-by-page from blocks, like the early Chinese books, appeared after 1470.

How far the adoption in Europe of the technique of printing from movable type was a result of reasonable inferences made from imported examples, how far from verbal descriptions and how far from immigrant craftsmen familiar with the technique must be a matter for speculation. What is certain is that by about 1448 Johann Gutenberg had a printing works in Mainz which used movable type. The first extant material known to come from this press dates from 1454: in the legal sense, the provenance of later work from this press is debatable as he was deprived of ownership of it after a legal dispute with one of his backers, Johann Fust. It is possible that some slightly earlier work was done in Holland and the south of France. For all practical purposes, however, the first organized printing works in Europe was that founded by Gutenberg about 1450.

That Mainz was the first European printing centre was not by chance: it was in an established metal-working area and from the beginning of printing there metal type was widely used. Gutenberg himself was a goldsmith, who would have been familiar with the coining of metal by stamping it in a die. Early type was made from a tin-based alloy, containing a little lead and antimony, cast in a lead mould impressed with the desired character by a steel punch. The typesetting procedure was strikingly similar to that practised in China. The type was picked out with tweezers from compartmentalized cases – an upper case for capital letters and a lower case for minuscule ones – and the assembled type was fixed firmly in its chase with wedges. It is difficult to believe that the basic technique was not directly inherited from China.

Intaglio printing

The principle of intaglio printing is that the writing or design to be printed is engraved with a burin (a chisel of tempered steel) on a soft metal plate, usually copper. The plate is then inked and the surface wiped clean, thus leaving ink in the incisions. If paper is squeezed hard against the plate, the inked design is transferred to it. In China, this technique was used from the latter part of the 12th century for printing banknotes, the delicacy of the strokes making it hard to counterfeit them. In Europe, it first appears about 1440, in Holland and northern Italy.

The European printing press

Whatever the debt of European printers to China, numerous innovations were quickly introduced. Apart from his introduction of metal types, Gutenberg was responsible for the equally important invention of an oil-bound ink that would adhere closely to them and was less liable to smudge. And

block printing 98–99 **Gutenberg** 102 **modern printing methods** 230–232 **Silk Road** 8–9, 21, 34

equally vital was the printing press, the basic design of which was to continue in use until the 19th century. It was adapted from the screw-presses used since classical times in such activities as pressing linen and expressing oil. It had a flat bed, on which the type chase was laid. A sheet of paper was fixed into the tympan, a frame hinged to the chase. When the type had been inked, the tympan was lowered over it. A flat plate, called a platen, was then forced down on to the paper by a screw, a device not used in this context by the Chinese. A sheet of felt in the tympan served to even out the pressure, compensating for slight irregularities in the height of the type. Up to 16 pages were printed at one time, and both sides of the sheet were used.

At an early date the screw was redesigned so that it would move a large distance for a small movement of the bar that worked it, thus speeding up the work. The price of this improvement was that the press exerted a weaker force. To compensate for this, a smaller platen, covering only half of the sheet, was used. Printing was accomplished in two stages, the chase being moved on rails between the impressions. The movable chase also permitted easier correction of the type, if necessary.

This illustration, from a French manuscript of 1537, is a representation of a contemporary printing press. The artist has been at pains to represent the equipment accurately and in detail. Note in particular the screw-press to apply the paper to the inked type; the composed type held in a chase; the printed sheets, carrying four pages of text on each side; and the bound volume.

For various reasons the impact of printing, which had been slowly but massively cumulative in China, was revolutionary in Europe. A major reason was that to print in Chinese a stock of many thousands of different characters had to be held. In Europe, where an alphabet had long been used, derived ultimately from that of the Phoenicians, less than a hundred were necessary, allowing for upper- and lower-case letters, numerals, punctuation signs and so on. In a sense, it can be argued that for a time the Chinese had printing but an unsuitable form of writing, while Europe had an appropriate form of writing but no printing process. A further reason is that knowledge of printing suddenly reached a highly civilized community poised to abandon the restrictive dogma of the past and embark on the freedom of thought and expression that characterized the Renaissance.

The printed word swept like a hurricane through Europe: by 1500 no less than 40,000 separate works – now collectively known as incunabula – had been printed. As editions were relatively small – rarely exceeding 500 copies – this probably represents a total of around 20 million books. Nevertheless, the Chinese lead was so formidable that it has been estimated that as late as 1800 more pages had been printed in Chinese than in all the world's other languages combined.

The advent of printing was not wholly advantageous in the West. China had a vast number of local dialects but the written language was the same everywhere: printing was, therefore, a unifying social influence. In Europe, however, it was arguably divisive: Latin, which was still the universal currency of scholars, began to be displaced by local vernacular publications.

Among the huge number of books published in the 16th century were many dealing with various aspects of technology, and the most important of these were translated into the principal European languages. Among them may be mentioned important works on mining and metallurgy, such as Agricola's *De Re Metallica* (1556) and Biringuccio's *Pirotechnia* (1540); military works like Macchiavelli's *Art of War* (1578) and Barwick's *Force of Fire Arms*; Lomazzo's *Art of Building* (1598); and Stevin's book on navigation, the *Haven Finding Art*. The printing press was thus a powerful instrument in disseminating knowledge in the technological field.

In Needham's great *Science and Civilisation in China*, Tsien Tsuen-Hsuin has succinctly stated the profound significance of paper and printing: 'Paper has proved to be the most satisfactory material on which human thoughts are committed to writing and when printing came to be allied to it, the ideas of one individual could be communicated to a multitude of others separated across great stretches of space and time. In short, the printed message has brought about changes in the intellectual mode of the human mind, and paper has provided the most economical and convenient means for its transmission.'

CHAPTER ELEVEN

Shipping and Navigation

Shipbuilding was a well developed industry during the times of classical antiquity. The almost tideless Mediterranean ports were linked by a complex system of shipping routes; boats plying from Red Sea ports had regular contact with others sailing from China and India; the Phoenicians, at least, had sailed beyond the Pillars of Hercules into the tidal Atlantic; and if Herodotus' tale of their circumnavigation of Africa is questionable, it is at least plausible. Ocean-going voyages far out of sight of land may have been hazardous but they were part of the regular pattern of trade. Carriage by water was by far the cheapest means of moving goods, and the great navigable rivers of the world too, augmented by canals, carried great quantities of freight.

By its end, the classical period had seen the development of most of the basic features of ships up to the advent of iron hulls and steam, in the sense that an experienced fifth-century mariner would find little that he would not readily understand in the 19th-century clipper-ships, the ultimate in sailing vessels. Yet there were many significant changes on the way. The nature of these is most easily understood in terms of systems – hull, steering, propulsion, navigation – rather than of complete vessels.

Although our knowledge of early ships is considerable, there are large gaps in it: we know surprisingly little, for example, of Byzantine ships. Artistic licence often makes pictorial descriptions of doubtful authenticity, and in the nature of things wooden vessels were broken up or decayed altogether. There are, however, some important exceptions to such generalizations. We are, for example, fairly knowledgeable about ships of the Vikings because of their custom of burying their chieftains with their vessels.

Recently, new techniques in underwater archaeology have greatly facilitated the location and investigation of wrecks: buried under mud or sand, they can be remarkably well preserved. Some early wrecks and their contents have been recovered largely intact; a notable example is the restoration of the *Mary Rose*, sunk with all hands off Portsmouth in 1545. And the investigation of ancient dockyards can sometimes throw light on the size and nature of the ships they served.

The hull

It has been argued that the earliest wooden ships of northern Europe, particularly those of Scandinavia, derived – by some kind of technological transfer – from those of the Mediterranean. But, apart from such a transfer being inherently unlikely at that time, we can clearly discern different lines of evolution which converge only when communication became much freer. That there are also similarities is not surprising, in that men independently facing the same practical problem naturally tend to find the same solutions.

Two important differences have already been noted. The keel, a longitudinal member on which the frame is built, was introduced into Mediterranean vessels at an early date, whereas a true keel – as distinct from a strong bottom plank, probably serving as a runner or skid for handling the boat on shore – cannot be firmly identified in Scandinavia before the Kvalsund boat of the sixth century. Further, Mediterranean ships were carvel-built (having edge-to-edge planking) in

In recent years, much has been learned about the construction of early ships by the recovery of well-preserved wrecks, some dating back to pre-Christian times. This is one of the best known, the English warship *Mary Rose*, which sank with all hands off Portsmouth in 1545. It was raised in 1982.

Another famous ship to be recovered from the sea is the *Vasa*, a 17th-century Swedish warship lost in Stockholm Fjord. This was raised in 1961, and its hull and a wealth of recovered objects have been restored and are now on permanent display.

Chinese printing 100–103 **early boats and ships** 49–53 **modern printing methods** 232–235, 310 **Renaissance** 98, 144 **shipbuilding** 172–176, 254–255

The Viking custom of burying their chiefs with their ships and personal possessions has provided modern archaeologists with a rich store of material. One of the earliest longships known is the Gokstad ship (above). Note the stump of the mast for a sail; the well defined keel; and the holes for oars, 16 on each side. Such ships often had elaborately decorated stem posts – analogous to later figure-heads – like that from a roughly contemporary ship recovered from the Scheldt in Belgium (below right).

The Gokstad longship (right) as she appeared when she was unearthed on Gokstad Farm, Oslo Fjord, in 1880 – about a thousand years after her burial. Powered by seated oarsmen, such longships were capable of making trans-atlantic journeys – as a replica of the Gokstad ship proved in 1893.

contrast to the clinker-built ships of the north (which used overlapping planking). Again, Scandinavian vessels, in sharp distinction to their Mediterranean counterparts, seem not to have carried a mast for a sail before the seventh century. This development went hand in hand with a true keel, which was necessary to provide a strong base for a mast. An early example is the Oseberg ship, dating from about AD 800, uncovered in Norway at the beginning of this century. She was 21 metres (69 feet) in length, was equipped for 15 rowers on each side and had a short mast carrying a square sail.

Perhaps the most famous of the Viking longships – so called because of their high length-to-beam ratio – was the Gokstad ship, of about AD 900, unearthed in 1880. It was in such a vessel as this, of around 50 tonnes, that the great Norse navigators such as Erik the Red and his son, Leif Ericsson, made their well attested voyages to Greenland – and almost certainly onward to the shores of North America – around AD 1000, predating Columbus by nearly five centuries. That such a voyage was feasible was demonstrated in 1893, when an exact replica of the Gokstad ship was sailed across the Atlantic and exhibited at the Chicago World's Fair. Scholars may argue about the direct evidence for this discovery, but when it is realised that the Vikings not only discovered Greenland, but for a time maintained and regularly supplied a colony of several thousand people there, it is difficult to believe that they did not indeed make the relatively small further step to the American mainland.

In northern waters longships were predominant for several centuries. They were the vessels that Alfred sent against the Danes – though by then his largest ship carried 60 oars – and William the Conqueror used for his invasion of England. Not until the 13th century, with the growth of the powerful Hanseatic League of trading cities in the Baltic with strong links with England and Holland, did a different kind of vessel appear. This was the cog: it had a straight stem and sternpiece fixed at an angle to a long straight keel, capable of

keeping the boat very stable even in rough weather. It had a single mast amidships. With an overall length of around 30 metres (100 feet) and a beam of about 8 metres (26 feet), these ships were quite tubby, like the merchantmen of the Mediterranean. A novel development was the appearance of two tall superstructures – a small one forward and a large one aft: initially they were built as an addition to the hull but later they became an integral part of it. These structures were called castles because – like their counterparts on land – they served both as observation posts and as platforms from which fire could be directed against enemy ships or boarders. A third castle, on the mast, was probably used mainly as a lookout, but some 13th-century prints show bowmen in it. In days when piracy was rife, this kind of defence was a very necessary precaution and it represents a convergence between merchantmen and warships. In England, a rather different type of cog evolved: this had the traditional curved stem and correspondingly shorter keel.

By the beginning of the 14th century cogs began to appear in the Mediterranean, mainly built to the orders of the great merchant traders of Italy, but in the regional tradition they were still carvel-built. A significant development was the introduction of a second mast behind the mainmast, and then a third, smaller one on the poop. Such vessels, at first particularly identified with Spain and Portugal, were known as carracks. From about 1400 this was the standard sailing ship of the Mediterranean. At the beginning of the century vessels of 250 tonnes would be considered large, but by its end 1000 tonnes, or even more, was not unusual. Henry V's

From about 1400 the carrack, successor to the smaller and simpler Baltic cog, was the standard merchant ship of the Mediterranean. (See the 16th-century examples above.)
The ships with which William the Conqueror invaded Britain in 1066 – depicted in the Bayeux Tapestry (right) – were descendants of the Viking longships.

Grâce Dieu, built in 1418, must have been exceptional for its time at 1400 tonnes, on the evidence of its wreck recovered in the Hamble River. A smaller shallow-draught version of the carrack was the Portuguese caravel, much used during the great voyages of discovery made in the 15th century and particularly identified with Prince Henry the Navigator. The shapes of ships continued to change. The tubby cogs began to be replaced by ships built to the so-called one-two-three formula, in which the length of the hull was three times the breadth, and the latter was twice the depth.

The recovered remains of the *Grâce Dieu* provide interesting evidence of the mode of construction. The overlapping planking was applied in three layers, with a total thickness of 10 centimetres (4 inches); it was fixed to the ribs with iron bolts. In the north, the carrack was not widely adopted, though caulked carvel construction was slowly substituted around this time for clinker. There the expanded cog was popular: the length was doubled and a three-masted rig adopted. Such vessels, originally known as hulks, developed into the full-rigged ship. Henry VIII's *Henry Grâce à Dieu* (the *Great Harry*) – built in 1512 and rebuilt in 1545 – was a fairly typical large ship of her day. Well over 1000 tonnes of timber went into her hull, constructed of oak some 15 centimetres (6 inches) thick. She was a four-masted ship with two large castles, designed for defence against any enemy who might invade the low waist; the crew totalled 700. We have little in the way of plans for such vessels: successive cross-sections were commonly marked out to scale in chalk on the floor of the shipyard's mould-loft. Very often, however, detailed scale models were made in advance, and many of these survive. A fifth-century Chinese archive contains an interesting passage: 'Wenzhou Prefecture has directed that it was in receipt of two volumes of drawings of ships from the Military Governor's Office. It is ordered that officials be dispatched to buy timber for the construction of 25 sea-going vessels in each magistrature.' It is possible, however, that the reference is to no more than sketches and purchasing requirements to guide the shipwrights. We cannot certainly date Chinese ship design plans before the 17th century.

In passing, it is interesting to relate shipbuilding to timber supply. A good oak might yield 2 tonnes of serviceable timber. Allowing 100 trees to the hectare (40 to the acre), a ship such as *Henry Grâce á Dieu* would require 6 hectares (15 acres) of woodland for her construction: this would take at least a century to regenerate, whereas the life of the ship might be no more than 20 years.

Some degree of decoration had long been a feature of ships, as, for example, the carved and gilded prows of the Viking longships and the shields displayed on the gunwales. In the 16th century painted embellishment was common, but there was little in the way of carving. In the 17th century, however, ostentatious decoration serving no functional purpose was widespread on important ships – sometimes to the extent of impairing their performance at sea. The best-known of such decorated additions are the elaborate figure-heads.

Steering

Boats propelled by paddles or oars can be steered by varying the strike on one side or the other or, if a sharp change of direction is needed, by back-paddling on one side. Nevertheless, some form of additional steering equipment

early boats and ships 49–53 **Hanseatic League** 90–91 **naval warfare** 140, 217–219 **shipbuilding** 172–176, 254–255 **Viking voyages** 97

was usually provided and the earliest form of this – depicted alike in Egyptian frescoes, on Hellenic vases and in carved representations of Roman warships – was a broad-bladed oar, trailed alongside the hull near the stern and controlled by a helmsman. Some vessels had two such steering oars, one on each side.

The now familiar stern-post rudder, acting centrally, and thus equally responsive to port and starboard, is a comparatively recent innovation, at least in Europe. Such rudders are clearly shown, however, in Chinese tomb-models of the Han Dynasty (202 BC–AD 220), but how widely they were used is another matter. In the West we cannot certainly date this very important innovation before the 13th century, though possibly Byzantine ships used it a little earlier. A single-masted sailing ship with a rudder of this kind is very clearly seen on a seal of the City of Elbing, Poland, dated *circa* 1242. Partly because of the high massive stern-posts, and partly no doubt because of long tradition, this kind of rudder was at first controlled, through a cranked rod, by a helmsman sitting conventionally at the quarter-rail near the stern as he had always done: such rudders can still be seen on some small Scandinavian boats. By the 15th century, however, the modern tiller-controlled rudder had appeared, controlled through a port cut centrally in the stern, now of square construction without the single stern-post.

The lateen sail, enabling ships to sail more closely into the wind than previously, derived from Arab dhows of the Red Sea (right). The Battle of Lepanto in 1571 (below) in which Don John of Austria defeated the Turks, is memorable as the last major encounter in which galleys were deployed. This picture gives a vivid impression of the battle and the size and appearance of the vessels engaged. Note in particular the massive rams in the bows, a tradition going back to the early Greeks: the ram thus remained in vogue for some 2000 years.

Propulsion

For all practical purposes, there were still only two methods of propulsion – oars and sails – during the period with which we are now concerned. Often, as in Venetian galleys, popular up to the 16th century, the two were combined, but for economic reasons the balance turned in favour of sail alone – save for heavy sweeps used only to manoeuvre over short distances in narrow waters. Even galley slaves involved some kind of capital investment and had to be fed, but increasingly paid labour had to be employed. As a large galley might require some 200 oarsmen, the cost was considerable.

In most parts of the world the winds follow a fairly regular pattern season by season and so far as possible voyages were planned to take advantage of a following wind. However, even if the wind was abeam, or nearly so, a square-rigged ship could still make progress by setting the sail aslant, when some of the force of the wind acting on it would be directed forward. There were, however, disadvantages in this. First, speed was necessarily slower than if the full force of the wind was exerted forward. Secondly, the residual lateral force of the wind would cause the ship to drift sideways off course. Thirdly, the ship would tend to twist and head up into the wind, and heavy work with the steering bar would be necessary to correct this. Aristotle, in a work on mechanics, described how this last problem could be overcome by reducing the area of the sail behind the mast by adjusting the brails, thereby diminishing the force of the wind on the sail. But doubtless he was only describing in simple physical terms a practice that was already long familiar to seamen.

Seamen were already well familiar, too, with another way of utilizing a wind on the beam. This was to progress by a series of tacks, the ship steering a zigzag course with the wind alternately on either quarter. Progress this way was necessarily very slow: not only was the ship's speed through the water much less than with the same wind blowing astern, but the distance travelled would be several times longer than the direct course. Moreover, at the end of each leg of the tack, the ship came virtually to a halt and the sails had laboriously

to be reset. In narrow waters, and in conditions of poor visibility and variable wind, it was very difficult, and often impossible, to steer the desired overall course.

Small wonder, then, that much attention was paid to the problem of sailing into the wind. Partly, this was a matter of skilled seamanship, but it was also a matter of the sails themselves. An important development – originating with the Arab dhows but quickly adopted in the Mediterranean – was the lateen (a corruption of Latin) sail. In this rig, the yard was set fore-and-aft and inclined at an angle to the mast instead of at right angles to it. The curve of the sail generated an aerodynamic force, as in an aircraft wing, and with this it was possible to make progress even if the wind blew considerably ahead of the beam.

This increased manoeuvrability encouraged the great Iberian voyages of exploration of the 15th century, most of which were made in caravels with lateen rig. Sea captains could overcome their very real fear of being unable to retrace their steps in the face of prevailing winds – a fear which deterred Portuguese mariners from rounding the western bulge of Africa until 1434 – and they could have much greater confidence in their ability to control their vessel among the hazards of uncharted landlocked waters.

Our understanding of the rigging of Chinese ships of this period is incomplete, partly because there were many different adaptations to local conditions and partly because colourful descriptive metaphors are freely used. We can readily interpret 'the turban on top which can lift the ship and make it light and fast' as some sort of topsail, but this does not in itself give an indication of its nature or how or when it was used. Nevertheless, the Chinese clearly kept pace with the West. The problem of drift when sailing with the wind in the beam was commonly solved by use of a leeboard which could be lowered into the water below the hull to increase lateral resistance to the water. A 13th-century writer clearly indicates that there, too, the problem of sailing into the wind had effectively been solved: 'Of all the eight quarters whence the wind may blow there is only one, the dead-ahead quarter, which cannot be used to make the ship sail.' The distinctive sail design of the junk was responsible for this. Each sail consisted of a series of panels, kept flat by a bamboo batten, so that the whole sail could be hauled close to the wind: the Western sail, by contrast, took up a curved form.

We have already mentioned briefly a fourth-century Latin description of a ship propelled by paddle-wheels powered by oxen, but dismissed it as probably a flight of imagination. Nevertheless, it embodied an important mechanical principle: the paddle-wheel represents continuous rotary action as compared with the intermittent action of rowing, in which time and energy are lost at the end of each stroke.

There is good evidence, however, of Chinese paddle-wheel ships centuries before they were adopted in the West with the advent of steam. Even if we discount an apparent reference to a paddle-wheel ship towards the end of the fifth century, Li Kao gives, about 782, a circumstantial description of 'a warship … which is propelled by two treadmill paddle-wheels. It flies through storms as fast as under sail.' This again may have been in the nature of a limited edition, but by the 12th century the situation was very different. Considerable fleets of paddle-propelled warships were then used on the Yangtze and on the large lake Tung-thing, particularly by Yang Yao, military leader of a peasant revolt in 1130. We know that some of these were large vessels up to 100 metres (330 feet) long with a crew of nearly 1000 men. These 'flying tigers' had multiple paddles, commonly as many as a dozen on each side. Such paddle ships have a continuous

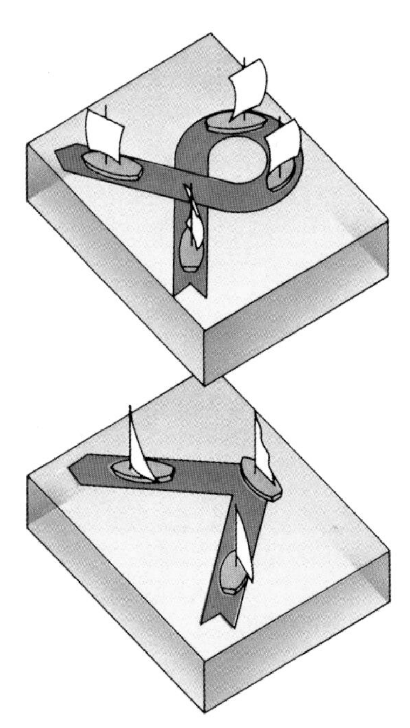

Sailing ships could maintain a course against the wind by tacking in a zigzag path so that the sail was kept at the same angle to the wind on each tack. However, at the end of each tack, square-rigged ships (top) had to make a downwind turn in which the ship turned completely around, a laborious and difficult manoeuvre that delayed progress. Ships with a triangular lateen sail (bottom) could change direction quickly and easily by turning into the wind. In addition, the lateen sail enabled ships to sail closer to the wind. Its adoption greatly improved manoeuvrability and fostered maritime exploration of the world.

early boats and ships 49–53 **modern ships** 254–255 **paddle-wheel ships** 68, 174–176 **steering oars** 53 **voyages of exploration** 90–91

The earliest seamen navigated simply by direct observation of the sun, moon and stars, but by the 16th century a variety of instruments, adapted from those long used by astronomers, were in common use at sea. One of the most important was the astrolabe (left): the example shown here is of German origin, made at Nuremberg in 1548.

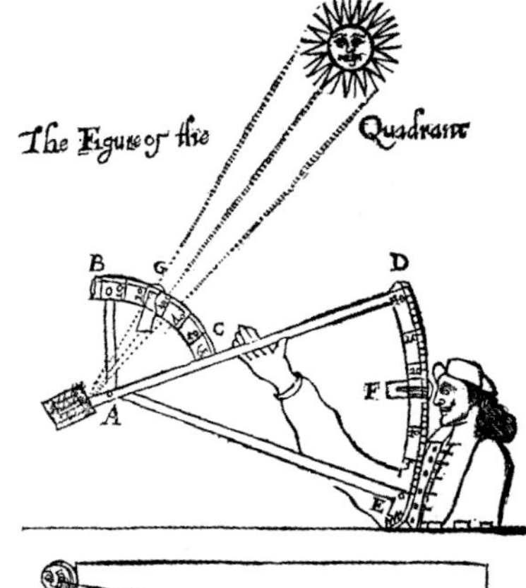

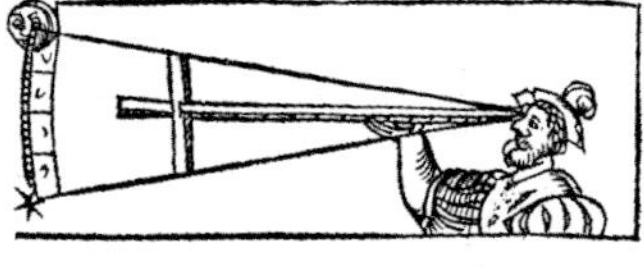

Right: Contemporary illustrations show early navigational instruments for measuring the altitude of the sun and other celestial bodies: a German astrolabe of 1493 (bottom), a Dutch cross-staff of 1538 (centre) and an English back-staff of 1669 (top).

For many centuries great trust was placed in Ptolemy's world map, prepared in the first century AD. This version (below right) was published at Ulm in 1486. Ordinary seamen required maps of more limited extent, such as this Portuguese map of the North Atlantic published in 1535 (bottom right).

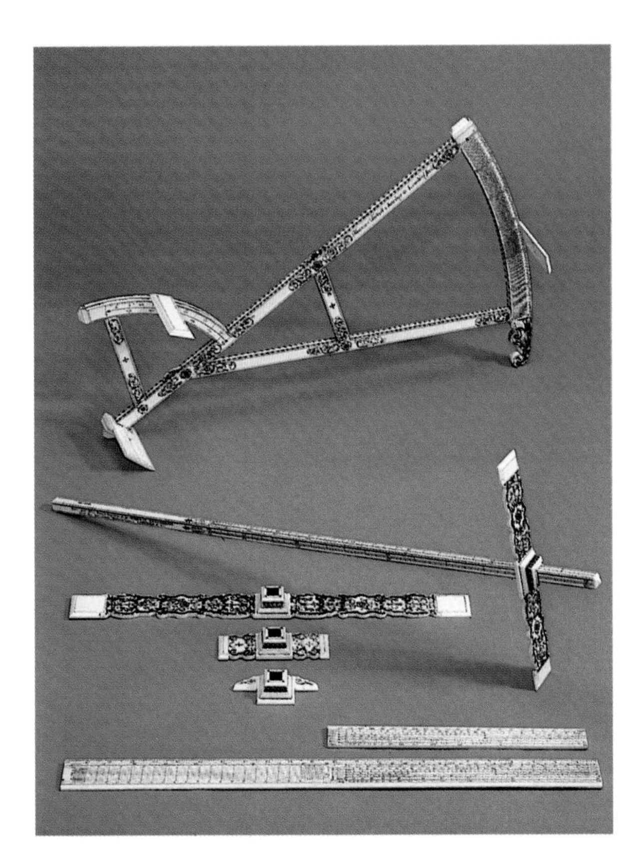

These finely finished navigation instruments, made about 1700, include a back-staff (top); a cross-staff (centre), with interchangeable crosspieces; and a pair of calculating rules (bottom).

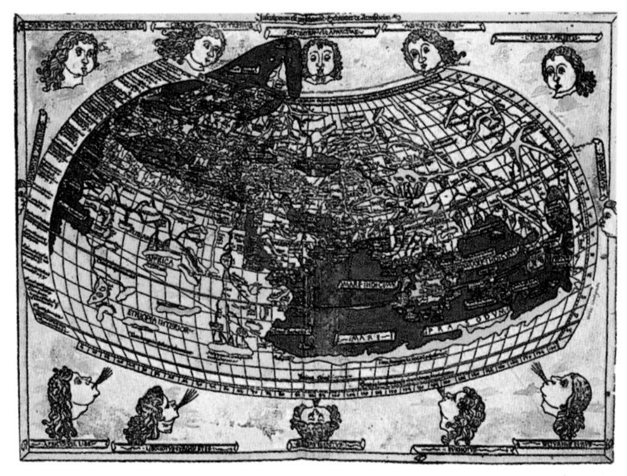

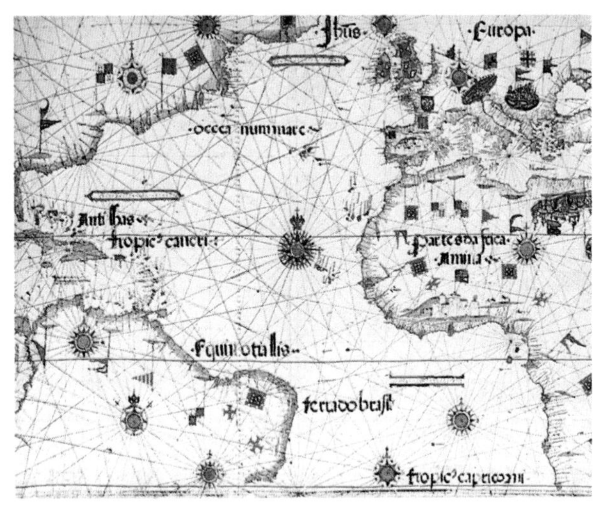

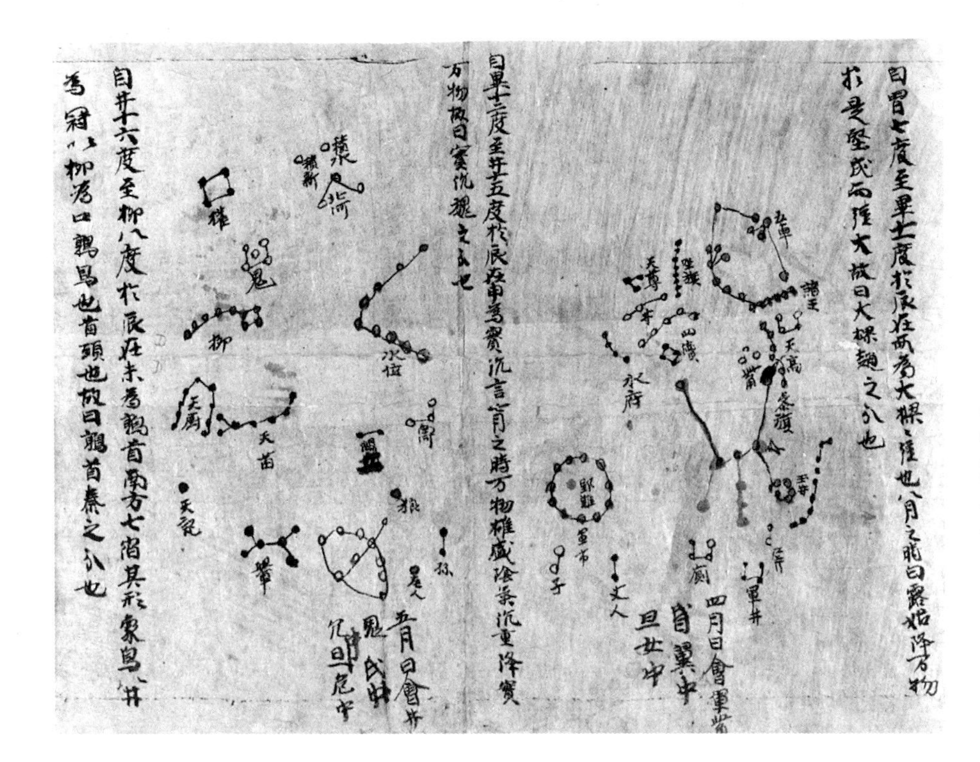

Chinese astronomers began to create catalogues and maps of stars from the fourth century BC. The earliest surviving example dates from AD 940, and was recovered from the Dunhuang caves in Gansu province. It shows various familiar constellations such as Canis minor and Cancer (left) and Canis major and Orion (right). The different columns denote the three great astronomers who had originally determined their positions more than a thousand years earlier.

history in China from the 12th century to modern times, but although considerable numbers were built they were never widely popular because of the high cost of the manpower: a 250-tonne vessel required 42 pedallers and some very large ones as many as 200.

Navigation

That the earth was spherical was well known at an early date, and in the third century BC the Greek scholar Eratosthenes, on the basis of astronomical observations, estimated its circumference to be about 46,000 kilometres (29,000 miles), which is not very different from the best modern value of 40,000 kilometres (24,800 miles).

On this globe early geographers, such as Ptolemy in the first century AD, located the main features of the ancient world with varying degrees of accuracy. In constructing their maps they used such astronomical observations as were available to them, but had also to rely heavily on travellers' estimates of distances (based on time of travel) and direction (which would never be very accurate). Naturally, the more familiar and accessible the region, the more numerous and accurate the sources of information. Europe, the Mediterranean area and the Middle East were surprisingly well charted but the geography of the Far East was much more vague and distances tended to be considerably overestimated. This had important consequences at a later date, as it suggested that eastern Asia could be more readily reached by travelling westward than it actually can be.

Nevertheless, on these early maps huge areas of the globe completely lacked any reliable information, though this did not prevent their makers from filling them with highly imaginative speculations. In particular, there was a strong belief that there must be a great southern continent – which the 16th-century Flemish cartographer Ortelius called the *Terra Australis nondum cognita* (the southern land not yet known) – to balance the great northern land mass of Asia. Ptolemy, indeed, showed the Indian Ocean as a landlocked sea. However, it is quite wrong to suppose that experienced navigators believed that the earth was flat and they might fall over its edge, though this might well be true of their crews. On the contrary, navigators were quite clear that, provided all the oceans were joined, it was perfectly feasible to circumnavigate the globe.

While the great cartographers concerned themselves with mapping the world, ordinary seamen had to rely on charts of more limited scope, such as the famous 14th-century map of the Mediterranean – the *Mappa Mundi* or the earlier *Carte Pisana* (*circa* 1275). Such maps embodied a great deal of practical navigational information of use to seamen – tides, shoals, prevailing winds, landmarks and the like. Sailing directions were incorporated into rutters (from the French *routier*) and it is interesting that the Chinese were producing similar guides at the same time. A Chinese manuscript in the Bodleian Library at Oxford – entitled 'Favourable winds escort you' – is reputedly derived from the first printed Chinese rutter of the 14th century.

When at sea, however, the mariner had to estimate his position by the method known as dead reckoning, in which every change of course and speed is recorded and plotted. If the information is accurate this gives a correct result, but in fact there are so many imponderables that great errors can accumulate: it is difficult, for example, to allow for the effects of currents and the sideways drift from a wind on the beam. The whole early history of Pacific exploration is clouded by territories being discovered and described, but not refound because they were wrongly located, often by hundreds of kilometres: others were discovered several times over before their identity was realized.

determination of longitude 146 **Marco Polo** 90–91 **paddle-wheel ships** 68, 174–176

These errors arose despite the fact that mariners had a range of navigational aids to assist them in determining their position, as expressed in terms of latitude and longitude. The former is measured in terms of degrees from the equator to the poles, and the latter in terms of degrees from some predetermined meridian, a great circle running from pole to pole. In principle, the determination of latitude is not very difficult, for it depends simply on measuring the altitude of some celestial body such as the sun or the Pole Star. In practice, however, it was often difficult on a pitching deck, while cloud or fog might make observations impossible for days or weeks on end, especially in northern latitudes.

The most common instrument for measuring altitude was the quadrant, a quarter-circle of metal or wood with sights along one edge. The circumference was graduated from 0 to 90 in degrees and a plumb-line was provided so that the angle of altitude could be measured on the scale when the star was aligned in the sights. A simplified version of the astronomer's astrolabe was also used for measuring altitude. Another simple device was the cross-staff. This consisted of a wooden rod about a metre (3 feet) long, along which slid a vertical crosspiece: one end of the latter was aligned on the horizon and the other on the sun or star being observed. An important improvement to the cross-staff was the back-staff, invented by John Davis in 1594, which enabled the observer to turn his back to the sun.

With such simple instruments latitude could be determined fairly accurately and easily provided the observational conditions were favourable. Longitude, however, was a very different matter, for it involved comparing local time with that at some fixed meridian. Not until the latter half of the 18th century, with the advent of the marine chronometer was this problem satisfactorily solved.

Two other simple navigational aids deserve passing mention. One was the lead-line, which served not only to measure the depth of water, but also – by means of a little tacky tallow fixed in its head – to indicate the nature of the seabed, a useful guide to the experienced mariner. The other was the log, a simple device for measuring the ship's speed through the water. It consisted of no more than a flat board weighted so that it would float upright in the water and thus resist movement. To it was attached a cord knotted at fixed intervals of about 14½ metres (48 feet). It was thrown overboard and the observer counted the number of knots that passed through his hand in a fixed period measured with a sand-glass. To this day, ships' speeds are stated in knots: a knot is now defined as one nautical mile (1.85 kilometres or 6080 feet) per hour. The first printed account of the log-line was given by William Bourne in 1574, but it was clearly already a well-established device. The importance of the log-line was such that its readings were entered systematically in a special 'log-book', which over the years became identified with a general day-by-day account of a voyage.

The compass

The most important innovation in navigational aids, however, was the magnetic compass. The attractive property of the lodestone (magnetic iron oxide) was known both to the Romans and to the Chinese of the contemporary Han Dynasty. Recognition of the ability of a magnet to orientate itself in a north–south direction is a different matter, however. In Europe, the first authentic account of this was given by Alexander Neckham about 1190. In 1269, Peter Peregrinus, in his *Epistola de Magnete*, described how a piece of iron wire, magnetized by being rubbed with a lodestone, would always come to rest in a north–south position if floated on a light piece of wood. Peregrinus – described by Roger Bacon as a 'master of experiments' – had more than an academic interest in this phenomenon. A practising military engineer, he constructed practical compasses with marked poles and graduated scales.

The mariners' compass – exemplified by this late 16th-century instrument (right) – was first used by the Chinese in the 12th century. Apart from difficulties arising from regional differences in magnetic declination, there were problems of relating compass bearings to directions on a chart. The basic problem was to translate positions on a three-dimensional globe – such as that of Mercator (below) – onto a two-dimensional sheet of paper.

That the phenomenon was put to practical use in China very much earlier is not to be doubted. As early as the fifth century BC a lodestone carved into the shape of a spoon, and known as *sinan*, was used for direction-finding. *The Book of the Devil Valley Master*, compiled in the fourth century BC, describes how jade-quarrying miners used *sinan* to guide them on their expeditions in search of the precious mineral. Thereafter there is constant reference to *sinan* in Chinese literature, but no major development until the utilization in the 11th century of magnetism induced in iron. A very thin sheet of iron was cut in the form of a fish, magnetized by rubbing with a lodestone and floated on water.

The Chinese used such 'fish' for various purposes. For example, much importance was attached to the proper orientation (feng shui) of houses, temples and other buildings lest ill-luck result. The first clear reference to the use of the compass at sea dates from a report of AD 1119 which states: 'The sailors are sure of their bearings. At night they judge by the stars. In daytime they tell by the sun. When it is cloudy, they rely on the south-pointing needle.'

Nevertheless, the compass was not without defects. On board ship, it was not easy to float the needle on water, and this led, in both Europe and China – though not until the 16th century in the latter – to the dry-mounted compass in which the disc is mounted on a needle point. From the 16th century in Europe compasses were mounted on gimbals to keep them horizontal, but in China this device appeared much earlier. More serious was the fact that the compass does not point to the true north – the axis of rotation of the earth – but to magnetic north. This declination, as it is called, was familiar to Chinese philosophers about 1050, even before the directive properties of the compass had been noted in Europe. Unfortunately, the declination is not a fixed quantity, but varies from place to place and year to year so that no simple correction is possible. Nevertheless, over large areas of the world the difference is small and changes only slowly.

How far, if at all, were these developments in East and West related? There is no evidence of an overland transmission from China to Europe, but there could well have been one by sea via the well established trade routes, with the Islamic world as the connecting link. Against this, however, is the fact that there is no known Islamic reference to the magnetic compass before AD 1232, rather later than its first description in Europe. Nor is there any known reference in Indian literature. At present, the question cannot be satisfactorily answered, but it may well be that here there were two independent lines of development.

When Magellan set out on the first circumnavigation of the globe in 1519 he carried with him 21 quadrants, seven marine astrolabes, 18 sand-glasses, 23 charts and 37 compass needles – probably a typical inventory for long open-sea voyages of the day. The combination of chart and compass was powerful, but in view of the limitations of both he might well have been more than 10° off course if both erred simultaneously in the wrong direction. And, in fact, by the time he reached the Philippines in 1521 his navigator had miscalculated his position by 5400 kilometres (3400 miles). One major problem with the charts was that they had to represent accurately on a flat surface geographical features which exist on a spherical one. Not until 1569, with the printing in Duisburg of Gerhardus Mercator's world map, using the mathematical system of projection which still bears his name, did this problem receive its first satisfactory solution.

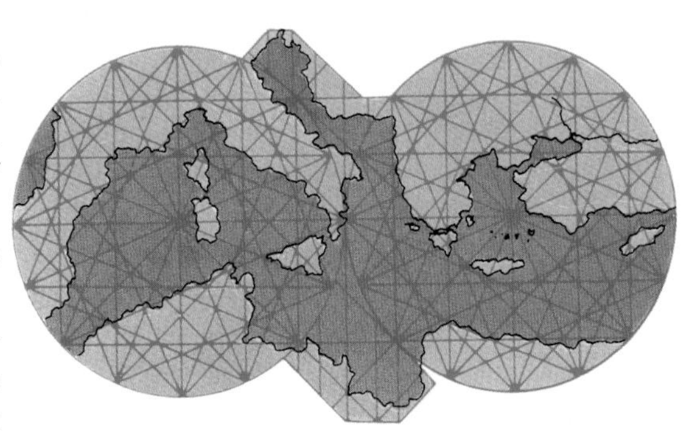

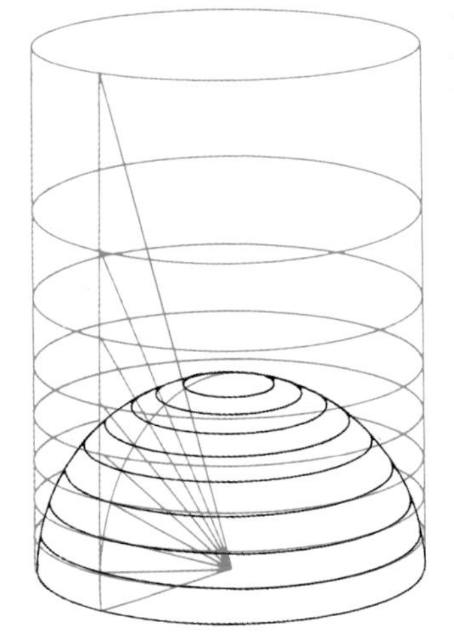

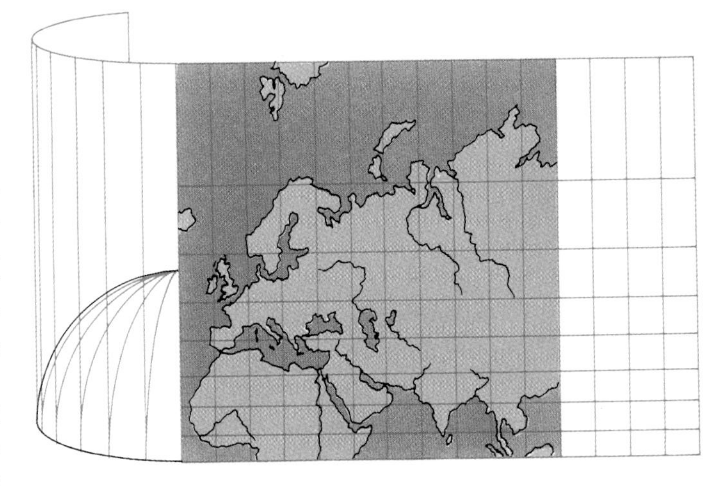

The *Carte Pisana* of *circa* 1275 (top) was the earliest surviving map to be constructed geometrically using compass bearings and distances. It shows the Mediterranean Sea but is not very accurate. In 1568, Mercator developed a cylinder projection method (centre) to produce a two-dimensional representation of the earth's curved surface. Lines of latitude get farther apart with increasing distance from the equator, while lines of longitude are parallel (bottom). Although Mercator's projection thus distorts shapes, it represents a bearing as a straight line and enabled mariners to plot a course with ease.

Magellan's voyage 90–91 **marine chronometer** 146 **navigational instruments** 110–111 **sonar** 324

CHAPTER TWELVE

The Beginning of Mechanization

As we have noted, the world of classical antiquity relied for its power almost entirely on that of men or animals, made more effective by the use of simple machines such as levers and pulleys. The windmill had yet to appear, though the power of the wind had long been effectively utilized in sailing ships, and such few water-wheels as existed had very limited output: their overall contribution was negligible, though locally they could be very important. About the time of the fall of the Roman Empire in the West, however, this situation began to change. Slaves were no longer so readily available and paid manpower could command increasing wages. The natural response was to replace men with machines, a response paralleled in America in the 19th century. It would, of course, be an over-simplification to see the development of power sources and machines as no more than a response to shortage of labour. There were other economic factors and technological ones as well. The accumulation of capital and the growth of banking, together with increasing technological capacity, encouraged the use of power and machinery. Again, the degree of change naturally varied from time to time and place to place: in Europe natural calamities such as the Black Death in the 14th century accentuated the labour shortage, while in large areas of Asia the pattern of life scarcely changed at all. When there was change, it was not without opposition: labourers reacted to the threat of machines long before the Luddites of 19th-century England. Nevertheless, if we take a global view we can clearly discern a gradual change, culminating in the 18th century in the Industrial Revolution. For the moment, however, we will limit ourselves to changes during the period of roughly a thousand years that preceded the Renaissance.

Windmills

The origin of windmills is obscure. On the one hand, it is plausible to relate them to the sails of ships, and one Mediterranean type, found mainly in Crete and the Aegean but occasionally as far west as Portugal, does have triangular canvas sails that can be furled or opened according to the state of the wind. On the other hand, it is plausible on the ground of similarity of construction to relate the earliest known windmills – those of Persia – to the Greek (or Norse) water-wheels. It does appear likely, however, that there were two independent lines of evolution, in the Middle East and northern Europe respectively.

The Persian mill cannot be dated earlier than the seventh century, but is possibly related to earlier wind-driven prayer-wheels used in Tibet and elsewhere in Asia. Its distinguishing feature is the vertical shaft, which lent itself, because no gearing was involved, to the grinding of corn. However, the earliest use was probably for the pumping of water, particularly in Seistan. Such mills commonly carried six to 12 sails, and were enclosed in shield walls that directed the wind to one side and were, apparently, fitted with shutters to control its force. Too high a speed would not only risk overheating the bearings but could also scorch the grain. These sails were mounted below the grindstones, an analogy with the Greek watermill which is perhaps significant. However, if we go further east, to China, we see from the 13th century a variant that again suggests a derivation from sailing ships. There the sails were not attached directly to the vertical shaft but to the six struts – or masts – of an hexagonal drum mounted on it: the sails bear a close resemblance to the traditional sails of Chinese junks. These sails empty and fill as the wheel rotates, and the shield wall of the Persian type is dispensed with. So far as the evidence goes, these windmills were built largely in

A windmill of a type found as far apart as Turkey, Crete and Portugal. The sails are furled according to the wind and the cap can be turned to face it, using a long pole. This Roman example is from Bodrum, Turkey.

Watermills and windmills were common features in the landscape, and from the 15th century were popular subjects for painters: this 17th-century example is by Jacob van Ruisdael. Inset is an example of wooden gearing from the mill shown on the facing page.

coastal areas, which was perhaps due to the availability of skilled sailmakers.

The consensus is that Persia was the source of the first windmills and that knowledge of them travelled east to China where they were built to a rather different design. This knowledge also travelled westward into Spain, through the influence of Islam, and thence, apparently, to the sugar plantations of the West Indies. There, vertical-axis windmills identical in design with those of Seistan were in use in the 17th century.

Long before this, however, there had appeared in northern Europe a different kind of windmill, which developed into an altogether more sophisticated device. Its distinguishing mark is that the shaft is horizontal – suggesting a derivation from the Vitruvian water-wheel – and carries four sails. This brings about a distinct improvement in efficiency over the Persian type: the wind acts continuously on the whole of the sail area instead of on just part of it. However, this advantage is obtained only if the sails are directed into the wind, and two devices were used to make this possible. In the wooden post-mill, dating from the late 12th century, the whole body of the mill is mounted on a central post and can be turned by pushing against a long horizontal arm or tailpole. Some two centuries later the tower-mill was introduced: in this the body of the mill, often of masonry, was fixed and the cap alone, carrying the sails, was turned to face the wind. In both types the mill machinery was located below the sails. In early mills, the shaft was horizontal but this imposed a great strain on the bearing immediately behind the sails. This problem was alleviated by inclining the shaft slightly upward, so that the main thrust was on a heavy bearing at its lower end.

Up to 1430 these European windmills were used only for grinding corn, but subsequently they became a power source widely used for various other purposes, notably land drainage: with their help, large areas of land were drained in the Low Countries, the English fenlands and elsewhere. We do not know the power of such windmills, but comparison with comparable Dutch ones of the 18th century suggests an output of about 5 horsepower.

The water-wheel

From Roman to medieval times water-wheels changed little in design and construction: the big development was in the extent to which they were used and the purposes to which they were applied. Originally, like windmills, they were used almost exclusively for grinding corn. But later they, too, were adopted as a simple prime mover for such diverse purposes as driving mine-pumps, sawing wood, forging metal, fulling cloth and operating bellows. We do not know the full extent of their use during this period, but the Domesday Book of 1086 records no less than 5624 watermills in England south of the Trent river alone. Their output was very variable, but an average value of 2 horsepower can reasonably be assumed. Their economic importance is indicated by the amount of legislation concerning the control of water on which they depended.

By medieval times, floating mills like those constructed by Belisarius at the siege of Rome in AD 537 were widely used in Europe. There, they were commonly moored downstream

early power production 64–70 **early sailing ships** 50–53 **engines and motors** 158–170 **Greek watermill** 68 **windmills** 166, 231, 277–278

This late 17th-century illustration of a forge shows several early devices for the application of mechanical power, here provided by an undershot water-wheel. The rotary motion of the wheel is translated by a crank and connecting rod into reciprocal motion to power a set of levers that work the bellows of the furnace. The wheel also turns a cam that raises and drops the hammer.

from the arch of a bridge: often they were so numerous as to obstruct the navigation channel, as in 13th-century Paris. Floating mills were also found in towns on the Tigris and Euphrates.

At appropriate sites, tidemills were built. In these, the water of a high tide was trapped in a reservoir and then allowed to escape, as the tide fell, through a narrow channel containing a water-wheel. Although their overall contribution was small until the 18th century, they were built before 1500 at places as far apart as Dover, Bayonne, Venice and Basra.

Machinery

The advent of water and wind power on a considerable scale made it possible to mechanize a number of basic processes formerly wholly dependent on manpower or draught animals. Limitations were set by the fact that although the prime movers were massive, their power output was low by modern standards. As we have seen, a windmill or water-wheel rarely produced more than about 5 horsepower. The machines they drove were equally cumbersome, constructed largely of wood, though often reinforced with iron. Frictional losses in crude bearings and gearing were very heavy. Some windmills incorporated large roller bearings in the early 18th century, but true ball bearings were not introduced until 1772.

The most common application was in the grinding of corn, a laborious task that had to be done day in and day out to feed a growing population. The next most common was probably the pumping of water – for land reclamation in northern Europe, for irrigation in the Middle East and to pump out mines in many parts of the world. Mining and metallurgy involved other tasks for which mechanical power was welcome – crushing ore with trip-hammers, working bellows for furnaces, raising ore from the bottoms of shafts, boring out tree-trunks to make water-pipes and so on. For many of these operations a reciprocating motion was needed, whereas the windmill and water-wheel provide continuous rotary action. One way of achieving this was provided by the crank and connecting rod, first clearly depicted in a German manuscript of about 1430. Another device was the cam, a projection from an axle which could raise and release a hammer or rod.

These machines, designed for heavy work, were massive and clumsy. By contrast, various kinds of mechanical instrument – especially those for astronomical observations and navigation at sea – were being made with a high degree of precision, and were often so elegantly ornamented that they are today collectors' pieces for this reason alone. By the 15th century the most important centre for this manufacture was the Bavarian city of Nuremberg, already noted for skilled metal-work. One reason was its strategic position astride the important trade route from Italy to the Low Countries and thence to Britain. Not until the 19th century, however, was a comparable degree of precision feasible, or indeed necessary, for the construction of heavy machinery.

Mechanical clocks

In China, in the second century, the astronomer Zhang Heng built a celestial globe whose movement was regulated by a clepsydra – or water clock – and corresponded with that of the natural celestial sphere. To this, in the eighth century, Yi Xing and Liang Lingzan added a mechanical clock, but the most famous development in this line was the water-driven armillary sphere, showing the movement of the heavenly bodies, built in 1088 under the direction of Su Sung, an imperial minister.

It fell into disuse in the middle of the 14th century, but such accurate contemporary records survive that a replica, on a 1:5 scale, was built in the 1950s and is now exhibited in the Museum of Chinese History in Beijing. The original clock tower was 12 metres (40 feet) high, and contained a 3-metre (10-foot) wheel fitted with a series of scoops which served as clepsydras. As each scoop became full the wheel turned one step. This movement was transmitted to the mechanism of the armillary sphere through a system of gearing.

The first mechanical clocks in Europe worked on a very simple principle. A weight was suspended from a cord wrapped many times round a driving shaft. As the weight descended, the shaft turned and the movement was transmitted to the hands – or, in the case of many early examples,

Right: An early mechanical clock with verge-and-foliot control (German, 16th century). The swing of the notched horizontal vanes (foliots) was regulated by adjusting the position of a pair of small weights. Even so, time-keeping was poor, rarely better than an hour per day: for this reason, a single hand was adequate.

Bottom right: A lantern clock of 1688 is regulated by a pendulum acting through a simple escapement. The plate on the back controls the striking.

Below: An escapement regulates a clock by releasing the energy of a spring or weight at a regular rate. The anchor escapement of a pendulum clock has an anchor that moves to and fro as the pendulum swings (A), releasing one tooth of the escape wheel with every swing (B). The pallets of the anchor alternately engage with the teeth of the escape wheel to check its motion and to give the pendulum a slight push that maintains the swing (C). The escape wheel then turns the hands.

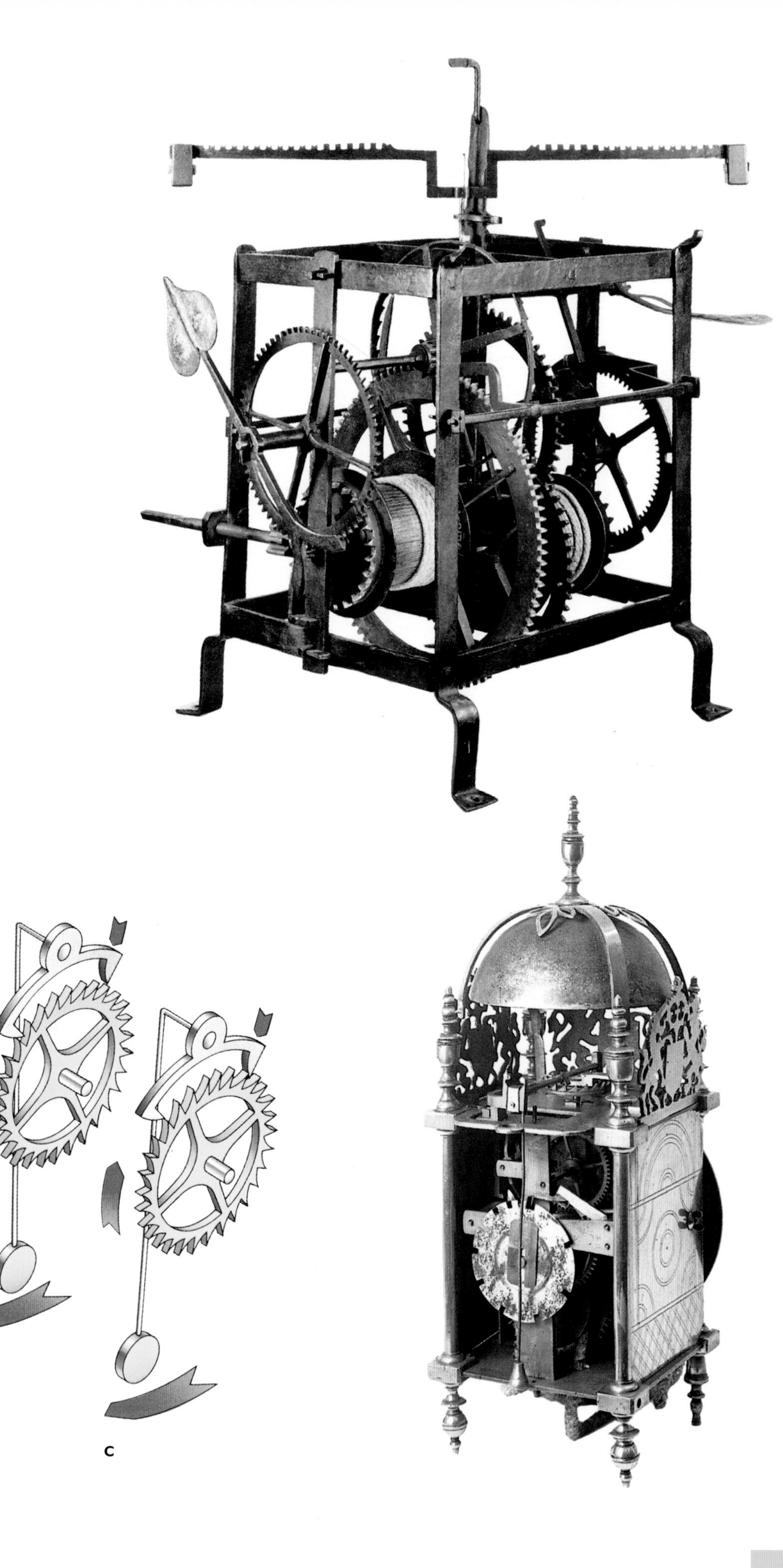

An astronomical clock showing the phase of the moon was installed in the royal palace at Hampton Court, London, in 1540. In 1649, Cromwell had it converted from verge-and-foliot to pendulum control, and added a handsome new dial (top) showing the time, the date and the position of the sun in the zodiac. Another famous clock is de' Dondi's elaborate 14th-century astronomical clock, shown here in a modern reconstruction (above).

the single hour-hand. The problem was to regulate the movement so that the hands rotated at a fixed rate.

Such clocks appeared around 1300, with an escapement – as the regulating mechanism was called – consisting of a pair of oscillating vanes (foliots) mounted on a vertical spindle (verge) carrying a protruding pallet which engaged with the teeth of a crown wheel. Some regulation of the rate of oscillation of the foliots was possible through a series of sliding weights on each arm; nevertheless time-keeping was poor. One of the earliest surviving examples of such a clock is that from Salisbury Cathedral: this dates from 1386, but it does not retain its original escapement.

Even before this, however, some very elaborate clocks were built, of which only descriptions survive. The most famous is the astronomical clock of Giovanni de' Dondi of Padua, built between 1348 and 1362. It showed the time in hours as well as the motion of the sun, moon and five planets, and also included a perpetual calendar.

The original clock was probably destroyed in a fire during the Peninsular War at the Convent of San Yste in Spain, where it had been taken by Charles V in the 16th century. However, de' Dondi, like Su Sung, left such a precise description of his masterpiece that several replicas have been made, one of which is in the Smithsonian Institute in Washington. This was a weight-driven clock – spring-driven mechanisms did not appear until the middle of the 15th century. They were first used in table clocks, but pocket watches had appeared by the end of the century.

Spring mechanisms presented the problem that its driving force grows weaker as the spring uncoils. The first satisfactory solution was the fusee, a grooved conical drum around which a cord or fine chain is wound. As the cord unwinds, under tension from the driving spring, it engages with an increasingly large radius of the drum and so exerts a greater turning force on it. The fusee was invented in the first half of the 16th century.

Hitherto day and night had been divided into 12 hours each. Daytime hours had therefore differed from night hours, and both varied with the seasons. But now the clocks that appeared on public buildings and in the homes of the well-to-do measured time in 'equinoctial' hours, which were of equal length.

Typically, the verge-and-foliot clock was accurate to no better than an hour a day, so that it had to be corrected frequently with the aid of a sundial, which therefore became more, rather than less, common with the spread of mechanical timepieces.

Not until the 17th century did Christiaan Huygens invent two devices that revolutionized time-keeping. These were the pendulum escapement (1657) for weight-driven clocks – first suggested by Galileo in 1581 – and the balance spring (1675) for spring-driven ones. Only then did it become worthwhile to put minute hands on timepieces.

CHAPTER THIRTEEN

The Birth of the Chemical Industry

For the most part, we have so far been concerned with processes involving only physical changes, such as shaping metals by casting and forging, working wood with hand tools, building in brick and stone. In all these the basic material – whether it be metal, wood, brick or stone – preserves its identity but changes its form. But in some processes the nature of the starting material can be radically and irreversibly changed, as when ores are smelted with charcoal to extract metals from them or when glass is made by fusing a mixture of soda, lime and sand. Such changes are chemical rather than physical, and although the underlying principles did not begin to be understood until the 18th century, many chemical processes, often quite elaborate ones, were carried out on an empirical basis from a very early date. This is a useful reminder that technology is not simply a matter of applied science. There was no revolution in man's use of chemical processes comparable with that in his use of power and machinery, and thus developments from ancient to medieval times can be treated together.

There is a good deal of truth in the assertion that chemistry, and thus the chemical industry, began in the kitchen. There, the focal point was the hearth or small oven, only one step removed from the industrial furnace. There, too, a variety of simple preparative processes were conducted as a normal part of daily life – pounding of grain and other foods with a pestle and mortar; boiling in pots and roasting in pans; straining to separate solids and liquids; fermenting sugar-rich fruit juices to make alcoholic beverages.

Salt and soda

Probably the earliest chemical to find general use was salt. Not only does it make food more appetizing when used as a condiment, but a small intake is necessary to preserve life, especially in hot climates where much water is lost in sweat: wild animals will travel long distances in search of a salt lick. Most foods contain a little salt, but if a salty flavour is desired it must be deliberately added. In many parts of the world this can be done only if salt is brought from a distance, for although it occurs very widely it is by no means a universal commodity. For this reason, salt is one of the most ancient of all articles of trade.

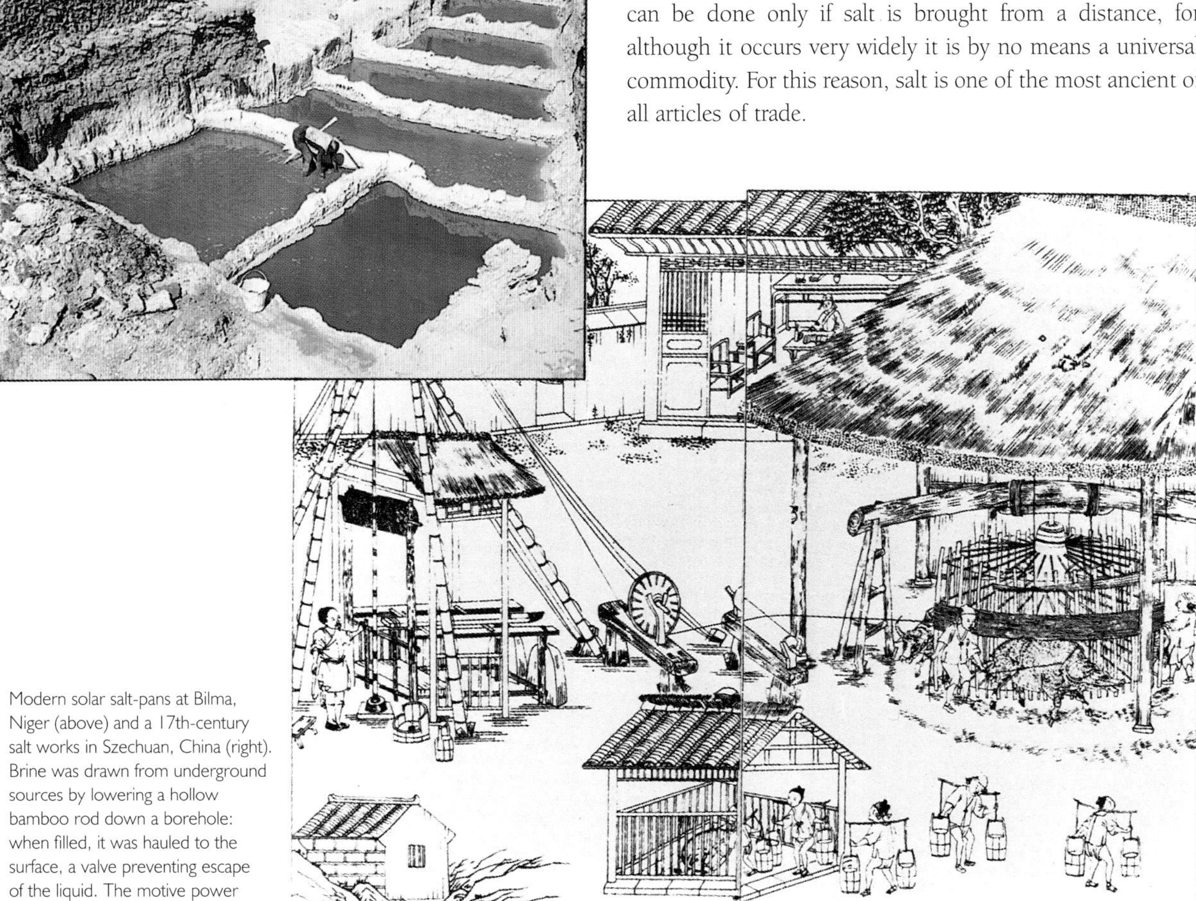

Modern solar salt-pans at Bilma, Niger (above) and a 17th-century salt works in Szechuan, China (right). Brine was drawn from underground sources by lowering a hollow bamboo rod down a borehole: when filled, it was hauled to the surface, a valve preventing escape of the liquid. The motive power was the winch on the right.

early food technology 43 **early smelting** 72–75 **glass-making** 41–43 **marine chronometer** 146 **sundial** 68

Ultimately, there is only one significant source of salt, namely the sea, but in fact it reaches the consumer by two quite separate routes. On the one hand it can be obtained, in coastal areas, by solar evaporation of sea-water in shallow ponds: naturally, this process is most productive in hot climates, such as that of the Mediterranean, but it is also used as far north as Britain. The salt content of sea-water varies enormously, from around 1 per cent in the Baltic to as much as 9 per cent in the Dead Sea. On the other hand, salt can be obtained from the enormous deposits that occur in many parts of the world, including Spain, Siberia and North America: essentially, these are the dried up remains of ancient seas.

These huge deposits were worked in two ways. In the first, the solid salt was simply mined like coal or any other mineral. In the other, water was pumped in and the salt was washed out as brine, which was then evaporated either in pans over a fire or by solar radiation in shallow ponds, as with sea-water. An exceptionally interesting variant was worked in China, where – as in other parts of the world – salt occurs underground naturally in the form of brine. These brine wells were exploited from at least the fifth century BC and by AD 300 wells over 100 metres (330 feet) deep were being sunk. These brine deposits were associated with accumulations of natural gas, which was tapped by sinking shafts alongside the wells: from about AD 200 gas, conducted through bamboo pipes, was used as fuel to evaporate the brine.

Most solar salt was very pure, but mined salt was often contaminated with clay and other impurities. It would, therefore, eventually degenerate into an earthy tasteless residue, perhaps the explanation of the biblical reference to salt that has lost its savour.

Salt was valued not only for flavouring food but also to preserve it. From the second millennium BC the Egyptians preserved great quantities of fish by pickling it in brine and this practice became universal throughout the ancient world. In the north, the abundant herring was the favoured species, and it was on salt-herring that the fortunes of the Hanseatic League – the developers of the sea-going cog – largely depended. Salt was used also as a flesh preservative in a more macabre context: it was widely used by the Egyptians in mummification. Salt was also used by early potters to form a glaze.

Today, the vast multiplicity of chemical substances can be clearly identified, but in the past even so simple a substance as salt was not always easily distinguished from other white crystalline minerals with a similar appearance. In particular, it was often confused with natron, another naturally occurring mineral. This does, in fact, often contain salt but its main constituent is familiar today as ordinary soda (sodium carbonate).

The Egyptians had three main sources of natron – a state monopoly – the most important of which was Wadi Natrun in the Western Desert, from which it takes it name. In altered form this very ancient term is today in universal use. The scientific name for soda in both French and German is *natrium*, from the Latin *nitrum*, and throughout the world the standard abbreviation for sodium, a constituent element, is Na. In Egypt, natron was used not only as a preservative but also in glasses and glazes for pottery and for cleansing textiles. Large quantities were exported to Europe for similar purposes, and even today – when most soda is manufactured from salt – huge quantities of soda are mined at Lake Magadi in Kenya and at a few other places.

The Egyptians attached great importance to the preservation of the dead, and common salt was among the chemicals widely used by embalmers. This picture shows the mummified head of Nefertari, wife of Ahmos I (*circa* 1500 BC).

Another naturally occurring mineral often confused with salt and natron was nitre (saltpetre): chemically this is potassium nitrate. The Akkadians made some use of it in glazing pottery, but it was never of great importance in the ancient world. As we shall see, however, it later became of immense importance as an essential ingredient, along with sulphur and charcoal, of gunpowder.

A third chemical traded from ancient times, and much prized by metal-workers as a flux, was borax. One of the most important sources was Lake Tengri-Nur, high up in the mountains of eastern Tibet, whence, under the name of tincal, it was brought down on the backs of sheep.

Mordants and dyes

Another chemical that has been in continuous use since antiquity is alum. Strictly speaking, this is a double sulphate of aluminium and potash, but in the distant past various natural forms of aluminium sulphate were also called alum. Much prized was the alum of Yemen, a reference to a particularly important source; the Romans obtained it from various volcanic regions, such as Sicily. Like natron and salt, it was an

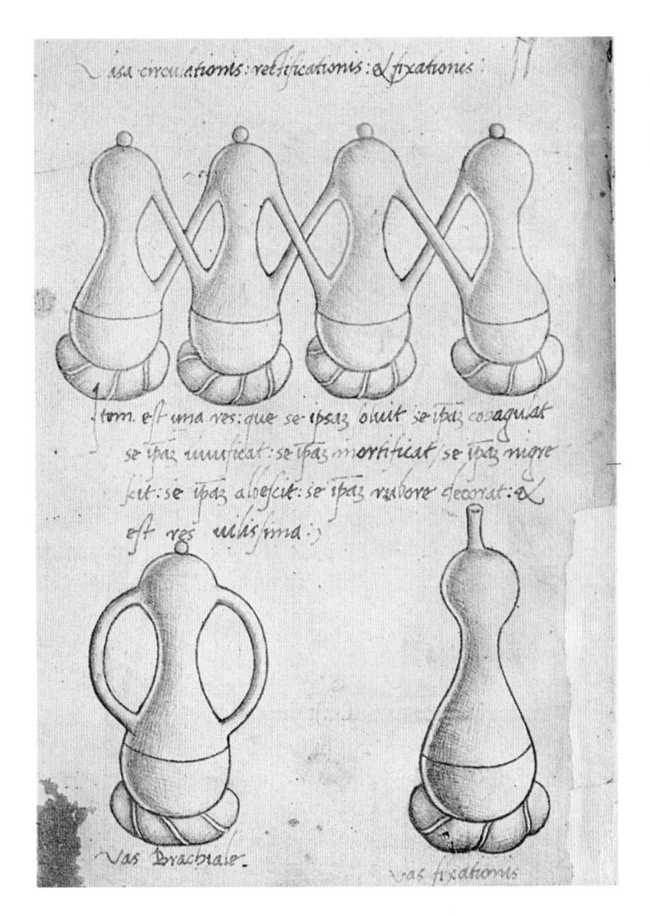

The separation of liquids by distillation was familiar to alchemists in pre-Christian times. For this purpose the traditional apparatus was the alembic, various forms of which are shown above. Vapour produced by boiling the liquids in the base condensed in the neck of the apparatus.

In many parts of the world, dyeing is still carried out by methods little different from those practised in antiquity, and vegetable dyes are still widely used. This picture shows dyed fabrics drying at a modern dyeworks in Lahore, India.

important article of trade, being used mainly in tanning leather and – as a mordant – for rendering dyes fast. Its importance became painfully apparent after the fall of Byzantium in 1453, for one effect of this was to cut Europe off from several major sources of supply. Happily, vast deposits were then found at Tolfa, in papal territory in Italy, where it is still worked.

Until the 19th century, dyers used only natural products, but from a very early date a wide spectrum of colours was available from plant and animal sources. Early Egyptian paintings typically show clothing dyed green, red and yellow. Some of the earliest accounts are too vague for the nature of the dye described to be precisely identified, but the general pattern is clear.

Among vegetable dyes the most important were the blue dyes indigo and woad. They are, in fact, chemically identical but they come from quite different sources. Indigo derives from the indigo plant (*Indigofera tinctoria*), which is of Indian origin: it has been identified in Egyptian fabric of the third millennium BC, but was not widely used there until very much later, in the fourth century BC. At about that time woad (*Isatis tinctoria*), the alternative source of the dye, began to be cultivated in both the Middle East and Egypt. In many places it grows as a weed and in pre-Roman times woad was used in Britain to dye the skin; by that time it seems to have been almost universally used, except in pre-Columbian America and Australasia.

The preparation of the dye-bath was a complex process, varying from place to place. Essentially, however, it was a fermentation process in which the plant was treated with some form of alkali – lime or wood-ash – together with stale urine and fruit juices. Like tanning it was in consequence an antisocial process, and the lot of the dyer was hard: the Papyrus Anastari states that 'the finger of the dyer has the smell of rotten fish; his eyes are red from fatigue'. Nevertheless, the dyers' work was so important that they had a recognized place in society and were organized in guilds at an early date: tradition dates the *Collegium Tinctorum* in Rome to the seventh century BC.

Other important vegetable dyes from the time of classical antiquity were safflower and weld (yellow) together with madder and archil (red). Greens were obtained by dyeing first with weld and then with indigo, and black by treatment with iron salts and extract of galls – essentially the same process as that long used to make ink. A medieval addition was the red dye brasil, extracted from the wood of trees of the *Cesalpinia* genus, which found its way to Europe first from the Far East, especially Ceylon (Sri Lanka) and the East Indies, but later (as its name indicates) from Brazil.

Important dyes were also obtained from animal sources. The most famous was Tyrian purple, the production of which was particularly identified with the Phoenician towns of Tyre and Sidon, though the first known mention of it was at Ugarit. It is secreted in minute amounts by a species of small

Mediterranean shellfish and was correspondingly expensive. In practice, only the very wealthy could afford it: the less affluent had to be content with paler purples achieved by dyeing successively with cheaper red and blue dyes. Some of this red dye – known as kermes – would have been obtained from an insect parasitic on a number of plants, but especially oaks. Much later, after the Spanish Conquest, a similar dye began to be imported into Europe from Central America: this was cochineal, still widely used as a food colorant, obtained from an insect parasitic on cactuses.

Pigments

The fermentation and other processes used in preparing the dye-bath had as their main purpose that of rendering the dye soluble so that it could be absorbed by the fabric. Pigments differ from dyes in being insoluble: they coat the surface of the material to which they are applied without being absorbed by it. For climatic reasons, we are familiar with the well preserved Egyptian paintings, which were done with water-based paints: these were often thickened with gum or egg-white to make them easier to apply. For protection, especially on wood, a final coat of varnish was often applied. This was prepared by dissolving resin in an oil that hardens on exposure to air and light: linseed and hempseed oils were those most commonly used. The varnish known as shellac derived, like kermes and some other red dyes, from a plant parasite.

A dozen or more pigments were in common use; these included: soot or charcoal, white lead, red ochre, vermilion (mercury sulphide), azurite (basic copper carbonate) and ultramarine (lapis lazuli). These raw pigments tend to be rather harsh in colour, and subtler effects were obtained by blending them. White lead was made by the action of vinegar – the strongest acid known to the ancient world – on lead sheeting.

The interdependence of artist and chemist is as old as civilization, for painting required the availability of a range of colours for blending. Arab artists – of whose vivid work this is a fine 15th-century example – devoted themselves to book illustration.

Early artists commonly displayed their skill on walls, either panelled in wood or plastered with gypsum. The surface was carefully smoothed and then often sized with gum. The artists of the Islamic world, however, specialized in producing beautiful illustrations for books on paper or vellum. Painting on canvas with oil-bound paints was a much later innovation: one of the first large paintings of this kind was Botticelli's *Birth of Venus*, dating from about 1485.

Ceramic glazes

The early potters used many chemicals other than salt and soda to produce glazes and colour effects. Among them was malachite, already referred to as an early source of copper, which gave a blue glaze; iron oxide, producing red or black glazes according to the conditions under which the pottery was fired; lead oxide, which produced a fine finish; and certain cobalt-containing minerals, referred to in Assyrian glass-making recipes of the seventh century BC, which give a rich blue colour.

Glazing was important not only for decorative reasons but because it made pottery impervious to water. If stored in unglazed vessels, water is slowly lost by evaporation from the surface – a phenomenon sometimes put to good use, however, to keep liquids cool.

Chemical processes

Chemical operations involved a number of basic processes, many of them designed to separate a desired product from impurities associated with it. One such process is the separation of solids from liquids, a common necessity in the kitchen. For coarse work the equivalent of the colander was used, but for finer separations the mixtures were strained through cloth. Vitruvius recommended that drinking water should be purified by filtration through beds of sand, a method still widely used by water engineers today; a porous mineral known as tufa was also used. An allied problem was the separation of oil from water: this arose as early as Assyrian times from the practice of extracting vegetable oils – such as linseed and sesame – by crushing the seeds and boiling them with water. Much of the oil was just skimmed off from the surface, but the final separation was done in a special vessel fitted with a tap so that the watery residue could be run off from beneath the oil. This kind of separation is still standard practice in modern chemical laboratories and in the chemical industry.

Another basic process, familiar to Alexandrian philosophers and traditionally attributed to the alchemist Mary the Jewess, reputedly the sister of Moses, was that of distillation. This depends on the fact that a liquid can be turned into a vapour by heating, and then condensed back to liquid form on a cold surface: it can thus be used to separate liquids from dissolved solids or volatile liquids from less volatile ones. The typical alembic consisted of a vessel which could be heated, fitted with a head from which the condensed liquid dripped

Alchemy, with its dual promise of worldly wealth by transmuting base elements into gold and of endless life through preparation of an Elixir of Life, had a wide appeal from the earliest times and even today belief in it is not wholly extinct. Its practitioners – here pictured in a painting by the 17th-century Dutch painter Adriaan van Ostade – achieved neither of these chimerical aims but nevertheless mastered many chemical transformations of considerable practical value. Their writings, however, contained a high content of mysticism and symbolism, partly to cloak their ignorance and partly to restrict their knowledge to the alchemical fraternity.

down into collecting vessels. Alcoholic drinks made by fermentation of fruit juices were made in Egypt and elsewhere as early as 4000 BC – though beer was not brewed until some 2000 years later – but the simple alembic was not sufficiently selective to make it possible to isolate pure alcohol: this was not achieved until the 12th century, in Europe.

Some substances can change straight from the solid to the vapour state without liquefying. One such is ice, which explains why snow slowly disappears even without a thaw. The sublimed vapour can be condensed on a cold surface, and this can be another useful means of purification. In the ancient world it was used for purifying sulphur and white arsenic. Another interesting example was in the preparation of the pigment cinnabar (mercury sulphide). This occurs naturally in the red state, but it was also made by grinding mercury and sulphur together: in this case the product is black, but if it is sublimed (purified) it condenses in the desired red form. Such different forms of the same compound are known as allotropes.

Sulphur occurs widely in volcanic regions, but although the Romans were familiar with it they made little practical use of it. Later, however, it became – like nitrate – very important in the manufacture of gunpowder and the various incendiary mixtures which preceded it.

The alchemists

Most of the processes involving chemical change that we have discussed – fermenting sugars, glazing pottery, smelting metal ores, making dyes and so on – were carried out by craftsmen on a purely empirical basis, utilizing knowledge passed down by word of mouth from one generation to another. But in parallel with this, more abstract studies of chemical processes were being carried out by alchemists. Their experiments were directed towards two principal goals – to transmute base metals into gold (by means of the Philosopher's Stone), and to achieve immortality (through the Elixir of Life). In a sense, the two goals were the same, in that both represented the attainment of perfection, in the one case material and in the other spiritual. Their writings were characterized by mysticism and symbolism, partly to obscure their own ignorance and partly to conceal their knowledge and techniques from the outside world. Inevitably, of course, many of their claims were simply fraudulent. Both these goals we now know to be chimerical, but in pursuit of them the alchemist did acquire some sound practical knowledge of chemical manipulations.

The aurigenic alchemy of Alexandria was supposed to have had its origins in the writings of Hermes Trismegistus ('Thoth the Very Great'), an Egyptian high priest who was believed to have flourished shortly after the death of Moses. However, the modern consensus is that this idea represented no more than overenthusiasm for ancient wisdom and that these writings date from no earlier than the second century AD. Greek alchemical texts were carried to Syria by Nestorian Christians in the sixth century and were there eagerly absorbed by Islamic alchemists such as Geber, who for a time was attached to the court of the caliph Harun-al-Rashid. There they were influenced also by notions of physiological alchemy – essentially the attainment of immortality – derived from China.

As we have noted earlier, Arab ideas gradually filtered back to Europe through North Africa, Sicily and Spain, first in Latin and then in the vernacular. Belief in alchemy persisted into the 19th century – and is, indeed, not wholly extinct even today – but chemistry began to emerge as a rational science in the 17th century, largely through the work of Robert Boyle. In his *Sceptical Chymist* of 1662, Boyle put forward a theory on which all modern chemistry is based: that the simplest form of matter is an element and that elements are immutable and cannot be changed into different substances. This replaced the old Aristotelian notion, which had been accepted for nearly 2000 years, that all matter is formed of fire, water, earth and air in varying proportions. His clear-sighted ideas earned for Boyle the title 'father of chemistry'.

early glazes 41 **early smelting** 72–75 **fermentation** 42, 285, 295 **gunpowder** 136–137 **modern paints** 304

CHAPTER FOURTEEN

Architectural Development

From the splendours of classical Rome, as exemplified by the Colosseum in the first century, there was little change in building materials or methods of construction until the mid-18th century. Manpower, whether that of slaves or hired labourers, was virtually the sole source of energy – though to a limited extent heavy work, such as the sawing of timber and stone, was increasingly done with the help of water-wheels or windmills. Simple cranes, pulley systems, wedges and inclined planes were still the common means of moving heavy objects into position. The key craftsmen – carpenter, mason and blacksmith – used tools which would have occasioned no surprise to their predecessors of pre-Christian times. The carpenter relied on adze, hammer, plane and chisel; for drilling, the traditional bow-drill persisted until the 15th century when it slowly gave way to the cranked brace-and-bit drill. For joints, iron nails were sparingly used, wooden dowels being the common device. Simple lathes for wood-turning were based on the same principle as the bow-drill, though treadle-driven lathes appeared in Europe in the 14th century. For the mason, mallet and chisel were still the basic hand tools.

The adaptability of these simple devices is demonstrated by the fact that they sufficed for the execution of buildings in a new range of architectural styles: the Romanesque style which in Europe gave way to the very distinctive Gothic about 1200; the graceful and intricately ornamented palaces and mosques of Islam, with their characteristic pointed arches; and the large ornate palaces and temples of China.

Alongside these buildings constructed of small units of masonry or wood, there persisted an old tradition of hewing buildings from living rock, which we may see as a logical extension of the primitive custom of living in natural caves. Indeed, in some parts of the world, as along the valley of the rivers Loire and Loir in France and in parts of Spain, cave dwellings still serve as homes.

In the ancient world the 'rose-red city' of Petra in Jordan, capital of the Nabataeans, carried the idea of the excavated cave dwelling to new heights of sophistication and beauty. Yet even Petra is outdone, if only in sheer scale, by the immense cave and rock-hewn temples of India and China.

The most famous of these are the chapels and schools that were cut into the cliffs at Ajanta, Hyderabad, to form the accommodation for the Buddhist university that existed there over some 800 years, down to the seventh century AD. The tradition found further development at Ellora, also in Hyderabad, where between the fourth and 12th centuries an immense complex of halls, galleries and cave temples were cut from the rock over a distance of some 2 kilometres (1¼ miles). Free-standing columns, decorative friezes, and superb three-dimensional sculptures combined to give the eerie effect of buildings erected underground. Among these marvels, the eighth-century Kailasantha temple is astonishing. It is no cave but a free-standing, two-storeyed structure almost 100 metres (330 feet) long and 30 metres (100 feet) high. Before work could begin on the temple itself, the masons had to excavate from the rock a pit, up to 50 metres (160 feet) deep in places, with, at its centre, a detached mass of rock from which the temple was, quite literally, sculptured. Other striking examples are to be seen at the Seven Pagodas, near Madras.

China

About the fourth century AD, the Indian Buddhist tradition of cave and rock-cut architecture spread, with the religion, to China. Among the earliest Chinese cave temples are those of the Thousand Buddhas at Tun Huang, begun by pilgrims though completed under official patronage. The Yun Kang

Artists did not paint with a view to informing future historians of technology, but their work can nevertheless be very informative.

This detail from Breughel's *Tower of Babel* translates into a biblical setting many building operations of mid-16th-century Europe.

Low-arch bridges appeared in China at an early date. The multi-span Jewel Belt Bridge (right) was originally built near Suzhou during the T'ang Dynasty (618–907). The three centre arches were later raised for the benefit of shipping. The Chinese were the world's greatest canal builders, and an already considerable system was greatly extended by the Emperor Yang Ti in the seventh century. Part of this was incorporated in the Grand Canal (above) largely completed in the 13th century, when its length exceeded 2000 kilometres (1240 miles).

caves near Ta-t'ung in Shansi province, began a century later, have colossal rock-hewn statues and relief carvings to rival those of India.

For all its revolutions and upheavals, China remains the world's most ancient continuous civilization; yet, surprisingly, there are very few buildings of real antiquity. The majestic stone enclosures and buildings at the centre of Beijing, known as the Temple and Altar of Heaven, date largely from the 15th and 16th centuries. But apart from military and some ritual and ceremonial buildings, for most of its long history the basic building material of China, as of its neighbour Japan, was wood, save for tiled roofs.

Many of the surviving stone buildings of China represent the extension of Buddhist influence during a period, continuing for some four centuries, when rival dynasties and petty states competed for hegemony while barbarian peoples from the north established independent power centres within the Chinese sphere. The traditional Confucian bases of Chinese society were under threat and the very idea of a single Chinese empire must have seemed, at times, a thing of the past.

This period of disintegration was brought to an end when an official from the northern Chou kingdom usurped the throne and asserted his authority throughout China as founder of the Sui Dynasty in 581. In barely 40 years the Sui not only restored the ancient unity of the empire and extended its conquests into central Asia and northwards into Mongolia, but laid the foundations of its continuation and development over the next 1300 years.

The Sui decades were a period of immense public works. Cities were beautified, and the Great Wall renovated and refortified at an immense cost in human life. More important still, communications between the northern and south-eastern regions of China were opened up and strengthened. Between 605 and 617 the Emperor Yang Ti extended the Grand Canal (parts of which had been begun centuries before) to such an extent that he may be regarded as its prime architect. With tree-shaded roads, posting stations and imperial pavilions along its banks, this immense trunk system linked the Yangtse, Huai and Yellow Rivers to give China the world's most extensive inland waterway network until modern times. It called for a large bridge-building programme, and the An-chi or 'Safe Crossing' bridge at Chou-hsien in Hopei province, built by the engineer Li Ch'un, is still in use today. It was a stone structure of great elegance and strength, taking the form of a single flat segmental arch pierced at the abutments on either bank by two spandrel arches which both lightened the design and provided overflow channels in case of floods. It is matched by the similar Lugon Bridge south-west of Beijing and the Jewel Belt Bridge south of Suzhou.

The immense construction programme which it undertook assures the short-lived Sui Dynasty of its place in Chinese history and a notable position in the story of technology, but it also, no doubt, contributed to its downfall.

building in antiquity 54–63 **early power production** 64–70 **Gothic building** 130–132 **Great Wall of China** 23, 87 **Romanesque architecture** 129–130

These tremendous undertakings were accomplished only by corresponding oppression. After a widespread peasant insurrection, the Sui were displaced by the T'ang in 618.

In many parts of the world where timber is plentiful and stone scarce, there has been a long history of constructing substantial and often richly carved wooden buildings. Some few at least survive from medieval times: inevitably the thatched wooden houses of ordinary people have all perished. In Scandinavia and Russia, for example, there was a strong tradition of building large wooden churches. In Britain wood was generally reserved for smaller buildings. One such which survives is Greenstead church in Essex: built of split logs, it dates from about 1015.

In the East, some wooden buildings, such as the little fifth-century shrine of the sun goddess at Ise in Japan, have been continuously rebuilt every generation and so are believed to present, even today, their original appearance. Still in excellent condition, the great mid-11th-century Yingxian pagoda in Mongolia, some 30 metres (100 feet) wide and 70 metres (230 feet) high, enables us to study an example of an early wooden building of great size and complexity. A complex cage of vertical and horizontal timbers, pegged with wooden dowels, surrounds a central space arranged on nine storeys. It tapers slightly so that the weight of the building tends to compress the framework and stabilize it. In structures of this kind a major hazard is the lateral and torsional effects of wind: to protect against the latter each floor consists of a rigid ring firmly bonded to the vertical cage. The effectiveness of the original design is demonstrated by the fact that when some diagonal ties were removed in the course of repairs the pagoda subsequently twisted slightly to the north-west, in accordance with the prevailing wind. In some early pagodas – though not in this particular one – the builders inclined them slightly into the prevailing wind to compensate for its force.

As a precaution against the rotting of the timber, the great pine or cedar columns providing the central structure of the building rested on stone plinths, and paints and varnishes were used as preservatives. Functionally necessary, they soon came to be selected for their colours, for both aesthetic and symbolic reasons. In the finest Chinese buildings the result was a structure integrated both as to form and colour, expressing ritual and symbolic concepts of deep traditional significance.

From the earliest times, timber was also the principal material in the architecture of Japan. But here it was rarely painted or even treated – hence the need to restore the structures generation after generation. The most primitive buildings were timber-frame huts, lashed with ropes. The principle of the frame building, with virtually no structural walls, remained the fundamental one of Japanese building in the classical period. The great castles of the shoguns were often founded on masonry platforms, often many metres high, which were generally sloped inwards, or 'battered', as a precaution against collapse in earthquakes, one of the several techniques adopted in a country plagued by seismic upheavals. The timber frame itself, which was flexible rather than rigid, was another factor favouring durability even under stress, and the use of oiled paper as a skin on room partitions and even exterior walls also helped lessen the dangers during catastrophe and the cost of reconstruction afterwards.

The 15th-century Temple of Heaven, Peking (Beijing), is a famous gem of Chinese architecture and includes several deliberately contrived acoustic structures to produce echoes (below). The 11th-century Friday mosque (right) and 17th-century Imperial mosque (far right) are both at Isfahan, Iran.

Islamic architecture

The characteristic building of the Moslem world, from Afghanistan and Persia in the east to Morocco and Spain in the far west, is the mosque. Nevertheless, during its centuries of ascendancy Islam and its rulers also raised palaces and administrative buildings of great splendour. One characteristic that distinguished this architecture from work in contemporary Europe was the use of water both for hygiene, with water supply and drainage systems far more elaborate than anything known in the Christian West, and for decorative effect. The pleasure gardens of the Alhambra in Granada, perhaps the most famous surviving example, bespeak a mastery of hydraulics then unequalled elsewhere in Europe.

From the seventh to the 10th centuries, a time when Christian Europe was only beginning to emerge from the Dark Ages, Islamic architecture proclaimed the advanced technology and cultural sophistication in this field of the Moslem over the Western Christian world. No doubt, much was learned from the achievements of the Byzantine architects who were beautifying Constantinople and its empire

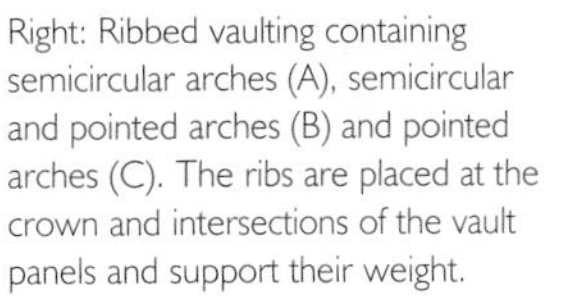

Right: Ribbed vaulting containing semicircular arches (A), semicircular and pointed arches (B) and pointed arches (C). The ribs are placed at the crown and intersections of the vault panels and support their weight.

with majestic churches, but Islam was soon making its own unique contributions to construction and building. From the early eighth century Islamic architects were making use of the pointed arch, low-pitched but embodying a principle which in Europe from the late 11th century was to give rise to the full glories of Gothic architecture.

Like some of their Byzantine contemporaries, Islamic architects based their ambitious and elegant structures on mathematical principles. This geometric approach to structural design, which came only much later in the Western tradition, is proclaimed in the great mosques of Islam in the immensely elaborate patterns of friezes and fretwork grilles which adorn the surface. Similar sophisticated calculation lay behind the design of the new capital of the Abbasid caliphs at Baghdad, laid out on virgin land in the mid-eighth century. For this we must rely on descriptions, for virtually nothing now remains of this immense exercise in city planning. Enclosed by a circuit of walls nearly 2½ kilometres (1½ miles) in diameter, it was entered by four gates at the cardinal points of the compass. It is said that these massive fortifications, with walls of brick bonded with reed matting, took four years to complete. At a time when the splendours of the Roman West were falling into ruins, or reduced to providing havens against the plundering raids of Vikings and even Saracens, Baghdad was a breathtaking achievement.

The Christian tradition

The earliest Christian churches were modelled on the Roman basilica, an aisled hall-like structure with a vaulted apse at the end, used for administrative business and judicial hearings by imperial officials. Many early Christian basilicas were roofed with timber but were based on Roman practices. The principal innovation was to come early in the fifth century when, as for example at Abu Mina church outside Alexandria, architects began to build domes on a new principle.

Impressive though it is, the vast dome of the Pantheon at Rome, built in the first century AD, is structurally unsophisticated in the sense that it crowns a circular building. It was only during the fifth century that architects solved the problems of raising a dome on a square base plan. The first answer was the squinch, a corbelled arch built up in the square corners to round them off. More flexibility and elegance of design was offered with the development of the pendentive arch, consisting of sections of vault rising from piers at each corner of the structure beneath.

The main structure of the magnificent church of Hagia Sophia, completed in Constantinople in 537, used Roman building techniques, though relying more on bricks than on masonry. But the great rectangular structure is surmounted by a technological splendour that no Roman architect could have envisaged. A drum, supported on pendentives and pierced with windows through which sunlight streams, carries the flat dome, spanning some 30 metres (100 feet). The walls beneath were covered by brilliantly coloured mosaics, an art – perhaps originating in the Middle East – at which the Byzantines excelled. It is said that when the Emperor Justinian viewed the completed building he exclaimed, 'Solomon, I have outdone thee!' The church of San Vitale, in

arch constructions 58–60 **Hagia Sophia** 129 **Pantheon** 55 **Roman architecture** 58 **Viking raids** 96, 97

As the Church could attract and extend wealthy patronage, many of the finest examples of Western architecture are to be found in ecclesiastical buildings. The soaring splendour of Gothic cathedrals owes much to the introduction of the flying buttress, as exemplified by Chartres (above left). In Britain, the hammer-beam roof, often richly carved, embellished the interior: the example illustrated here is from St Mary's, Woolpit, Suffolk (above right).

The basilica of St Mark's, Venice (right) illustrates the lavish use of gilded mosaic for decoration. The largest cathedrals involved structural work that would tax the resources of modern contractors.

The elegant octagon on the top of Ely Cathedral (below), completed in 1346, has a span of over 23 metres (75 feet): it incorporates eight oak beams each weighing six tonnes.

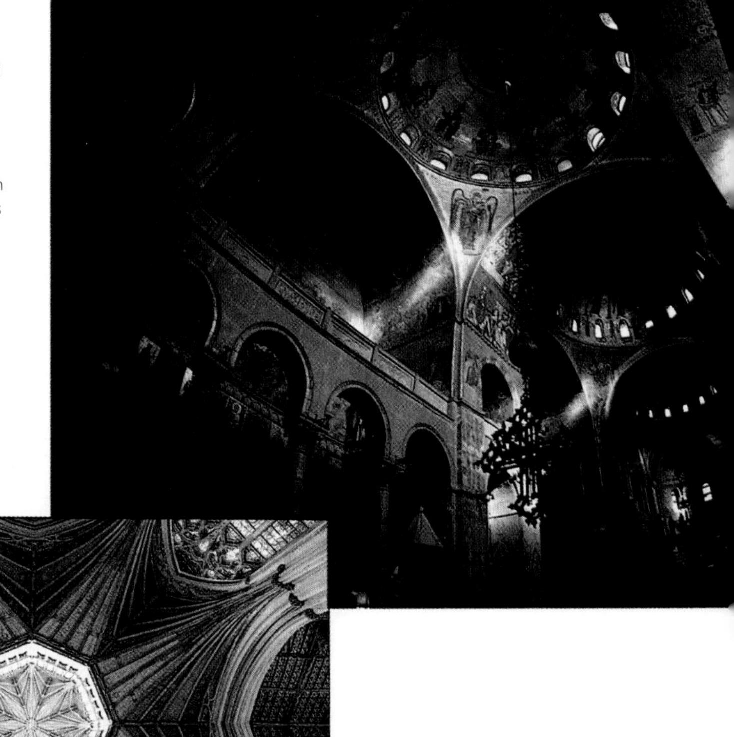

The fact that there are very few basic architectural structures tends to be obscured by the fact that they are so highly ornamented as to mask their true function. This capital pillar in Hagia Sophia, Constantinople (above), is in typical Byzantine style: note also the strengthening iron tie-bars. Windows, originally purely functional to admit light, also became major decorative features, as exemplified by this glorious northern rose window of Notre-Dame, Paris (right).

Ravenna, completed 10 years later – and reflecting Byzantine influence after Justinian's reconquest of Italy – is interesting in that, for lightness, the central dome was built not of the customary solid brick, but of hollow earthenware pots.

Byzantine influence persisted in Italy, so that St Mark's in Venice, built on the shape of a Greek cross and surmounted by five domes, belongs much more to the Eastern than the Western tradition. It was modelled on a now destroyed church at Constantinople and like other Byzantine-type churches is remarkable for its mosaics, also a familiar decoration in many Roman buildings. The making of mosaics was an elaborate and painstaking technique in which patterns and often lifelike pictures were built up from tiny cubes of coloured stone and rock (tesserae); it called on the skills of the stonecutter and the plasterer as well as those of the artist. The peculiar brilliancy of many Byzantine mosaics derives from the use of coloured glass. Still richer effects were achieved by the use of plain glass tesserae backed with gold leaf, which to this day retain their lustre.

The blank, almost box-like structure of Byzantine churches, pierced by few windows at ground level and from the outside enlivened only by their domes, reflect a building tradition suited to countries of hot sun. Within, the sparkling mosaics picked up and reflected the little light admitted by the carefully placed window openings. In the West glowing colour, equally important in church design, was to be achieved with stained glass, a considerable achievement in terms both of chemical and glass-firing technology.

Romanesque architecture in Europe

Byzantine buildings contained certain unusual features because they were constructed in an area prone to earthquakes. In a masonry tradition, the precautions taken were necessarily different from those evolved for the lightweight buildings of Japan. Columns were strengthened with bronze bands, and layers of soft lead were put at their tops and bottoms to absorb shocks. Additionally, cross-bonds of wood or iron were freely used to give additional strength.

As we have noted, brick was a common material in Byzantine building. In western Europe stone was generally preferred. In areas where flint was abundant, small shaped flints set in courses and bonded with mortar were used not merely for decorative effect but even for bearing loads. Because of transport difficulties, where wheeled vehicles had to be used on poor, muddy roads, small blocks were used; when possible, stone was moved by sea or river. This practice is in marked contrast to the use of huge units favoured by the Egyptians, except in their earliest days. As with the Egyptians, however, stone was so finely dressed from the beginning of the 12th century that the joints were almost

buttresses 130 **Egyptian building** 65 **hammer-beam roof** 132 **Japanese building** 126 **stained glass** 131

invisible. To support the heavy superstructures, walls of cathedrals and castles alike were massive, their interiors being generally filled with rubble. Some of the late 11th-century walls from the original cathedral at Winchester, for example, are well over 2 metres (6½ feet) thick.

Being built of brick or stone, and cared for by generations of worshippers, many of these early churches survive, though often they have been substantially reconstructed over the centuries. Some of the finest are those of Spain before the Islamic invasion, such as that of San Pedro de la Nave, with fine stone carving, dating from the seventh century. For the medieval pilgrim, Santiago de Compostela, in the north-west corner of Spain, was a powerful magnet; well-established pilgrim routes converged on it from many points in France, such as Rouen and Paris in the north and Le Puy and Arles in the south. The similarity of architectural style of many of the great churches built along the route provides an interesting example of how ideas were spread by generations of travellers following the same path.

We still know little about the way in which the architects' initial conceptions were translated into the reality of these magnificent buildings. Very few original plans survive, but an interesting exception is that for the building of the Benedictine monastery of St Gall in Switzerland about 820. This shows a whole complex of buildings for the accommodation of monks and pilgrims, on a 5-hectare (12-acre) site, with a typical Roman-style basilica as the central feature. We know rather little, too, of how or in what numbers labour was deployed on medieval buildings, but Caernarvon Castle in Wales (1283) called for 1200 labourers and 400 masons. Many of the latter were itinerant, going where the work called them.

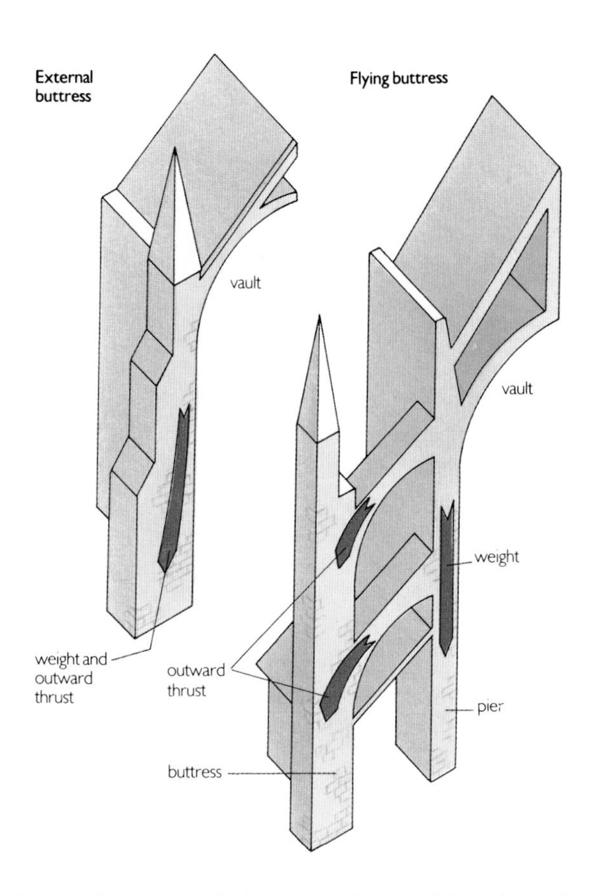

An external buttress was built against a wall in order to support a vault inside a building by taking most of its weight and outward thrust. Flying buttresses were designed to take up the outward thrust of the vaults, enabling the intervening walls to be pierced with windows.

The massive structure of the Tower of London, commenced in 1078 to secure the Norman conquest of Britain, is typical of the Romanesque tradition.

The Gothic revolution

The style of Romanesque buildings can fairly be epitomized as massive. The walls were thick and relieved only by small windows and doors, and the interiors correspondingly dark. These limitations were imposed by the design and building techniques of the day. Around the middle of the 12th century, however, two new concepts combined to give more graceful and much lighter buildings. These were respectively the pointed arch and the rib vault.

As we noted earlier, the Roman arch was semicircular and a direct consequence of this was that its height was determined by its span. Apart from the monotony of proportion that this imposed, there were problems in joining vaultings of different span. These problems were resolved by the introduction of the pointed Gothic arch. Not only could the arch be made higher if desired but cross-vaults could be cleanly joined with a common keystone by varying the angle of the arch. Moreover, the use of the Gothic arch reduced the lateral thrust on the walls, so that these could be of less massive construction. Much more significant in this respect, however, was the development of a design in which a light vault is supported by slender piers, reinforced by external buttresses: what originally had been the thick load-bearing wall could be freely pierced with windows to admit far more light to the interior, and the usable internal area, free from encumbrance, could be increased to a considerable extent.

As is to be expected, these developments did not emerge fully fledged, but developed over a period as designers gained confidence and builders became familiar with new techniques. The earliest example is the Abbey of St Denis, rebuilt in Paris about 1140. Here for the first time we see the possibilities opened up by the lighter and more spacious style of building and, above all, the flood of light let in through large windows glazed with stained glass. As early as the seventh century, glass had begun to replace shutters or fabric in church windows – in 675 the Venerable Bede had sent to France for craftsmen to glaze the windows of his church at Monkswearmouth, Durham. But coloured glass, soon to become the glory of the great Gothic cathedrals, was not used in Europe until the 12th century. Glass-painting began to be practised in the 14th century, with important centres in London and York.

Coloured glass was not the only form of decoration to be widely used: very elaborate carving and sculpture were also prominent features. If St Denis marked the beginning of the Gothic phase, the great cathedral at Chartres, built only some 50 years later, represents its full flowering.

Such buildings, towering upward for 50 metres (150 feet) or more, were triumphs of the mason's skill, but still the necessity for good foundations was not always recognized and some spectacular collapses occurred. One such was at Beauvais in 1284, when a substantial part of the soaring vaulting, completed only 60 years earlier, fell to the ground. Generally speaking, however, the risk was increasingly understood and guarded against. A common device was to drive in massive timber piles: even in wet soils these were surprisingly durable in the absence of air. When buildings for Eton College were erected in Britain in 1453, however, great loads of stone were brought from Yorkshire and laid in mortar for the footings.

Iron continued to be used, as it had been in Greek and Roman times, to strengthen buildings: the masonry of these graceful Gothic edifices often conceals a surprising amount of this metal. It was used, for example to reinforce mullions and to strengthen stonework, especially on slender external features such as pinnacles. The vaulting ribs of the chapter-house at Westminster Abbey have been found to contain a complicated radiating system of iron ties.

Gothic buildings also contained enormous amounts of timber, mainly in the roofs; these were commonly secured with wooden pins, though iron was sometimes used. Such timbers could be very massive: the lantern of Ely Cathedral, erected in 1322, contained eight oak beams 16 metres

The elegance of Gothic building, made feasible by new techniques of building, is typified by the soaring spires of Chartres Cathedral, France (above), which dates from the 12th century, and the vault of the Lady Chapel of Ely Cathedral, England (top). At the time of its building in 1349, the latter had the largest span of any chapel in Britain: it was no more than 45 centimetres (18 inches) higher at its centre than at the sides.

Chartres Cathedral 128 **fortified buildings** 133–135, 138–139 **vaulting** 58

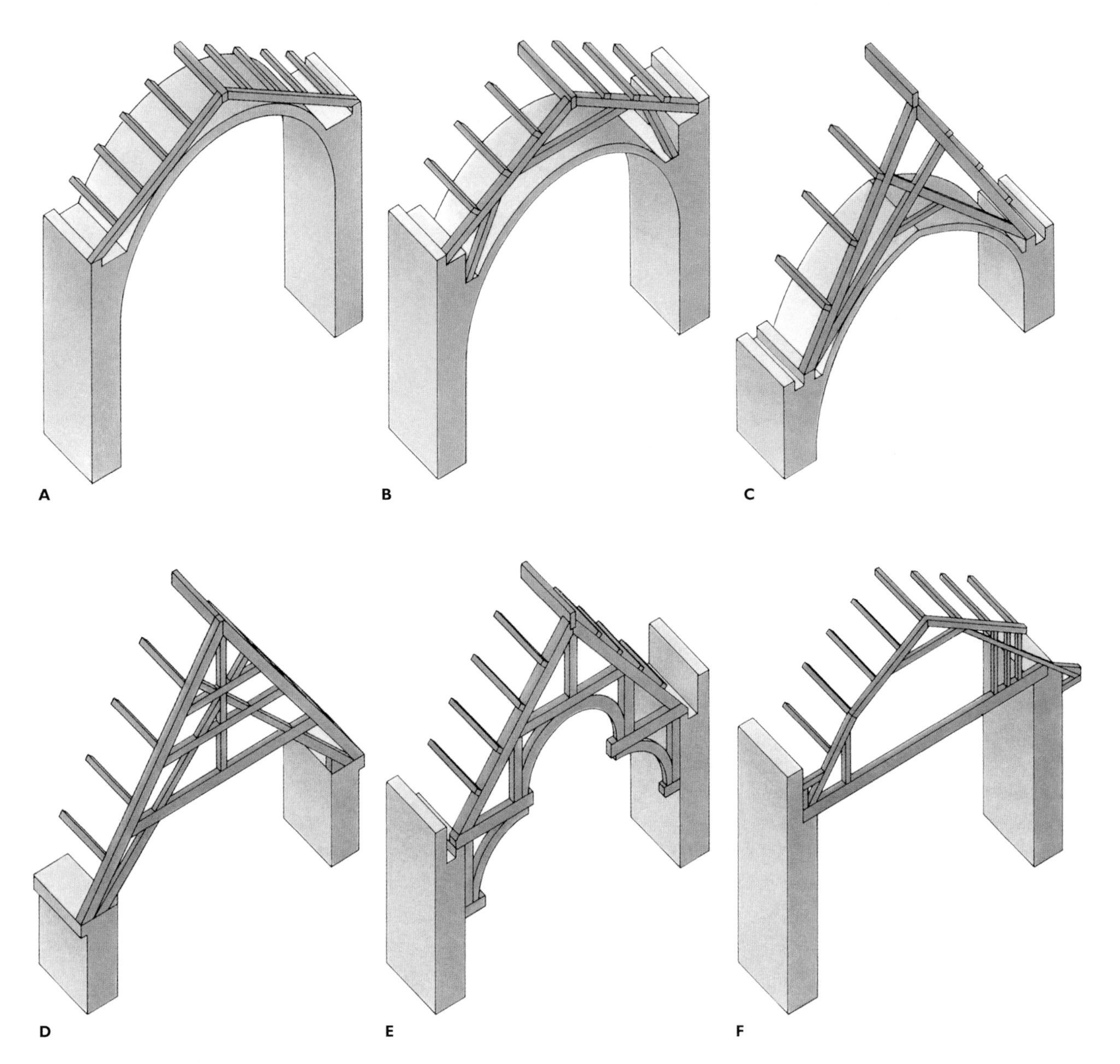

Early Romanesque roofs (A) were not self-supporting but built on a wooden truss resting on top of a vault. In late Romanesque roofs (B), a cross-beam was incorporated so that the truss supported the roof above the vault and the piers of the vault or walls of the building took the weight of the roof. Gothic roofs employed more elaborate trusses with diagonal braces placed over a vault (C) or above a wooden ceiling (D). The late Gothic (English) hammer-beam roof (E) had an elaborate truss of curved struts and short beams that could be left visible and the spaces filled with lattice work. The mansard roof (F), which dates from the 16th century, had a truss with a large space in which an attic room could be built. The roof is shown with a parapet on the left side and eaves on the right.

(53 feet) long and 80 × 50 centimetres (30 × 20 inches) in section. Such a beam would weigh about six tonnes, making a total of nearly 50 tonnes for this woodwork alone. Hauling them into position far above the ground must have been an impressive sight. The pitch of roofs varied considerably: in Germany and other countries prone to heavy falls of snow they were often very steep. To prevent the legs of the rafters spreading under the load they carried – which often included sheets of roofing lead ½ centimetre (¼ inch) thick – some form of cross-tie was necessary. This was most effective if it joined the feet at their lowest point, where they rested on the walls. In this arrangement the ties are obtrusive, however, and various devices, such as diagonal ties, were used as a palliative. One, peculiar to England, was the hammer-beam truss, projecting inward from the foot of a rafter and supporting on a bracket which in turn supported the main roof timbers.

Finally, we must mention a form of medieval construction, favoured in northern Europe, which represented a compromise between wood and masonry. The timber-framed house consisted, as its name implies, of a framework of heavy timber: the spaces between were filled with brick, masonry, or – in the cheapest construction – wattle and daub. Few such houses now exist, having perished through normal processes of decay or by fire. The latter was a particularly serious hazard with all timber structures: in 1666 more than 13,000 timber-framed houses were destroyed in the Great Fire of London.

CHAPTER FIFTEEN

The Arts of War

In military, as in other branches of technology, the Dark Ages were a period of relative stagnation. The walls of Constantinople, built in the fifth century by the Emperor Theodosius II, stood unbreached for a thousand years. The only new siege weapon to appear, about 1100, was the trebuchet, a device for launching stones. Effectively, however, this was little different from the ancient onager – though its recoil was less – except that the pivoted arm slinging the shot was activated by a heavy counterpoise weight, not by torsion. Nevertheless, although there was little in the way of innovation, the pattern of warfare changed radically in Europe. With the collapse of the Roman Empire in the West, the concept of defence based on the control of distant frontiers came to an end, leading to a rash of fortresses built for local tactical, rather than strategic, purposes: such, for example, were the castles built by the Normans for the subjugation of England and later by English kings for the subjugation of Wales. Although these castles were not different in kind from their predecessors – and, indeed, resemble that built at Sinjerli in Syria in the seventh century BC – they did incorporate a number of important modifications.

This period of quiet evolution came to a dramatic end in the 14th century with the introduction of gunpowder. This quite literally revolutionized warfare: in a very short space of time new kinds of weapons had to be made and men trained to use them. Almost overnight, fortifications which had seemed impregnable became highly vulnerable. Military engineers, after a period of strengthening existing fortresses as best they could, responded by introducing new designs.

Pre-gunpowder castles

Although the Franks began to build stone castles around the middle of the ninth century as a defence against Norse invaders, the common European fortress of the time was the so-called motte-and-bailey castle developed in Normandy. Basically, it consisted of a wooden strongpoint erected on a motte, or mound, of earth: most of the material for the construction of this came from the excavation of a surrounding ditch. The area around the motte was the bailey, which, except in times of danger, housed the garrison, domestic staff, cattle, stores and the like. Around this was a second ditch, the material from which formed a rampart, which was surmounted by a palisade. The size of the bailey was such that the whole of it was within bowshot of the motte. As the construction was of wood, such castles could be built very quickly and were thus very suitable as a first step in consolidating the hold of invaders. After their conquest of England in 1066, the Normans built a hundred castles of this kind before the end of the century. On the other hand, they were vulnerable to sudden destruction by fire and slow disintegration by rotting, nor could they long resist the siege weapons of the day.

Once new territory was subdued, more permanent buildings were called for and a stone keep was substituted for the wooden one. This raised building problems, in that the hastily constructed motte was not nearly strong or stable enough to act as a foundation for a mass of masonry. A common solution was to build up the motte in layers of locally available material rammed hard and waterproofed with clay. At the same time, the wooden palisade was replaced by a stone curtain wall. The keep, or donjon, was at first usually round, but the Normans in England favoured square structures, such as the White Tower of the Tower of London, commenced in 1078. This was a massive building, 36 × 33 metres (120 × 110 feet) in area, rising to a height of nearly 30 metres (100 feet): the walls are 4½ metres (15 feet) thick at the base.

Crenellation has already been mentioned as a device whereby bowmen standing on the rampart could fire with

Works of art are often fruitful sources of military history. A 13th-century stained glass window in Canterbury Cathedral (above) commemorates the siege of the city by the Danes in 1011. Another siege – of Dinant – is depicted in the Bayeux Tapestry (below).

minimum exposure of their bodies. This protection was useless, however, if the enemy had reached the foot of the walls, when the bowman could effectively fire only by leaning outwards, thereby making himself an easy target. To overcome this difficulty it became common practice to build a hoard. This was a timber catwalk built out from the parapet, and protected by more timber in front and at the top. Here bowmen could lurk in safety, removing floorboards in order to fire vertically downwards or drop missiles on the enemy below. Hoards were often put up only when danger threatened and were very vulnerable to fire or a lucky shot from a siege weapon. Gradually, therefore, similar stone structures, known as machicolations, came to be built as permanent features of castles.

Potentially the weakest point in any castle is the main entrance, and elaborate devices were introduced to make it as secure as possible. The first line of defence was the approach, which was made narrow so that men could come near only in small numbers. It was protected by one or two towers, from the embrasure of which arrows could be fired at invaders. In a state of siege, the massive drawbridge over the surrounding ditch was hinged up by a windlass to close the entrance passage, which was further secured with a portcullis that could be lowered and one or two heavy wooden doors. As a further deterrent, the ceiling of the entrance was pierced with what were aptly called 'murder holes', through which the defenders could drop missiles or injurious mixtures on attackers who had got so far. Apart from incendiary mixtures, a variety of other unpleasant substances were used. These included hot sand and powdered quicklime, which would penetrate chinks in armour, and blinding mixtures of soft soap and lime.

Some of these devices – machicolations, murder holes and portcullises – were features of Byzantine forts long before they were used in the West, but how far the ideas were transferred, and how far they were independently invented, is not altogether clear. One obvious means of transfer was the Crusades, the first of which – beginning in 1095 and comprising 100,000 men at the outset – led to the capture of

Jerusalem in 1099. Subsequently, the Crusaders consolidated their hold by building a series of strongholds in the feudal style in the Holy Land from Antioch in the north to Kir Moab in the south. Some were taken over from the Saracens – whose defensive system was no doubt closely studied – and modified, but others were new constructions. The most impressive of those which survive is that of Krak des Chevaliers in Syria, the building of which began about 1110: during the 12th century it had a garrison of 2000 men.

Increasingly the defence of the Holy Land became the responsibility of two powerful religious orders, the Knights Hospitallers and the Knights Templars, especially after the Crusaders' defeat by Saladin at Hattin in 1187. They quickly became powerful in Europe also and in 1178 the Knights Templars were made responsible for strengthening the castle of Ponferrada on the pilgrim route to Santiago de Compostela. Here we have a clear example of a direct technological link between East and West.

In the first flush of success the Crusaders built some simple castles of the Norman motte-and-bailey type, but as the opposition became more organized, and manpower from Europe became less plentiful, more massive strongholds like Krak became necessary. In a naturally inhospitable land,

From the earliest times, siege warfare has been comparable in importance to pitched battles. This imaginative picture of the siege of Jerusalem (1099) from a 15th-century manuscript (above) contains many anachronisms but is nevertheless full of interesting detail. Subsequently, the Crusaders consolidated their hold on the Holy Land by building many massive fortresses of which the great Krak des Chevaliers (1110) is the best known (left).

A wall walk and machicolation from the 13th-century castle at Beersel, Belgium (above). The contemporary castle at Angers (right), built 1223–38, well illustrates the vogue for round towers in France at that time.

which the enemy was ready to devastate in order to cut off all possibility of local supplies, such fortresses had to be designed to stave off long sieges and provisioned accordingly. Krak des Chevaliers, for example, was kept constantly provisioned for five years and nearby Saone had vast subterranean cisterns holding 9 million litres (2 million gallons) of water.

As religious fervour waned and commercial interests increasingly took command, the Crusaders were forced on the defensive and castle after castle was lost to the Saracens. Finally, in 1291, their last stronghold, Acre – which had been taken and retaken in the past – fell after a six weeks' siege. Although one cause of the defeat of the Crusaders was loss of morale and lack of support from home, a major factor was that their adversaries were themselves skilled engineers well versed in the strengths and weaknesses of fortifications. For example, the defences of Cairo – begun in 1087 – equalled anything the West could then construct. The great Bab al-Futuh gateway was defended by two round towers – similar to that of Harlech in Wales (1285) – well equipped with embrasures for archers. An interesting feature of the towers, and one not encountered outside the Arab world, was the inclusion of massive stones set horizontally to give added strength against mining operations.

Greek fire

Although the first use of cannon in warfare can be fairly precisely dated to the early 1320s, the history of gunpowder, and of military pyrotechnics generally, is far longer and still not entirely clear.

In the ancient world, wood was so abundantly used as a structural material, for both buildings and ships, that it would have been surprising if incendiary materials had not been used in warfare at a very early date. There is, indeed, an abundance of references to their use from Assyrian bas-reliefs of the ninth century BC onwards. Blazing mixtures of pitch, sulphur, resin, bitumen and the like were delivered by arrows and firepots and even, on occasion, by fireships, as at the siege of Tyre in 332 BC. Pliny even describes fire-bombs containing quicklime which ignited themselves from the heat generated when they were wetted.

This kind of warfare – inflicting terrible injuries as well as destruction of defences – is particularly identified with what is known as Greek fire. It seems to have first been used by the Byzantines about 671 for the defence of Constantinople against the Arabs. This is plausible, for there were alchemists in the city who had inherited the chemical knowledge of the Hellenistic Alexandrian school; but if this is true, 'Greek fire' is a misnomer. The Byzantines never called themselves Greeks, whom they despised, but Romans.

The nature of Greek fire is a matter for dispute, but the likelihood is that it was a liquid derivative of petroleum, which occurs naturally in surface pools at many places in that part of the world. Its composition is, in any case, of no great consequence – and it was probably variable according to local circumstances – for it is its mode of delivery that is important. The evidence is that a jet of the burning liquid was forced out of a bronze tube by some form of force-pump and literally sprayed over the target. A 10th-century Byzantine manuscript includes a detailed picture of six men in a boat enveloping an opposing boat in a cloud of flame.

The making of Greek fire is said to have been a Byzantine state secret but – like that of the atom bomb in modern times – it was not a secret that could long be preserved. The Saracens certainly penetrated it quite quickly and used it against the Crusaders with great effect. The fact that they

Crusaders 136 **guns** 215–216 **high explosives** 215 **missiles** 322–324, 326 **tanks** 216–217

Although Greek fire is commonly identified with the defence of Byzantium in the seventh century, the use of highly flammable mixtures in warfare – sometimes containing crude petroleum – is certainly of a far earlier date, not only in Europe but also in the East. This well known picture from a 10th-century Byzantine manuscript shows the use of Greek fire in a naval engagement: the blazing liquid is apparently being discharged from some sort of pump.

called their specialist flame-throwing soldiers 'naphtha troops' (*naffatun*) supports the view that the main ingredient of Greek fire was petroleum. The Crusaders brought back to Europe tales of the terrifying powers of this incendiary device, but little use was made of it there. Under the name 'wild fire' it was apparently used by Edward I at Stirling in 1304, but its general adoption was forestalled by the advent of a far more deadly mixture – gunpowder.

Gunpowder

Despite some assertions to the contrary, the weight of the evidence is that Greek fire contained no saltpetre and essentially it is this that distinguishes it from gunpowder. The significance of saltpetre is that it decomposes to yield oxygen when heated, and thus the powder can burn with explosive rapidity and without need of an external air supply.

For the origin of gunpowder we must once again turn our attention to China where, as in the West, there was a long tradition of using incendiary mixtures in warfare. The Chinese were also skilled in practical chemistry, and by the Han Dynasty were quite familiar with the three basic ingredients of gunpowder – charcoal, sulphur and saltpetre. Just when these began to be combined to form gunpowder is not certain, but it may have been as early as the seventh century. At any rate, a book (*Collection of the most important military techniques*) published in 1044 gives a precise formula: six parts saltpetre, three parts sulphur, one part charcoal. (This compares with a typical 4:3:3 mixture for early European gunpowders.) Initially, such powders, mixed with inflammable materials such as hemp and tar, were apparently used as an exceptionally destructive form of Greek fire, and the explosive force generated by igniting them in a closed space was not then exploited. By the 13th century manufacture was on a considerable scale; the *Tearful records of the battle of Qizhou* (1221) states: 'On the same day were produced 7000 gunpowder crossbow arrows, 10,000 gunpowder ordinary arrows, 3000 barbed gunpowder packages and 20,000 ordinary gunpowder packages.'

The first description of gunpowder as an explosive dates from the first quarter of the 12th century, when iron fragmentation bombs were used with devastating effect: 'When it went off, it made a report like sky-rending thunder. An area more than half a *mu* [about 300 square metres or 3200 square feet] was scorched on which men, horses and leather armour were shattered. Even iron coats of mail were riddled.'

Open-ended bamboo tubes packed with gunpowder were used in 1132 as fire-spitting lances, like gigantic Roman candles, and sometimes the powder was mixed with round bullets, thus making a primitive musket. However, these must have been very inefficient, as bamboo could withstand only very moderate pressures without bursting, and there would be little force behind the bullets. Not until bronze or iron barrels were introduced could the full explosive force of gunpowder be used. This had certainly been done by 1332, the date on a cannon in the Museum of Chinese History, Beijing. At about the same time, the Chinese began to use gunpowder as a rocket propellant, and from contemporary accounts it is clear that some of these carried anti-personnel bombs detonated in the air by time fuses.

In the event, the West quickly outstripped China in the development of artillery – and it is just possible that firearms were independently invented in Europe – but there is little doubt that knowledge of gunpowder itself was transmitted from the East. What route it took is another matter, but the probability is that once again it was via the Arab world. There, saltpetre was known as 'China snow' (in Persia as 'China salt') and an Arabic military treatise of about 1290 gives careful directions for its purification. Some versions of

Marcus Graecus's *Liber ignium* (*Book of fires for the burning of enemies*) which can be dated to about 1300 gives recipes for making gunpowder. It is possible, too, that the Arabs had direct experience of early firearms through Mongolian expeditionary forces.

In any event, it is certain that knowledge of gunpowder reached the West around the end of the 13th century: the popular belief that it was invented by Roger Bacon, who died in Oxford in 1294, is without foundation, though he may have been aware of it. Once its properties were known, it was put to use with surprising speed all over Europe. Cannon were used at Metz in 1324, at Florence in 1326, and in Scotland in 1327; a manuscript in Christ Church, Oxford, dated 1326, depicts an early cannon – apparently firing an arrow rather than a round shot. By the middle of the 14th century every European army of consequence included cannon as a matter of course.

Early cannon were forged from metal, with the seam sealed by brazing or welding, but very soon they were being cast from bronze. This presented no great difficulty, for the craft of bell-founding was well established: bells up to a tonne in weight were cast in the 11th century. Iron guns were originally made by welding rods longitudinally round a core which was later removed: the barrel was then bound with iron hoops and mounted on a heavy wooden framework. For firing, the lower part of the barrel was filled with powder through the muzzle and the stone or iron shot or other missile placed on top of it. The powder was ignited through a touch-hole by a red-hot iron or a fuse made from rope impregnated with saltpetre. The mode of construction followed no set rules and inevitably there were many accidents resulting from burst barrels. The first cannon were quite small, weighing around 70 kilograms (150 pounds), but by the end of the 14th century 300 kilograms (650 pounds) was not exceptional. Some much larger weapons were also made: an iron cannon made at Caen in 1375 weighed a tonne and another made two years later for the Duke of Burgundy's artillery train could fire shot weighing 200 kilograms (440 pounds). However, this was only the beginning: in 1453 the Turks were using a monstrous 19-tonne gun in the siege of Constantinople.

Initially, the power of cannon was limited not only by the quality of the powder – reasonably pure saltpetre was hard to come by – but also by the way in which it was prepared. When rammed home hard, the fine powder was so compacted that it burned rather slowly after the original ignition. About 1425, however, this defect was remedied by mixing the powder wet, which was safer, and then passing it through a sieve so that it formed granules, which were then dried. Such 'corned' powder almost doubled the power of artillery.

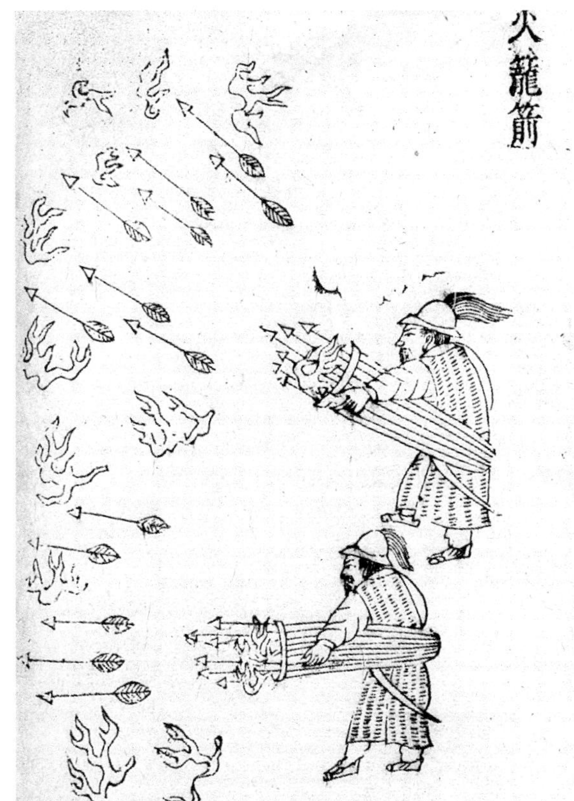

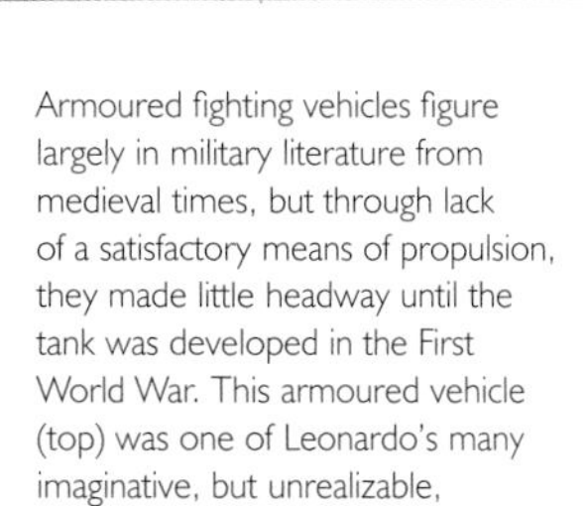

Armoured fighting vehicles figure largely in military literature from medieval times, but through lack of a satisfactory means of propulsion, they made little headway until the tank was developed in the First World War. This armoured vehicle (top) was one of Leonardo's many imaginative, but unrealizable, inventions.

Gunpowder was invented by the Chinese, who probably used it as early as the seventh century. A Sung Dynasty print (centre) shows a *tuhuoqiang* (fire-spitting lance) which first appeared about 1250. It discharged arrows, and perhaps bullets, from a bamboo tube packed with gunpowder and was thus a primitive musket.

The earliest cannon were fired by sighting along the barrel but very soon the gunner adapted various surveying instruments, such as the quadrant, as shown in a 16th-century print (bottom).

artillery 216 **Byzantine Empire** 25 **Ottoman Empire** 90–91 **siege warfare** 85–88, 133

In siege warfare, it was often necessary to lob projectiles high in the air so that they fell almost vertically behind defences. The picture above shows typical 16th-century mortars and the mechanisms and instruments for laying them.

Post-gunpowder castles

Although cannon were generally developed for use in the field, they were heavy and cumbersome and thus difficult to deploy. By the end of the century, however, Charles VIII of France, in his campaign into Italy, had mounted his cannon on carriages in such a way that they could be lowered and elevated, and also traversed. In the event, their strongest impact was in siege warfare, both in attack and defence, and this resulted in radical redesign of fortresses.

From the attacker's point of view, the great merit of cannon was that even massive masonry walls and entrance gates could be steadily pounded to pieces from a safe distance, by direct low-trajectory fire. As early as 1377, at the siege of Ardres, cannon balls went right through the walls. Additionally, short-barrelled mortars could lob explosive projectiles in a high trajectory over castle walls to drop close behind them.

Initially, defenders could do little more than strengthen walls on an *ad hoc* basis, often by banking earth up against them inside. They could reply with cannon of their own, but the location of these was a problem, as was the safe storage of large quantities of powder. One feature of a cannon is that it recoils after firing and it must have sufficient space behind it to allow for this. For this reason large guns could not be mounted on narrow ramparts nor in small turret rooms: in practice, the only suitable sites, where gunners could also be protected, were the tops of towers.

In the longer term, the answer lay in redesigning fortresses, and in doing so two principles were followed. First, walls were made even more massive, to stop penetration. Secondly, flat surfaces were avoided as far as possible: rounded ones improved the chance of a shot being deflected. Rounded towers therefore came back into favour and even the tops of the merlons on the battlements were rounded instead of being cut square.

Contemporary illustrations are not always to be taken at their face value. The illumination (right) from the *Chronik* (1483) of Diebold Schilling – Swiss chronicler and theologian – represents an impossible situation. Nevertheless, it is interesting for showing – within about 50 years of the introduction of cannon – that there were already ideas about troops being transported in armed, and armoured, ships. The representation below of a very early gun is from a 14th-century manuscript.

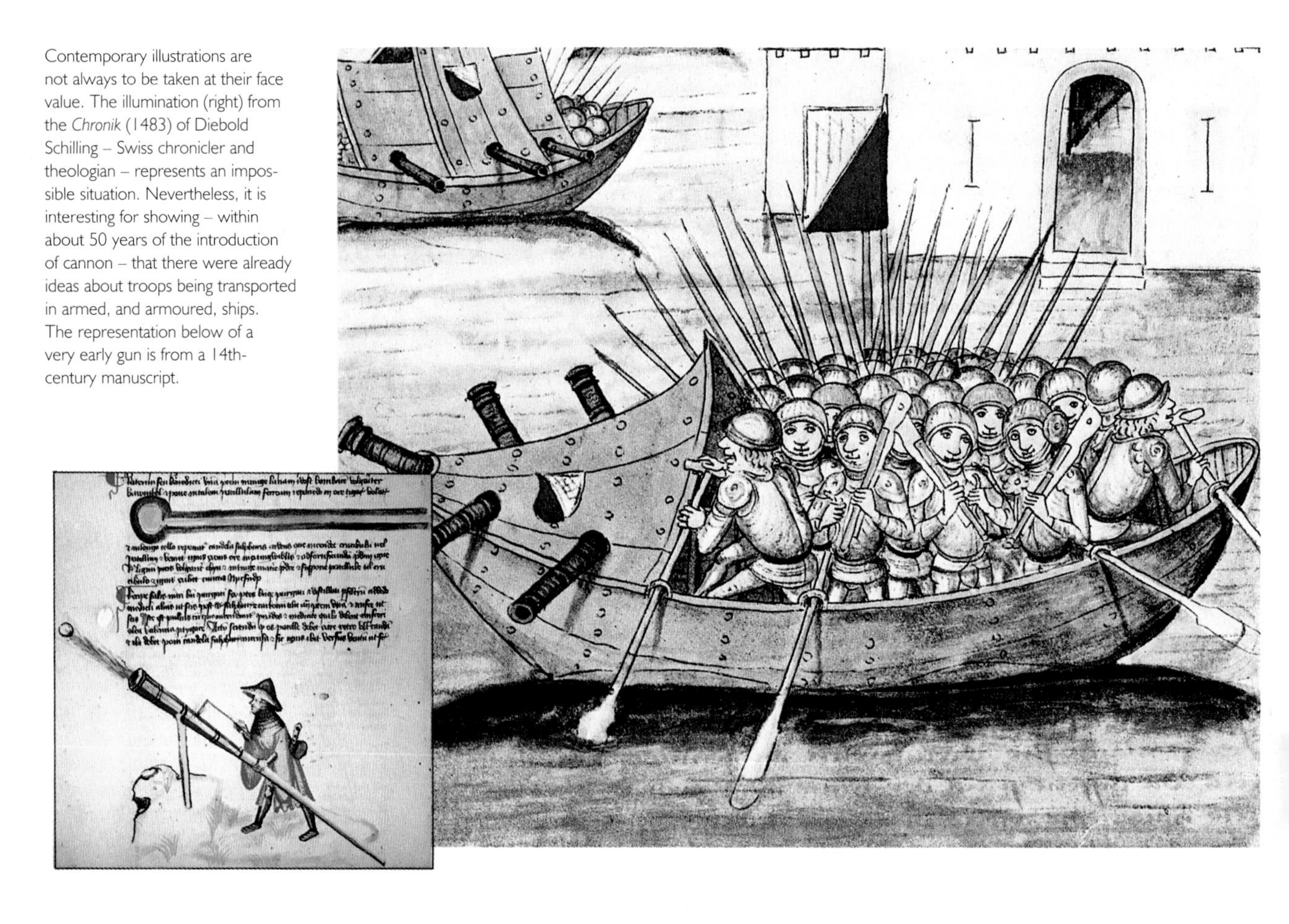

The fortress at Naarden is one of the most perfectly preserved in the world. It achieved its final form after rebuilding by William of Orange in the late 17th century.

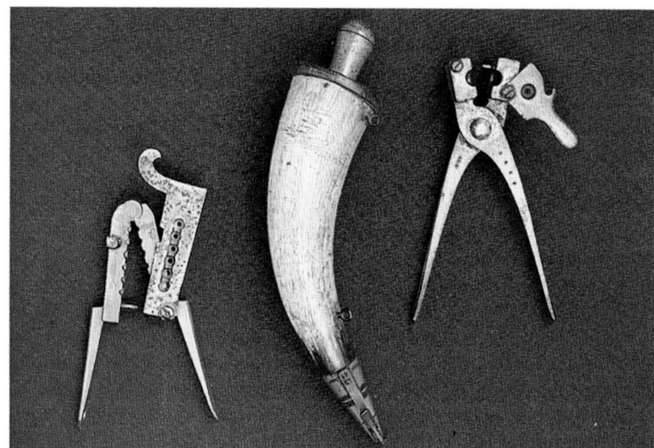

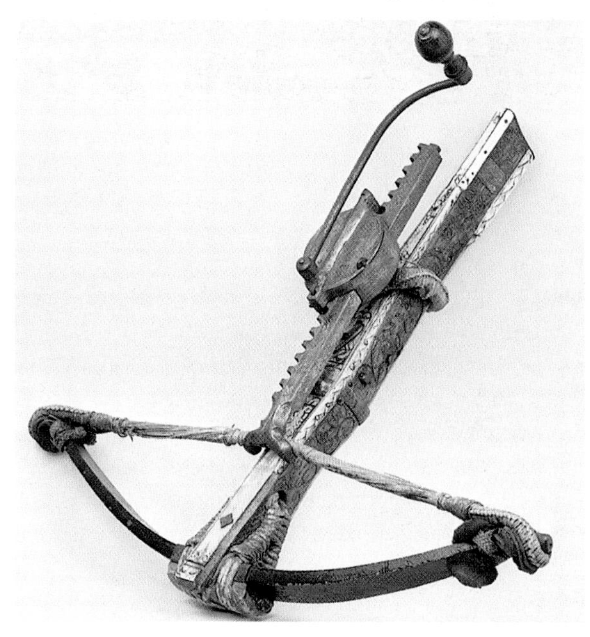

The bullets used in all early muskets were made in hand moulds (top left and right), and the powder was carried in a horn (top centre).

Because it was silent, the crossbow (above) was used by hunters long after it had been abandoned as a military weapon. This bow is late 16th century, from Poland.

The destruction of defensive walls by mining underneath them is a very ancient military device, but it became very much more deadly when the excavated space could be filled with gunpowder and exploded. Much attention was therefore given to designing fortresses so that there was no 'dead ground' near the walls where fire could not be brought to bear on potential sappers. The geometrical and architectural problems of arranging lines of fire to cover all situations were considerable, and engaged the attention of such powerful Renaissance thinkers as Roberto Valturio, Leonardo da Vinci and Albrecht Dürer.

One of the main structural developments was the replacement of the curtain-wall tower by the bastion, a larger structure which could house several cannon and from which the stretch of wall on each side could be protected by covering fire. One of the main exponents of this geometric design was Sebastien Vauban, who in 1678 was appointed Commissary-General of Fortifications in France, where he built many forts of distinctive polygonal shape. For attack he developed the use of parallel trenches giving besieging forces continuous protection as they came within striking distance. He also developed the technique of ricochet fire, and pioneered the use of the socket bayonet, first used at Bayonne about 1650.

Hand-held firearms

Although hand-held firearms were introduced at an early date, they did not have the revolutionary effect on field warfare that cannon had in siege warfare. Early muskets were all muzzle-loading and the charge had to be fired by applying a glowing match to some powder in a priming pan. Thus the rate of fire was very slow and – in bad weather – uncertain. The longbow, by contrast, was a quick-firing weapon and was important up to the end of the 17th century. Where firearms made the most progress was in competition with the crossbow, which also had a slow rate of firing and required an expensive mechanism and skilled users. The crossbow was necessarily largely a defensive weapon, used when the bowman could operate from a protected position. As a military weapon it became obsolete in the 16th century, though it remained popular as a hunting weapon.

Many early firearms were richly ornamented, like this German wheel-lock of 1540.

hand weapons 82–84 **long bows** 83, 133–134 **machine-guns** 216 **naval warfare** 217–219 **rifles** 215–216

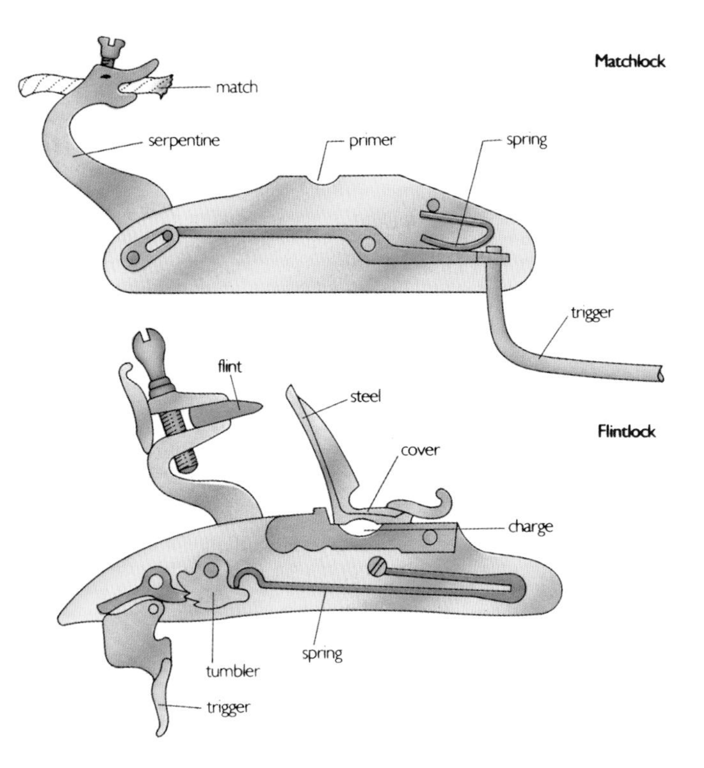

In a matchlock musket, the trigger moved a serpentine that applied a glowing match to the primer. The spring returned both serpentine and trigger. In the flintlock, the trigger released a tumbler that moved the flint against the steel to make a spark. The cover fell back and the spark ignited the charge.

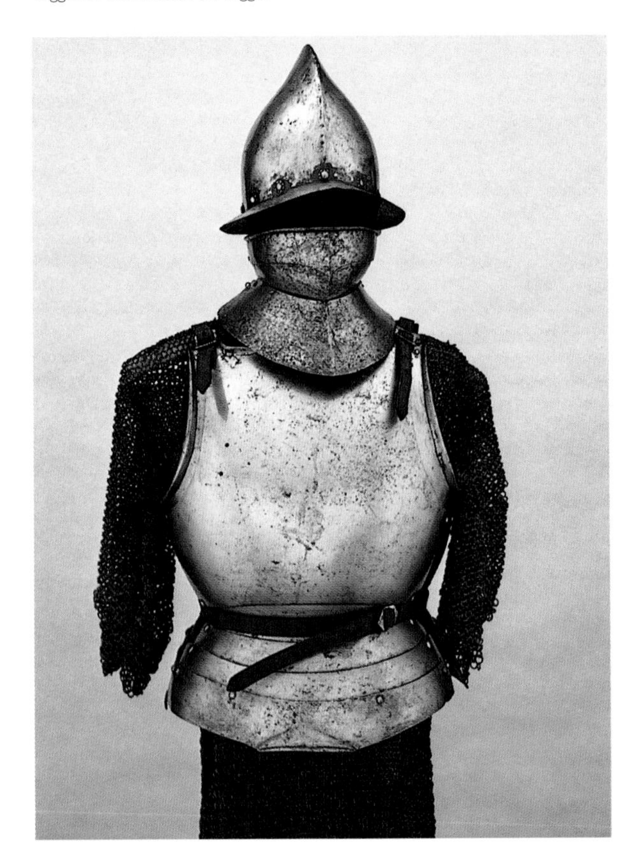

This fine suit of half-armour is most probably of German origin. It was made during the last decade of the 15th century.

That the smooth-bore musket was not very accurate mattered little, for it was used mainly at close quarters. The main developments were in the design of the lock. In the matchlock, the glowing match was applied to the primer when the trigger was pulled, leaving the musketeer with both arms free to steady his weapon. The wheel-lock – probably an Italian invention – appeared early in the 16th century: in this a spring-loaded friction wheel rubbed against a piece of pyrites, generating sparks which fired the priming. It was followed about 1630 by the flintlock, which worked on a similar principle but was more easily cocked. The French army adopted it around 1660 and shortly afterwards it was incorporated in the British army's famous Brown Bess.

Early in the 16th century, possibly earlier, it was realized that accuracy could be much improved if the bullet was made to spin by rifling the musket barrel. Such weapons were favoured by sportsmen but did not come into military use until the 17th century. One deterrent was that the soft lead bullet had to be hammered home to make it a close fit, so loading was even slower than with the smooth-bore musket.

Body armour

Feudal knights were protected with suits of chain mail weighing up to 40 kilograms (90 pounds). This consisted of hundreds of small steel rings welded together to form a flexible sheet; steel plates gave added protection to vulnerable parts. Protection of the head was afforded by an iron helm or a chain-mail covering resembling a balaclava. For jousting and ceremonial occasions a full suit of steel plate, skilfully jointed and burnished, would be worn. With their elaborate ornamentation, these were fine examples of the contemporary metal-workers' skill.

By the 15th century full armour was no longer appropriate, for a suit thick enough to stop a bullet at even moderate range would have been very heavy. There was still much fighting at close quarters with hand weapons and protection in the form of comparatively light plate armour was deemed necessary. By the mid-17th century this was reduced to little more than back and breast plates and a helmet. Today all that remains of the old body armour is the steel helmet.

Naval warfare

In battles at sea two tactics had long been practised: either the enemy vessel could be rammed, as by the Greeks, or she could be boarded, as by the Romans. With the advent of cannon, a third possibility emerged. The enemy could be pounded from a distance and perhaps sunk: during the bombardment a lucky shot with a red-hot ball might explode the magazine. In fact, such methods were not much used before the 16th century, up to which time reliance was placed mainly on small cannon and musket fire to weaken the enemy before boarding. Much later, however, the heaviest warships were to become floating, heavily armoured gun platforms.

PART

3 The Birth of Industrialization

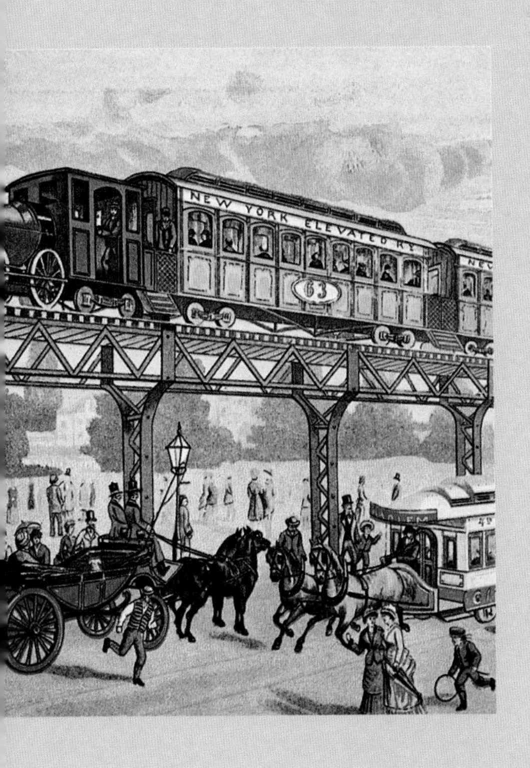

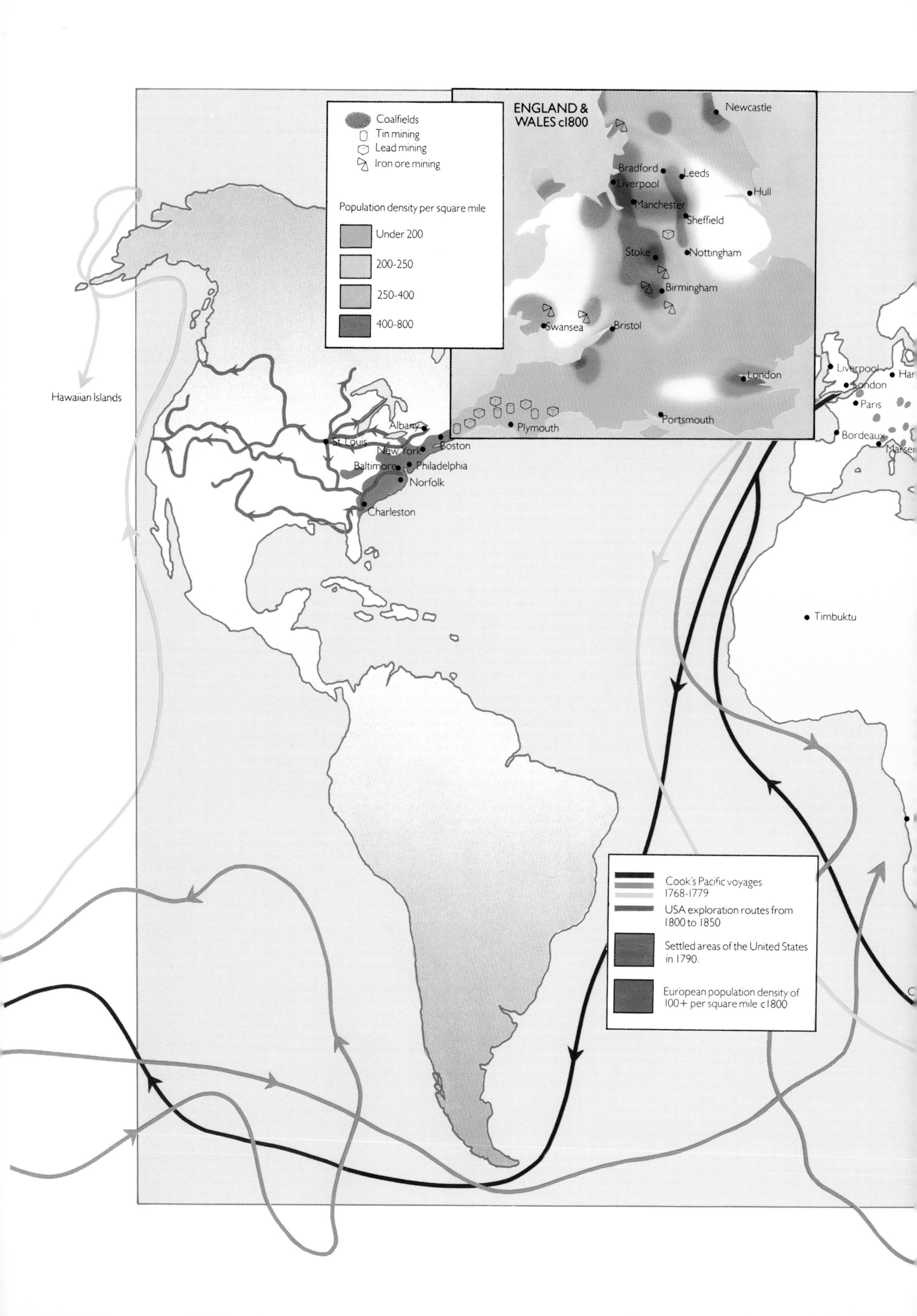

ENGLAND & WALES c1800
Coalfields
Tin mining
Lead mining
Iron ore mining
Population density per square mile
Under 200
200-250
250-400
400-800
Newcastle
Bradford
Leeds
Liverpool
Hull
Manchester
Sheffield
Stoke
Nottingham
Birmingham
Swansea
Bristol
London
Plymouth
Portsmouth
Hawaiian Islands
Albany
St Louis
Boston
New York
Baltimore
Philadelphia
Norfolk
Charleston
Liverpool
London
Paris
Bordeaux
Timbuktu
Cook's Pacific voyages 1768-1779
USA exploration routes from 1800 to 1850
Settled areas of the United States in 1790.
European population density of 100+ per square mile c1800

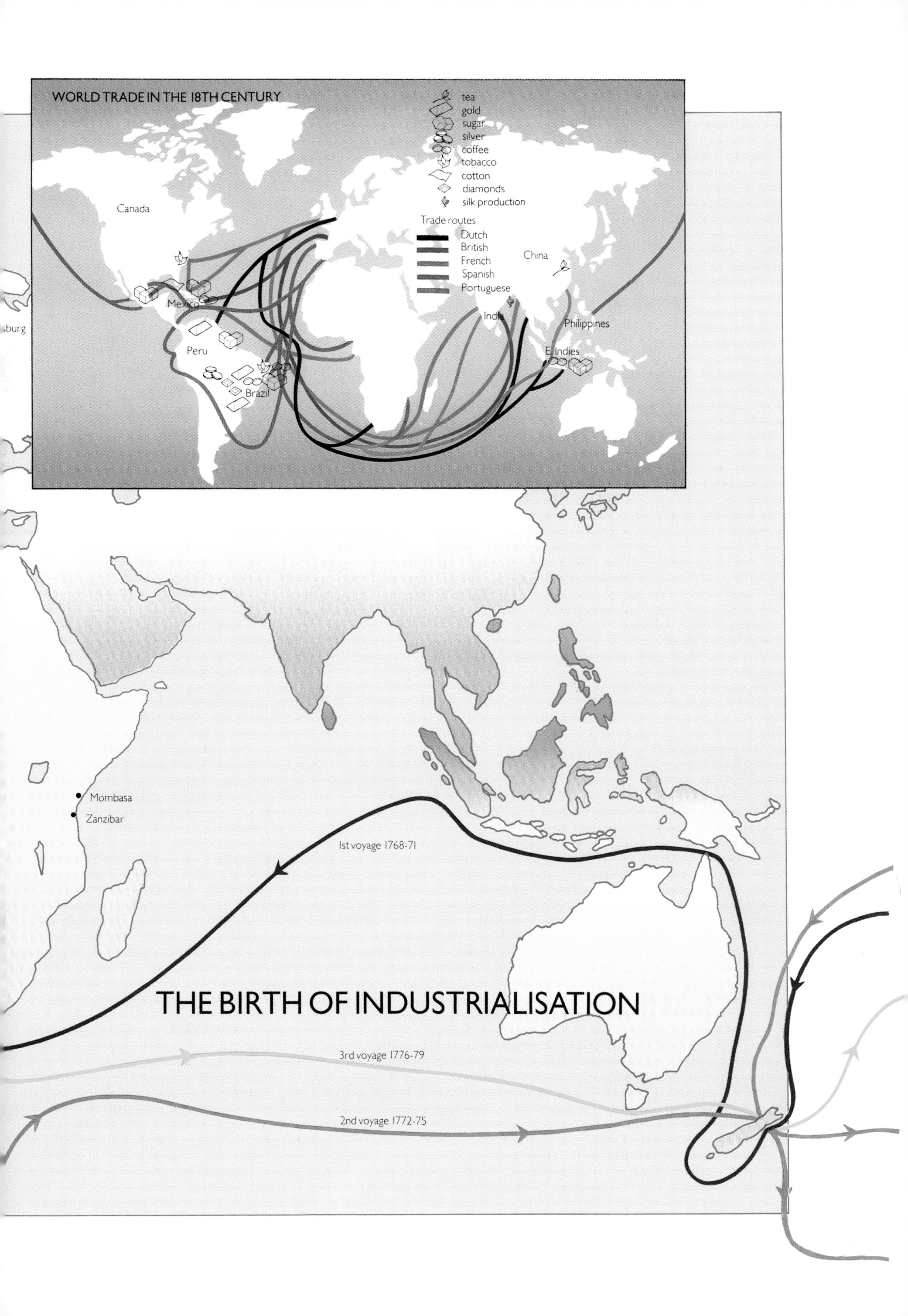

THE BIRTH OF INDUSTRIALISATION

CHAPTER SIXTEEN

The Dawn of the Modern World

The Renaissance, with its centre in the city states of Italy which had grown rich on trade with both East and West, was essentially a cultural movement. Wealthy patrons could encourage writers and painters, architects and sculptors, and craftsmen skilled in working in silver and leather. Yet it was founded on a great increase in commercial, rather than industrial, activity. Although the end-products might be more sophisticated, the basic processes of building and agriculture, weaving and dyeing, shipbuilding and road-making were all carried on in much the same way as previously. Even mining, in which there was a great increase in activity, was hardly receptive to new ideas. Although exploding mines under enemy fortifications was one of the earliest uses of gunpowder, the first recorded use of gunpowder in the European mining industry, in what is now the Czech Republic and Slovakia, was in 1627, three centuries after the first cannon was fired.

Scientific societies

The Renaissance nevertheless fostered the developments that led on to the Industrial Revolution and the science-based industry of the 19th century. The intellectual milieu of Italy in the 16th century was favourable to the conduct of independent scientific inquiry through experiment. The *Academia Secretorium Naturae* was founded in Naples in 1560 to provide a means by which a new breed of natural philosophers could exchange views: membership was conditional on making some discovery in natural science. One can see a tenuous link here – through 'many-tongued Sicily', one of the great centres of translation from the Arabic – with the Brethren of Sincerity, an Islamic society founded about 983 for the promotion of scientific knowledge. The Academia was short-lived, and was followed in 1603 by the *Accademia dei Lincei*, which in turn was followed by the *Accademia del Cimento*, under the patronage of two science-loving Medici brothers, in 1657. These last included among their members Galileo Galilei, his assistant and successor Evangelista Torricelli and the physiologist Giovanni Alfonso Borelli.

The pre-eminence of Italy, however, did not last for long. In northern Europe, the Royal Society – now the oldest surviving scientific society in the world – was founded in London in 1660 and the Paris Academy of Sciences in 1666. Both were dedicated to the promotion of science and useful arts. In Germany, too, similar societies were founded in the first quarter of the 17th century but not until 1700 did a national academy, the *Societas Regia Scientiarum*, finally emerge in Berlin. Elsewhere, the courts of Sweden, Denmark

1500 1550 1600 1650 1700 1750

INVENTION AND DISCOVERY
- Florentine soft-paste porcelain. isochronous swing of pendulum observed
- Accademia dei Lincei
- telescope invented
- pendulum clock
- Royal Society of London
- Paris Academy
- Meissen porcelain
- Kay's flying shuttle
- American Philosophical Society founded
- spinning mule
- water frame

PRIME MOVERS
- von Guericke's atmospheric experiments
- Papin's digester
- Savery's fire engine
- Newcomen's beam engine
- Watt's improved steam engine

TRANSPORT
- first circumnavigation of globe
- Mercator's map projection
- Hakluyt's *Voyages*
- Snell introduces triangulation to cartography
- Greenwich Observatory
- École des Ponts et Chaussées
- Board of longitude established
- iron rail tramways
- marine chronometer
- sextant
- Cugnot's steam locomotive

MINING AND METALS
- Agricola's *De re metallica*
- 'Calamine' brass
- large-scale blister steel manufacture, Nuremberg
- gunpowder used in mining
- iron smelted with coke
- Polhem's gear-cutting machine
- crucible steel
- Cort's pud process

DOMESTIC LIFE
- Harington's water closet
- first regular newspaper in Germany
- smallpox inoculation introduced in England
- lightning conductor
- Argand oil lamp
- food preservatic

AGRICULTURE
- Fitzherbert's *Boke of Husbandrie*
- potato introduced to Europe
- rubber brought to Europe from S. America
- tea first shipped to Europe
- English fenland drained
- new forage crops
- new crop rotation systems
- seed-drill in Europe
- curved mould-boards
- threshing machine
- self-sharpen ploughshare

The 17th century was notable for the foundation of a number of national academies of science, through which the new generation of natural philosophers could exchange views. Outstanding among them were the Royal Society of London (1660) and the Paris Academy of Sciences (1666). This picture shows a group of French scientists meeting in their new accommodation, with the tools of their trade as a background.

and Hungary also encouraged scientific inquiry. Natural philosophy became a reputable and fashionable pursuit throughout Europe.

For the moment, therefore, Europe was the leader of the world in the prosecution of scientific inquiry. But new challengers, though not immediately apparent, were already emerging. The idea of the Royal Society was conceived in Britain in the uncertain years of the Civil War, and at one time it was proposed that it should be established in America, as the Society for Promoting Natural Knowledge, under the Governor of Connecticut, John Winthrop. With the Restoration, this plan was abandoned, and Charles II became the Royal Society's patron, but many Americans, including Winthrop himself and Benjamin Franklin, were among the early Fellows. Nevertheless, as early as 1683 a short-lived Philosophical Society was founded in Boston, and it was followed by other comparable American societies. The American Philosophical Society was founded in 1743.

In the present century, technology has become almost synonymous with applied science and the whole edifice of science is built on foundations laid in the 17th and 18th centuries. Nevertheless, the men who brought about the Industrial Revolution were in the main not scientists – the word itself was not even coined until 1840 – but practical men of affairs with an eye for profit and little or no formal education. Where they were men of position and substance – like the early chemical industrialist Archibald Cochrane, ninth Earl of Dundonald – their education was almost invariably in the classical tradition. Indeed, the intellectual approach can be counter-productive, as the 17th-century philosopher Francis Bacon remarked: '[Great technical discoveries] were more ancient than philosophy and the

1820 1840 1860 1880 1900

natic
ic
ry
first photograph (Niépce)
photographic negative
electromagnetic induction
anaesthesia
electric telegraph
safety lift
celluloid
oil well USA
reinforced concrete
antiseptics
telephone
radio waves
electron
X-rays
radioactivity

principles of thermodynamics.
water turbine
electric motors
compound steam engine
internal combustion engine
accumulator
steam turbine
petrol engine
ac electric motor
diesel engine

n
ge
team
ocomotive
steamship
railway tunnelling shield
screw-driven ship
telegraph
Cayley's glider
Great Eastern
underground railway
bicycle
Suez canal
electric trams and trains
pneumatic tyres
oil tanker
turbine ship
motor car
submarine

um vessels
ventilation
miners' safety lamp
safety fuse
steam hammer
electroplating
turret lathe
Bessemer converter
open-hearth process for steel
pneumatic drills
dynamite
chromium steels
electrochemical aluminium extraction

t
hting
food canning
rubberized fabric
gas cooker
sewing machine
synthetic dyes
typewriter
margarine
phonograph
electric power station
electric cooker
cinema
gramophone

beet
ished
superphosphate fertilizers
harvesting machines
steam ploughing
milking machine
pasteurization
rubber plantations
cream separator

Benjamin Franklin 171 **Cook's voyages** 142–143 **Galileo** 118 **gunpowder** 136–137, 186 **Industrial Revolution** 153–156

intellectual arts; so that ... when contemplation and doctrinal science began, the discovery of useful works ceases.' This observation was all the more telling because Bacon himself was a distinguished scholar and sought to prepare a 'grand instauration of the sciences' – a systematic conspectus of all human knowledge.

Clocks and the longitude problem

However, this is far from saying that the new scientific researches had no immediately useful results. Astronomy, for example, bore fruit of great value to seamen exploring the remoter parts of the world, especially the Pacific, as well as to those who were making regular long trading voyages from Europe to the East Indies via the Cape and to California via the Horn. On these journeys they were still handicapped and endangered by the difficulty of determining longitude at sea. In 1675 the Royal Observatory was established at Greenwich, with John Flamsteed as the first Astronomer Royal. Its task was to 'perfect the art of navigation' by preparing lunar tables so that longitude could be measured by observing the movement of the moon across a given fixed star. The Royal Observatory of Paris had been founded several years earlier, in 1667. There, in 1676, the young Danish astronomer Ole Rømer first measured the velocity of light – a physical constant of fundamental importance in the atomic physics of the 20th century – on the basis of discrepancies in the eclipses of Jupiter's satellites. His value of 225,000 kilometres (140,000 miles) per second is not too far off the accepted modern figure.

At the heart of the determination of longitude was the need to be able to determine exactly, anywhere in the world, the time relative to a fixed meridian, now that of Greenwich. In the early 1580s Galileo had discovered that the time of swing of a pendulum depended only on its length and not on the amplitude of the swing – a 1-metre (39-inch) pendulum takes two seconds. But only shortly before his death in 1642 did he attempt, unsuccessfully, to apply this principle to the strict regulation of a mechanical clock. In the event, the first such clock was not invented until 1656, by the Dutchman Christiaan Huygens. In theory, this solved the problem of determining longitude anywhere, but in practice pendulum clocks were quite useless on board a ship rolling and pitching at sea. Huygens' later invention of the balance-spring regulator still failed to achieve the necessary accuracy: on a long voyage great accuracy might have to be maintained for months on end. So serious was the problem that in 1714 the British Government set up a Board of Longitude empowered to award a prize of £20,000 – a huge sum in those days – for a solution. This was finally won by John Harrison with his famous H-4 Marine Chronometer, based on a spring-driven mechanism controlled by a balance-wheel. Subsequently, efficient chronometers of simpler design were introduced. It is a salutary thought that mariners have been able to navigate accurately for barely two centuries.

The first chronometer to tell the time accurately, and thus determine longitude, was John Harrison's H-4 (1759). This replica was made by Larcum Kendall and accompanied James Cook on two Pacific voyages in *Resolution* (1772–8).

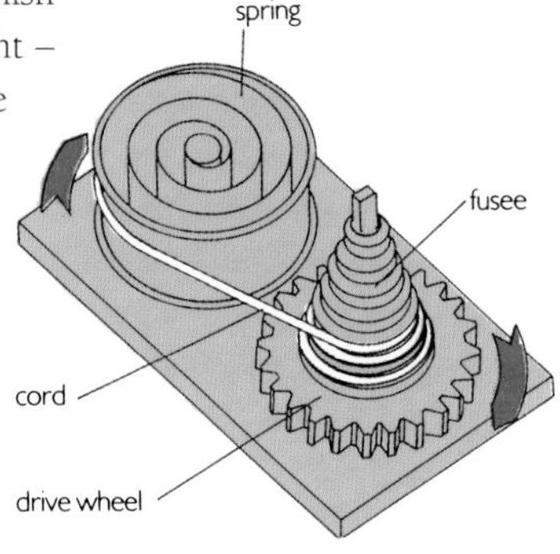

Early spring-driven timepieces contained a fusee to compensate for the slowing of the spring as it unwound.

The advent of steam

One of the first to test Harrison's chronometer was Captain James Cook whose great voyages of discovery between 1768 and 1779 – when he died violently in the Sandwich (Hawaiian) Islands – transformed European knowledge of the geography of the Pacific. Cook, and generations of seamen after him, made their long lonely voyages in wooden sailing ships, more sophisticated but not different in kind from those in which the Portuguese had sailed down the coast of Africa 300 years earlier. Long before his death, however, an invention had been made that was destined to revolutionize not only transport, on land as well as at sea, but the whole basis of industry. In 1698, Thomas Savery had devised an engine for raising water 'by the impellent Force of Fire'. In the hands of Thomas Newcomen, and later James Watt, this was translated into the crude but effective

Although conditions in many factories were poor, some owners adopted a paternalistic attitude towards their workers. Saltaire was a model textile factory and town founded by Titus Salt in 1851. Built on the Leeds and Liverpool Canal for ease of transport, it included shops and libraries. Its style, like that of Boulton and Watt's Soho Works (rebuilt 1795), was modelled on that of the great country mansion of the aristocracy.

steam engines which were a central feature of the Industrial Revolution. And only eight years after Cook's death John Wilkinson, the English ironmaster, launched the first iron-hulled ship on the Severn. By the beginning of the 19th century, these two developments had been linked and the iron steamship had arrived. By mid-century, Brunel had completed his huge *Great Eastern*, designed to carry 4000 passengers on the Australian run, in the wake of Cook who had plied the same route less than a century earlier. A commercial failure, as a result of excessive fuel consumption, she was nevertheless an outstanding technological achievement: later, she was used to lay the first transatlantic telegraph cable, which was completed in 1866, and marked a new era in long-distance communication.

Only later, as it became more compact, was the steam engine effectively applied to transport on land. Steam carriages for public use appeared on the roads about 1830 but achieved only limited success in competition with the railways. The world's first public steam-hauled railway was opened in England, between Stockton and Darlington, in 1825. By the end of the century railway track in Britain alone exceeded 30,000 kilometres (20,000 miles), a billion passenger journeys were being made every year and the weight of goods carried was 400 million tonnes. By that time, of course, railway transport was commonplace everywhere: the biggest system in the world was in the United States, with a total of over 300,000 kilometres (200,000 miles) at the end of the century.

Textiles

As we shall see in the next chapter, the first impact of the steam engine on industry was in mining, where it was used to pump floodwater from deep levels. Its popular identification with the start of the great expansion of the textile industry in the second half of the 18th century is not in accordance with the facts, however. Here the power source was for many years the water-wheel, and so far as steam engines came into the picture their role was in pumping, to ensure a good head of water to keep the wheels turning. Only later, with the availability of Watt's more compact and efficient engines, was steam used directly in factories.

In the series of inventions that revolutionized the British textile industry in the course of half a century, the first significant one was the flying shuttle invented by John Kay of Lancashire in 1733. Up to that time, a single weaver could make only a narrow fabric: he had to throw the shuttle through the shed of fibres with one hand and catch it with the other. Wide cloth required two operators, one on each side of the loom, to throw the shuttle to and fro. Kay's device returned the shuttle automatically, which both speeded up the process and made it possible for a weaver to make broad cloth single-handed, with a corresponding saving in labour costs.

This speeding up of the weaving process demanded a similar acceleration in the spinning of fibre. Here the first in the field was probably James Hargreaves, who is said to have invented his hand-operated eight-spindle spinning jenny in 1764, though he did not seek a patent until 1770. In this he was unsuccessful, though having previously sold some machines. The credit is, therefore, usually given to Richard Arkwright, who patented his much heavier spinning machine in 1769: it came on the market two years later as the water-frame machine. This title was misleading: although it was too heavy for a human operator, it was originally designed to be driven by a horse rather than water-power. In about 1770, Arkwright entered into partnership with two local hosiers, Samuel Need and Jedediah Strutt, and opened a water-driven mill at Cromford. This may be regarded as the start of the factory system: while machines were hand-driven, textile workers could operate from their homes, but once power was introduced they had to work at its source. Sales of fabrics consisting entirely of cotton were subject to a high tax, but in

Cook's voyages 142–143 **early weaving** 37 **mechanical clocks** 116–118 **railways** 177–182, 252–253 **steam engines** 158–165

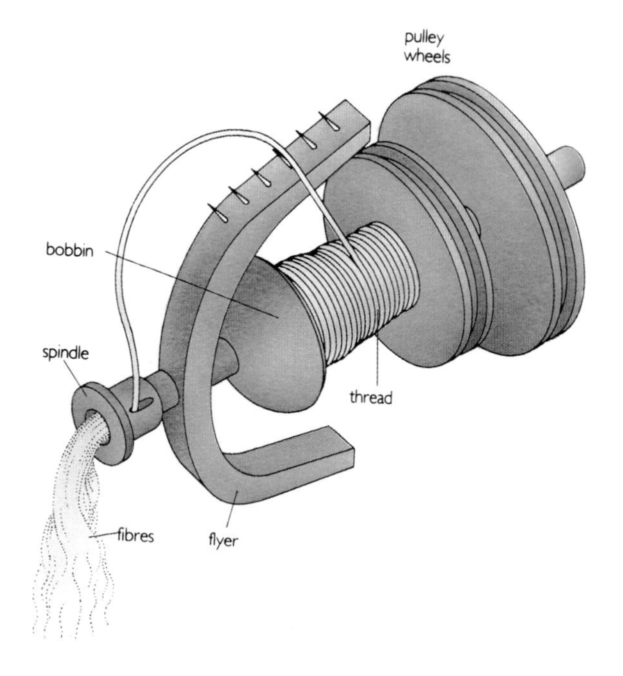

In the thread-maker of a spinning-wheel (above left), loose fibres are twisted as they pass through the hollow spindle and over the rotating flyer, which is driven by a belt from the main wheel to the large pulley wheel. The fibres are then wound on to the bobbin, which is driven by the smaller pulley wheel so that it spins faster than the flyer, pulling the fibres taut to produce thread.

The mechanization of the textile industry – and the use of water, and later steam power in place of manual labour – was at the heart of the Industrial Revolution in Britain. These contemporary prints show the drawing-in process (top) and mule spinning (above). The new mills gave much employment to women.

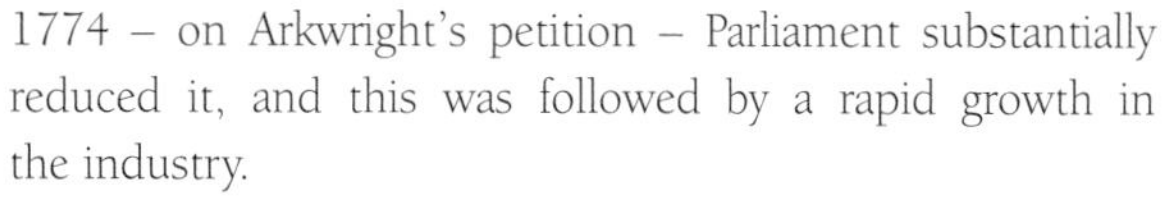

1774 – on Arkwright's petition – Parliament substantially reduced it, and this was followed by a rapid growth in the industry.

Between 1774 and 1779, Samuel Crompton devised another spinning machine: this was called a mule because it embodied features of the machines of both Hargreaves and Arkwright. In 1785, Arkwright's patent was declared invalid, and the use of the water-frame was open to all. In the same year a Boulton and Watt steam engine was installed in a mill at Papplewick, and with this the modern system of factory production may be said to have been firmly established.

There followed a tremendous expansion of the British cotton industry, especially for export. In 1751, the export of cotton goods was valued at only £46,000: by 1800 it was £5.4 million, and in 1861 £46.8 million. By contrast, wool – for long the mainstay of the British textile industry – became relatively unimportant: exports of woollen goods in 1861 were only £11 million.

The increased speed of weaving effected by Kay's flying shuttle was, as we have seen, an incentive to the development of faster methods of spinning, but the output of machines such as Arkwright's soon threatened to outstrip the capacity of hand-workers to weave it. This led Edmund Cartwright to experiment with a power-loom, and he set up his own works in Doncaster, with a rather unsatisfactory prototype, in 1787. Improvements were only made slowly, however, and even at the turn of the century there were probably no more than 2000 power-looms in the whole of Britain. But by 1825, the tide was turning decisively: there were then 75,000 power-looms and 250,000 hand-looms. Output was divided roughly equally between them, however, as the output of the power-looms was roughly three times that of the hand-looms.

The rapid introduction of so much machinery caused widespread unemployment, and despair led to rebellion. In Leicestershire in 1782 Ned Ludd, an idiot, destroyed some stocking-frames, and from him the Luddites took their name. They were involved in some serious destructive riots between 1811 and 1818, and a number were tried and executed.

The expansion in weaving had far-reaching effects on the source of raw cotton. Until 1795, British imports came mainly from the East and West Indies, Egypt and India, with only a little from the United States, where it was introduced into Georgia about 1786. Then, in 1793, Eli Whitney invented his cotton gin, a simple machine which greatly speeded up the separation of the cotton from the pod and its cleaning. This gave a tremendous impetus to the growth of cotton in the southern states and, incidentally, helped to prolong the institution of slavery there for many years. Slaves in the United States numbered 700,000 in 1790, two million in 1820 and over four million in 1860. In 1795, British cotton imports from the USA amounted to around 2000 tonnes; by

Left: Eli Whitney's cotton gin for separating the fibre from the pod (1793) gave a powerful stimulus to the American cotton industry. By the 1880s exports to Britain exceeded 500,000 tonnes annually. This print depicts the busy scene on the New Quay at New Orleans, a great centre for the cotton trade, in 1884.

Left: Arkwright's original spinning machine (1769) was known as the water-frame. In fact, it was designed to be driven by a horse and only later was water-power used.

Top: This crank mill built in about 1790 at Morley, Yorkshire, was one of the first mills to be powered by steam.

Above: The introduction of the steam engine to the cotton industry marked the beginning of the factory system: workers had to congregate at the seat of power.

clothing industry 196–197 **early loom** 37 **Jacquard loom** 150 **synthetic fabrics** 268, 304–305

1830 they had risen to 96,000 tonnes, and by 1860 to over 550,000 tonnes. The 1860 figure was not exceeded again until 1881, the hiatus being caused by the disruption of the Civil War and increased competition from production in India and Egypt.

With the new machinery, Britain established a commanding position in the new cotton industry and maintained it until well after the middle of the 19th century. It is estimated that in 1846 there were in the world some 27.5 million spindles: of these 17.5 million were in Britain, 7.5 million in Europe and 2.5 million in the United States. By the end of the century, however, the pattern was very different. Worldwide, the industry had multiplied nearly fourfold, with a total of 110 million spindles, but of these less than half were in Britain, whose share of the market was then roughly equal to that of the Continent and the USA combined. Much of the remainder came from mills in Asia, especially India.

Although the 19th-century textile industry was dominated by the spectacular growth of King Cotton, other branches of it were similarly mechanized and expanded. Allowing for annual fluctuation, the wool industry grew steadily: in Britain, long famous for its lustrous wool, exports of woollen cloth roughly doubled between 1850 and 1900. As with cotton, however, the sources of supply changed greatly: whereas in 1842 some 6000 tonnes were imported from Australasia, by the end of the century this had risen to 175,000 tonnes. South Africa and South America were also important new sources.

As a luxury item, the production of silk has never rivalled that of cotton and wool and, not surprisingly, mechanization was comparatively slowly introduced. The industry was particularly identified with France and there a very significant development occurred in 1801. In that year, Joseph Marie Jacquard devised a new form of loom which is still named after him. The weaving of high-quality silk involved the production of elaborate patterns by complicated interaction of the warp and weft. Traditionally, this was effected by a draw-boy acting on the instructions of the weaver, but in the Jacquard loom the weaver controlled the whole operation with the aid of a treadle. This is not in itself of great interest, for it was part of the general trend to save labour by mechanization. Its significance was that the movement of the strings, normally pulled by the boy, was controlled by a punched-card device in which the holes were arranged in accordance with the pattern desired. This system was widely used later to program the early computers.

Pottery

Pottery-making is a technology which reached an advanced state of development very early in the history of civilization, as is testified by surviving examples of early Greek, Chinese and Islamic ware. In an earlier chapter, we noted that the Chinese made a hard, impervious stoneware by using vitrifying clay mixtures different from those available in the West. In the eighth century AD, in the T'ang Dynasty, the Chinese began to make the much finer ware known as porcelain – a hard, white, translucent material. The firing technique was similar to that for stoneware, though a slightly higher kiln temperature – about 1300°C (2370°F) – was required: the main difference was in the material used. This was a mixture of kaolin (China clay) with a mineral known as petuntse, a weathered granite containing quartz and feldspar. From the early 14th century, porcelain began to find its way to the West, via the Silk Road, where it was highly prized for its elegance and delicacy.

The high quality of porcelain led to many attempts to imitate it. In the 1570s, Florentine potters produced what is known as soft-paste porcelain, so called because it could be fired at a much lower temperature than true porcelain. In this, the petuntse was replaced by powdered glass. From

During the 19th century the pattern of the cotton industry changed radically. At mid-century Britain was still dominant, but by 1900 the USA – jointly with the Continent and new mills in Asia, especially India – was a strong challenger. This spinning machine – at a factory in Newton, North Carolina – was typical of the early 20th century.

Florence, the manufacture spread to France, especially in the neighbourhood of St Cloud in Paris and Rouen. Not until the early 18th century, however, was a porcelain produced in Europe that was equal to that of China. Around 1707 Johann Böttger, alchemist to Augustus II – Elector of Saxony and King of Poland – experimented with the addition of various calcareous fluxes to the normally infusible Saxony clays. Eventually he produced a fine white porcelain and devised a satisfactory means of glazing it. This led, in 1710, to the foundation of the famous Meissen pottery, near Dresden, with Böttger as its manager. In France, too, much effort was devoted to the production of hard-paste porcelain, and a pottery at Vincennes was established under royal patronage in 1738. It moved to Sèvres in 1756. At first it manufactured soft-paste porcelain, but by 1770 a hard-paste product equal to that of Meissen was being produced.

In Britain, too, the search for a true porcelain was assiduously pursued. The first to achieve success was William Cookworthy, a Quaker preacher in business as a chemist in Plymouth. About 1745, he began to experiment with local clays and discovered an ideal source of both kaolin and petuntse nearby at St Austell: to this day the St Austell mine is one of the world's most important sources of fine China clay. In Plymouth, Cookworthy had no local supply of coal for his kilns. This led him to move to Bristol, where financial difficulties obliged him to make over his patent rights to his partner, Richard Champion. Unfortunately for Champion, the patent was successfully challenged by a group of Staffordshire potters including Josiah Wedgwood, and from 1781 hard-paste porcelain began to be made at New Hall, near Shelton. Wedgwood had himself experimented widely with the addition of various new minerals to the local clay and was famous for his stoneware – especially his jasper which could be tinted to any colour and polished like stone. In 1790, he made a fine copy in black jasper of the famous Portland Vase. In the wider market he is remembered for his creamware known as 'Queen's Ware', a name consonant with his title of Royal Potter. In 1769, he established his world-famous pottery at Etruria, whence his fragile products were transported safely on the smooth waters of the Grand Trunk Canal.

Meanwhile, however, other British potters had been pursuing a different line of development. In experiments to improve soft-paste porcelain, a variety of additives had been investigated, including bone-ash, and about 1750 the manufacture of bone china was established at Stratford-le-Bow in London. At the turn of the century, Josiah Spode began to incorporate bone ash in hard-paste porcelain too, with excellent results. At potteries in Worcester, trials had been made with the addition of a local mineral known as soapstone or steatite, and from about 1751 this, too, gave a distinctive form of porcelain. Thus by the end of the 18th century the manufacture of porcelains comparable with, and sometimes excelling, those of China a thousand years earlier had been established on the Continent at Meissen and Sèvres, and at several places in Britain.

Whatever its source, such fine porcelain – thin and translucent and usually richly ornamented with coloured glazes – was very much for the carriage trade, like the fine silks of the textile industry. For the ordinary citizen cheaper,

The Chinese made fine porcelain from the eighth century, but the art was mastered much later in Europe. In the 18th century good porcelain began to be manufactured at Meissen and Sèvres: this bleu-celeste jardinère (above) is a fine example of Sèvres ware. In Britain the lead was taken by Josiah Wedgwood, who began manufacture in Burslem in 1759. This traditional blue willow pattern china (top left) has been made by the Wedgwood firm since 1806.

ceramics 303 **early pottery** 38–41 **glazes** 122 **punched cards in computing** 307 **Silk Road** 10–11, 23, 36

but still attractive, ware was required and the heavy demand for this made necessary some sort of mass-production system. In the 1740s Ralph Daniel, in Staffordshire, introduced the technique of slip moulding in which very plastic clay – or slip – was poured into a mould. When the clay had dried sufficiently to handle safely, the article was fired in the usual way. Originally, metal moulds were made but it was soon found that ones made of plaster were preferable because they speeded the drying process by absorbing moisture from the slip. A second method, introduced a little later, was that known as jigging. In this, one side of the object was shaped by putting the clay on a rotating plaster mould; the other side was then shaped by lowering a template on to it. For many everyday purposes a cheap, plain glaze was sufficient to finish the article, but for others some sort of decoration was called for. In mass production, the old technique of painting each item by hand was far too slow and expensive, and this led, in the early 1750s, to the design being transferred to the article from a suitably inked plate, via a sheet of paper. Sometimes, for better-class ware, this transfer pattern would be enhanced by hand painting.

Agriculture

Agriculture has never been a fast-evolving industry and it is fair to say that over much of the world – notably Africa and the Middle and Far East – traditional methods of subsistence agriculture prevailed in this period. Even in Europe, a relatively small part of the globe, change was slow. But new systems of rotation of crops were followed and the move towards larger areas of enclosure encouraged the mechanization of the basic processes of cultivation – ploughing, seed-sowing, reaping, threshing and so on. So, too, did the rapid growth of population during the Industrial Revolution. In the United Kingdom, the population in 1483 was about five million; in 1760, nearly three centuries later, it had risen to around nine million; but by 1830, the population had topped 24 million, almost a threefold increase in 70 years.

From the latter part of the 18th century, increasing use was made of iron for the construction of farm implements, but up to the end of the 19th century the chief source of power remained draught animals and human strength. Overall, the contribution of steam – mainly in ploughing and threshing – was small and when, in the 20th century, engine power became widely adopted, it was the newly developed internal combustion engine that prevailed. The importance of animal manure for keeping the soil fertile had long been recognized but with the advent of artificial manures in the 19th century purely arable farming became more practicable.

In contrast to the situation in the Old World, change in the newly settled territories of North and South America and Australasia was rapid. There the vast areas of virgin land and the lack of manpower forced the pace of mechanization. At the same time in Europe, the growth of population and the drift of labour from the countryside to the town created a new pattern of demand for food. It also created increased demand for cotton and wool for the increasingly mechanized textile industries. New affluence encouraged expansion of luxury crops such as tobacco, tea, coffee and sugar: with the exception of the last, which could be grown in Europe as beet, all were tropical crops.

Developments in food technology opened up immense new possibilities. From François Appert's first experiments in the preservation of food by heat sterilization in 1795, a vast new canning industry grew up. Similarly, from about 1870 margarine became a substitute for butter for the great standing armies of the European powers and for the urban poor. Finally, with the advent of large-scale refrigeration at the end of the 19th century, even meat could be imported in bulk into Europe from as far afield as Australia, New Zealand and Argentina.

Expansion of the canals and waterways

From the earliest times the most economical and simplest way of moving heavy goods within a country had been by water, using inland waterways and canals. Despite the advent of railways, canals not only remained important but attracted much new investment. In Britain a 6500-kilometre (4000-mile) network, culminating in the Manchester Ship Canal, had been built in the century following the opening of the Sankey Warrington in 1757, but thereafter little was done. Elsewhere, however, things were different. In China, the vast network of rivers and canals begun by the Emperor Yang in the seventh century served the double purpose of conveying tax grain to the capital and facilitating the movement of troops. In France, where the Canal du Midi joining the Mediterranean and the Atlantic had been completed in 1681, much new work was done in the 19th century, including a system linking the Somme, Oise and Seine on the one hand with the Marne, Rhône and Rhine on the other. The last year of the century saw the opening of the 103-kilometre (64-mile) Dortmund–Ems Canal in Germany.

Across the Atlantic, where there was no tradition of canal-building, several great enterprises were embarked upon. Chief among them was the great Erie Canal, 586 kilometres (364 miles) long, opened in 1825, which provided a water route to bring grain from the Great Lakes to New York. The Michigan–Mississippi Canal was opened on the first day of the 20th century.

In a different category, but immensely significant in the context of world trade, were the great canals built to shorten the sea routes. Two examples were outstanding. The Suez Canal, linking the Mediterranean and the Red Sea, was completed in 1869; and the Panama Canal across the Isthmus of Panama, which shortened the voyage from Europe to the West Coast of America by 4800 kilometres (3000 miles), was commenced in 1881 but not completed until 1914.

Britain's extensive canal network was largely completed by the middle of the 19th century. It was designed for navigation by long, narrow boats, as portrayed below right in an engraving (1830) of City Road Lock on the Regent's Canal. Elsewhere, however, some very large enterprises were embarked upon, some designed for use by ocean-going ships. The Suez Canal (below left), linking the Mediterranean and the Red Sea, was completed in 1869: this picture shows the opening ceremony. The Panama Canal, across the Isthmus of Panama, was dogged by ill fortune and not completed until 1914. It involved the building of six massive locks (left) and an 11-kilometre (7-mile) cutting through the Culebra Hills.

Road-building, in contrast, made rather little progress so far as the mode of construction was concerned, though military needs often prompted improvement and expansion of road systems, as in France under Napoleon. The main reason was the lack of incentive: the products of the heavy manufacturing industries could readily be moved by water or, later, the railways. It was not until the 20th century, with the advent of the car and motorized transports of all kinds, that they came into their own.

In the last quarter of the 18th century, a new mode of transport was born. In 1783, the Montgolfier brothers in France reached the climax of their experiments with hot-air balloons by achieving a flight of 12 kilometres (7½ miles) with two passengers. Ever since, there has been an unbroken interest in lighter-than-air flight. In the event, though, the future lay with heavier-than-air craft, but that story is strictly part of the 20th century.

Machine tools

Apart from adopting steam as a new source of power, the Industrial Revolution was characterized by the increasing use of machinery – often of a quite novel kind – to carry out the work once done by hand. But this new machinery itself had to be manufactured and a variety of machine tools were developed to carry out such basic engineering operations as turning, planing and milling. At one end of the scale, the makers of clocks and instruments needed precision lathes for cutting screws and arbors. At the other, the makers of the early steam engines needed accurately bored cylinders up to 2 metres (6½ feet) in diameter and 3 metres (10 feet) long. The need to forge massive parts led to James Nasmyth's steam hammer of 1839, so powerful that it could forge a steel ingot nearly a metre (3 feet) in diameter – as for the paddle-shafts of the new steamboats – yet so delicate that it could crack an egg.

canals 63, 261 **development of agriculture** 200–206, 291–301 **early agriculture** 26–33 **first balloons** 171–172

The new machine tools were capable of accurate and fast repetitive work. An early example was the machinery installed at Portsmouth Dockyard by Marc Brunel in the first decade of the 19th century to make wooden pulley-blocks for the Royal Navy. A total of 43 machines, driven by a 30-horsepower steam engine, converted logs into pulley-blocks of three sizes at the rate of 130,000 per year. Such precision mass-production made possible the manufacture of interchangeable parts, apparently first practised by a French gunsmith about 1785 for making locks for firearms. Each part was so accurately made that any one, picked at random from stock, would fit perfectly with the other components.

Despite its undoubted Continental origin, this technique became known as the 'American system', because it was enthusiastically adopted in the USA with its expanding industry and lack of skilled craftsmen. Among the first to practise it were Eli Whitney, Simeon Worth and Samuel Colt, for the manufacture of muskets and pistols for the US Government. By 1853 Colt, famous as the inventor of the revolver, was employing 1400 machine tools in his armaments factory. The clock trade, too, used it to stamp out the gearwheels required to meet the growing demand for mechanisms. Only after 1851, when this mass-production system was demonstrated at the Great Exhibition, was the American system adopted in Britain, first by the Royal Small Arms Factory at Enfield.

Gas from coal

The most important of the early manufacturers of steam engines was the firm of Boulton and Watt in Birmingham, who supplied them to many of the tin mines in Cornwall for pumping out water. The installation was supervised by their local manager, William Murdock. At Redruth, he experimented with the production of lighting gas from coal, and early in the 19th century Boulton and Watt began to manufacture the necessary equipment on a commercial basis, first supplying one of the large Lancashire textile mills. This was the basis of a great new industry which had far-reaching practical and social consequences. By the middle of the century, city streets and public buildings were brightly lit by gaslight and poor town-dwellers (problems of distribution prevented penetration of country areas) enjoyed illumination of a brilliance previously affordable only by the rich. In a time of growing literacy, when improved methods of printing and paper-making were releasing a flood of cheap literature, the effect on leisure habits was considerable.

Before the end of the 19th century, coal-gas had found another domestic use as a source of heat as well as of light. Coal-gas lighting was soon widely adopted in Europe and to some extent in the USA, but there huge reserves of natural gas were exploited for the same purpose at an early date: in western Europe, this source of gas was not developed – and indeed, not discovered – until after the Second World War.

Increasing mechanization in all branches of industry led to increasing demands for precision engineering. This gear-cutting machine (top) was built in Madrid in 1789 by Manuel Gutierrez. The importance of these developments was reflected in the great 28-volume *Encyclopédie ou Dictionnaire Raisonné des Sciences, des Arts et des Métiers* (*Pictorial Encyclopaedia of Trades and Industries*) compiled by Denis Diderot 1751–1772, from which this picture of lathes (above) was taken.

Before the end of the century America had begun to exploit its enormous oilfields, which were often associated with natural gas. But as the main use of petroleum was as a fuel for internal combustion engines, this is essentially part of the history of the 20th century. So, too, is the story of rubber – required in great quantities for the tyres of cars – though it was used in relatively small amounts for waterproofing fabric and other purposes from the early part of the 19th century. Rubber was encountered in Peru in the early 17th century by the Spanish *conquistadores*, but not seriously investigated until it was brought back to France more than a century later.

The age of steel

The Industrial Revolution is popularly identified with the heat, glare and smoke of furnaces, especially those firing pottery and smelting iron. There is much justification for this. The use of iron for making machinery of all kinds and rails for the new railways, as well as for construction, expanded rapidly, particularly after Abraham Darby showed, in the middle of the 18th century, that coke could replace charcoal in the smelting of iron ore. Between 1740 and 1850 the production of iron in Britain alone rose from 17,000 tonnes to 1.4 million tonnes annually. Some little part of this was converted into steel – essentially iron containing 0.5 to 1.5 per cent carbon – which had been used in very limited quantities from pre-Christian times. It was a difficult manufacture, lending itself only to small batches, but at mid-century production suddenly rose remarkably. The reason was the advent of cheap new production processes: one was developed independently by William Kelly in America and Henry Bessemer in Britain, while on the Continent the Siemens-Martin open-hearth process was introduced. Between 1870 and 1900, a mere 30 years, world production of steel rose from 0.5 million tonnes to 28 million tonnes, with America the largest producer, followed closely by Germany. Much of this went into steel rails, which were far superior to iron, for the rapidly expanding railway network. Special alloy steels were developed for use in machine tools, enabling them to cut faster and last longer.

If the course of history is to be punctuated in terms of dominant metals, the Age of Steel logically follows, and ranks with, the Bronze and Iron Ages. Much steel went into productive industry, but a great deal was used to make armaments too. At sea, this was increasingly the age of heavily armoured battleship, protected with thick steel plates, and on land as well as sea the size and power of guns grew rapidly. A major factor here was the appearance of high explosives, pioneered by the Swedish engineer Alfred Nobel as the first improvement on gunpowder for more than 500 years.

The rise of steel was thus a technological development of great importance. So, too, was the advent of a new metal, aluminium. In chemical combination, as alum, it was one of the earliest of all commercially traded chemicals, but it was not

In Britain, mail-coaches for conveying letters were first established at Bristol in 1784 and a national network was soon in operation. This characteristic print shows the departure of the Royal Mail from Lombard Street, London, in 1827. From 1838, however, the mail began to be sent by rail and the old mail-coach system was one of the casualties of the new form of transport.

isolated as a metal, and then in only very small quantities, until 1827. In 1886, however, a method of producing it cheaply and in large quantities was invented independently in France and America. At first demand was modest – production in 1900 was no more than 7000 tonnes – but the metal came into its own in the 20th century as a major constituent of the light alloys required for the nascent aircraft industry.

Science-based industry

The developments we have so far considered support the general thesis that the Industrial Revolution was brought about by ambitious men with a keen practical streak rather than by scientists. If we take the steam engine as one of the central inventions of the age, it is difficult to dispute the familiar aphorism that science owed more to the steam engine than the steam engine did to science. Of the pioneers, Thomas Newcomen was an ironmonger's apprentice and James Watt a carpenter's son; Matthew Boulton, Watt's partner, was the son of a buckle-manufacturer. Thomas Telford, the great builder of roads, bridges and canals, was a shepherd's son who had no more education than a village school in Scotland could provide. James Brindley, another great canal builder, was never more than semi-literate. Josiah Wedgwood, who transformed the pottery industry, began work at the age of nine. Richard Arkwright, pioneer of mechanical spinning, began life as a barber. Henry Cort, the 'great finer' of iron, was the son of a brickmaker. So it goes on, yet it would be quite wrong to assume that such men were divorced from the scientific developments of their time. Their success was based largely on an ability to see the practical significance of new discoveries and by dint of immense energy and enthusiasm make them the basis of great new

coal-gas engines 166–167 **explosives** 185–186, 215, 264–265 **metal technology** 183–191 **oil industry** 223–227 **paddle-wheel steamships** 174–176

industrial enterprises. That they did not lack in intellectual capacity is shown by the fact that in the later days of their success many were elected to Fellowship of the Royal Society. For almost 25 years many of the great industrialists of the Birmingham area met regularly as members of the highly regarded Lunar Society to discuss the significance of new developments in science: it was so called because for convenience of travel they met when the moon was full.

A main reason why men of this kind of background could achieve such success was that the Industrial Revolution was at first concerned largely with mechanical and civil engineering – steam engines and textile machinery, canals and railways, forges and lathes. These were fields in which practical experience and mechanical flair were far more important than an understanding of basic principles. A surprising amount could be, and was, achieved by rule-of-thumb methods. Nevertheless, these were not all-sufficient; purely technical considerations became increasingly important.

Chemicals

A rudimentary chemical industry had to be expanded to meet the needs of the rapidly growing textile industry: it has been said that the chemical industry grew up in the shade of the textile industry. But here native wit was no alternative to grasp of principles, even if those principles were themselves still only partially understood by the chemists of the day. To make soda from salt and alum from shale; to manufacture chlorine for bleaching as the output of textiles outstripped the capacity of traditional bleachfields, where they relied on the slow action of sunlight; and to make synthetic dyes to complement those found in nature – all involved problems that increasingly demanded the skill of a new breed of professional chemist.

Electricity

The 17th century saw the first serious investigation of a phenomenon that in the 19th was to be the basis of a wholly new industry and was profoundly to influence many others. From experiments on friction-generated electricity came Michael Faraday's crucial experiments on electromagnetic induction at the Royal Institution in London. These demonstrated in the early 1830s that mechanical power could be converted into electricity and that electricity could be a source of mechanical power. Well before the end of the century electricity was being supplied from central generating stations for public consumption and the incandescent filament lamp was challenging the long-established supremacy of gas. Electric trains and automobiles appeared in the 1880s. No less important, electricity was pressed into service for telegraphy and telephony: by 1900 there were already a million telephones in the United States. Even before then, the possibility of wireless telegraphy had been demonstrated by Heinrich Hertz, and in 1901 Guglielmo Marconi transmitted a signal across the Atlantic.

The Great Exhibition of 1851

With the Renaissance, Europe became the leader of the world – not yet in the military sense of being able to impose her will on Eastern powers like China or Japan, who still successfully pursued a policy of keeping foreigners at arm's length – but in the sense of exploring new ideas in the arts and sciences and accumulating wealth through commerce. Within Europe the leadership eventually passed to Britain, which emerged in the middle of the 18th century as a new kind of state, whose wealth and power were based on manufacturing industry, a state which was literally the 'Workshop of the World'.

The period of the Industrial Revolution cannot be defined, in that no single event began or ended it, but it is generally regarded as being completed during the years 1760–1830. To many it must have seemed by the latter date that Britain's commanding position was unassailable, and this seemed to be borne out by the Great Exhibition mounted in London, under the patronage of the Prince Consort, in 1851. Here hundreds of different manufactured goods – from locomotives to sewing machines, from carpets to pianos, from cooking stoves to electroplated items – were judged by international juries. At its conclusion, Britain swept the board.

Education for a technological age

Despite this brilliant success, the reality was that Britain was already in decline relative to other European countries, as was demonstrated by the Paris Exhibition of 1867, at which it gained no more than a dozen awards. The shock to the nation was considerable and triggered off many inquiries into what had gone wrong and how it could be put right. Various explanations were offered, but the consensus was that the English educational system was totally unsuited for the needs of the day. Lyon Playfair, a distinguished chemist who had been closely associated with the Prince Consort, wrote bluntly: 'The one cause upon which there was unanimity of conviction is that France, Prussia, Austria, Belgium and Switzerland possess good systems of industrial education for the masters and managers of factories and workshops, and that England possesses none.'

That this was true cannot be doubted. The old and prestigious universities of Oxford and Cambridge virtually ignored science, as did the public schools from which most of their students came. Even in the old Scottish universities, where more practical views prevailed, the teaching of engineering subjects was limited and half-hearted. The plain fact was that the ruling classes saw no need for technically trained men.

On the Continent, the situation was very different. There the rich rewards of industrialization were apparent and there was a natural desire to share in them. Yet the British policy of deliberately obstructing the export of machinery and the engagement of her skilled artisans abroad made it difficult for

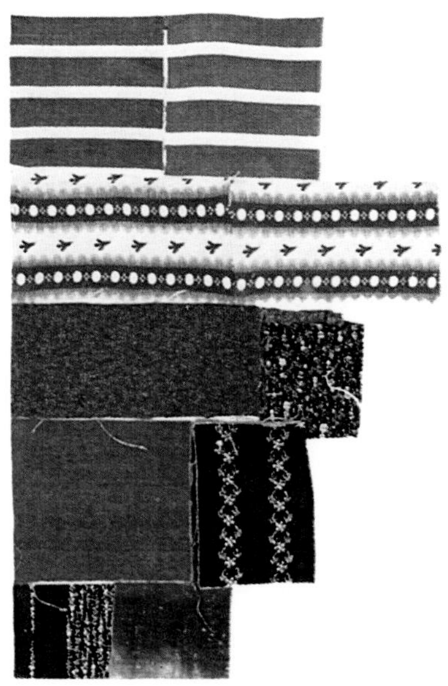

At the Great Exhibition of 1851, Britain's wide range of manufactured goods (above) were unsurpassed by any other nation. In reality, this supremacy was crumbling. Unlike Britain, European countries and the USA recognized the importance of technical education and spent liberally on technical colleges, such as the Munich Polytechnic (right), founded in 1868. Systematic research in chemistry made Germany the leading manufacturer of synthetic dyes before the end of the century: a leading firm was BASF, some of whose 19th-century dyes are illustrated (above right).

them to participate. The natural result was that the leading countries of Europe set up technical education systems of their own, to enable them to compete with each other and, above all, with Britain. The *École Centrale* was established in Paris, and Polytechnic Schools in Germany, Holland and Switzerland. To staff them men were sent to study English industrial methods. An important feature of these new institutions was the close links they established with industry. Recognizing that higher education was possible only after a good grounding in the schools, much attention was paid also to primary education. While compulsory schooling was not established in Britain until 1870, it had been enforced in Prussia since 1826, up to the age of 14: in the year of the Great Exhibition, Prussia already had 26 technical schools.

The German universities excelled in chemistry and through their research laboratories – headed by such internationally famous men as Justus von Liebig at Giessen, Robert Wilhelm Bunsen at Heidelberg, and Friedrich Wöhler at Göttingen – provided a stream of highly qualified men for the burgeoning chemical industry. Although Britain got off to a flying start in the field of coal-tar dyes, with the discovery of mauve by William Perkin in 1856, much subsequent work was done by visiting German chemists who soon returned to their own country to found a dyestuffs industry there. By 1884, Germany's production of dyes was four times Britain's and 80 per cent of it was exported.

In the United States, where labour shortage encouraged mechanization and a flood of European immigrants brought new ideas about education with them, land grant colleges were endowed from 1862. The now world-famous Massachusetts Institute of Technology (MIT) was founded in 1865 'for those who seek administrative positions in business, where a systematic study of political and social relations and familiarity with scientific methods and processes are alike essential.'

In the face of this determined opposition Britain reacted too slowly and inadequately: the lead which had seemed unassailable at mid-century had been irretrievably lost by its end. Not until 1889 were technical schools formally established and it was 1902 before secondary education became compulsory. By that time, the specialist had decisively ousted the amateur in science and technology. On the Continent and in the USA, industry had deliberately set out to train specialists: in Britain, industry had painfully to discover their value. At the turn of the century, a new international pattern of technology was emerging. Britain was in decline, Germany was in the ascendant and across the Atlantic America was gathering the strength that was to make it the greatest power of all after the Second World War.

chemical industry 263–268 **development of telephone** 236–238 **electricity production** 165, 169–170, 228–231 **Marconi and radio** 238–240

CHAPTER SEVENTEEN

Prime Movers

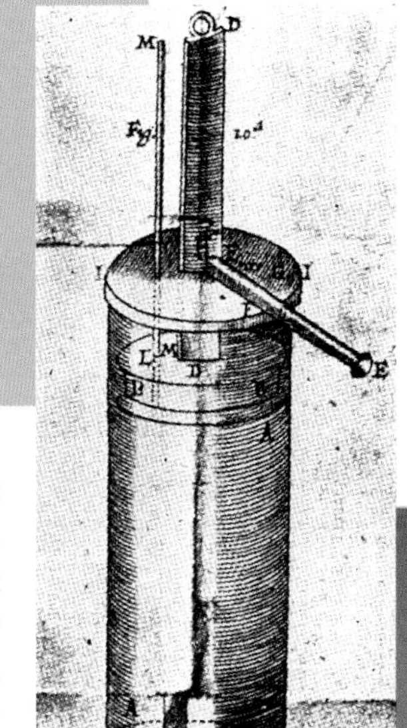

By the 17th century air had become a tangible substance in the minds of the new school of experimental philosophers. From Torricelli's research developed the barometer, a fine early (1705) example of which is shown (below left). With his famous experiment with evacuated hemispheres at Magdeburg (1654) von Guericke demonstrated the great potential power of atmospheric pressure (below right). This led on to Papin's atmospheric engine of 1690 (left), a predecessor of the steam engine.

Even in medieval times, when both windmills and water-wheels were in common use, the main sources of power were still either men or draught animals. When heavy work had to be done, they might be used in numbers, with gangs of men tailing on to a single rope or teams of oxen harnessed to a heavy wagon, or their task might be simplified by the use of some form of gearing – multiple pulley-blocks or simple systems of cogs. For certain kinds of repetitive work, such as grinding corn, wind and water power were quite satisfactory in that some interruption in the supply was of no great consequence: reserves of flour could be drawn on when rivers were low or the wind was light. For certain tasks, however, an intermittent source of power was not satisfactory: in pumping water from mines, for example, extraction had to keep pace with the inflow or flooding would occur. For this reason, machines operated by human power – such as Archimedean screws or treadmills – were essential, if only as a standby. Better still, however, would be some sort of engine that could be put into motion at will.

The development of the steam engine was closely related to the problem of mine drainage, which by the 17th century had become acute as veins of ore near the surface became exhausted and shafts became deeper. Mining and metal-working attracted huge investment, encouraged by the rise of great banking families such as the Fuggers of Augsburg. For reasons not then understood by mining engineers, suction pumps could not raise water above a height of 9 metres (30 feet). In 1643, Galileo's assistant, Evangelista Torricelli, discovered why this should be: the pressure of the atmosphere is equal to that of a column of water of that height and is the limit of what it can support. Using mercury, a liquid 13½ times denser than water, Torricelli noted that a column 76 centimetres (roughly 30 inches) high could be supported. He predicted, and subsequent experiments proved, that with increasing altitude the atmospheric pressure would diminish and the length of the column would be correspondingly shorter; thus was born the barometer. It became clear, therefore, that the seemingly intangible atmosphere could exert surprisingly large forces, as Otto von Guericke strikingly demonstrated in a famous experiment at Magdeburg in 1654.

When two close-fitting hemispheres only 50 centimetre (20 inches) in diameter were evacuated, two teams of eight horses, hauling in opposite directions, could not pull them apart. More directly significant, he carried out a similar experiment with a piston working in a vertical cylinder of about the same diameter. A rope attached to the piston was passed over a pulley and held by a team of 50 men – yet when von Guericke pumped out the air from beneath the piston the men could not restrain it. It thus became clear in principle that if a vacuum or partial vacuum could be alternately created and released beneath a piston working in a cylinder, considerable power could be generated. How this could be achieved in practice, however, was still far from clear.

The first, and as it turned out impracticable, proposal was Christiaan Huygens' gunpowder engine of 1680. He exploded the charge in a cylinder beneath a piston, driving it upwards: then, as the hot gas cooled, the piston was driven down again by atmospheric pressure. This had two defects. First, only a moderate degree of vacuum was created below the piston, and second, repeatedly recharging with gunpowder was slow and laborious.

Ten years later Huygens' assistant, Denis Papin – who had worked with Robert Boyle on the development of the air pump – put forward what proved in the long run a workable alternative. This was based on the fact that water when boiled yields 1300 times its own volume in steam. In his first device – called not a steam engine but an atmospheric engine because it utilized the pressure of air – he placed a little water in the bottom of a tube 7½ centimetres (3 inches) in diameter, closed by a piston. The tube was heated to boil the water, driving the piston up the tube: when the tube was cooled, the steam condensed back to its original volume and the piston fell back. By alternate heating and cooling, the piston could be made to move up and down repeatedly – until there was no working fluid left. From his own account, Papin had a very clear idea of the basic principles: 'Since it is a property of water that a small quantity of it turned into vapour by heat has an elastic force like that of air but upon cold supervening is again resolved into water, so that no trace of the said elastic force remains, I concluded that machines could be constructed wherein water, by the help of no very intense heat, and at little cost, could produce that perfect vacuum which could by no means be obtained by gunpowder.'

Papin's small device was extremely simple. It combined in a single unit boiler, cylinder, piston and condenser, and followed the cycle of operations characteristic of later versions. He himself did not pursue the invention very far, though he spoke of a machine with 'a two-foot bore and a four-foot stroke that might raise 8000 pounds of water four feet in one minute'. On the rating used by engines later this would correspond to 1 horsepower (defined as 550 foot-pounds per second, which is equal to 746 watts). He was deterred by the difficulty of making so large a cylinder. In 1679, he did, however, make another invention of lasting value. This was his digester, which relies on the fact that the boiling point of water depends on the pressure to which it is exposed; the higher the pressure, the higher the boiling-point. This device is familiar today as the domestic pressure cooker and the hospital sterilizer.

The Miner's Friend

In the event, the first practicable steam engine was that devised by Thomas Savery and patented by him in 1698, the patent running for 14 years but later extended. It was: 'A new invention ... for raiseing of Water, and occasioning Motion to all sorts of Mill Work by the Impellent Force of Fire, which will be of great Use and Advantage for Drayning Mines, serveing Towns with Water, and for the Working of all Sorts of Mills where they have not the benefit of Water nor Constant Windes.'

Savery's 'fire-engine' worked on the principle of raising water as high as atmospheric pressure would allow and then forcing it still higher by the use of superheated high-pressure steam such as Papin used in his digester. This use of high pressure was ambitious and not very successful, simply because the constructional techniques of the day were inadequate and boiler explosions were a hazard.

In practice, Savery's engine – which rated about 1 horsepower – had to be situated no more than 9 metres (30 feet) above the lowest level in a mine and – supposing 3 atmospheres excess pressure on the high-pressure side, about as much as was feasible – could pump up another 27 metres (90 feet) or thereabouts, making a total lift of 36 metres (120 feet). For deep mines it was, therefore, necessary to have more than one engine, at successive levels.

Savery's pump had no piston. Steam from a boiler was led into an oval vessel filled with water, driving the water up through a vertical pipe. A valve then closed this, and the vessel was cooled, creating a vacuum inside it: the vessel was then connected with the low level, from which it sucked up mine water. In practice, two vessels were used, one filling as the other emptied.

Savery's engine had some success in supplying water to towns and large residences – in 1712 he supplied one to the York Buildings Waterworks Company in London to pump water from the Thames – but in mines, where the need was greatest, it failed to live up to its popular name of 'the Miner's Friend' and found little application. Its fatal defect was that, to get large lifts, higher pressures had to be used than were really safe and even then several pumps often had to be installed in series. Generally speaking, mine-owners thought the danger and expense too great and stuck to their bucket-chains worked by horses.

early power production 64–70 **mine drainage** 81 **water-power** 68–70, 115–116, 165–167 **windmills** 114–115, 166, 231, 277–278

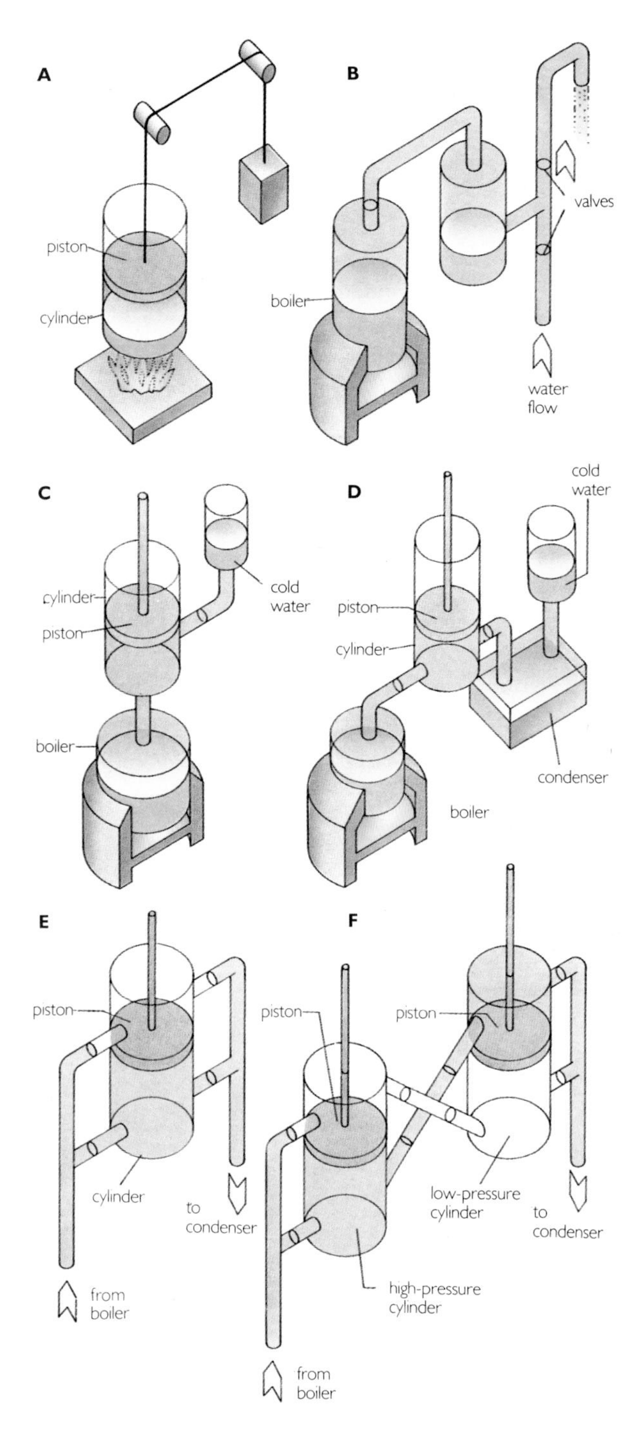

The development of the steam engine began with Papin's engine (A), in which steam was produced to force the piston upwards: it fell under atmospheric pressure when the steam condensed. Savery introduced a separate boiler producing steam that forced water up a pipe; condensing the steam sucked up more water (B): effectively it was a pump rather than an engine. Newcomen's engine used a spray of cold water to condense steam in the cylinder (C), and Watt invented a separate condenser (D) so that the cylinder remained hot. Watt also introduced a double-acting closed cylinder in which steam drove the piston from both sides (E), and Arthur Woolf refined this by using two cylinders (F), one at high pressure and one at low pressure. Watt also invented the centrifugal governor for automatic speed control.

The Newcomen engine

Thomas Newcomen's engine was the next to appear. It had the great advantage that it was designed to work with steam at atmospheric pressure – that is, ordinary boiling water. It was thus very easily constructed by the kind of craftsmen generally available. It consisted of a heavy beam pivoted at its centre, at one end of which was hung the rod that worked the pumps in the mineshaft. The other was attached to the top of a piston working in a vertical cylinder placed directly above the boiler. With the piston at its lowest point, steam was admitted to the cylinder, forcing the piston to rise and the far end of the beam to fall. When the piston was at its highest point, the steam in the cylinder was condensed with a spray of cold water and the piston fell back. Initially known as a lever-engine, by mid-century the term beam-engine was more common.

While it is apparent that Newcomen's engine was very different from Savery's, not only in the use of steam at atmospheric pressure, but also in reverting to Papin's piston, Savery's patent for 'Raiseing Water ... by the Impellent Force of Fire' was so widely drawn that Newcomen had to come to an accommodation with him. In view of the lack of commercial success of Savery's engine, this would not have been difficult.

It is said that Newcomen had no knowledge of Savery's invention when he made his own. This may be true, but we know little about Newcomen's early life or when he first conceived the design of his engine. Savery's patent was granted in 1698 and in 1702 he was advertising his engine for sale from his workshop in Salisbury Court in London, and Newcomen, with similar interests, cannot have been unaware of this. In the event, the first Newcomen engine was not erected until 1712, at Dudley Castle in Worcestershire. Its purpose was to extract water from a coal-mine, and – working at 12 strokes a minute – it raised 540 litres (120 gallons) of water 46 metres (153 feet). This corresponds to an effective output of a little over 5 horsepower.

Just as Savery's engine presented a serious practical problem – the mastery of high steam pressures – so, too, did Newcomen's. In his case, it was that of manufacturing cylinders of the required size. Cannon could then be bored very precisely up to 18 centimetres (7 inches) in diameter, but Newcomen engines needed cylinders far larger than this. By 1730, cylinders 74 × 274 centimetres (29 × 108 inches) in size had been made and by 1765 this had risen to 188 × 320 centimetres (74 × 126 inches) corresponding to a weight of 6½ tonnes. Many of these cylinders were cast by Abraham Darby at Coalbrookdale, where he had introduced smelting with coke in place of charcoal: by 1758, he had manufactured more than a hundred. At this sort of size a close steam-tight fit between cylinder and piston simply could not be achieved: even in the 19th century the best mechanical engineers were talking in terms of 'a bare sixteenth' of an inch or 'a full thirty-second'. Newcomen's solution, possible because his steam was at atmospheric pressure, was to cover the top of the piston with a leather disc and seal this with a layer of water.

Newcomen's engine was very successful and, especially after the patents expired in 1733, was widely used in the mining industry, particularly the tin mines of Cornwall and the deep flood-prone coal-mines of the north of England. Before Newcomen died in 1729, he saw it in use in Hungary, France, Belgium and other European countries. Nevertheless, his engine did not have the same degree of success in France and Germany that it enjoyed at home. On the Continent mining and water engineers had developed a crude but effective system of push-pull rods (*Stangekunst*) which made it possible to transmit the power of water-wheels for distances up to 1½ kilometres (1 mile). They remained wedded to this even though the steam engine enabled all the machinery to be installed where it was required.

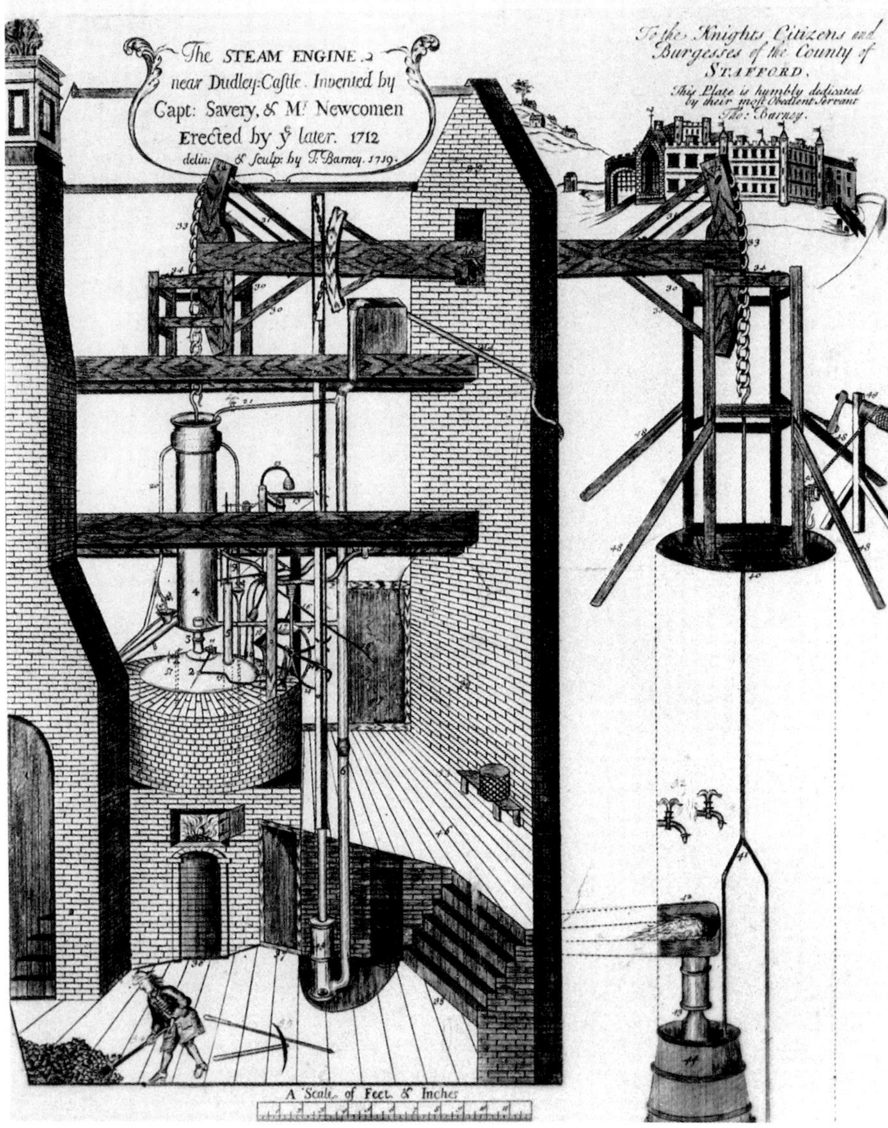

Papin's atmospheric engine of 1690 illustrated an important principle but was of no practical value. The first to make effective use of the power of steam was Savery's Miner's Friend (1698) which had limited success in draining mines and pumping water for public supply. The first true steam engine – with a reciprocating piston working in a cylinder – was Newcomen's lever (beam-) engine (left), the first example of which was erected at Dudley Castle, Worcestershire, in 1712. Cumbersome and inefficient though it was, it was not effectively challenged for half a century, until James Watt invented a much improved steam engine (1769) incorporating a separate condenser for spent steam (above).

Abraham Darby 187 **blast furnace** 184 **Hero's steam engine** 70 **Newcomen engines in America** 163 **trip-hammers** 186

For 50 years there was little improvement in the Newcomen engine, not because this was impossible, but because there was little incentive. Inefficient though it was, it was easily built and maintained, and the coal with which it was fuelled could be of very low grade. It injected fresh life into the mining industry, just as metals – and coal for smelting them – came into increasing demand. Apart from its use in drainage, it was used to work trip-hammers for crushing ore and bellows for blast furnaces, and for other heavy tasks.

Around 1770, the engineer John Smeaton – who had gained fame 10 years earlier as the builder of the Eddystone lighthouse – much improved the design and construction of beam-engines. From a careful survey of more than a hundred existing engines, he discovered that for every bucket of coal – 38 kilograms (84 pounds) – burned they would on average raise 2540 tonnes of water 30 centimetres (1 foot). By means of a new boring machine, which gave a more truly circular section to the cylinder and other improvements, he succeeded in almost doubling this performance. He built a number of huge and highly successful engines – including one with a 2-metre (6½-foot) cylinder for pumping out the docks at Kronstadt. Even so, the thermal efficiency was no more than a miserable 1 per cent.

Watt's improved beam-engine

In 1763, James Watt – then working as an instrument-maker at Glasgow University – was asked to repair a model Newcomen engine used for demonstration purposes. He was struck by its poor performance, and experiments led him to develop the very important concept of carrying out the condensation of the steam in a condenser separate from the cylinder: the latter could then be kept constantly hot. This great improvement he patented in 1769 as 'A new Method of Lessening the Consumption of Steam and Fuel in Fire Engines'. Watt could not afford to exploit his patent himself and eventually entered into partnership with Matthew Boulton, who had an engineering works at Soho near Birmingham. Thus was born the world-famous firm of Boulton and Watt. It was, however, some years before all the practical problems were solved, and they were fortunate in being granted, in 1775, a 25-year extension of Watt's original patent. On the strength of the success of two large engines – one for pumping out a coal mine and the other for blowing air into blast-furnaces – with an efficiency three times greater than that of the best Newcomen engines, orders poured in. During the years of their virtual monopoly, Boulton and Watt manufactured about 500 steam engines.

As it stood, Watt's engine was a notable advance, but it was quickly improved. It was made double-acting, so that the downward as well as the upward stroke delivered power. Steam was admitted only during the first part of the piston's motion, its own expansive force completing the stroke. No less important, an effective mechanism was devised for converting the reciprocal motion of the beam – satisfactory for moving the plungers of pumps – into rotary motion. The obvious device of connecting the end of the beam to a crankshaft was not open to Watt, as it had already been patented. Instead, he devised his famous sun-and-planet gearing, which was used until the expiry of the rod-and-crank patent in 1794. Finally, in 1787, he introduced a centrifugal governor to maintain the speed constant under varying loads.

By 1790, with all these improvements, the steam engine had become a powerful and reliable source of power, including rotary power, even though it was still clumsy and, in terms of fuel consumption, inefficient. Its great merit was that – though making the cylinder and piston was a skilled task – erection and maintenance were well within the capacity of local labour in the form of masons, blacksmiths and carpenters. For protection from the weather, the engine itself was enclosed in an engine-house, but the end of the beam stuck out through a slot in the side. The many ruins of these buildings still standing in the old mining areas are a reminder of the dramatic effect of beam-engines on the industry.

High-pressure steam

Watt was born 20 years after Savery died, but he would have been well aware of the many explosions associated with the Miner's Friend. Prudently, he himself always worked with steam at, or only a little above, atmospheric pressure. Yet the attraction of high-pressure engines was great, for they offered the prospect of much more compact engines in terms of weight per unit of power developed and higher efficiency in terms of fuel consumed. For mines, these arguments were not considered compelling, but for small factories and, above all, for land transport, great prizes might be won if high pressure could be mastered.

This was fully realized by the German engineer Jacob Leupold, whose magnificent nine-volume *Theatrum Machinarium Generale* (1723–39) – the first systematic treatise on mechanical engineering – included a design for a high-pressure non-condensing steam engine. There is no evidence, however, that this was ever built, and it is doubtful whether the materials and techniques of the day would have made it possible.

The first to achieve demonstrable success was Richard Trevithick, whose father was manager of Dolcoath Mine in Cornwall, where a Newcomen engine had been installed. In 1802, he built a steam engine at Coalbrookdale which worked at 10 times atmospheric pressure, to sustain which the boiler was 4 centimetres (1½ inches) thick. The cylinder was only 18 centimetres (7 inches) in diameter and the piston had a stroke of 1 metre (3 feet). By the standards of the day it was a remarkable advance, and within two years he had built 50 engines on the high-pressure principle. An important innovation was to divert waste steam up the chimney to improve the draught. Unfortunately, he never

The steam engine only slowly displaced draught animals as a source of power, as illustrated below by an 18th-century print of the surface workings of a coal-mine, where both are prominent. By the middle of the 19th century, however, steam engines had grown to a size which no other source of power could rival, as exemplified (right) by this 300-horsepower example (Germany, 1865).

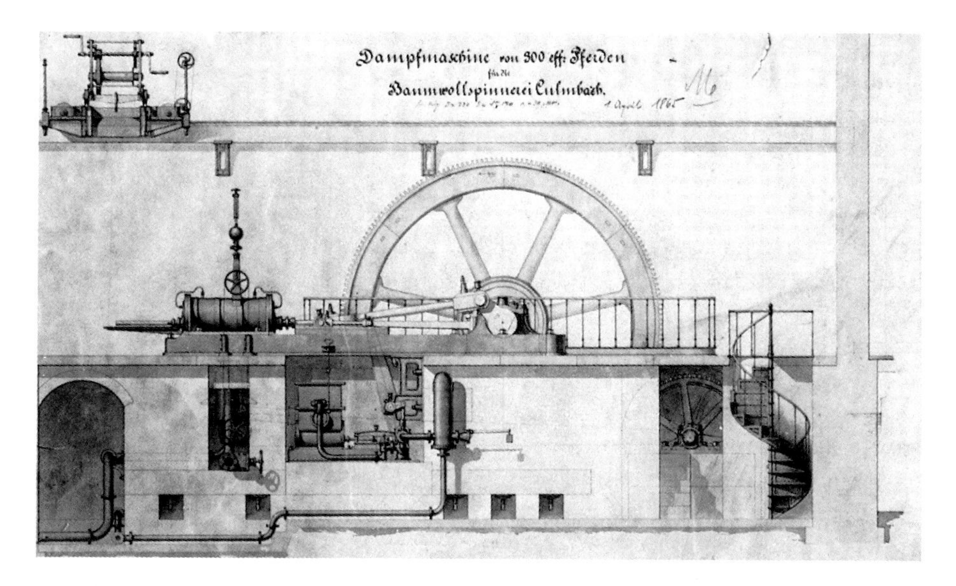

patented this: had he done so, he might have been in as powerful a position as Watt had been 30 years earlier with his patent for the separate condenser.

In 1810, on the eve of Trevithick's success, Britain had had tolerably efficient steam engines for nearly a century; high-pressure engines still had a bad reputation and there was little inducement to change. Across the Atlantic different considerations applied. At that time, in America, there were no more than half a dozen steam engines – the first Newcomen engine was imported in 1753, to drain a copper mine at Newark, New Jersey – and there were no preconceived ideas to inhibit experiment. There, in the first years of the 19th century, Oliver Evans, who began life as an apprentice to a wainwright, began to construct engines working at about 8 atmospheres pressure, much the same as Trevithick's. One, with a double-acting cylinder of only 15 centimetres (6 inches) in diameter and with a stroke of 20 centimetres (8 inches), ran at 30 revolutions per minute. A novel feature of Evans's engine was that he put the cylinder and the crankshaft under the same end of the beam, instead of opposite ends, as in the conventional Watt engine. This meant that a much lighter beam could be used. He also invented a new straight-line linkage for the upper end of the piston, which became world famous. In the course of 15 years, until his death in 1819, he built some 50 engines but did not achieve the success he deserved. Evans's work was roughly contemporary with that of Trevithick, and the latter must certainly have been aware of it, as it was described in a *Treatise of Mechanics* written in 1807 by Olinthus Gregory, who taught mathematics at Woolwich. How far the two men influenced each other is not clear.

Evans set a pattern for the future in America, where high-pressure engines were widely used from the outset, whereas in Britain Watt-type engines remained popular until mid-century. The main change was the increasing use of iron instead of timber in its construction, especially for the beam. From 1803 Arthur Woolf, another Cornish mining engineer, developed a Watt engine in which a high-pressure cylinder was placed alongside a low-pressure one. Because of its efficiency, it was popular in France, where coal was relatively expensive. In 1845, the engine gained a new lease of life when John McNaught 'compounded' it in a different way by adding a high-pressure cylinder with a short stroke.

Transmission of power via the beam was appropriate for working pump rods and very convenient in that only simple pin-joints were required. Where compactness was sought, however, it was desirable to dispense with the long beam, as in the so-called 'grasshopper' engine, patented by William Freemantle in 1803.

In all early steam engines, the cylinder was set vertical in the belief that to arrange it otherwise would result in uneven wear caused by the weight of the piston. From about 1825 this was discovered to be a fallacy and horizontal engines were increasingly used.

Up to this point the beam-engine had been developed almost entirely empirically – there was no understanding of

Boulton and Watt 147, 155, 160, 174 **Richard Trevithick** 176–177, 179 **Savery's Miner's Friend** 159 **Watt's steam carriage** 176

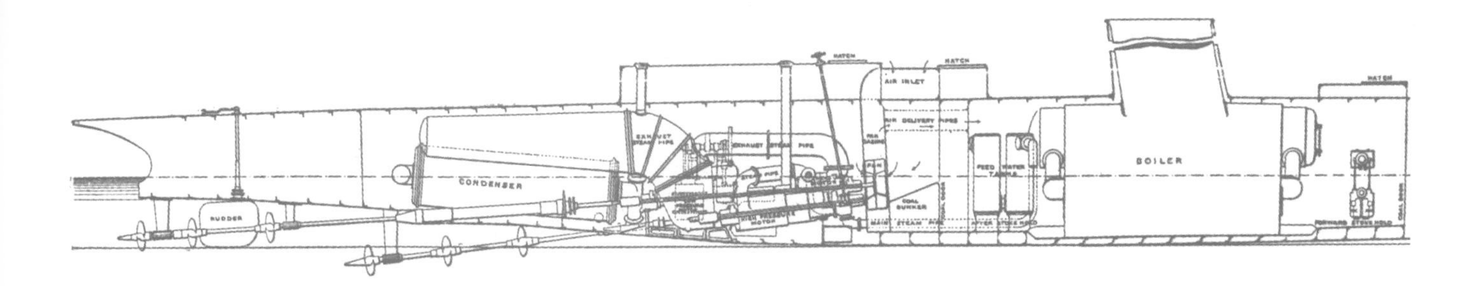

the theory that underlay its working. This situation changed in 1824 when Sadi Carnot – noted for his contribution to the new science of thermodynamics – published his *Réflexions sur la puissance mortrice du feu* in which he demonstrated that there was a maximum theoretical limit to the efficiency of any heat engine. However, despite its very important practical implications, Carnot's work made little immediate appeal to practical engineers. It was not until 1849, when William Thomson (Lord Kelvin) made use of it in formulating the Second Law of Thermodynamics, that its brilliance was appreciated.

But Carnot was no mere theoretician, for he had a very clear idea of the importance that the steam engine had assumed in the British economy. In his *Réflexions* he wrote: 'To rob Britain of her steam-engines would be to rob her of her coal and iron, to deprive her of her sources of wealth, to ruin her prosperity, to annihilate that colossal power.' Carnot was substantially right, though even 10 years after he wrote Lancashire and the textile area of Yorkshire still had nearly 900 working water-wheels as against some 1400 steam engines.

The steam engine had a symbiotic relationship with the coal industry: the mines supplied the coal for the steam engines which, among other duties, pumped them dry. In 1700, at the birth of the Newcomen engine, coal consumption in Britain was 3 million tonnes, mostly for domestic use: by 1800 it had doubled and by 1850 it had risen to 60 million tonnes. By the end of the century it had risen to an astonishing 225 million tonnes.

Cornish beam-engines working at around 3 atmospheres pressure continued to be made up to the end of the 19th century and, as they were virtually indestructible, some continued in use until well into the 20th. As late as 1903, an engine with a 173-centimetre (68-inch) cylinder was built for use in the Dorothea slate quarry in North Wales. Increasingly, however, the horizontal engine was favoured, a trend assisted by the development of machine tools capable of making the moving parts with the necessary precision at an economic price. By the end of the 19th century a tolerance of 25-thousandths of a millimetre (one-thousandth of an inch) was being attained. Typically, such engines of British manufacture would develop up to 180 horsepower at rather less than 100 revolutions per minute. In America, however, higher speeds were favoured, particularly as the electrical industry developed and dynamos had to be driven. As early as 1863, 168-horsepower engines working at 350 revolutions per minute were being built in Philadelphia and some 20 years later in England, under licence. In the latter part of the 19th century some exceedingly powerful engines were built, some models developing nearly 3000 horsepower.

The *Turbinia* opened a new era in marine propulsion when Charles Parsons staged a dramatic exhibition of the power of his steam turbine at the Spithead Naval Review in 1897. The pictures (above and opposite) show the appearance of the ship at sea and the internal layout of the machinery. The efficiency of Parsons' turbine owed much to the fact that the steam pressure was lost in stages. Gustav de Laval's turbine (1887) lost pressure in a single stage, but its relatively simple construction made it popular for light work, such as driving cream-separators and other dairy machinery (opposite, below).

The steam turbine

By 1900 the steam engine had come, in nearly 200 years, a long way from Newcomen's first clumsy beam-engine. It was a magnificent example of the design and constructional skill of the mechanical engineer, and its thermal efficiency had risen from about 0.5 per cent to over 20 per cent for the big triple-expansion engines. Nevertheless, for certain fields, notably electricity generation and ship propulsion, a different kind of steam engine, the turbine, was found to be superior.

The conception of the turbine was not new for, as we have noted, Hero of Alexandria, in the first century AD, had described how a hollow 'sphere of Aeolus' could be made to rotate at high speed by feeding it steam and allowing it to escape through tangential jets. In 1629, Giovanni Branca described how a wheel could be turned by allowing jets of steam to impinge on vanes attached to its circumference. In his early work on high-pressure steam, Trevithick constructed a 'whirling engine' in which steam at 7 atmospheres pressure escaped from the ends of a 4½-metre (15-foot) arm pivoted at its centre.

However, these early attempts came to nothing and in the event the first successful designer of steam turbines was Charles Parsons. A major factor in his success was that the nascent electrical industry required for its generators engines running faster than the 1000 revolutions per minute which was about the best that reciprocating steam engines – such as those built by Peter Brotherhood in England about 1870 – could achieve. Parsons' first turbine, built in 1884, achieved an astonishing 18,000 revolutions per minute.

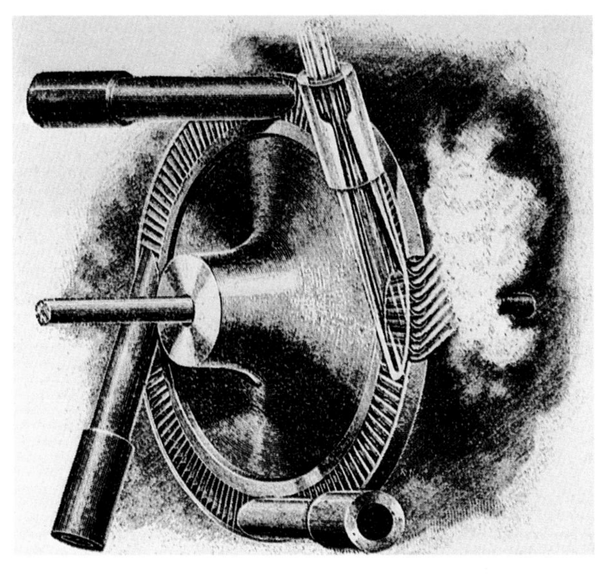

This turbine was of the axial-flow type, and consisted of two parts. The fixed part or stator had rows of stationary vanes on the inside of its casing. The rotor carried rows of vanes alternate to those on the stator. When high-pressure steam was admitted to the assembly at a central point – to balance the thrust – its impingement on the rotor blades caused the shaft to revolve rapidly. From 1889 to 1893, because of patent difficulties, Parsons devoted his attention to an alternative design in which the steam flow was radial instead of axial. When the patent problem was resolved, he reverted to the axial turbine, and in the first year of the new century supplied two 1500-kilowatt turbogenerators – then the biggest in the world – to Elberfeld, Germany. In 1897, Parsons extended his success by demonstrating a turbine-propelled ship, the *Turbinia*, at the Spithead Naval Review, where it achieved a spectacular speed of 34½ knots and opened a new era of steam propulsion at sea.

Although Parsons was the most successful of the turbine pioneers, he was by no means the only one. In Sweden, for example, Gustav de Laval – who was interested in high-speed engines for cream-separators for the dairy industry – in 1887 devised a turbine in which jets of steam impinged on vanes set on the rim of a wheel: in effect, this was a practical realization of Branca's idea of 1629. It had the defect that, unlike Parsons' turbine, the steam pressure was lost in a single stage, but it proved very successful for the relatively small machines in which he was mainly interested.

Water-power

The history of the steam turbine was also closely related to developments in the water-wheel. As we have noted, this was the main source of power at the beginning of the Industrial Revolution, though it was superseded by steam as the years went by. Generally speaking, there was little change in the traditional wheel, though John Smeaton effected some improvements by better design and by making more extensive use of iron in its construction. As late as 1854, a giant iron wheel, 25 metres (82 feet) in diameter, was erected at Laxey on the Isle of Man.

In France, however, important developments took place in the early part of the 19th century. There, Claude Burdin gave a practical course on water-wheels in the Technical College at St Étienne and in 1824 he coined the word turbine, from the Latin *turbo*, a spinning top – based on the radial outward flow of water from a vertical-axis rotor. Thus encouraged, one of his students, Benoît Fourneyron, constructed a 6-horsepower turbine of this novel type. He then moved on to build a larger engine of 50 horsepower. These machines worked under high pressures, around 100 metres (330 feet) of water pressure, and at high speeds, upwards of 2000 revolutions per minute. They were thus eminently suitable, like Parsons' steam turbine, for generating electricity and were in fact installed at the end of the century in the great hydroelectric plant at Niagara Falls. Unlike the steam turbine, however, they were limited to situations where suitable high heads of water were available. Such conditions are not widely encountered, and electricity-generating plants must be located reasonably near the consumer because of power losses in transmission. Many potential sites were therefore too remote to be useful. For the big slow-flowing rivers of Europe an axial-flow water turbine – in which the water flowed parallel to the horizontal shaft, as introduced by N. J. Jonval in 1841 – was more suitable, and it was also used in America from 1850.

Hero's steam engine 70 **hydroelectric power** 231 **John Smeaton** 162 **machine tools** 153–154 **water-wheels** 68–70, 115–116

Both types found a serious rival, however, in a third type of turbine – in which the water flows radially inwards – proposed in 1826 by another French hydraulic engineer, Jean Victor Poncelet. Such a machine does not seem to have been built, however, until 1838 and then in small numbers, not in France but in New England, by Samuel Howd. These prototypes seem to have been rather crude, and the design was much improved by another American engineer, James Francis, working in Massachusetts. His employers were locomotive manufacturers, but his success led them to abandon this in favour of hydraulic engineering. His turbine eventually combined radial and axial flow, and for this reason is generally referred to as the mixed-flow turbine.

Windmills

Windmills remained important sources of power long after the use of steam was widespread. In Holland, industry developed on the banks of the Zaan to the west of Amsterdam and here alone there were nearly 1000 windmills at the end of the 17th century, used not only for grinding corn but also for pumping water, sawing wood, working trip-hammers and a variety of other industrial purposes. At one time the United Provinces had 8000 windmills in operation and even as late at 1900 there were still 2000.

Although no new principles emerged, the windmill continued to be refined, and some of the later ones are magnificent examples of mechanical engineering in heavy timber. As they became more massive, mechanical aids were invoked to turn the sails into the wind, including, from about 1750, fantails which acted automatically through gear chains. Also, as in most other machinery, increasing use of iron was made from about the middle of the 18th century, particularly for casting gear-wheels. However, unlike the water-wheel – which fathered the water turbine – the windmill was, for the moment, the end of the line, though even in the 20th century it remained important, especially for local use in remote areas. Lately, with worldwide searches for new energy sources, 'aeolian' power has again been investigated. This has led to some novel designs – notably vertical rotors somewhat like the original Persian windmills – but no significant overall contributions.

The internal combustion engine

Papin's concept of a steam engine stemmed from Huygens' experiments in 1680 with a gunpowder engine, in which the gas generated in the explosion drove a piston up a cylinder. The practical difficulties and hazards of this arrangement led to its abandonment, but the principle was sound enough, and inventors reverted to it a century later. In 1794, for example, Robert Street lodged a patent in England for a piston engine fuelled with gas derived from tar-oil or turpentine and ignited with a flame. In 1801 Philippe Lebon, pioneer of the coal-gas industry, took out a patent in Paris for a double-

As its name implies, the post-mill could be turned on a vertical pillar to face the wind. From about 1750 this could be done automatically by use of a fantail, as in this example from Hotton, Suffolk, England.

acting coal-gas engine with electrical ignition: a remarkably advanced conception, for Alessandro Volta had invented his electric battery only a year earlier.

In the event, however, no real success was achieved until 1859 when Étienne Lenoir, a French engineer, built a piston-engine, modelled on the steam engine, in which a mixture of coal-gas and air was ignited by a spark from an induction coil. By that time the coal-gas industry was firmly established and gas was widely and cheaply available in urban areas. However, this engine was very heavy in proportion to the power it developed, and although a few were sold no real progress was made until 1862 when another Frenchman, Alphonse Beau de Rochas, realized that efficiency could be much improved by compressing the gas/air mixture before ignition. This led to the development of a really successful gas engine by Nikolaus August Otto in partnership with Eugen Langen. Together they founded the firm of Otto and Langen, and in 1867 their gas engine won a gold medal at the Paris Exhibition. These original engines were essentially 'atmospheric' engines of the Newcomen type. The explosive in the vertical cylinder drove the piston upwards and as the gas cooled atmospheric pressure forced it down again.

Otto's main contribution was the four-stroke cycle, which he announced in 1876. The first stroke draws the explosive mixture into the cylinder and the second compresses it; the mixture is fired at the beginning of the next stroke – when the piston is near top dead centre and compression is at maximum – and the burned gases are expelled on the fourth stroke. (See the illustration on page 251.) There seems no doubt that Otto independently invented this cycle, which he patented in 1876, but it proved to have been described already by Beau de Rochas and in 1886 his patent was invalidated.

At first sight, the success of the Otto cycle was surprising. As we have seen, in the double-acting steam engine – recognized as an important advance on the original Newcomen engine – every stroke was a power stroke. In the Otto cycle, however, only one stroke in four was a power stroke, which on the face of it was wasteful and would lead to uneven running. The explanation is that the cycle secured the all-important compression of the mixture and the advantage of this outweighed the disadvantage of three non-power strokes in four.

Otto's original atmospheric engines had perforce to have a vertical cylinder, but this new development made it possible to design horizontal ones, which proved very popular. In 1869, Crossley Brothers of Manchester became licensees of Otto and Langen and were highly successful in manufacturing the engine in Britain, selling 40,000 before 1900: by the mid-1880s few British factories of any consequence lacked an Otto engine. It was an excellent engine – commonly developing power in the 1- to 10-horsepower range, though a small number of much larger ones were built, up to 1000 horsepower, for driving individual machines, or small groups of machines. It also found use on farms for driving barn machines such as chaff-cutters. In Britain, the Crossley gas engine was guaranteed to cost less than one (old) penny per horsepower-hour: if we assume that a 5-horsepower engine was employed on a farm for five hours daily, its annual overall cost would have been around £40.

Bearing in mind that great numbers were made, and that in their developed form they were cheap and efficient, gas engines have been strangely neglected in the history of technology. By 1900 some 200,000 were in use in a dozen countries but mainly in Germany, Britain and America. However, they enjoyed only a limited popularity, for about 30 years, before being supplanted by the petrol engine, which had the advantage of being independent of a fixed source of fuel and so was very suitable as a power unit for transport.

The petrol engine

In 1869 Otto and Langen built a factory near Cologne. There Otto looked after the business side, leaving Langen to develop the engineering side. Langen was helped by two young men, Gottlieb Daimler and his assistant Wilhelm Maybach, who were thus very well placed to see at firsthand the nature and problems of building four-stroke internal combustion engines. In 1882 Daimler left, taking Maybach with him, to set up on his own and try to develop an engine light enough but powerful enough to be used to propel road vehicles. It was a brave move, for apart from the purely technical difficulties there were also doubts whether there could be much demand for such an engine even if it should prove successful. In the event, within the space of 20 years, they succeeded in developing the automobile engine to something closely resembling its present form, and there was soon no question of the demand for it for mechanical road transport.

The first effective internal combustion engines were fuelled by coal-gas, as in this horizontal Otto engine of about 1878 (left). Although tens of thousands of such engines were made, they were eventually displaced by the petrol-fuelled engine, which was peculiarly suitable for the propulsion of road vehicles. The pioneer was Daimler, one of whose early engines (*circa* 1890) is shown above.

four-stroke cycle 251 **Huygens' gunpowder engine** 159 **trip-hammers** 186 **Volta's battery** 169 **windmills** 114–115, 231, 277

The diesel engine is similar in principle to the conventional petrol engine but is intrinsically more efficient. It works at a much higher compression ratio, and the heat generated by the compression ignites the fuel, so no ignition system is needed. Inset into Diesel's original sketch is an engine of about 1920.

Some attempts had been made to make the gas engine mobile by using cylinders of compressed gas, such as had been introduced in Germany in 1871 for lighting railway carriages, but the weight of these and the problems of replacement proved too great. To use transportable gas generators was also dismissed as impracticable, so the choice of fuels limited itself to some volatile liquid that could be vaporized and drawn into the cylinder mixed with air.

Here several alternatives offered themselves, such as alcohol or ether, but for cheapness and ready availability in quantity the obvious choice was petrol (gasoline), then an unwanted and dangerous product of the new petroleum industry. The ease of ignition of petrol, which caused so much anxiety in petroleum refining, was its main virtue in the eyes of engineers such as Daimler. Nevertheless, the problem of preparing and igniting just the right mixture proved very difficult.

Even before leaving Otto and Langen, Maybach had successfully run a stationary gas engine on gasoline vapour, but to do this on a moving vehicle was a different matter. Initially Daimler and Maybach used an evaporation system in which air was drawn over liquid petrol, but by 1893 Maybach had devised a float-feed carburettor in which a fine spray of petrol was injected into the airstream. Ignition was also very difficult, as this had to be done at engine speeds of around 600–900 revolutions per minute. Their solution was a tube inserted into the cylinder and kept constantly hot by a petrol-fuelled flame. Working independently in Mannheim, Karl Benz reintroduced in 1885 the electrical system of spark ignition which Lenoir had used: this was universally adopted and is still standard. Benz's electricity came from an accumulator, but this was not reliable and for a time, therefore, engine-driven magnetos were popular. However, Benz's engine was generally less satisfactory than Daimler's in that its speed was much lower, and its performance was limited by a surface evaporation carburettor.

In 1889 Daimler patented a V-type two-cylinder petrol engine and in the same year he demonstrated a car equipped with this at the Paris Exhibition. This attracted little attention, however, and a car exhibited by Benz suffered the same indifference. Daimler eventually gained some backing in France, but from 1895 he worked again as an independent manufacturer in Germany. Subsequently he sold considerable numbers of his V-2 engine to Continental car manufacturers. The further history of the petrol engine is so closely related to the history of the car that it is best discussed later, in the context of transport, though mention must be made of another variant, the diesel engine.

The diesel engine

The early history of the internal combustion engine supports the general thesis that the Continental industrial strategy of investing heavily in technical education paid good dividends. Lenoir, it is true, was self-taught, but both Daimler and Benz had benefited from the training offered by the new German technical schools, in Stuttgart and Karlsruhe respectively. Rudolf Diesel was a graduate of the Technische Hochschule in Munich. In Britain, where Parsons' turbine had produced a new and much-needed variant of the traditional steam engine, there was no pioneer work at all on the new engine. Not until 1897, when Frederick W. Lanchester introduced a more evenly balanced engine with a carburettor in which air was drawn over a series of wicks dipping into a reservoir of petrol, was there any significant British contribution, and by then German engineers had fully evolved the basic design.

In Munich, Diesel studied thermodynamics under Carl von Linde, later famous for the manufacture of liquid air. In 1880, he went to Paris. By this time, the originally overlooked pioneer work on thermodynamics done by Carnot had received recognition and Diesel realized the significance of the Carnot heat cycle for the design of an efficient engine. Indeed, even before starting to build his engine, but not before he had secured his position with a patent, he published in 1893 a description of it in a memoir entitled *The Theory and Design of a Rational Heat Engine*.

The diesel engine works at very high compression – in his first experiments Diesel thought in terms of 250:1 – and initially it is air and not an air/fuel mixture that is compressed. This high compression heats the air so much that when fuel is injected it ignites spontaneously, and there is thus no need for an external ignition system. It is interesting to recall that some primitive people used heat generated by air compression as a means of making fire. As the mixture burns while the piston moves downward, combustion proceeds at constant pressure, as distinct from the Otto cycle where it takes place at constant volume. Thermodynamically this is a more efficient arrangement and this, coupled with the fact that the engine will run on heavy oil, which is cheaper than petrol, gave it distinct advantages. Diesel attracted worldwide interest when he exhibited his new engine at the Munich Exhibition of 1898 and he soon became a rich man. But the practical advantages of his engine were realized only after several further years of intensive development work, and then only partially. A disadvantage was that the engines ran rather rough and had therefore to be of very solid construction. This was of no great consequence for stationary engines, but had to be overcome before the diesel engine could be used for road transport.

The electric motor

Although the Industrial Revolution was initially based on the use of water-power, steam quickly became dominant and remained so into the 20th century. By then, however, it was being seriously challenged not only by the internal combustion engine but also by the electric motor. Like the gas and petrol engines, this had the powerful appeal that it could be made in small units suitable for driving individual machines, thus avoiding the noisy power-wasting systems of belts and overhead pulleys that were a particular characteristic of factories in the 19th century.

Up to the beginning of the 19th century virtually the only sources of electricity were frictional machines such as that described by von Guericke in 1672. In 1800, Alessandro Volta invented the voltaic pile, consisting of alternate discs of zinc and silver or copper separated by felt soaked in brine: this gave a continuous supply of electricity.

In 1866, Georges Leclanché introduced a battery – suitable for bells and other intermittent uses – in which rods of zinc and carbon were dipped into a solution of ammonium chloride. In the 20th century, it was developed as the

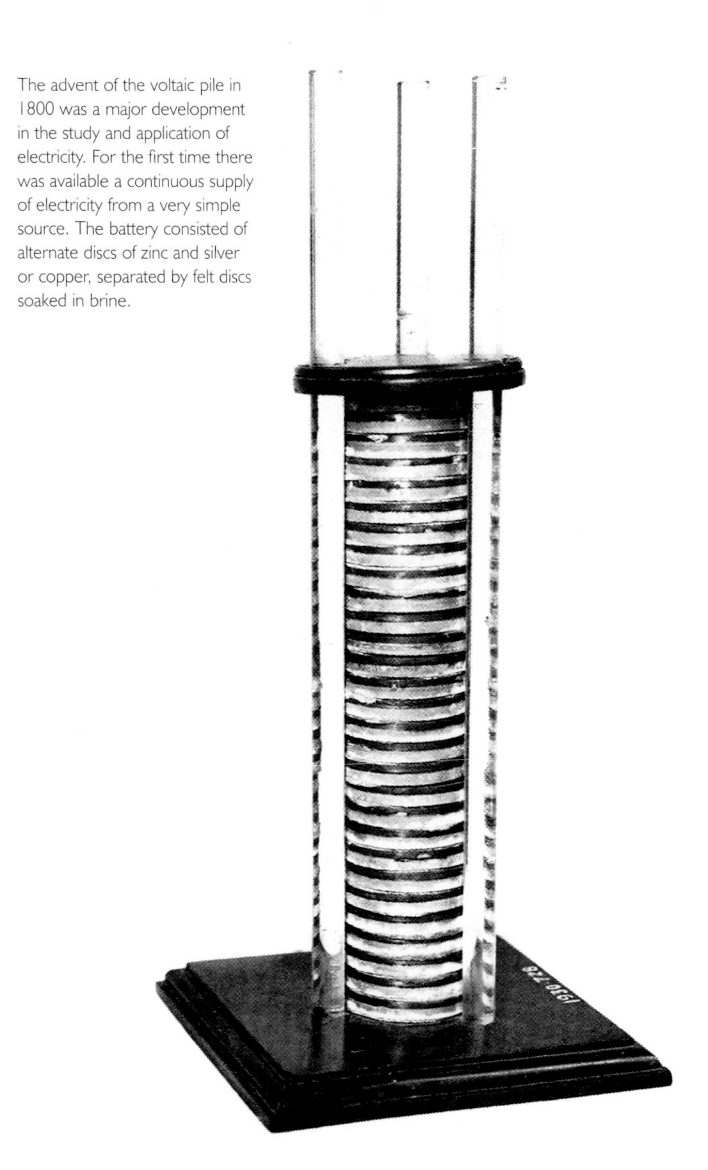

The advent of the voltaic pile in 1800 was a major development in the study and application of electricity. For the first time there was available a continuous supply of electricity from a very simple source. The battery consisted of alternate discs of zinc and silver or copper, separated by felt discs soaked in brine.

development of automobile 248–250 **diesel engine** 250, 251 **Drake's oil strike** 226 **Parsons' steam turbine** 164–165 **Sadi Carnot** 164, 223

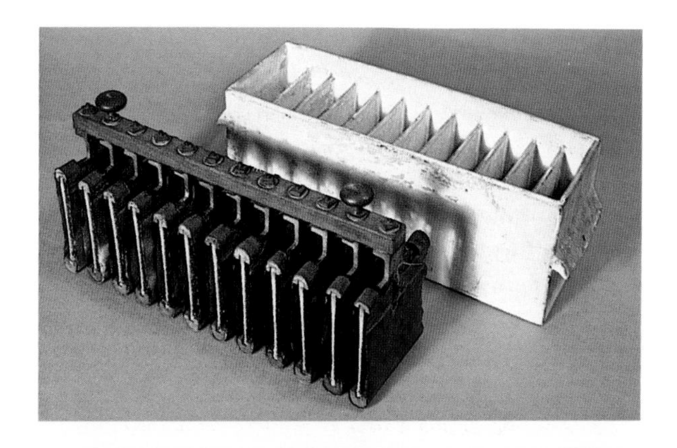

Volta's simple battery soon appeared in more sophisticated forms, as in this 12-plate zinc-copper-sulphuric acid trough battery (top) devised by William Hyde Wollaston about 1805. The coil (above) devised by Nikola Tesla in the 1880s, was a convenient means of transforming low voltage current from a battery to a much higher voltage; it was popular in the form of the 'shocking coil'.

familiar dry-cell battery used in billions throughout the world. Such batteries are called primary batteries, because they have a limited capacity and become exhausted with use. Secondary batteries (accumulators) are batteries which can be repeatedly recharged by passing a direct electric current through them. Although such batteries were devised as early as 1803, the prototype of modern accumulators was the lead acid battery invented by Gaston Planté in 1860. Such a battery gives about 2 volts, but higher voltages are easily obtained by coupling them in series, and large amounts of electricity can be stored by making them large. Such batteries were much used in the electric supply industry – which initially generated direct current – so that electricity could be stored in slack periods to meet peak demands. With the widespread adoption of alternating current, this outlet diminished, but in the 20th century the car industry created an enormous demand for batteries to power starting motors and lights.

In 1820, the Danish physicist Hans Christiaan Oersted discovered a direct connection between electricity and magnetism by demonstrating the existence of a magnetic field surrounding a wire carrying a current, and in Paris André Marie Ampère quickly quantified the relationship. But it was in London that the really significant experiments were carried out. There, at the Royal Institution, Michael Faraday showed in 1831 that a wire carrying an electric current would tend to move if placed in a magnetic field, and conversely a current would be generated in a wire moved through a magnetic field. Thus electricity could be converted into motion and motion into electricity.

Very quickly electrical generators were devised based on the rotation of a wire coil relative to a permanent magnet: the first such seems to have been that exhibited by Hippolyte Pixii to the *Académie des Sciences* in Paris in 1832. Permanent magnets were soon replaced by electromagnets activated by batteries. Then, in 1886, Cromwell F. Varley discovered and patented the principle of self-excitation, whereby the electromagnets were activated by electricity provided by the generator itself. By that time, too, multiple coils had been introduced, forming an armature, so that a fairly steady, rather than fluctuating, voltage was produced. Thus the main features of the dynamo had emerged, and a steady and dependable supply of electricity could be provided by coupling it to a suitable motor. As we have seen, high-speed motors were desirable for this purpose and this encouraged the development of turbines.

The electric motor is essentially a dynamo working in reverse; if electricity is supplied to the field coils, the rotor will turn. In 1873, Zénobe Théophile Gramme, a pioneer manufacturer of dynamos, showed that two of his dynamos could be coupled together to act indifferently as generator or motor. However, Gramme's motor could be used only with direct current, whereas from 1882 the electricity supply industry was increasingly turning to alternating current, the voltage of which could be more readily stepped up or down by transformers to meet the varying requirements of power transmission and industrial or domestic use. The difficulty was resolved in 1881 by Nikola Tesla, a Serbian electrical engineer working in Budapest. However, it was not until 1888, after he had emigrated to America, that he was finally able to market his alternating current motor. Five years later his system attracted wide publicity when it was used to illuminate the Chicago World Exposition of 1893.

CHAPTER EIGHTEEN

New Modes of Transport

Until well into the 19th century, the pattern of transport was very much what it always had been. In all cases the propelling force was the power of men, draught animals or the wind. These were to remain important – and in many parts of the world dominant – up to the present day, but at last a revolutionary new factor was introduced: on both land and sea, engines – especially steam engines – began to provide the motive force. No less important in the long term, but of no great practical significance until the 20th century, was the first conquest of the air. For the first time, man was no longer confined to the surface of the globe but could soar into the air: transport had quite literally gained a new dimension.

The beginning of flight

From the earliest times, birds had demonstrated the feasibility of flight, and the legends of many races have references to beings endowed with this ability – Mercury, messenger of Jupiter, with his winged heels and helmet, is a typical example. The reality came very late in human history, however.

Aerial transport must be taken to imply flight by vehicles with no connection whatever with the ground. While this excludes kites, which are anchored by a rope, their early history is nevertheless interesting. Although Western tradition ascribes the invention of the kite to Archytas of Tarentum in the fourth century BC, it was in China and the Far East that they were most widely used and developed. Apart from being a very popular pastime, kite-flying has a long history of military use. While the story that man-carrying kites were used for observation during a siege of Sung in the fourth century BC is of doubtful authenticity, kites were certainly used for signalling in the sixth century AD. At the siege of K'ai-feng in 1232, kites were used to distribute propaganda leaflets over the city to encourage defection by Chin soldiers. In America, in a classic experiment conducted in 1752, Benjamin Franklin used a kite to demonstrate the electrical nature of lightning.

The first to achieve free flight, with no ground contact whatever, were almost certainly the Montgolfier brothers who began their experiments with hot-air balloons in the early 1780s. Their interest is said to have been aroused by the possibility of breaking the blockade of Gibraltar (1779–82). Their first model flew in 1782, and on 21 November 1783 two passengers made a flight of 9 kilometres (5½ miles), remaining airborne for half an hour and reaching a height of 100 metres (330 feet). The Chinese, it is true, have a long

Balloons had a long history of use for military observation. This picture shows the inflation of an observation balloon during the French invasion of Fès, Morocco, in 1911.

tradition of 'flying dragons', which may have been similar hot-air balloons, but equally may have been paper lanterns attached to kites. In any case, there is no suggestion that these were of a size to carry passengers.

From 1783 there is a continuous history of ballooning up to the present day, partly for pleasure and partly for transport. Nevertheless, it was never of more than marginal success: the future lay with heavier-than-air machines – though not until the 20th century. Nevertheless, achievement was not negligible, especially after the introduction of the newly discovered gas hydrogen as an alternative to hot air only a week after the epic Montgolfier flight. By 1785, the English Channel had been crossed and before the end of the century the armies of Britain, France and America were using balloons for observation. Not until 1984, however, was the Atlantic first crossed by a manned balloon, and the first voyage around the world was completed in 1999.

The great disadvantage of the balloon was that it was at the mercy of the wind, and attempts to overcome this by means of an engine and propeller were unsuccessful through lack of an engine with a high enough power:weight ratio. The steam engine was quite unsuitable, but in 1884 a French airship, *La France* – cigar-shaped to lower wind resistance – made an 8-kilometre (5-mile) flight under the power of an electric motor run by batteries. Such early airships were pressure-ships: that is, they were kept in shape by the pressure of the gas inside them. At the end of the century, the Germans introduced the rigid airship, in which a light metal framework enclosed a number of gas bags, the contents of which could be individually regulated. These were developed by Count Ferdinand von Zeppelin, who launched his first airship in 1900.

The ship

In 1760, the availability of the steam engine and Darby's invention of iron-smelting by means of coked coal instead of charcoal were two important elements that paved the way to the dominance of the steam-propelled iron ship in the 19th century. Yet the day of the great wooden sailing ships was far from over and it was estimated that as late as 1890 as much as 90 per cent of the merchant shipping of the world was of wooden construction.

The Battle of Trafalgar on 21 October 1805, in which the Royal Navy defeated the combined fleets of France and Spain, was the last great battle between sailing ships. But perhaps the finest ship afloat that day was the Spanish vessel *Santísima Trinidad*, representative of a class of three-masted sailing ships known as *navios*. Some 220 of these were built in various 'ratings' according to size and armament. They were so designed that certain structural elements were common to all: this element of mass production favoured speedy building and kept the cost down.

About 70 of these *navios* were built in the Spanish dockyard in Havana, largely from the locally available mahogany, which was much more durable than the oak commonly used in French and British shipyards. The *Santísima Trinidad* itself was a 1900-tonne vessel, with a crew of 1200 sailors and marines, and with walls 60 centimetres (2 feet) thick. The whole was held together by iron bolts up to 2 metres (6 feet) long. A third-rate ship of the line – carrying about 80 cannon on two decks – required the timber from 3000 trees.

The three-masted *Anglesey* (left), 1150 tonnes, was typical of the 19th-century clipper ships designed mainly for the Chinese tea trade and the African slave trade. Sailing barques, such as the iron-hulled *Garthsnaid* (above), survived into the latter part of the 19th century, but the large crews necessary to set the great spread of sails proved an increasing disadvantage.

Such vessels were essentially floating gun platforms. At the other end of the line were the long slim clippers built mainly for the tea trade with China and the slave trade between Africa and the southern states of America. These were capable of making the run from Shanghai to London in less than 100 days. On some days a clipper might sail more than 400 nautical miles (1 nautical mile is equivalent to 1.85 kilometres or 1.15 land miles). These 'ocean greyhounds', with a length:beam ratio of about 6:1, were of American origin, the first being built in Virginia and Maryland around 1812. Apart from their slim design, they were unusual in being built not of oak or other hard woods but of the soft woods abundant on the American coast. These soon became soggy and warped, and the life of the hull could be as little as five years. Nevertheless, they were so much cheaper to build, and their skippers such notoriously hard drivers, that they still earned a handsome profit for the owners.

From mid-century these clippers were particularly identified with Donald McKay of East Boston, and some of his ships made spectacular runs. The discovery of gold in Australia in 1848 started a rush of traffic and several clipper ships were ordered from McKay by James Baines' Black Ball Line of Liverpool. One, the *James Baines*, sailed the 14,000 nautical miles from London to Melbourne in 1854 in 63½ days, carrying 700 passengers and 1400 tonnes of cargo. In 1868, the Aberdeen-built *Thermopylae* set a new record, which still stands, of 59 days for this run. Top speeds of 20 knots were attainable: this is little less than the cruising speeds of the great Atlantic liners built just before the First World War. With the opening of the Suez Canal in 1869, steamships took over the lucrative tea trade, but the growth of sheep-rearing in Australia provided work as a new and profitable cargo for the clippers until the end of the century. There still remained other competitive runs for sailing ships, notably to the Pacific coast of South America, via Cape Horn, where no coaling stations had been established. This required much bigger vessels, and the largest sailing trader ever built was launched in France at late as 1911: this was the 8000-tonne, five-masted *France II*. Such ships carried an enormous spread of sail – up to 5000 square metres (54,000 square feet) – requiring a correspondingly large crew: this, too, was a disadvantage in the losing battle with steam.

The iron ship

The later sailing ships had hulls of steel, an innovation that began with a 21-metre (70-foot) iron barge that the ironmaster John Wilkinson launched on the Severn in 1787, but did not make much impression on the shipbuilding industry for nearly half a century. The first iron ship in America, the 18-metre (60-foot) *Codorus*, was built in 1825. In Europe, Britain had become a great shipbuilding centre, but even one of the biggest firms, Cammel Laird, did not start building iron ships until 1829, seven years after the first iron steamship crossed the English Channel. It was a development that went hand in hand with the adoption of steam in place of sail for propulsion.

In anything but a calm sea the hulls of ships are subject to considerable strains because they are not equally supported at all points as they plough through the waves. Although shipbuilders from Egyptian times were aware of this liability to 'hogging' and 'sagging', they could combat it only by empirical means: in practice, this limited the length of wooden ships to about 100 metres (330 feet). By the middle of the 19th century, however, marine architects had gained an understanding of the underlying theory and could apply it effectively to the design of iron hulls of much greater length.

One source of weakness in wooden hulls was that they had of necessity to be constructed from relatively short lengths of timber, simply because that is the way trees grow. This practice was so ingrained that when iron was first used it was still followed, so that the great strength of iron was not fully realized. By the middle of the 19th century, however, the main frame members could be rolled or cast in a single piece. The skin of the vessel consisted of iron plates of standard size held in place by rivets driven into drilled holes while red hot and then hammered flat: not until the end of the century did pneumatic riveters begin to appear. With the advent of the Bessemer and Siemens processes in the 1850s, steel became increasingly available as a constructional material. It was sporadically used in shipbuilding by 1880, but 10 years later its use was almost universal.

The *Great Eastern*, here seen under construction, was five times bigger than any ship before her. She was a technological triumph but a commercial disaster, because of a serious underestimate of her fuel consumption.

Abraham Darby 187 **sailing ships** 50–53, 105–109 **steam engines** 159–165 **steel production** 187–189 **Zeppelin airships** 221, 256

Propulsion

The steam engine had obvious attractions for ships' propulsion: it could provide a steady output of power even in a flat calm. Its initial disadvantages were its great weight and size in proportion to the power produced, and its appetite for coal. The latter could be easily satisfied for ships working from a fixed base, or coastwise, but long ocean voyages were possible only if either large supplies were carried on board – with corresponding loss of cargo or passenger capacity – or a regular system of coaling stations could be established. It was this problem of fuelling and the natural conservatism of the military that made the world's navies hesitant about dependence on steam long after merchant fleets on fixed runs had adopted it. Naval vessels had to be able to undertake duties at short notice in remote waters, often for long periods.

A basic problem was that of converting the rotary motion of the steam engine into forward movement of the ship, for which there was no precedent. For two reasons, the paddle-wheel was adopted. First, it was essentially a long-familiar water-wheel acting in reverse. Secondly, the idea was not new, as we have noted: a Roman author had proposed a paddle-boat driven by oxen in a treadmill and the Chinese – of whose achievements the West had become increasingly aware since the Jesuits established themselves in China in the 17th century – had built many such ships driven by manpower.

The French were the pioneers of the steamship. Their first attempt, on the Seine in 1775, was a failure, but in 1783 the 185-tonne *Pyroscaphe* gave a promising demonstration on the Saone. Only four years later, the *John Fitch*, named after its inventor, sailed on the Delaware at Philadelphia; this was a 15-metre (50-foot) stern-wheeler powered by a steam engine built by Henry Voight, a local clockmaker. Ambitiously, Fitch then sought to establish a regular steamboat service between Philadelphia and Burlington, 30 kilometres (20 miles) away, but it proved too unreliable. The scene next shifted to Scotland, where Patrick Miller had been experimenting with manually operated paddle-wheels mounted between the twin hulls of a catamaran. In 1788, he commissioned William Symington to install a Watt-type steam engine in it, but although this achieved a speed of 6½ kilometres per hour (4 miles per hour), Miller seems to have lost interest and reverted to manpower. However, this trial had been noticed by Lord Dundas, Secretary for War, and he commissioned Symington to build a steam tugboat to replace the horses used to tow barges on the Forth and Clyde Canal, of which he was a governor. The outcome was the famous *Charlotte Dundas* (named after his daughter), an 18-metre (58-foot) wooden boat fitted with a 12-horsepower single-cylinder engine driving a stern wheel. This was in regular use for a month and is generally regarded as the first successful steamship; she was withdrawn from service only because her wash was damaging the banks.

One of the first successful steamboats was the *Charlotte Dundas* (top), here seen undergoing trials in 1802. The remarkable engineering drawing (above) is one of a series of tinted lithographs of the *Great Eastern* prepared by John Scott Russell, the architect, in 1860 to vindicate his reputation after the ship's failure.

Thereafter progress was rapid. One of those who had travelled on the *Charlotte Dundas* was an American engineer, Robert Fulton. After trials on the Seine, he returned to America to build the 100-tonne *Clermont* to ply on the Hudson River. As no suitable engine could be built locally, he ordered one from Boulton and Watt. This drove two side-paddles. This ship was an immediate commercial success from its maiden voyage of 1807, and after two seasons Fulton ordered a sister ship, the *Phoenix*. As this was built at Hoboken, New Jersey and had to steam 150 kilometres (100 miles) south to the mouth of the Delaware, it was the first ocean-going steamship. During the next few years regular services were widely established – between Glasgow and Fort William, Brighton and Le Havre, London and Leith. These were all relatively short hauls with small vessels, but already owners had their eyes on much longer and potentially more lucrative runs. The main problem was no longer mechanical, for engines were much more reliable, but that of storing sufficient coal, and the first compromise was to combine sail and steam. In 1819 the full-rigged American ship *Savannah*, equipped with paddle-wheels that could be lifted out of the water when not needed, crossed the Atlantic in 27 days 11 hours, but used her engine for only 85 hours. More ambitiously, the British ship *Enterprise* reached Calcutta in 103 days in 1825, on more than half of which the engine was used.

But the principal goal was the Atlantic crossing and this was the occasion of a famous race in 1838. In that year the Great Western Railway Company launched the 1300-tonne *Great Western* to provide a regular service between Bristol, then their main-line terminal, and New York. Their rivals, the British and American Steam Navigation Company, responded with the promise of a larger ship, the *British Queen* and, when completion of this was delayed, chartered the *Sirius*, from the London–Cork service. The outcome was that the *Sirius* sailed into New York from Cork just ahead of the *Great Western*; but her fuel was virtually exhausted, while the *Great Western* had not only made a faster run but arrived with 200 tonnes of coal still in her bunkers. To *Sirius* goes the palm of being the first ship to cross the Atlantic solely under steam, but the *Great Western* demonstrated that transoceanic voyages were commercially profitable.

The paddle-wheel had the disadvantage of lifting out of the water in rough seas. The Archimedean screw for pumping water had been in use for more than 2000 years, and screw propulsion for ships had been experimented with even in the late 18th century, but its practicability was not convincingly demonstrated until 1838. In that year the Swedish engineer John Ericsson successfully employed it in a small ship aptly named *Archimedes*. Among those who witnessed the trials was Brunel, who had then embarked on building a 4000-tonne paddle-steamer, the *Great Britain*, at Bristol. He promptly halted work on this and redesigned her for screw propulsion. This, like other steamships of the day, was fully rigged for sailing, but during scores of Atlantic crossings sails were only once put into service.

The superiority of the screw for marine propulsion was convincingly demonstrated in 1845 in a contest between the warships *Rattler* (propeller) and *Alecto* (paddle-wheel). By that time screw propulsion had been tested for more than half a century, using a great variety of designs (below).

In 1854, Brunel embarked on a new venture that was to be as great a commercial failure as it was a technological success. This was the *Great Eastern*, five times bigger than any ship then afloat. Her total weight was nearly 19,000 tonnes, of which the iron hull accounted for 6350 tonnes: 30,000 iron plates and three million rivets went into her construction. She was equipped for screw and paddle propulsion. She was designed for the Indian and Australian run, to carry 4000 passengers. In fact, she never made this run, as her fuel consumption proved to have been fatally underestimated. After a brief period on the Atlantic run, she was put into service in 1865 to lay transatlantic telegraph cables and, later, one from Aden to Bombay.

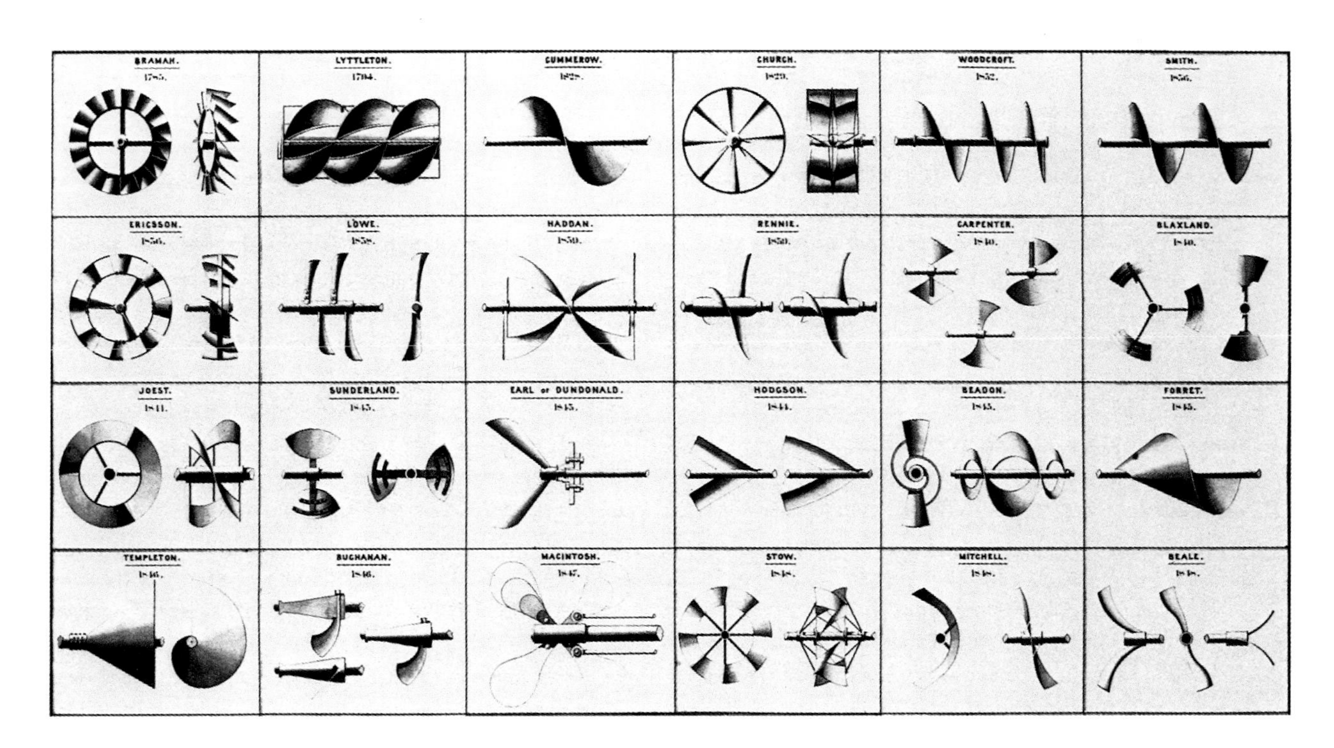

Archimedean screw 30 **early paddle-wheel ships** 68, 109–111 **modern ships** 254–255 **Watt's steam engine** 162

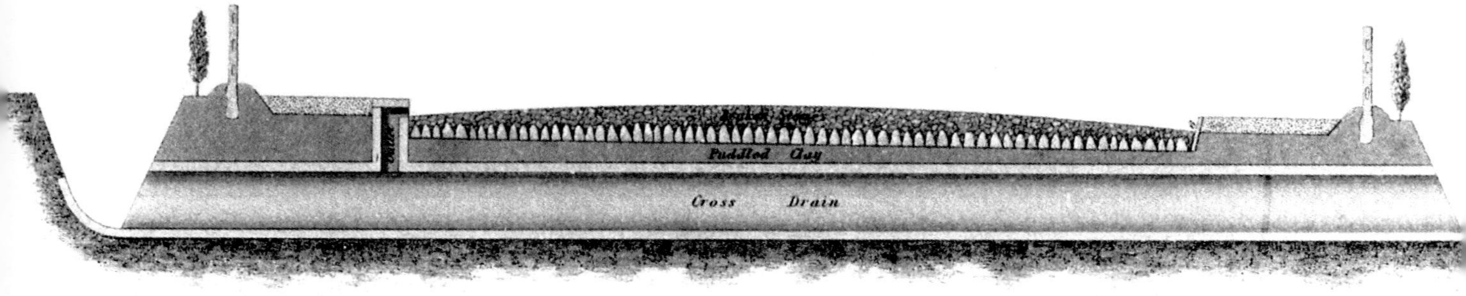

Puddled Clay

Cross Drain

The latter part of the 18th century saw a general improvement in the technique of road construction, using a stone bed and a cambered surface to throw off rainwater to lateral ditches. In Britain, one of the pioneers was Thomas Telford, famous builder also of canals and bridges. His method is illustrated in this section of the improved Glasgow–Carlisle road, for which Parliament voted £50,000 in 1814.

The superiority of the screw was only slowly accepted, one problem being that as it needed to turn faster than paddles some form of gearing between engine and propeller shaft was necessary. The Royal Navy was among the sceptics but in 1845 staged a convincing comparison. Two identical frigates, *Rattler* and *Alecto*, were fitted respectively with a four-bladed propeller and a pair of paddles. In a 150-kilometre (100-mile) race *Rattler* won by several kilometres and then, linked to *Alecto* by a hawser, pulled the latter backwards at nearly 3 knots. At the very end of the century Parsons' steam turbine provided a new form of marine engine, fast-turning and with a very favourable power:weight ratio. In 1899, turbines were fitted in HMS *Viper*, the first warship to use them. Before the First World War, diesel engines, too, had been applied to marine propulsion.

Steam transport on land

As a source of motive power for land vehicles, the steam engine had obvious attractions in being a self-contained unit. Whereas the development of the ocean-going steamship was hampered by the need to carry disproportionately large supplies of coal; land vehicles could be refuelled at frequent intervals. The main obstacle to be overcome, apart from public prejudice and the opposition of vested interests, was the lack of a suitably compact, efficient and reliable engine. While beam-engines were used in Mississippi and Ohio steamboats – 5000-tonne vessels with paddle-wheels up to 12 metres (40 feet) in diameter – land vehicles required more compact prime movers. Equally, they needed good roads to run on, and these were rare in the 18th century. In the event, steam made its great impact on land transport through the railways – which provided their own low-gradient, low-friction iron roads – but historically the steam carriage was the first in the field.

Steam carriages

The first steam wagon was that built by Nicolas Cugnot in 1769. That it should have appeared in France was not a matter of chance, for it was instigated by Jean-Baptiste Gribeauval, a general of artillery, and demonstrated to representatives of the French army in 1770: Cugnot was himself an army engineer. In contrast to Britain, a great naval power, France had a large standing army and thus a strong interest in mobility on land. After the end of the Seven Years War in 1763 there was much military interest in technical improvement and this coincided with Pierre Trésaguet's development of an improved technique of road construction using graded stones. Thus a steam wagon, or a tractor for artillery, was an attractive proposition.

As it turned out, Cugnot's locomotive was not adopted as no financial support was forthcoming after the fall of the enlightened minister Étienne-François Choiseul, but it was nevertheless a very interesting machine and still survives. It was a three-wheeler steered by a double-handled tiller and powered by two cylinders supplied with high pressure steam from a 1800-litre (400-gallon) boiler. The reciprocating action of the pistons was converted to rotary movement of the single front wheel by a crank movement. The bigger of the two machines – the first was little more than a model – could pull a three-tonne cannon at walking pace. In Switzerland, Isaac de Rivaz experimented with steam carriages from 1785 and he had one on the road in 1802. In America, Oliver Evans drove one on the streets of Philadelphia in 1800, but – like Boulton and Watt – his real interest was in stationary engines.

In Britain, James Watt patented a steam-driven carriage in 1784 according to the design of his assistant, William Murdock. The latter built a working model of this carriage in 1786, but no more was heard of it, probably because of lack of enthusiasm on the part of Boulton and Watt. In the early years of the 19th century, Richard Trevithick built several road steamers and in 1801 he achieved a speed of 13 kilometres per hour (8 miles per hour) at Camborne in Cornwall. In 1831 another Cornishman, Sir Goldsworthy Gurney, introduced a regular service between Gloucester and Cheltenham, covering the 14 kilometres (9 miles) in three-quarters of an hour. In the same year, Walter Hancock – brother of Thomas Hancock, pioneer of the rubber industry – established a public omnibus service in London that lasted for five years. In 1833, there was a regular London–Birmingham service. After that, steam vehicles of one kind or another were seen on the roads until the middle of the 20th century, but achieving success – and that only limited – for very heavy loads such as quarry stone or as tractors for pantechnicons or fairground equipment. In Britain, they were discouraged by the imposition of punitive tolls – on the grounds that they damaged the roads and frightened horses – and by the Red Flag Act of 1865. The latter restricted the

Cugnot's road locomotive (right), built in 1769, was the first effective steam vehicle to appear on the road, but half a century elapsed before steam carriages made any serious impression. By 1818 the idea, at least, had gained some popular recognition, as the French caricature below indicates.

speed of all mechanical road vehicles to 6½ kilometres per hour (4 miles per hour) – 3 kilometres per hour (2 miles per hour) in towns – and demanded a crew of three, one preceding the vehicle with a red flag to give warning. The effect of this was largely to limit steam vehicles in Britain to use as mobile power sources, particularly on farms for ploughing; an important exception was the steamroller, invented in France in 1859.

One disadvantage of the steam engine in transport was the time taken to rise a sufficient head of steam. On the Continent, free from Britain's restrictive legislation, there was greater interest and an important development was Leon Serpollet's flash boiler, in which steam was generated by pumping water through red-hot steel pipes. This was applied both to steam carriages and, in 1887, to a steam tricycle on which he rode from Paris to Lyons. He then went on to build steam cars in Paris, and in 1903 one of these – equipped with pneumatic tyres – attained a speed of more than 130 kilometres per hour (80 miles per hour). At the turn of the century the Stanley brothers of Newton, Massachusetts, began to manufacture steam cars, but although these had a certain vogue the future of land transport lay with the internal combustion engine.

Growth of the railways

The use of rails to guide vehicles long antedates the advent of steam traction. As we have noted, ruts in the stone-paved streets of some Greek and Roman cities seem to have been contrived to fit a standard wheel base, rather than being the result of long wear, and the Chinese certainly used this system of guiding vehicles as early as the third century BC. In the 16th century, mining industry trucks were hauled by horses along wooden tramways in which a groove was cut to engage with a pin on the bottom of the truck: not later than 1550, wooden rails and flanged wooden wheels were in use in Transylvania.

Wooden rails wear and rot quickly, and iron rails were introduced at Whitehaven Colliery in Britain in 1738; they were more extensively used after Darby began to cast them at Coalbrookdale. One such tramway was that which linked the Pen-y-Darran ironworks in South Wales with the Glamorgan Canal, 16 kilometres (10 miles) away. There, in 1803, Trevithick had installed a stationary steam engine to drive the rolling-mill. In 1804, this engine was equipped with wheels to serve as a locomotive and successfully pulled five trucks – carrying 70 men and 10 tonnes of iron – the full distance at 8 kilometres per hour (5 miles per hour). The project was short-lived, however, as the cast-iron rails would not stand up to the traffic, and the steam engine was converted back to stationary use. However, a second locomotive was built in 1805 for a colliery near Newcastle.

Trevithick is also remembered for his 'Catch-me-who-can' exhibit in Euston Square, London in 1808, in which a steam locomotive pulled passengers in a carriage round a small circular track. As the passengers paid, this might be called the world's first public railway enterprise, but it was in the nature of a diversion rather than an event on the main line of progress. Trevithick soon turned his attention back to stationary engines. Pen-y-Darran was a convincing demonstration, but the track was nearly level and left unresolved one important question: was the adhesion between an iron wheel and an iron rail sufficient for steam locomotives to be used on the quite stiff gradients on which horses were used? Doubt about the adequacy of grip of an iron wheel on an iron rail led John Blenkinsop to devise, in 1811, a form of traction in which a cog on the locomotive engaged with a toothed rail – a device still used on steep mountain railways all over the world, but not necessary on normal gradients.

development of railways 247, 252–253 **Parsons' steam turbine** 164–165 **steam engines** 159–165 **steam transport** 250, 252

The early days of railway travel were bedevilled by differences of gauge between systems. A picture (top left), of the 1840s, colourfully illustrates the chaos inflicted on passengers.

Although the railways were mostly for long-distance travel, compact urban networks soon appeared. New York's Elevated Railway (top right) was originally designed for steam traction, but was converted to electric in the 1880s. Note the competing horse-drawn tram.

Railway engineers, obliged to lay a nearly level track, had many difficulties to overcome. In Britain, one of the major achievements was the crossing of the water-logged land of Chat Moss in the building of the Liverpool and Manchester Railway (centre).

In the USA the railways played a special role in opening up vast new tracts of virgin territory. The transcontinental link was completed in 1869 (bottom).

By 1820, there were a number of small industrial railways powered by steam in Britain, and George Stephenson had begun to manufacture locomotives at Killingworth. Almost all were for private use but an Act of Parliament of 1801, sanctioning a horse-operated railway from Wandsworth to Croydon (Surrey Iron Railway), had opened the way to public systems. This railway had no rolling stock of its own, but was open for use, at a price, by any carrier providing his own wagons. In 1821 a further Act was passed approving a public railway to transport freight and passengers between Stockton and Darlington. Originally planned as a wooden tramway using horses, the enthusiasm of Stephenson – appointed engineer in 1823 – led to the adoption of an iron railway and provision for some steam traction. The railway was finally opened on 27 September 1825 and Stephenson's *Locomotion* drew into Stockton 12 wagonloads of freight, 21 wagonloads of ordinary passengers and a coachload of directors. The superiority of steam over horsepower was quickly demonstrated, and the way was paved for the world's first public railway relying wholly on steam and using only its own rolling stock. This was the Liverpool and Manchester Railway, opened with great ceremony on 15 September 1830 by the Duke of Wellington, then Prime Minister. The first train was hauled by Stephenson's *Rocket*, which had been chosen on the basis of its fine overall performance at the competitive Rainhill Trials of 1829. It embodied the essential features of all steam locomotives: separate cylinders on each side driving the wheels through short connecting rods and a multi-tubular boiler for efficient heat transfer. The latter had been invented – but patented only in France, so it was open to use in Britain – by Marc Seguin, pioneer of the St Étienne–Lyons Railway, the first in France.

These projects demonstrated not only the feasibility of steam traction on railways, but also that a great deal of money could be made from it. The Stockton and Darlington Railway was paying shareholders a dividend of 14 per cent after five years and the Liverpool and Manchester made £14,000 in its first year. The railways were demonstrably a unique form of fast cheap long-distance transport for passengers as well as goods, and after a difficult start money became freely available. This started what has been called the railway mania, which reached its peak in Britain at mid-century. By then the paid-up capital exceeded £250 million; by 1900 it was well over £1000 million. At the peak of the mania, in 1847, 250,000 navvies were employed in railway construction work. The social consequences were enormous. Henry Booth, a great railway projector, wrote: 'What was slow is now quick; what was distant is now near, and this change in our ideas pervades society at large.'

The Great Exhibition of 1851 illustrates the scale of the change: in that year the railways brought six million visitors to see the wonders of Victorian technology. Additionally, new opportunities arose for holidays for the population at large:

Trevithick's 'Catch-me-who-can' circular railway set up in Euston Square, London, in 1808 was no more than a public entertainment: the engraving (top) by Thomas Rowlandson is far from accurate. More significant was the opening of the Stockton and Darlington Railway in 1825 (above).

Thomas Cook organized his first 'package' holiday in 1841. In Britain, the greatest expansion was completed by 1860, when the national network amounted to 16,000 kilometres (10,000 miles); by the end of the century it had no more than doubled, and there had been little increase since 1885.

British railway pioneers quickly came into conflict with the canal owners, already firmly established conveyors of heavy freight: in the 19th century only one major work was completed after 1831 – the Manchester Ship Canal of 1894. In America, the situation was very different, for there was no established transport network. The 584-kilometre (363-mile) Erie Canal in New York – from Albany on the Hudson to Buffalo on Lake Erie – had been commenced in 1817 and posed a threat to the trade of Philadelphia. The citizens of the latter, conscious of their total ignorance of the subject, sent a representative to England to learn all he could about railways. This led to the opening of the Philadelphia and Columbia Railroad in 1831. Originally designed for public use with horse-drawn vehicles, it acquired its own rolling stock and locomotives in 1834. The Baltimore and Ohio Railway had a similar background and history. Initially locomotives had to be imported – the first was the *Stourbridge Lion* for the

development of railways 247, 252–253 **Great Exhibition** 156, 157 **Richard Trevithick** 162, 176–177, 179, 186

Delaware and Hudson Railroad in 1829 – but manufacture was soon started at home. By 1838 some 350 locomotives were running on 2400 kilometres (1500 miles) of track, and of these no more than a quarter had been imported. This was roughly the same mileage as in Britain at the time, but whereas the British network grew to only 35,000 kilometres (22,000 miles) at the end of the century, the American system already covered 48,000 kilometres (30,000 miles) in 1860, and exceeded 320,000 kilometres (200,000 miles) by 1900. This corresponded, of course, to the opening up of vast tracts of virgin territory: one of the great events was the completion of the transcontinental railway linkage on 10 May 1869. The American Civil War of 1861–5 was the first to be fought with railways making a significant contribution to transport; the Union had a marked superiority, as it did in other fields of technology, especially the mass production of arms.

In Russia, railways got off to a slow start as a national, rather than private, enterprise. The first considerable railway in Russia, between St Petersburg and Moscow (650 kilometres, 400 miles) was opened in 1851. The Trans-Siberian Railway was started in 1891 but not completed until 1916.

The lessons of the American Civil War were learned in Prussia. As a preliminary to the invasion of Bohemia in 1866, the Prussian chief of staff, Moltke, had drawn up most detailed schedules for the swift movement of troops and equipment by rail. In the Franco-Prussian War, too, a major factor in the defeat of Napoleon III, whose army was deemed the best in Europe, was very carefully planned use of rail transport.

As the railways grew, the demand was for higher speeds and power and greater fuel efficiency. In Britain, the Great Western Railway was running some scheduled services at a mile a minute (96 kilometres per hour) in 1847, and in 1904 the same company's *City of Truro* reached the magic 100 miles per hour (161 kilometres per hour). Greater rail adhesion was achieved from 1859 by using coupled driving wheels: the greater length of boiler was commonly accommodated by free-wheeling bogies. This led to the general adoption of an international notation. In Britain and America, a 4-6-2 steam locomotive was one that had four bogie wheels at the front, six driving wheels and two trailing wheels. On the Continent, it was the number of axles that counted, and the same locomotive would be described as 2-3-1. A prefix T denoted a tank engine, in which water was carried on the locomotive rather than in a tender. In America, and some other developing countries, the poor state of the tracks made weight-spreading particularly important and some monster locomotives had as many as 24 wheels. In such situations, reason of economy often dictated sharp curves to avoid the cost of building tunnels and bridges, and this originally necessitated locomotives with short wheelbases. In 1861, however, Robert Fairlie devised a solution to this problem by means of a locomotive fitted with power-driven bogies at front and rear: this was later developed in the huge locomotives designed by Garratt.

The second half of the century saw three important technical developments in locomotive design. The first was the invention of a much improved valve gear by the Belgian engineer Égide Walschaerts in 1844. It was quickly adopted in many countries, though in Britain not until 1878. The second, introduced in France in 1874 by the Swiss engineer Anatole Mallet, was the compounding of engines, introducing steam first to a small high-pressure cylinder and then to a larger low-pressure one. The third, adopted in Germany in 1898 by Wilhelm Schmidt of Kassel, was the use of superheated steam in order to reduce condensation losses. This was quickly adopted also in France and Britain.

Electric traction

The electric motor had obvious possibilities for land transport and at the 1879 Berlin Exhibition Werner von Siemens – an electrical inventor and founder of the Siemens and Halske company – displayed a narrow-gauge railway on which up to 30 passengers were carried on a circular track in an open truck drawn by a 3-horsepower locomotive. The speed was about 6½ kilometres per hour (4 miles per hour) and low-voltage electricity was drawn from a centre rail. It is very reminiscent of Trevithick's 'Catch-me-who-can' display 70 years earlier but was more immediately productive. In 1884 Siemens and Halske established an electric tramway between Frankfurt-am-Main and Offenbach, drawing current from an overhead cable, but a year earlier Magnus Volk had opened a narrow-gauge line along the seafront at Brighton, a popular English seaside resort.

Also in 1884, Frank Julian Sprague electrified the bogies of cars on the New York Elevated Railway. More significantly, he devised in 1897 a system in which a whole train of motorized cars could be controlled from the leading one: this was the first example of the multiple-unit train.

For safety reasons, and with no smoke to disperse, electric traction had an immediate appeal for mining engineers. The same advantage applied on the underground public

Railway enthusiasm soon spread to Continental Europe. France's first railway, from Paris to St Étienne (right), opened in 1829. By the end of the century, electric traction complemented steam: Italy's first electric railway (opposite, below), from Milan to Monza, opened in 1899.

railways, which were a product of urban expansion. The Metropolitan Railways in London – not using, strictly, a tunnel but a covered cutting – had been built in 1863 to link some of the principal main line railway stations, but the smoke and dirt of the steam locomotives which were used were often intolerable. When the 5½-kilometre (3½-mile) City and South London line (1887–90) was planned, electric traction was decided on from the outset: passing beneath the Thames, this was the world's first deep-level electric railway. The advantage of electric traction was recognized in 1895 in America when a 6½-kilometre (4-mile) tunnel on the Baltimore and Ohio Railway was converted from steam to electricity. However, main line electrification was essentially a 20th-century development.

Track and rolling stock

The Pen-y-Darran experiment described above had been terminated not because of mechanical difficulties but because the track collapsed under the pounding of the locomotive. As railways developed, the brittle cast iron was replaced by the more flexible wrought iron, but the biggest development was the increasing use of steel from about 1860. A particular point of weakness was the junction between two lengths of rail. Initially, they were joined by simply putting them end-to-end in the same 'chair', or seating, but in 1847 fishplates to join them were introduced, and safety was much improved. To some extent the pounding action of the driving wheels on the rails was mitigated by fitting them with counter-balancing weights.

From the beginning, railway gauges varied greatly, even within countries. In Britain 143½ centimetres (4 feet 8½ inches) was widely adopted – and eventually became standard gauge – for no better reason than that this was the gauge of the original North Country colliery tramways. When Brunel built the Great Western Railway, however, he favoured a gauge of 214 centimetres (7 feet ¼ inch) on the ground of safety at the high speeds he had in mind: this caution proved unnecessary but the broad gauge was not finally phased out until 1892. In America gauges were chaotic: many railways were built to the 4 feet 8½-inch standard but some, such as the New York and Eire, used gauges as large as 183 centimetres (6 feet). Not until after the Civil War were steps taken to adopt the British standard and the change was not completed until 1885. Such differences were, and are, a great hindrance to through traffic when different systems join. Where cost was an important consideration narrow-gauge railways, down to 60 centimetres (2 feet), were popular, but rarely for major systems.

A major factor in laying out the route of a new railway was the need to avoid stiff gradients. In the earliest days, Stephenson thought 1:330 reasonable, but with greater understanding of the causes of adhesion much more severe climbs proved feasible. In 1848, locomotives tendered for the Semmering Pass line in Austria were required to demonstrate their ability to climb a gradient of 1:40 at 19 kilometres per hour (12 miles per hour) towing 80 tonnes.

Even accepting stiff gradients, tunnelling, an extremely expensive operation, could not always be avoided and many major works were undertaken in the second half of the 19th century. These included the 13-kilometre (8-mile) Mont Cenis tunnel (1870) and the rather longer St Gotthard (1882): the latter shortened the rail journey between Italy

'Catch-me-who-can' 177, 179 **development of steel** 187–189 **electric motor** 169–170 **electric trams** 252 **electrification of railways** 253

and Germany by nearly 40 hours. Unbridgeable rivers were another major obstacle: in Britain the 6½-kilometre (4-mile) tunnel under the Severn, completed in 1886, was a major achievement, accomplished only after overcoming severe flooding from an unexpected encounter with underground springs.

Tunnelling techniques had been well developed by miners, but these later works had the advantage of some technical advances, particularly the pneumatic drill. This was specially developed by Germain Sommeiller for the Mont Cenis project, speeding up the rate of progress threefold. In building his tunnel beneath the Thames (1825–43) Marc Isambard Brunel used a shield which was pushed forward as the working face was excavated: he is reputed to have based this on his observations of the wood-boring mollusc *Teredo navalis*, which infests the hulls of wooden ships. This device was much improved upon by James Henry Greathead, who invented a steel shield advanced by hydraulic rams, while water was held at bay by compressed air. This was used in the building of the Hudson River tunnel in New York (1874–1908) and in the underground City and South London Railway.

As the railways grew, signalling devices became necessary for safety reasons and, under various guises, the system of military semaphores devised by Claude Chappe in 1793 was widely adopted. On busy routes 'block' systems were introduced by which no train could enter a block before the previous one had left it. In America and other large sparsely populated countries with long lengths of single-track line, much reliance was placed on the electric telegraph, first used in this context in Britain between Paddington and West Drayton in 1839. A further safety measure was the interlocking of signals and points, introduced in 1859.

Another important aspect of safety was the braking system, initially very primitive and limited to the locomotive and the special brake-van at the rear of the train. Communication between them was limited to whistling or waving. Train-long systems of air-brakes – using either pressure or (later) vacuum – were introduced by George Westinghouse, supposedly inspired by the use of compressed-air drills in building the Mont Cenis tunnel.

The main role of the railways was originally seen as the carriage of freight, and the great popular demand for conveyance of passengers came as a surprise. Initially, the gentry simply sat in their own carriages securely lashed to flat cars, while the cheapest fare-paying passengers sat on wooden benches in open wagons. However, as the potential of the trade was perceived, better amenities were provided. Four- or six-wheeled closed carriages – faithfully modelled on road carriages of the day – appeared in the 1830s, the comfort of the accommodation varying according to the fare. By modern standards, however, comfort was meagre until after mid-century, especially in Europe, and much of it had to be provided as an extra by the passengers themselves. Lighting was by candle and heating by foot-warmers filled with hot water. The frequent stops acted as a disincentive to the provision of lavatory accommodation, which entailed wasting seating space on corridors: lavatories did not appear until the 1860s. When lighting was first provided, it was by oil but in Germany compressed-gas cylinders were introduced in 1871 and the Metropolitan Railway in London used them from 1876. By that time, however, electric lighting was challenging the long supremacy of gas, and by the end of the century dynamos driven from the coach wheels were in use. A third category of goods, small in volume but exceedingly important, was the mail. The railways greatly speeded up inland distribution, as steamships did internationally.

That rail travel would be supremely comfortable, at a cost, was not in doubt, however. Queen Victoria, undeterred by a mounting catalogue of fatal accidents – including one at Versailles in 1842 in which 53 august personages had lost their lives – was a great devotee of rail travel, especially from London to Windsor, and the directors of the Great Western Railway provided a sumptuous royal train. It was as nothing, however, compared with that commissioned in the 1860s by the notoriously eccentric Ludwig II of Bavaria or one built for Pope Pius IX, at about the same time, which included an ornate throne room in a three-room suite.

In America, whose very size encouraged thinking on a grand scale, George Pullman conceived the idea of offering a much enhanced degree of comfort – including sleeping and dining facilities – to ordinary fare-paying passengers. His first 'hotel on wheels' was no more than an adapted open carriage, but its success encouraged him to build the far more lavish *Pioneer* in 1864. This was so successful that Pullman found it impossible to meet the demand for his conversions – including export orders from Europe – and in 1880 he had to set up his own manufacturing facilities on the outskirts of Chicago.

In transit the traveller got comfort in proportion to his fare, but at railway stations the main concourse was available to all, though there would be some segregation in refreshment and waiting rooms. Very often, too, there would be hotel accommodation for the better-off. In their heyday, the railways had plenty of money and they spent freely on their principal termini, often employing leading architects. King's Cross, built in London in 1852 to the design of Lewis Cubitt, was remarkable for two great barrel-vaults each spanning 32 metres (105 feet). The vast and incredibly ornate Victoria Station in Bombay, terminus of the Great Indian Peninsular Railway, is a memorial to its architect, F. W. Stevens. But even these, and the great Continental termini, were eclipsed by the enormous Grand Central Station in New York, built between 1903 and 1912. Built on two levels, it accommodated 67 tracks. Such buildings are today regarded as outstanding examples of the architecture of their time.

CHAPTER NINETEEN

Mining and Metals

It is a rather astonishing fact that from the days of classical antiquity to the end of the 19th century only one new metal came into general use. This exception was aluminium, but even in 1900 world production was no more than about 10,000 tonnes: not until after the Second World War was its annual production to be measured in millions of tonnes. However, this generalization must not obscure the fact that a number of new metals found application in very small amounts for special purposes. Such, for example, were osmium for making electric light filaments; chromium and tungsten for high-duty steels; platinum for chemical-resistant vessels; and nickel for electroplating. The great changes were in the volume and pattern of use. During the 19th century world annual copper production rose from about 10,000 to 525,000 tonnes – about half this increase being ascribable to the demands of the new electrical industry, which in turn provided new electrochemical techniques for metal extraction and refining, as in the case of aluminium. In the last half of the century, annual production of tin rose from 18,000 to 85,000 tonnes, reflecting the rapid growth in demand for tinplate for the new canning industry. Iron, increasingly used in the form of steel, remained by far the most important of the metals, world production always greatly exceeding that of all others put together. This reflected the many new uses to which it was put – in machinery of all kinds, shipbuilding, railway lines and the frames of buildings and bridges, among others.

Overall, the increase in demand was too great to be met from traditional mining sources in the Old World and, increasingly, fresh ventures – often on a vast scale – were located in newly developed territories. Chile, for example, became a major source of copper, Malaysia of tin and Australia of lead and zinc. The exploitation of these new sources depended upon parallel advances in other fields of technology, especially in transport: they were often on a scale so large as to effect significant shifts in population. The need to minimize transport costs led to smelting and refining being conducted where feasible at the mine itself. In 1800, for example, three-quarters of the world's copper was smelted in South Wales, especially in Swansea where alone there were several hundred furnaces. Before the end of the century, however, Swansea's share of world production was negligible and by 1921 the once thriving industry was extinct. Even when smelting was not feasible at the mine site, some preliminary concentration of ore might be done there.

Increasing emphasis on productivity led to the development of new mining techniques to supplement traditional pick-and-shovel work. Gunpowder seems not to have been used for blasting underground until the 17th century, but Alfred Nobel's dynamite, introduced towards the end of the 19th century, was originally sold for mining and not military purposes. Heavy earth-moving machinery, such as was used on the Panama Canal and other great civil engineering projects, was also appropriate for open-cast mining. Hydraulic methods, too, were well established before 1900. In one form, large mechanical dredgers worked on artificial lagoons; in another, powerful jets of water were used to wash down masses of exposed ore.

From the earliest times until the end of the 19th century, the miners' work was heavy and dangerous: picks and shovels were the main tools (left). However, the advent of the safety lamp greatly reduced the risk of explosion of firedamp: the three examples (above) were made by Humphry Davy about 1816.

chromium and tungsten 213 **dynamite** 186, 215, 265 **early mining and metal-working** 71–81 **undersea minerals** 341

Blast furnaces at Priestfield depicted on a trade token of 1811.

The blast furnace was at the heart of the iron-smelting process. In principle, there is little difference – except in the important matter of size – between those at Priestfield (top left) and that above at Dowlais, *circa* 1870.

Working conditions in foundries deteriorated. Even allowing for artistic licence, pictures of a British foundry of about 1805 (left) contrast with the grim conditions in a German one of the 1870s (top right).

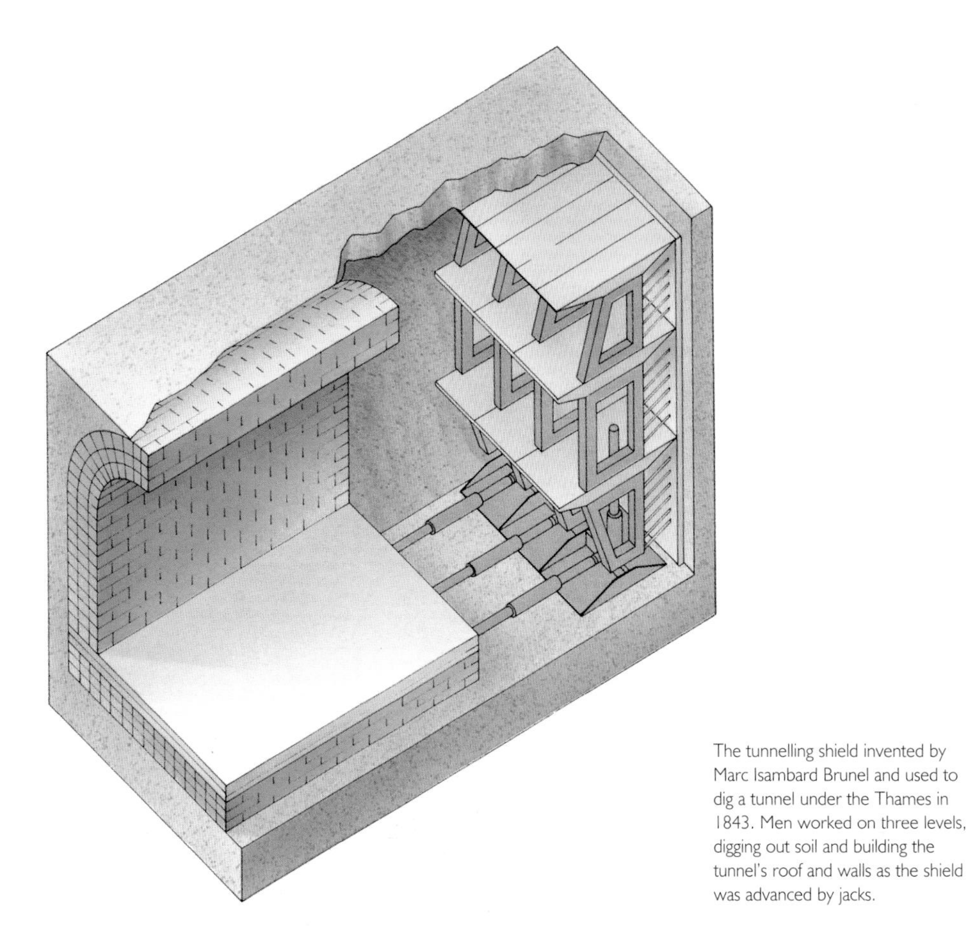

The tunnelling shield invented by Marc Isambard Brunel and used to dig a tunnel under the Thames in 1843. Men worked on three levels, digging out soil and building the tunnel's roof and walls as the shield was advanced by jacks.

Advances in technology made possible totally new methods of extraction and refining. While the first large-scale aluminium production was achieved in the 1880s by a traditional smelting process, this was almost immediately superseded by an electrochemical one. Electrolytic processes were used also in refining copper and other metals. However, the older methods were still capable of substantial improvement. An outstanding example was the invention in 1875 by Thomas Gilchrist of a satisfactory method of smelting the previously intractable phosphoric iron ores, including those of Alsace-Lorraine: by the end of the century half the world's steel came from such ores.

Mining techniques

The first stage in the manufacture of metals is, of course, to extract the crude ore from the ground, and to reach and follow the veins large quantities of rock or other non-productive overlay have to be removed. In this sense, mining is, therefore, analogous to other civil engineering enterprises, where large quantities of material have to be shifted – such as the building of railway and road tunnels and cuttings. Two important roles of steam have already been noted – in pumping floodwater from deep mines and transporting minerals away from the mine. Underground, the smoke and heat of steam engines precluded their use at the working face, but from the 1840s centrally situated stationary engines, discharging their exhaust up a shaft, were used to haul trucks on endless ropeways. So far as pumping is concerned, the civil engineering aspect is exemplified by the Severn railway tunnel. When finally opened after a disastrous flood in 1879, pumps had to be installed capable of handling up to 270 million litres (60 million gallons) of water daily.

Flooding was particularly a problem of deep mines and these, with their extensive system of horizontal galleries, also created difficulties of ventilation. A common device was to have two main shafts – after 1862, following a major disaster, this was obligatory in Britain, to provide an alternative exit in case of accident – and to light a fire at the bottom of one to provide a strong up-draught. Horizontal galleries were divided by brattices to encourage two-way flow of air. In coal mines, and to a lesser degree in other mines, fires or any other naked flames were potentially dangerous in the presence of firedamp (methane). In 1807, John Buddle introduced a crude wooden air pump to improve ventilation in Wallsend Colliery and about 1840 steam-powered pumps were installed by John Nixon in South Wales. By the end of the century, huge fans capable of extracting 6000 cubic metres (211,725 cubic feet) of air a minute – 25 times more powerful than Buddle's first air pump – were in use.

The candles and lamps used by individual miners were a greater explosion risk than fires lit for ventilation, however. In the early part of the 19th century this became so serious that in Britain the Sunderland Society appealed for technical aid. This resulted in the development of a variety of safety lamps, the best known of which was that invented by Humphry Davy in 1816. In this, the flame is protected by a metal gauze.

The carpenter's toolkit had included manual drills from pre-Christian times, but the much harder task of drilling rock

aluminium extraction 189–190 **coal mining** 228 **early mine drainage** 81 **invention of gunpowder** 136–137 **steam engines in mining** 159–164

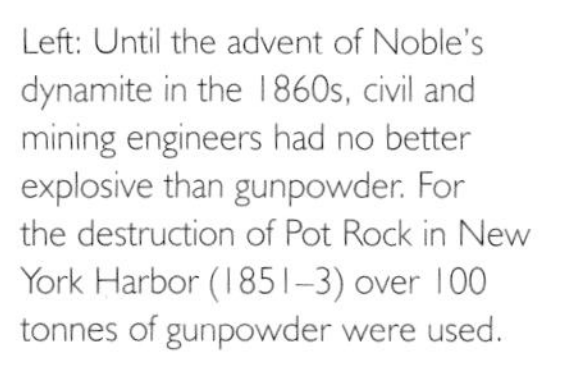

Left: Until the advent of Noble's dynamite in the 1860s, civil and mining engineers had no better explosive than gunpowder. For the destruction of Pot Rock in New York Harbor (1851–3) over 100 tonnes of gunpowder were used.

Right: Forge with tilt-hammer; from a penny trade token issued by John Wilkinson, Ironmaster, in 1790.

Below: For heavy forging increasing use was made of mechanical power, as with these trip-hammers at Abbeydale, Yorkshire.

really demanded mechanical power. A steam drill is said to have been invented by the versatile Trevithick, but the first effective device of this kind was not put into use until the building of the Mont Cenis tunnel, where it was shortly superseded by a pneumatic drill. An unexpected bonus was that the spent compressed air substantially contributed to better ventilation. The earliest machines drilled holes either by continuous rotation or by a hammering action: later, both movements were combined on one hammer-drill machine.

A major use of drills in mining and civil engineering was in drilling holes for shot firing. Until the latter part of the 19th century, however, the only explosive available, after more than 500 years, was gunpowder. By modern standards this is a weak explosive and for major works large quantities were required. In 1851–3, for example, more than 100 tonnes were used in New York Harbor to destroy Pot Rock. The safe use of large gunpowder charges was promoted by the invention of the safety fuse by William Bickford in Cornwall in 1831. In 1845, Christian Schönbein, in Basle, discovered that an exceedingly powerful explosive could be made by treating cotton and other forms of cellulose with nitric acid. In Italy, a year later, Ascanio Sobrero discovered the same effect with glycerine. Initially, however, these new products proved too unstable and dangerous for general use and not until the 1860s was the Swedish engineer Alfred Nobel able to offer an acceptable, but still hazardous product. He set up his first nitroglycerine works at Winterwick, near Stockholm, in 1865: two years later he marketed dynamite, a much safer product in which nitroglycerine was mixed into a dough with kieselguhr, a kind of clay. Dynamite was five times more powerful than gunpowder, and within ten years he was selling 3000 tonnes a year for rock blasting. Not until later did Nobel direct his attention to military explosives.

Ore enrichment

As the demand for metals accelerated, the richer ores in known deposits became exhausted and poorer ores had to be exploited. Enrichment by hand-picking then became impracticable and various mechanical methods were adopted. With heavy ores, such as those of tin and lead, a good deal of gangue (the rock in which the ores are embedded) could be washed away with water, as in gold panning. This was assisted by mechanical agitation, but in 1895 A. R. Wilfley introduced a sloping vibrating table with parallel grooves or riffles, in which the ore particles became concentrated before being discharged at the side.

In the special case of magnetite (magnetic iron oxide) the ore particles can be separated by means of powerful electromagnets. One of the pioneers of this method was Thomas Edison, who built a huge plant at Ogden in the Appalachians in the 1890s with a capacity of 5000 tonnes daily. It was a technological success but a commercial failure because of

Abraham Darby, pioneer of the use of coal (as coke) in iron smelting, was the first of a dynasty of ironmasters at Coalbrookdale, Shropshire. He is also remembered as the builder of the world's first iron bridge, crossing the Severn in the vicinity of his works.

unforeseen changes in the geographical pattern of the American steel industry.

The most important ore-enrichment process is that known as flotation. This was first used in 1865, when it was discovered that if low-grade sulphide ores such as copper pyrites were crushed and treated with oil, the oil attached itself to the sulphide particles but not to the gangue, which could then be washed away. However, this needed a lot of oil and was not very efficient. At the very end of the century, a much improved method was devised in which the ore/water mixture was treated with a little oil, and air was then blown through it: the ore particles are then carried up to the surface by bubbles of air. This technique was first used at Broken Hill, Australia, in 1901 to concentrate zinc blende.

The Age of Steel

Until the 18th century, iron was smelted with charcoal, but by then this was becoming increasingly difficult to sustain. One reason was that the iron-smelting areas were becoming denuded of wood. Another was that charcoal was easily crushed and so the height of blast furnaces could not easily be increased to accommodate a bigger charge. On the face of it a solution was ready to hand: coal, with a high carbon content, was cheap and abundant and might be substituted for charcoal. Although Dud Dudley claimed that he had successfully smelted iron with coal in 1619, this is very questionable, as all raw coal contains some sulphur which weakens the metal. The solution was to remove the sulphur by first coking the coal by roasting it in coke ovens. Abraham Darby, founder of a dynasty of ironmasters, first smelted iron with coke in 1709 at his works in Coalbrookdale, in Shropshire. He owed his initial success partly to the fact that the local coal, and consequently the coke made from it, had a very low sulphur content. Other manufacturers were less fortunate, and coke-smelted iron was not widely made until the second half of the 18th century, when a much improved product was made by re-melting the iron before use, when further impurities separated. A further problem with coke was that it burned less freely than charcoal, necessitating a more powerful draught in the furnaces.

Cast iron served many purposes: for example, in 1777 Darby cast the members for the world's first iron bridge, crossing the Severn at Coalbrookdale. In the century from 1740 to 1840 production in Britain rose to nearly two million tonnes annually. To obtain the more malleable wrought iron, however, this had to be decarburized and in 1784 Henry Cort – the 'Great Finer' – introduced his very important puddling process. In this, cast iron was melted in a reverberatory furnace – in which the metal did not come into direct contact with the fuel – and stirred with long poles to decarburize it. Cort's initial success owed much to his using as his starting material high-quality iron made in Sweden from ore from the famous Dannemara mine near Uppsala. Later, however, he was able to supply the Royal Navy with chain anchors and similar large items made from old mast hoops and other scrap – all equal in quality to the Swedish product. This process gave Britain a worldwide lead in the iron industry. Cort also introduced in Britain the use of steam-driven grooved rollers to finish iron bars in place of the traditional forge hammer: this greatly reduced the cost. Sadly, Cort was among those whose commercial ability did not match his inventive genius: an ill-judged financial transaction led to his ruin. The grooved roller was not Cort's own invention, but that of the Swedish engineer Christopher Polhem in 1745. Sweden had a long tradition of high-quality iron manufacture and one of the best descriptions of the mining and metal-smelting industries in the first half of the 18th century was the *Regnum subterraneum* published in 1734 by Emanuel Swedenborg – better known as the founder of the Swedenborgian religious sect.

Abraham Darby 161, 172, 177 **early smelting techniques** 72–74 **explosives** 215, 264–265 **Henry Cort** 155 **iron-working** 20–21, 75, 78, 94

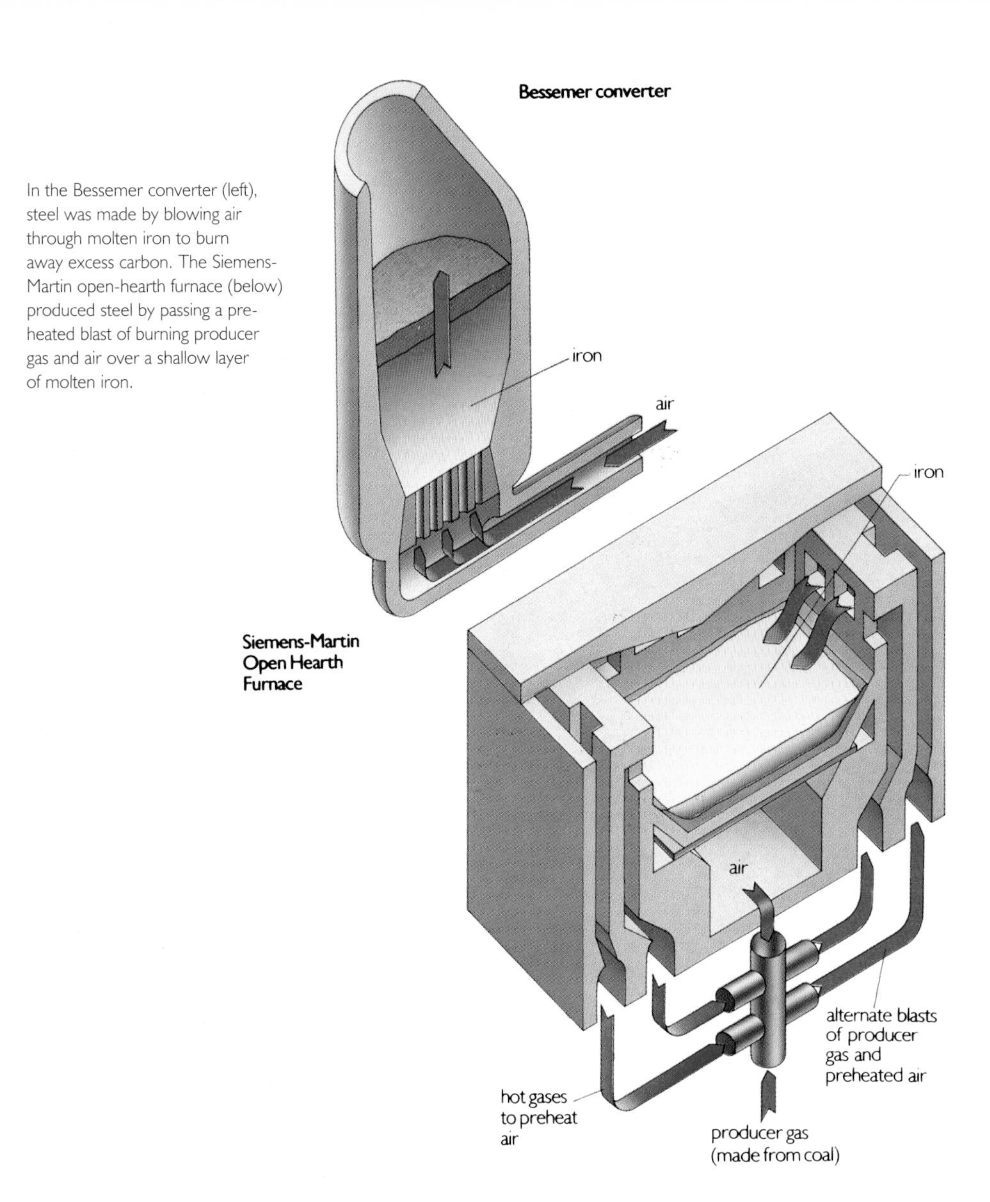

In the Bessemer converter (left), steel was made by blowing air through molten iron to burn away excess carbon. The Siemens-Martin open-hearth furnace (below) produced steel by passing a pre-heated blast of burning producer gas and air over a shallow layer of molten iron.

As we have noted, steel is virtually as old as iron, in the sense that it is the natural product of smelting certain kinds of iron ore in a particular way. To make any considerable quantity of steel deliberately was another matter. From about 1300, steel bars began to be made by a scaled-up version of the cementation technique used from Roman times to 'steel' small iron articles such as sword blades. Blister steel – so called from the pitted appearance of its surface – was made by heating iron bars at red heat for long periods in a bed of charcoal: the resulting steel ingots were then subjected to further heat treatment and forging. This process – as distinct from treating articles already fabricated – seems first to have been employed at Nuremberg, a great metal-working centre, in 1601. Later, however, the north-east of England was to become the centre of the steel industry for many years. There, according to an account written in 1767, a steel furnace might hold as much as 11 tonnes of iron bars and the heating went on for five days and nights.

After forging blister steel was suitable for many purposes, such as making cutlery, but it was not of the highest quality. An improved product was made by Benjamin Huntsman who, using a coke-fired furnace, succeeded in generating heat sufficient to melt steel in small crucibles: entrained impurities then separated. Huntsman perfected his crucible process – which relied entirely on high-quality Swedish iron as a starting material – in Sheffield during the period 1745–50, but the local cutlers at first refused to buy his steel, complaining that it was too hard to work. He therefore exported his entire output to France and it was only when demonstrably superior cutlery began to find its way back to Britain from there that the value of his product was acknowledged.

Despite its valuable properties, the use of steel was necessarily limited as long as it was made by such a slow, expensive batch process. In the middle of the 19th century, the output in Britain, by far the largest producer, was only about 60,000 tonnes. Quite suddenly the situation changed dramatically as a result of a single technological innovation. In 1870 world production had risen to about 500,000 tonnes and by 1900 it had soared to almost 28 million tonnes, with the USA, followed by Germany, the biggest manufacturer. The reason was the independent invention – by William Kelly in America and Henry Bessemer in Britain – of a process for decarburizing steel by blowing a blast of air through or over the liquid metal. The revolutionary feature of this process was that no additional fuel was needed, a possibility never previously contemplated in the steel industry and initially regarded with incredulity. From the 17th century the Chinese had 'improved' the quality of iron by running it molten through the open air, but this was not necessarily a comparable process.

First in the field was Kelly, proprietor with his brother of the Suwanee Iron Works and Union Forge, manufacturers of sugar kettles. Around 1847, he noticed that if an air blast was directed on to molten iron not covered with charcoal its temperature rose to white heat. What was happening was that the carbon contained in the iron was burning in the oxygen of the air blast. Between 1851 and 1856, he built seven experimental converters to develop his 'air-boiling' process and in 1866 – learning of Bessemer's work – lodged a patent with the US Patent Office.

Bessemer was an inventor whose versatile ingenuity is not well known. Born in 1813, the son of a typecaster, by the mid-century he had invented a machine to cancel stamps by perforation, an improved graphite pencil, a process for making imitation Utrecht lace and – one so successful that he had no further financial cares – a machine to make brass dust which was universally used to manufacture 'gold' paint. Not until the Crimean War did he have any serious interest in steel. Then, reversing long-accepted practice, he worked on a field gun in which the barrel was smooth and the shell rifled. Cast iron proved too weak for the barrel and experimenting with fusion of a mixture of iron and steel in a crucible he observed that the metal could be decarburized by blowing air through it. This process he patented in October 1855; five years later he patented a tilting converter that was eventually widely adopted. However, he had to endure several troubled years because licensees soon discovered that the steel they produced could not be forged when hot and was brittle when cold. The trouble proved to be due to the fact that Bessemer had been using phosphorus-free iron from Sweden, whereas most British ores contain a good deal of phosphorus. Eventually the difficulty was overcome, especially after the discovery that the oxygen content of steel could be controlled by adding manganese and the invention of the dephosphorizing process by Thomas Gilchrist in 1878. Bessemer became a rich man; Kelly, in contrast, eventually went bankrupt and had to come to terms with the Bessemer interests in America. In Sweden, G. F. Göransson, who had access to very pure iron ore, discovered that varying grades of steel could be made by strictly controlling the force and duration of the air blast.

Bessemer's invention was timely, for his steel was very suitable for structural purposes, plate and railway lines. It was not ideal, however, and it found a powerful rival in the product of the Siemens-Martin open-hearth process developed in Germany. By the end of the century more steel was being made by this process than by Bessemer's, and an important factor in this was its adoption by the American steel magnate Andrew Carnegie.

Aluminium, first isolated in 1824 but not available on a substantial commercial scale until some 60 years later, was only slowly adopted. One very familiar example of its use, in alloy form, is in the statue of Eros in Piccadilly Circus, London.

Although the advent of huge quantities of cheap steel was the outstanding metallurgical event of the century, there were important developments in other fields. Old metals found new uses and some new metals were introduced though only one, aluminium, was used in quantity.

Electrochemical processes

The invention of the voltaic cell in 1800 transformed the science of electricity: for the first time a simple continuous source of electric current was available. One early consequence of this was Humphry Davy's isolation of sodium and potassium – and later several other new metals – by passing electricity through their fused salts. But it was not until the last quarter of the century – when mechanical generating systems made large quantities of electricity cheaply available – that this method was applied industrially, initially to the manufacture of aluminium.

In 1824, the Danish scientist Hans Christiaan Oersted isolated aluminium by treating aluminium chloride with sodium, and it was later found that potassium could be used in place of sodium. Attempts were made to develop this process in Germany and France in the 1850s, but the cost was prohibitive. A few aluminium articles were shown at the Paris Exhibitions of 1855, 1867 and 1878; a set of cutlery was made for Napoleon III and Tiffany's made some items of jewellery, but generally interest was negligible. Nevertheless, its lightness and resistance to corrosion still made aluminium attractive, and in 1879 an American chemist, Hamilton Young Castner, devised a much cheaper method of making sodium. In 1888, he set up a works at Oldbury, in England,

aluminium 302–303 **Bessemer and typesetting** 233 **electrolytic refining** 185 **Niagara power plant** 165 **voltaic cell or pile** 169

to manufacture 50 tonnes of aluminium annually, intended primarily for alloying with other metals. For two years, this was a success, but in 1886 Charles M. Hall in America and Paul Louis Toussaint Héroult in France had independently invented a process for making aluminium at much less cost – subject to the availability of cheap electricity, as in the new hydroelectric schemes at Niagara and elsewhere – by electrolysing a molten mixture of bauxite (aluminium oxide) dissolved in cryolite. As soon as this began to be operated commercially, Castner was put out of business so far as aluminium was concerned. Even so, demand for aluminium rose only slowly and it was destined to be very much a 20th-century metal.

Castner's only remaining asset was his cheap sodium and he turned this to good account by manufacturing sodium cyanide for the flourishing gold-mining industries of America, South Africa and Australia. There, potassium cyanide had lately come into use as a 'wet' method of extracting gold from crushed ore but Castner had some difficulty breaking into a conservative market: finally he succeeded by resorting to selling his product as '130 per cent potassium cyanide', which it was indeed strictly equivalent to in chemical terms.

Zinc

Brass, an alloy of copper and zinc, was made in the first millennium BC by co-smelting copper (or copper ore) and zinc ore (calamine). Like copper, it is a bright attractive metal, but zinc itself is one of the dullest. The Chinese appear to have been the first to prepare the pure metal – simply by heating its ore with charcoal in crucibles – and it was used there in coinage as early as 1402. Small quantities were brought to Europe by the Portuguese – used mainly for small ornamental castings – and were commonly known as speauter, which may be a corruption of pewter, an alloy of lead and tin rather similar in appearance. Later this changed to spelter, the trade name for zinc until recent times. In nature, zinc is commonly associated with lead and in the early 17th century it was recovered in the Harz region of Germany as a valueless by-product of lead smelting, known as *conterfeht*. The first person deliberately to manufacture zinc in the West seems to have been William Champion, of Bristol, about 1743: his purpose was to make brass, in which his family business had long had an interest, by directly melting together copper and zinc in appropriate proportions.

Apart from brass, zinc was used to make a number of other useful alloys. By the middle of the 19th century, German silver, a bright copper-nickel-zinc alloy, was being widely used. Muntz metal, invented by C. F. Muntz in 1832, soon displaced copper for sheathing ships' bottoms: it was a brass with a very high (40 per cent) zinc content. One of zinc's most important uses, however, was in 'galvanizing' iron to prevent corrosion: in exposed positions, the zinc is preferentially decomposed before the iron. For this purpose, iron

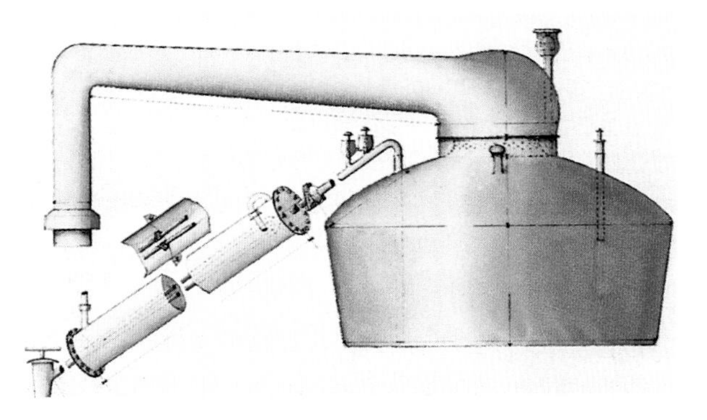

This platinum still made in 1867, could handle 5000 kilograms (11,000 pounds) of sulphuric acid daily for the chemical industry.

sheet was dipped into baths of molten zinc. Huge quantities of corrugated galvanized iron sheets were used for roofing farm and factory buildings, especially in developing countries where cheap, quick construction was essential.

Platinum

There is no real evidence that platinum was known at all to the ancient world, nor indeed in Europe before the 16th century. For all practical purposes, platinum was first encountered in the New World by the Spaniards, who called it platina, meaning 'little silver'. There the native metal had been used for several centuries to make small pieces of jewellery, many examples of which still survive.

Platinum is an extremely intractable metal because of its very high melting point and hardness. How, then, did the natives of Ecuador, Colombia and surrounding regions, with extremely limited resources, make so many delicate articles from it? The answer appears to be that they used a version of the modern technique of powder metallurgy. Small granules of platinum were mixed with a little gold dust and then heated sufficiently for the gold to melt and bind the whole together. This material was then subject to long heating in the flame of a blowpipe, when a little of the gold diffused into the platinum, and vice versa. By prolonged heating and hot forging a homogeneous mixture can be made, and then worked up into the desired article.

During the early 19th century platinum was found also in Europe, especially in alluvial deposits in the Urals, but little use was made of it because of the difficulty of fabricating it into useful articles. This problem was solved by William Hyde Wollaston who, in effect, rediscovered the powder metallurgy technique of the Central American Indians. After a series of chemical reactions, he precipitated the metal in a very fine powder from a solution of ammonium platinichloride: this he compressed, heated and hammered, finally obtaining platinum in the form of a malleable sheet.

Wollaston set up his own manufactory in London and found ready customers in the chemical industry. There the highly resistant platinum was very welcome for making vessels for the concentration of sulphuric acid, which was in growing demand. The first such vessels he supplied in 1809 to a London works. Later, platinum was linked with sulphuric acid in quite a different way, as a catalyst in the so-called contact process for manufacturing the acid. After Wollaston's death in 1828, the leadership of the industry was assumed by the British firm of Johnson Matthey. At the Paris Exhibition of 1867, they created a sensation by showing platinum equipment weighing in all over 500 kilograms (1100 pounds), including crucibles and two large boilers.

Electroplating

Proverbially, beauty is but skin-deep and workers in precious metals – and in rare woods, too – soon realized that they could extend their market by encasing base material. A simple method, of great antiquity, was to apply gold leaf – that is, gold beaten out as thin as fine paper. In the 16th century, Vannoccio Biringuccio describes in his *Pirotechnia* an elaborate technique for gilding iron and other metals by means of gold amalgam. In 1742, Thomas Bolsover of Sheffield first made plate by fusing or soldering a thin plate of silver to a thick plate of copper and then rolling it out while hot. From its place of origin, this was known as 'Sheffield' plate but the main centres of manufacture were Birmingham and London. It was used for making buttons and buckles, Bolsover's own trade, but was later extended to many small domestic articles such as candlesticks.

Sheffield plate was much in demand for nearly a century – the many fine pieces are now collectors' items – but it was displaced by the system of electroplating invented by the Elkingtons of Birmingham in 1840. In this, the article – commonly made of German silver – was made one electrode of a cell containing a solution of silver or gold dissolved in cyanide. When current was applied, the metal was deposited on the suspended article. The first to operate the process commercially seems to have been Thomas Prime of Birmingham, but by the end of the century Germany was the main manufacturer. A similar process was used to plate with nickel and, later, chromium and cadmium.

Mechanical fabrication

A notable feature of the Industrial Revolution was the growing use of metals – especially iron – for a great variety of purposes. There was an enormously increased demand for massive items such as girders for bridges and bases for machinery as iron displaced heavy timber, as well as for small items, such as nails and bolts. Apart from questions of speed and cost, many of these big items could not be forged at all by the traditional methods of hammer and anvil.

There was, therefore, growing emphasis on the development of heavy machinery which could shape iron and other metals into basic shapes such as rods, bars and sheets. Such stock could be further worked up by more specialized fabricators using a new generation of machine tools. These machines, such as grooved rollers and slitting mills, were able to turn sheet steel quickly into bars and rods. Another was Nasmyth's steamhammer of 1839. Much use was made of massive rolling mills, which could roll red-hot ingots weighing many tonnes. Even by mid-century 20-tonne ingots could be handled in this way, but by 1900 Alfred Krupp could roll a 130-tonne steel ingot into naval armoured plate 14 × 3½ metres (45 × 11 feet) in area and 30 centimetres (12 inches) thick.

An important area of increased demand was for wire of varying gauges. Iron wire, including barbed wire from the 1870s, was needed by the mile not only for established purposes but to fence the great new cattle and sheep ranching areas that were becoming a major source of food supply for the Old World. It was needed too for netting, mattresses and hawsers, all made with ingenious new machines. Additionally, the advent of the telegraph and telephone and the new electrical industry, in the second half of the 19th century, called for enormous lengths of copper wire because of its excellent conducting properties.

Another important area of growth was in the manufacture of metal tubing for a variety of purposes, but especially copper tubes for the boilers of steam engines.

In the 18th century copper was plated with silver (Sheffield plate) by rolling sheets of metal together, but from 1840 a new process of electroplating came into use. These electroplated articles were illustrated in a catalogue of Thomas Dixon of Sheffield in 1901.

CHAPTER TWENTY

Household Appliances

Home is the focus of life, the operational base for making a living and the retreat for rest and sustenance. We have already considered developments in methods of building, how utensils are made, how cloth is woven and dyed and other technologies that relate directly to daily life. Generally speaking, the relationship is self-evident; nevertheless we have spent much of our time seeing what was going on in the outside world, rather than looking inward to see how advances in technology directly affected the domestic life of individuals at the family level. It is, therefore, relevant to consider this aspect for a moment, particularly in view of the great changes that occurred in the 18th and 19th centuries, making the usual proviso that at all times there were great variations, depending on social and geographical differences.

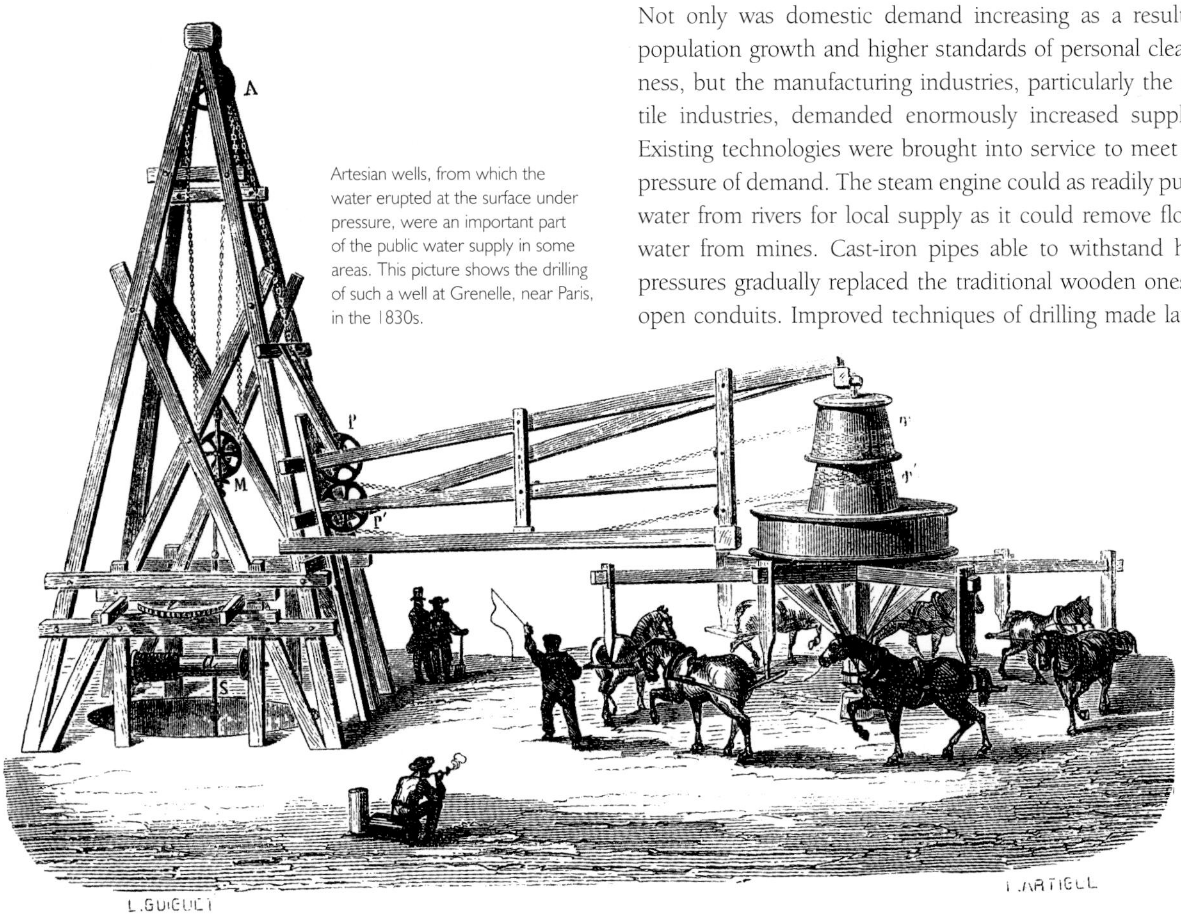

Artesian wells, from which the water erupted at the surface under pressure, were an important part of the public water supply in some areas. This picture shows the drilling of such a well at Grenelle, near Paris, in the 1830s.

Public utilities

'Public utilities' is a term of quite recent origin, but it is a convenient umbrella under which to consider a variety of communal domestic services. The most important of these is water. The whole pattern of human settlement is determined by the availability of adequate supplies of potable water, either immediately to hand or at no great distance. In this respect, there is a great difference from food. Even in the ancient world food could regularly be imported over many hundreds of kilometres – as Rome did from Egypt and Peking from its provinces – and by the end of the 19th century Europe, with the rise of Australian agriculture, was importing large quantities of food from literally the other side of the world. Even today, however, water is essentially a local supply – there is nothing comparable to the grid systems for electricity – and with respect to distance we still have to think in terms not very different from those of antiquity. Only in certain special circumstances – such as the importation of water in tankers to wealthy communities in hot, arid zones – are there any major exceptions to this.

For the countryman in the 19th century, the sources of water remained unchanged: he necessarily depended on local streams, springs and wells, and additionally might collect rainwater in a cistern against times of drought. For the town-dweller, the ultimate sources were, of course, basically the same but they had to be exploited much more intensively. Not only was domestic demand increasing as a result of population growth and higher standards of personal cleanliness, but the manufacturing industries, particularly the textile industries, demanded enormously increased supplies. Existing technologies were brought into service to meet the pressure of demand. The steam engine could as readily pump water from rivers for local supply as it could remove floodwater from mines. Cast-iron pipes able to withstand high pressures gradually replaced the traditional wooden ones or open conduits. Improved techniques of drilling made larger and deeper wells possible. Pumping from great depths was often impracticable, but it had long been known that some underground water was at such pressure that it would spontaneously rise to the surface and spurt out; such wells are known as artesian wells because the first recorded example was drilled in the French province of Artois in 1216. Another French well, sunk to a depth of 550 metres (1800 feet) at Grenelle in the 1830s, yielded 36 million litres (8 million gallons) a day.

Getting the necessary quantity of water to the towns was one thing; to distribute it to householders was another. In the 19th century water-carriers, selling water at so much a bucket, were a familiar sight, but the usual source of supply was a standpipe in the street, serving a number of houses. Only better-class houses would have their own supply and those would rarely exceed a single pipe in the kitchen.

Both in Europe and in the United States, the 19th century saw a great population explosion with a corresponding increase in demand for potable water. In Britain, for example, the population rose only from 7 to 11 million between 1760 and 1800 but in 1900 it exceeded 42 million. The greatest increases were in the urban areas. New York, for example, grew from under a million inhabitants in 1860 to almost 3½ million in 1900.

Sanitation

The provision of food, water and housing for these densely populated areas was a major problem: no less so was the disposal of excreta. In country districts, it was no great problem for it could be buried or mixed with farmyard manure and spread on the land. In the congested cities of Europe – where manure was available from cows and hens kept to supply local needs and horses were a main mode of transport – the contents of outdoor privies were discreetly removed by night-soil men and much of it was sold as an agricultural fertilizer. Even in London, this was done up to the end of the 19th century.

Such a primitive system is still normal for some of the world's population, but with increasing availability of piped water a more convenient system became possible. In 1589, Sir John Harington of Kelston in Somerset installed in his house a water-closet which could be flushed once or twice a day from an overhead tank. Harington possessed a Rabelaisian wit which often got him into trouble at court: he chose to describe his invention in 1596 in a literary work punningly entitled *The Metamorphosis of Ajax* ('jakes' was then a term for a lavatory). However, his water-closet was an isolated instance and not until the last quarter of the 18th century were such toilets at all widely used. Two important developments were the U-bend water-seal (1782) and the automatic flushing cistern of roughly the same date.

While the water-closet was a relief to the squeamish, it did nothing for the disposal problem. Drainage was commonly into a cesspit, which might be emptied once a year, and the contents tipped into the nearest river or sea. Meanwhile it, and the connecting pipes, would probably have been leaking slightly and contaminating the surrounding soil, from which many households still drew well-water. If local by-laws permitted – and often if they did not – water-closets emptied straight into sewers, which discharged into

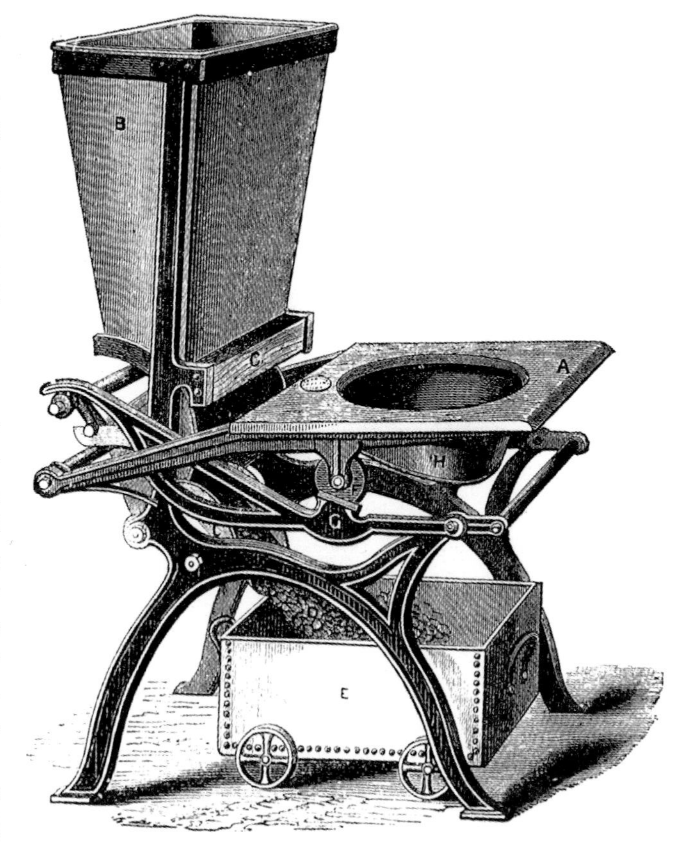

Although the water-closet was invented in 1589, bucket sanitation was the rule until the end of the 19th century. In the example above (*circa* 1870), soil was automatically shot into the tray when the weight of the body was lifted from the seat.

Water-carriers (top) were a familiar sight in city streets until well into the 19th century. The engraving above shows such a carrier in London in 1808.

domestic life in antiquity 34–44 **early water supply and distribution** 60–63 **public health measures** 289 **steam-powered pumps** 159–162

the nearest river, itself often used as a source of drinking water. In crowded areas, the conditions became not only very unpleasant but a serious menace to health. This risk was recognized in a general sort of way, though not until Louis Pasteur established the relationship between microbes and disease in the 1870s did the precise nature of the connection between contaminated water and diseases such as cholera and typhoid become clear.

Around the middle of the 19th century, the great conurbations recognized that some radical measures were needed. The general solution was to adopt a dual system of drainage – one to carry away storm-water and the other to deal with sewage. The latter could be pumped away to discharge at a point comfortably remote from the city centre. One of the first cities to adopt this system was Hamburg, following a fire which devastated it in 1842 and made extensive rebuilding necessary. In London, too, it was a disaster – in this case, a serious outbreak of cholera in 1854 – that led to reconstruction of the entire drainage system.

In large fast-flowing, turbulent rivers, direct discharge of sewage from small towns was acceptable, but in London and other very large cities located on estuaries, the discharged sewage drifted up and down with the tide, becoming more and more noxious. Around the end of the century, pre-discharge treatment began to be adopted. In this, the sewage was first passed through coarse filter beds, to remove the largest suspended solids, then treated chemically to precipitate the remainder. The sludge obtained was taken out to sea and dumped, and the residual innocuous liquid was allowed to go into the river.

Lighting

Until the 19th century, the sole sources of artificial light were candles or oil lamps. They are similar in the sense that both depend upon the capillarity of a wick to suck up liquid fuel: in the case of the candle the heat of the flame melts the solid wax and provides a little cup of liquid fuel at its top. Candles were certainly used in ancient Egypt and were made by dipping a cotton or flax wick into liquid animal fat or beeswax repeatedly. Moulded candles came into use in the 17th century, supposedly invented by a Parisian candle-maker named Brez. The wicks of such candles had to be trimmed as the wax burned away, and candle-snuffers were a standard piece of domestic equipment. A small but important innovation was made by J. J. Cambacères in 1824: he introduced a plaited wick which curled outward beyond the flame and so gradually consumed itself.

In certain areas whales were quite familiar to ancient man, for they are frequently stranded on the shore. Bones found in middens prove that they were appreciated as food and the largest bones were even used for building. At an early date, whales were regularly hunted by driving them ashore, and by the Middle Ages the Basques had evolved the

The simple candlestick was capable of almost infinite elaboration, as this Meissen candelabrum – reputedly the largest in the world – demonstrates.

technique of spearing them with a barbed harpoon attached to a line: when they were exhausted, they were stabbed with a lance. This method was used until the advent of Sven Foyn's harpoon gun in the 1860s. From the 17th century onwards whaling was a regular industry in Arctic and, from the late 19th century, Antarctic waters: the whalers commonly worked from remote shore stations where the blubber could safely be boiled down and barrelled, a hazardous process on board a ship. Whale oil was valued as an illuminant for lamps and as a lubricant; the whale also yielded a wax known as spermaceti, which made excellent candles.

In the latter half of the 19th century, however, a serious rival appeared: the rapidly growing petroleum industry provided both a good lamp oil and a hard, paraffin wax suitable for candles. Meanwhile, however, another important illuminant had appeared in the form of coal-gas. At first, this was just burned as a flare, which gave a smoky weak light, but improved burners were soon devised. One of the most popular was the fish-tail, in which two jets of gas impinged on each other. More important was the adaptation to gas of the Argand oil burner. This had a circular wick through which air was drawn up a glass chimney, providing a strong draught and a brighter flame. One of the best known designs was Sugg's London Argand, adopted as a parliamentary standard for luminosity in 1868.

In the 1880s, the Austrian chemist Auer von Welsbach was investigating a group of very similar metals known as the rare earths, which include thorium and cerium. He noticed that when asbestos fibre impregnated with these metals was strongly heated they emitted an intense white light. From this chance observation, he developed the Welsbach gas mantle, containing 99 per cent thorium and 1 per cent cerium, which he patented in 1885. It was a timely invention for the gas industry, which then was being threatened by the appearance of the incandescent filament electric lamp, invented independently by Joseph Wilson Swan in Britain and by Thomas Edison in America. In this a carbon, or later metal, filament is heated to incandescence in an evacuated glass globe. A key invention in its development was Hermann Sprengel's mercury air pump of 1865, which made it possible to attain easily the high vacuum needed. Swan and Edison came into conflict over patents, but eventually settled their differences and formed the Edison and Swan United Electric Light Company in 1883. Since that time billions of such lamps have been manufactured and they are used in every part of the world in a great variety of sizes.

Thanks to Welsbach, gas-lighting held its own against electricity until the First World War and is to be found in a few streets and houses even today. Latterly, it has found a new use in portable lamps, fuelled from bottled gas cylinders, in boats, caravans and tents.

Heating

For almost the whole span of human history, the sole source of artificial heat was solid fuel – initially wood, sometimes in the form of charcoal, or later coal. The Greeks and Romans made some use of coal, but only where they encountered it as outcrops. Weight for weight it gives much more heat than wood and for this reason it was favoured by smiths for their furnaces and by lime-burners and brewers. Its unpleasant smoke counted against it, however, and for a time its use was forbidden in and near London. However, by 1400, it was being regularly shipped there – hence the name sea-coal – from the Newcastle area. Open fires were the main form of domestic heating with coal-fired ranges in the kitchen. Large houses might have dozens of them, with very convoluted chimney systems. The employment of small children to sweep these was a great social scandal until regulated by legislation in the 19th century, in Britain and elsewhere.

The 19th century saw the advent of gaseous and then liquid fuels for heating. Coal-gas began to be used for cooking in the 1820s, and became more popular after the famous chef Alexis Soyer adopted it for the Reform Club in London in 1841. Fears about food cooked by gas being injurious to health were dispelled and smokeless burners were invented by Robert W. Bunsen in the 1850s. In this context, the relatively high cost of gas was outweighed by its convenience. Space heating was a different matter, however, and not until the 1880s were gas stoves attractive and efficient enough to be at all widely used.

The elaborate Dobereiner lamp of 1823 (top left) was self-igniting. It was fuelled by hydrogen gas generated in the base by the action of sulphuric acid on zinc, and self-ignited by letting the hydrogen play on a piece of platinum serving as a catalyst.

In the early days of electric lighting, architects paid great attention to the design of fittings – as at Cragside (top), home of the inventor William George Armstrong, built by the Victorian architect W. N. Shaw.

The coal-fired kitchen range (above) served not only for cooking, but also as a source of hot water for the whole household. This example is from a trade catalogue of 1903.

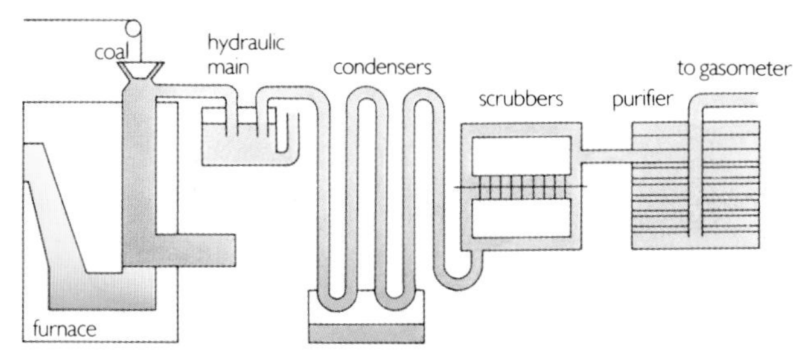

One of the earliest gas-making installations (above) was that built by Samuel Clegg in 1812 for Ackermann's Depository of Fine Arts in the Strand, London.

Coal-gas (see diagram at top) is made by roasting coal in a closed retort to give coal-gas, coke and coal-tar. The gas flows through a hydraulic main, a liquid seal which separates some coal-tar, and condensers that cool to deposit the remaining tar.

Today, the availability of hot water all over the house is commonplace, but in the 19th century it was carried in cans from a boiler attached to the kitchen range. In 1868, B. W. Maughan's 'geyser' offered hot water on demand in the bathroom, and by the end of the century it could be available all over the house from a central boiler.

Electric fires, like gas fires, suffered from the disadvantage of high running costs. Additionally, however, there was the technical problem that the iron wires used for the heating elements lost their strength at red heat and readily oxidized. Although electric fires were available around 1890, they made little headway until Albert L. Marsh invented nichrome – nickel–chromium – alloy for the elements in 1906. Rookes Evelyn Bell Crompton exhibited electric cookers at a Crystal Palace exhibition in 1891 and offered them for sale three years later, but they were not widely used until after the First World War.

Clothing

While the Industrial Revolution affected virtually all aspects of life, its biggest impact was in the textile industry, where mechanization of all the basic processes enormously increased production and lowered prices. Leaving aside the richest and most elaborately designed garments, in the course of the 19th century the ordinary citizen could aspire to a wardrobe as varied as that of the wealthy of earlier generations. Although no new fibres came into use – nor did they until the advent of man-made fibres such as nylon and Terylene in the middle of the 20th century – new processes were introduced that substantially changed the properties of old ones. Charles Macintosh perfected his rubberized fabric in 1823 and within ten years 'macintosh' was the generic name for any kind of waterproof coat. In 1844, John Mercer discovered that cotton could be given a lustrous sheen by treating it with caustic soda, and 'mercerized' cotton became available from 1890. He discovered, too, that wool could be more easily dyed if it was first treated with chlorine; later it was realized that this also made it crease-resistant. In 1892, Charles Frederick Cross and Edward J. Bevan discovered the viscose rayon process for making 'artificial silk' from cellulose, and this was developed by Courtaulds in the early years of the 20th century. Perhaps the most dramatic development of all, however, was that of a great variety of synthetic dyes, following William H. Perkin's discovery of mauve in 1856.

Although the first synthetic dyes were made in England, following William Perkin's discovery of mauve in 1856, they initially found little favour there. They became popular only when taken up by leaders of fashion in Europe. Prominent among these was the Empress Eugénie, wife of Napoleon III, painted here in a gown dyed with aldehyde green.

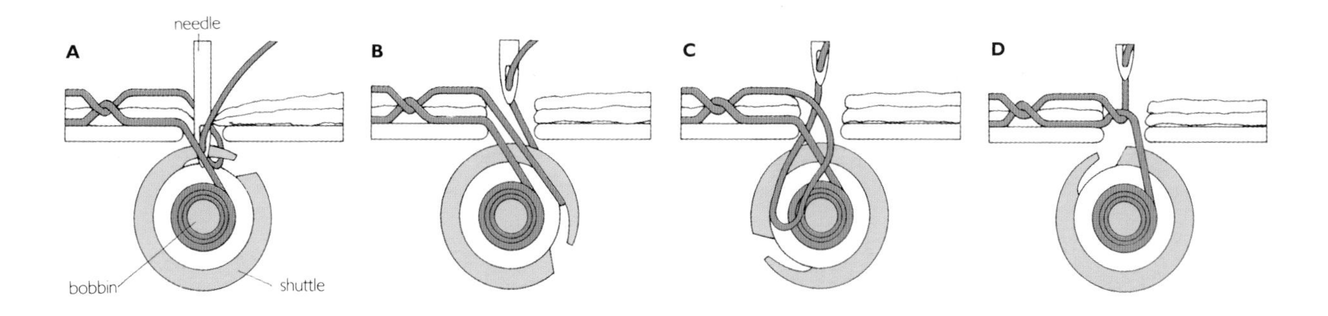

A machine stitch (above) is made with two threads, one wound around the bobbin and the other threaded through the point of the needle. The shuttle rotates around the bobbin, hooking the needle thread as the needle pierces the fabric (A). It then pulls a loop of the needle thread around the bobbin (B), so that it slips off the hook and under the bobbin thread (C). The needle thread is then pulled tight to form the stitch (D).

The first successful sewing machine was that of Isaac Singer in 1851 (right).

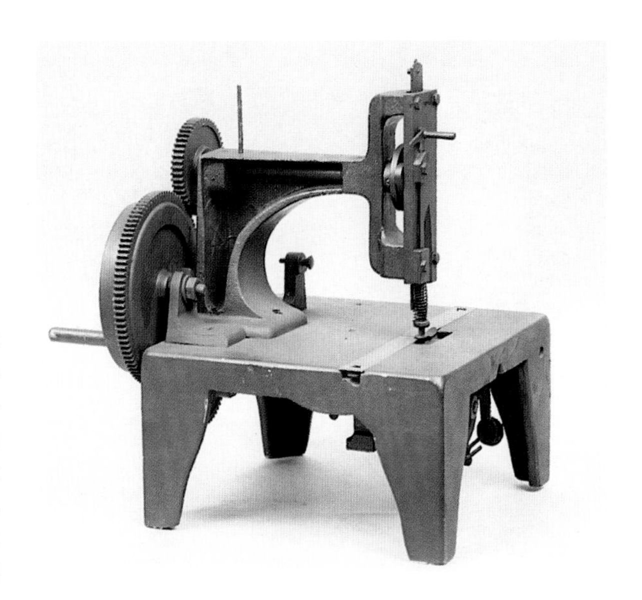

These developments were manifested in the home not only in clothing but in all other goods in which textile fibres were used – such as carpets and soft furnishing. They resulted from the utilization of machinery in distant factories, but at mid-century this invaded the home itself. In 1851, Isaac Singer patented the first practical lock-stitch sewing machine, successor to a number of earlier devices – beginning with Balthazar Krems' chain-stitch machine of 1810 – which had failed for a variety of technical and commercial reasons. It was cheap and efficient – sewing ten times faster than by hand – and appealed strongly to housewives, first in America and quickly in Europe, especially after the introduction of the lightweight 'family' machine in 1858.

In 1874, Levi Strauss in San Francisco introduced metal rivets in working clothes to reinforce weak points. So were born the modern jeans, so called because the twill used to make them came from Genoa (in French, Gênes).

The sewing machine had two important repercussions. First, it led to the development of the ready-made clothing industry, especially in alliance, in the 1880s, with G. P. Eastman's reciprocating knife which made it possible to cut through 50 thicknesses of fabric at once. Secondly, Singer introduced the convenient, but expensive, system of hire-purchase, his door-to-door salesmen offering the machines for a small down payment and a succession of monthly ones. Not surprisingly, Singer died a multi-millionaire.

Food

The focal point of every home is the kitchen where, as we have seen earlier, gas cookers began to displace the old-fashioned range from the middle of the 19th century. By then, a food-processing industry had become well established, especially for the long-term preservation of food. To an extent, some of the everyday culinary tasks were transferred to the factory, just as the spinning and weaving of wool had already been.

Here, the major development was in the increasing availability of preserved food. The establishment of the frozen meat industry in the latter part of the 19th century had the advantage of low cost rather than long life: until the advent of domestic refrigerators in the 1920s such meat had to be eaten very soon after it came out of store. Much more significant was the rise of the canning industry, for this allowed ready-prepared food to be kept in the store-cupboard for months or even years.

At the end of the 18th century – in response to a prize offered by Napoleon for improving the supply of provisions for the French army – François Appert developed a heat-sterilizing process. The food was heated in boiling water in glass jars, which were then hermetically sealed. At that time nothing was known of the microbial causes of food putrefaction and it was not until 1810 that, working empirically, he developed a reliable method and opened his first food-preserving factory at Massy, near Paris. In that same year Peter Durand, in England, patented the use of tin-plate canisters instead of glass. This was quickly taken up by Bryan Donkin, who by 1812 was supplying the Royal Navy from a cannery in Bermondsey, London. The industry grew rapidly but was limited by the fact that the cans had to be made individually

early mordants and dyes 120–121 **food industry** 291–301 **food technology in antiquity** 43 **man-made fibres** 268, 304–305
William Perkin and mauve 265

The food canning industry expanded enormously during the 19th century, both in total production and the variety of foods treated. This trade card for canned Frankfurter sausage was printed in 1895. The multilingual label indicates a wide export market.

by hand. From 1847, however, their manufacture began to be mechanized. By the end of the century, with the introduction of rubber seals, the process was entirely automatic.

The canning process was introduced to America from Britain in 1817 and led, among other things, to the development of the huge meat canning concerns in the Chicago area, particularly associated, from 1880, with the name of P. D. Armour. By the latter part of the century, a great range of canned meats, condensed milk, fruit and vegetables were cheaply available throughout the world: already production was measured in tens of millions of cans a year.

Food preservation is, of course, a very ancient technique. From time immemorial fish and meat have been preserved by drying, salting or smoking; quickly perishable milk has been turned into cheese and butter; vegetables have been pickled in vinegar. Increasingly, however, these processes were adapted for large-scale factory production. For example, cheese-making, traditionally a local farm process, was being transferred to factories in Australia, Canada and America well before 1900. Apart from the canning of milk condensed by evaporation, dried milk in powder form began to be made from 1855: this was followed by malted milk in 1883.

Butter, too, was increasingly factory-made and the invention of de Laval's cream-separator in 1878 made for much greater efficiency: it was widely used in the larger dairies throughout the world from 1880. But after mid-century butter found a serious rival in margarine, invented by Hippolyte Mège-Mouriés in 1869. Initially, it was a rather crude emulsion of beef suet, skimmed milk, cows' udder and pigs' stomach, but it was soon improved in taste and texture by blending with better quality vegetable oils. Initially, it was sold in bulk but from about 1890 it began to appear in America in packaged form, like butter. It posed so serious a threat to butter that in Britain the original trade name of butterine was forbidden in 1885. Nevertheless, by 1900 Britain imported 50,000 tonnes of margarine. Today sales of margarine in America and Europe exceed those of butter, partly because of a much improved product and partly because its unsaturated fats were believed to be less conducive to heart disease than the saturated fats of butter. By 1960, world production had risen to over 2½ million tonnes annually.

Domestic appliances

During the 19th century a number of labour-saving devices began to appear in the houses of more prosperous families. While many originated in Europe, still not lacking in inventive genius, it was in America, with its shortage of labour, that they were most readily adopted.

The laundering of clothes stands high on the list of routine domestic tasks. Traditionally, and to this day in many parts of the world, this consisted simply of scrubbing in clean water, removing as much water as possible by wringing with the hands and hanging out to dry. From the earliest days a mixture of oil and potash might be used to help shift the dirt, but soap as we know it is essentially an 18th-century commodity, though for personal hygiene it was used as a luxury from the 14th; that of Castile was particularly prized.

Some English country houses had clumsy washing machines in the 18th century and there was even a patent application as early as 1691, but effectively the domestic washing machine, manually operated, dates from the middle of the 19th century. This followed an American patent of 1846 for a machine that imitated the action of the human hand on a ribbed scrubbing board. The following year saw the appearance of a wringer which also duplicated the traditional method of expelling water by twisting the fabric, but this was soon superseded by mangles with rollers.

The last stage was ironing, a tedious task then aggravated by a predilection for frills. By the end of the 19th century, gas

The latter part of the 19th century saw the beginning of the mechanization of domestic work, though cheap labour offered little incentive. Early devices, like the 1874 hand-operated mangle (right) were primitive and clumsy. Vacuum cleaners for carpets were originally operated by travelling contractors, but from the turn of the century individually owned machines – such as the 1906 example shown far right – began to appear in larger households.

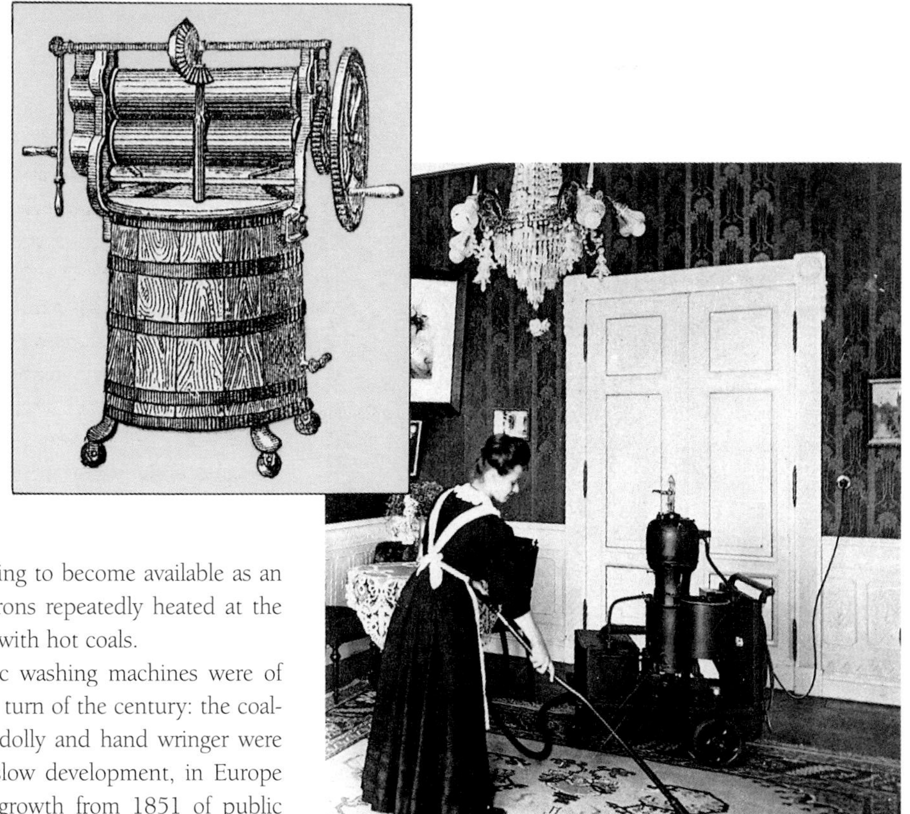

and electric irons were beginning to become available as an alternative to the simple flat-irons repeatedly heated at the range or the hollow iron filled with hot coals.

Even in America, domestic washing machines were of marginal importance up to the turn of the century: the coal- or wood-fired boiler, wooden dolly and hand wringer were normal. One reason for this slow development, in Europe as well as America, was the growth from 1851 of public laundries which would collect household linen and quickly return it washed and ironed. In these, the clothes were vigorously swirled around in large tanks of hot water by injecting steam.

Some stains, especially greasy ones, cannot be removed by the most thorough laundering. Grease dissolves readily in organic solvents such as benzene, however, and dry-cleaning establishments using this fluid appeared in Paris in the 1850s. The high fire risk led to chlorinated hydrocarbon solvents being introduced around the turn of the century.

As carpets became popular, periodically beating the dust from them became another considerable chore. Carpet-beating machines, often located in laundries, appeared in America about 1860 and were followed by compressed-air machines to blow the dust out of railway carriage upholstery.

In London, it occurred to Hubert Cecil Booth that removing the dust from carpets by suction would be more satisfactory than blowing it around. About 1901, he introduced a mobile suction machine which cleaned domestic carpets. This used a long tube passed in through the window: similar machines were in use in Paris in 1903. From these devices developed the variety of domestic vacuum cleaners in use today.

Thus by the turn of the century a considerable range of domestic appliances was available to those who could afford them. Apart from the disincentive that domestic labour was still relatively cheap, especially in Europe, their disadvantage was that they were mostly manually operated. The situation changed rapidly in the 20th century with the availability of small electric motors.

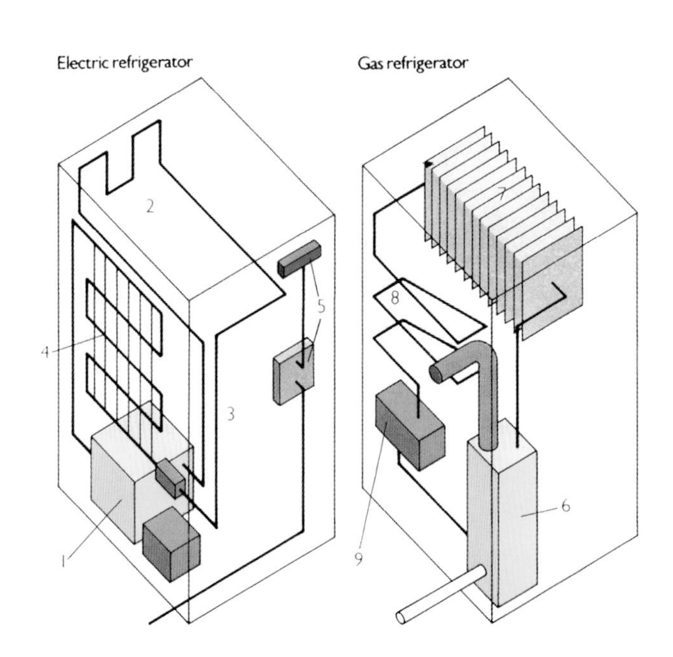

In a compression (electric) refrigerator, refrigerant is circulated through a pipe by an electric compressor (1). It evaporates, absorbing heat from the freezing (2) and cooling (3) compartments. The refrigerant then flows to the condenser (4), where it liquefies and loses heat to the air, before returning to the compressor. Thermostats (5) control operation. An absorption (gas) refrigerator uses ammonia produced in a generator (6) by heating ammonia solution and liquefied in a condenser (7) to lose heat. It then goes to an evaporator (8), where it takes heat from the refrigerator. The absorber (9) dissolves the ammonia vapour and returns the solution to the generator.

CHAPTER TWENTY-ONE

The Changing Pattern of Agriculture

Agriculture is the most basic of all technologies, for on its success the life of the community depends. For this reason it has always tended to be conservative: with so much at stake there was little incentive to experiment when traditional methods would provide an adequate yield. In the main, these methods were handed down from generation to generation by word of mouth and this, too, tended to perpetuate the existing order rather than prompt innovation. Nevertheless, written accounts of farming practice appeared at an early date. Thus Varro – 'the most learned of Romans' according to Quintilian – wrote an agricultural treatise *De re rustica* in the first century BC: from this, one of his very few works surviving nearly intact, we have a good account of Roman farming techniques, based partly on his own experience. Almost exactly contemporary was a detailed Chinese work by Fan Sheng-Chih: this has long since been lost, but enough can be deduced from later quotations from it to reconstitute much of the original. From this, it is clear that it was a comprehensive guide to practical agriculture under Chinese conditions somewhere in the Middle Yellow River region: for example, the reader is told how to plant mulberry trees and how many silkworms a given acreage of trees will support. Much of his advice is as sound today as it was then: 'Winter wheat may be sown 70 days after summer solstice. Earlier sowing gives solid ears and firm stands: late sowing gives smaller ears and less plump grains .... When it thaws in spring time plough the soil that has become mellow and sow spring wheat. After sprouting and good growth of the root system, hoe thoroughly as with winter wheat.'

The reference to the summer solstice is interesting. We have already remarked on the very early importance of astronomical observations as a means of compiling a calendar as a guide to the proper seasonal activities in agriculture.

Much of our knowledge of the original text of Fan Sheng-Chih comes from a Chinese agricultural encyclopaedia of the sixth century, the Chi'i Min Yao Shu. This is remarkably comprehensive: apart from detailed advice on growing the ordinary cereal crops, it discusses fibre crops, timber trees, the principles of animal husbandry and the brewing of beer. As the author, Chin Ssu-hsieh, puts it: 'Beginning with ploughing and cultivation, down to the making of vinegar and meat-paste, any art useful in supporting daily life is jotted down.' Fan Sheng-Chih advises his readers that: 'The basic principles of farming are: choose the right time, break up the soil, see to its fertility and moisture, hoe early and harvest early.'

We can do worse than follow this sequence in continuing our survey of the history of agricultural techniques, but first we should take a look at changes in the overall pattern of agriculture.

The changing pattern

For more than 1000 years after the fall of Rome there was little significant change in agricultural practice. Experience had shown that after repeated cropping with the same crop, land became impoverished but could be improved by allowing it to lie fallow for a season, by varying the crop and by applying animal manure. In much of western Europe, this found expression as a three-field system. Individual farmers held strips of unenclosed land in multiples of three: each was cropped in turn with winter corn and spring corn, and then left fallow. The long narrow strips facilitated ploughing because they reduced the number of times the plough had to be reversed. Cattle, the source not only of meat and dairy products but also of the all-important manure, grazed the surrounding open land or the strips themselves immediately

In the Middle Ages agricultural practice differed little from that of classical times, as this German book illustration of 1515 – not to be taken as accurate in all its details – clearly indicates. Seed is broadcast, as in ancient civilizations, and the square harrow is essentially Roman.

after the crops had been harvested. Around the Mediterranean a variant was practised in which small enclosed fields were as a whole put down to a succession of crops and then left fallow for a season.

By the 17th century, however, this age-old pattern began to change. With increasing acreage falling into the hands of landlords, there was a move to improve the productivity – and profitability – of farmland. This was done by introducing new forage crops such as clover and alfalfa: the Spanish name of the latter, a kind of lucerne, is a reminder that these were not strictly novel crops but ones that had found a limited use since the classical systems described by Varro. New crops – such as turnips – were also introduced for winter fodder to supplement hay. This involved four-course or more elaborate rotations: in Flanders a seven-course rotation was practised. Basic farm implements, such as the plough, were improved and new ones devised. On the animal side, increasing attention was paid to livestock improvement.

These changes were not compatible with the working of small open units of land and led increasingly to enclosure of larger areas. In Britain, this had been advocated as early as 1523 by Sir Anthony Fitzherbert in his *Boke of Husbandrie*. That this, the first textbook on farming in English, should have appeared at all is symptomatic of the new attitudes that were beginning to emerge. It may be taken as heralding the start of the new Agricultural Revolution though this, like most of the revolutions beloved by historians, was in reality a leisurely affair: it is generally taken to have started in 1600 and to have been completed in 1850. Inevitably, the enclosure of land was bitterly resented by many of those directly affected, but it was facilitated by appropriate legislation. From the late 18th century it was possible in Britain to apply for private Acts of Parliament to permit land enclosure, and between 1727 and 1844 more than 4000 such Acts were approved, involving more than 7 million acres. A million acres more were enclosed by the end of the century under the Inclosure Act of 1845.

In the field of livestock much attention was paid to the creation of improved breeds, though the meat producers found it hard to keep pace with changing public taste and demand. Of particular interest was the merino sheep, famous for its fine wool. This was a jealously guarded Spanish monopoly, but it proved impossible to prevent some beasts being smuggled out: during the second half of the 18th century breeding stock had been established in most European countries. It did not do well in Britain, where the most important development was the new Leicester sheep bred by Robert Bakewell, who also did much to improve breeds of cattle and horses. He and other breeders kept their methods secret, but it is clear that their basic technique was out-breeding to develop and in-breeding to stabilize desirable characteristics.

Such, in brief, was the development of agriculture in Europe – a transformation, certainly, but a very gradual one.

Although American agriculture accepted mechanization with enthusiasm, because of the large areas and scarce labour, progress was uneven. This picture of maize cultivation in the Midwest in the 1870s shows primitive equipment and a house built from turf sods.

In the New World the situation was very different: there, too, a transformation took place, but a far more dramatic one. Naturally, the early settlers brought with them ideas about agriculture conceived in the European context – particularly Britain, Germany and north-western Europe until the end of the 19th century – but in the different climate and soil conditions of America, especially in the hot southern states, these ideas were not always applicable.

The American situation was different, too, in another respect. Whereas Europe was densely populated and much of the best land was already under cultivation, in America the population was small – only 23 million even in 1860 – and there were vast tracts of virgin territory available at little cost, much of it open prairie. In 1800, Thomas Jefferson remarked succinctly that: 'In Europe the object is to make the most of their land, labour being abundant: here it is to make the most of our labour, land being abundant.'

Initially, the eastern seaboard was the main area of settlement, but with increasing population and better transport the frontier was pushed steadily westward. Cattle flourished on the grassland, which had once been the home of vast herds of bison and in the second half of the 19th century these increasingly supplied the great meat-packing areas around Chicago. The quality of meat at first was poor, but improved breeds and the fattening of cattle in grazing areas – such as those of Iowa and Illinois where the prized 'blue' grass grew – gave a more acceptable product. West of the Mississippi huge areas of wheat were grown for the home market and increasingly for export. Maize, too, became a important grain crop, especially for fattening cattle. By modern standards much of this farming was not only primitive but often reckless. The general separation of arable farming from cattle-raising made it difficult to restore exhausted land

calendar 31–32 **development of ploughs** 203–205 **early agriculture** 26–33 **modern agriculture** 291–301

The latter part of the 19th century saw a great expansion of the rubber industry and of the variety of its products: plantations were established in the Far East. The advertisement for rubber boots (below) appeared in Boston in 1898. Another important new crop was sugar-cane. The picture (right) shows the millyard of a plantation in Antigua in 1823.

by manuring. Worse, ill-advised clearance of land in hot, low-rainfall areas was beginning to cause the soil erosion that was to prove so disastrous in some areas in the 20th century. That so much was produced reflects the fact that the original soil fertility was high and the acreage so vast that low productivity could still be profitable. North of the 49th parallel, the picture was very similar, though the emphasis was even more strongly on wheat production.

Although the economic importance of cattle and grain was very great, other crops were grown on a very substantial scale. These included tobacco, cane sugar and cotton – the latter finding a ready market in the mills of Lancashire. In both America and Canada, the extensive forests were a source of timber for both construction and, from the 1880s, of wood pulp for paper manufacture.

When the Declaration of Independence was signed in 1776 America already had a population of 2½ million people. Twelve years later the township of Sydney was founded in Australia, with a population of 1030: even by 1843 the European population of Australia was no more than 200,000. Mineral wealth was potentially very great, but agriculturally conditions were adverse: the soil was difficult to work and in many areas drought was a constant menace. Nevertheless, the young colony soon became more than self-sufficient. The merino sheep did well and wool was its first export, followed closely by wheat.

Again, however, it was abundant land rather than good husbandry that brought results: productivity per acre was comparable with that in Europe 500 years earlier. The more interesting statistic – productivity per person – is not reliably ascertainable. The world's first commercial plant for meat refrigeration was established at Sydney in 1861 and this led to the shipment of lamb and beef to Europe: the beef market, however, was lost to Argentina by the end of the century. Refrigeration also encouraged the export of butter and other dairy products. Apples and grapes were also grown, the latter providing the basis for what was to become a flourishing wine industry. New Zealand, too, became an important exporter of lamb and dairy products after 1890.

South Africa had been practising agriculture modelled on that of Europe, but adapted to local conditions, since 1652, when the Netherlands East India Company first used Table Bay as a port of call on the way to Java. From the 17th century viticulture and wheat growing have been important within the areas of winter rain. Later, merino sheep did well and the sale of the wool-clip in Europe became an important part of the settlers' income.

In the temperate zones, farming could be conducted on lines resembling those of Europe, but in tropical and semi-tropical countries different crops were important. Here, too, a notable difference was in the scale of operation: plantations tended to replace the small holdings of the Old World. Improved transport made it possible to export new products to distant markets. Among such new crops was sugar-cane – though from the start of the 19th century this was in competition with sugar-beet grown in Europe at the instigation of Napoleon – tea, coffee, cocoa, bananas and palm oil. Tea is a typical example of the expansion of trade in these exotic products. Tea-drinking became particularly popular in

A selection of farming and woodworking hand tools used in America from about 1750 to 1850: an axe, sickle, pitchfork, plane, adze, brace and bit and compasses.

Britain: in 1830 three-fifths of the world's consumption was drunk there. This amounted to an annual import of 30 million pounds annually: by 1900 this had risen to 300 million pounds. We have already mentioned the development of the fast clipper ships especially for this trade.

Coffee-drinking, too, became widely popular in Europe following the establishment of the first plantation by the Dutch in Suriname in 1718: previously, its use had been largely limited to the Middle East, where it was popular among the Arabs.

Another important new crops was rubber. Rubber, made by coagulating the latex of forest trees, was familiar to the Incas and Aztecs and small quantities were brought back to Europe by the Spaniards. At first little more than a novelty, it gradually found practical uses, particularly for making waterproof clothing and, later, tyres for road vehicles. By the latter part of the 19th century, the supply of wild rubber from South America was beginning to be insufficient to meet the demand and attempts were made to establish plantations in the Far East with seeds smuggled from Brazil. After early failures, seedlings raised at Kew Gardens in England were established in Ceylon (now Sri Lanka) in 1877. From this arose a huge plantation industry in the Far East and parts of Africa, notably Liberia, supplying, in particular, the demands of the burgeoning car industry.

Implements and machinery

In medieval Europe, farm implements were made of wood and wrought iron and assembled by local craftsmen: the carpenter and blacksmith played a dominant role in rural life. The blacksmith provided not only the parts of implements subjected to the hardest wear, such as the shares of ploughs and the tines of forks, but also tyres for wagon-wheels and shoes for horses. Construction of all kinds of farm equipment tended to be simple, but there were regional variations dictated partly by local climatic and soil conditions and partly by the traditions of design. In Europe, this continued until well into the 20th century and in the developing countries of the world it is still normal. Nevertheless, from the middle of the 18th century a significant change can be seen: increasingly, implements were made not individually but in batches and small manufacturers began to supply regional, if not national, rather than purely local markets.

The plough

The first stage in cultivation is to break up the impacted soil as a preliminary to producing a fine weed-free tilth in which seeds can germinate and the seedlings can establish themselves. For this, the basic implement is the plough. In its simplest form this does no more than make a shallow scratch on the surface: such scratch ploughs, or ards, were widely used and one dating from 400 BC has been recovered from a peat-bog in Jutland.

Ards were normally drawn by one or two oxen, but peasants who had not even this source of power would use a breast plough, in which the ploughman pushed against a crosspiece to force a blade through the ground: in stony ground in Scotland a *cas chrom* – literally 'crooked spade' – was used: in this the blade was pushed through the ground with the foot.

Such primitive implements are still to be found, but by Roman times the main features of the modern plough had emerged: a coulter to cut roots in front of the share and a straight mouldboard to deflect the sod as it was cut. Such a plough could be made entirely of wood but it was soon common to protect the working surfaces with iron: this was particularly necessary when working the wet heavy soils of northern Europe. By Saxon times such swing ploughs were beginning to be complemented, but not supplanted, by ploughs with wheels which supported the front of the implement and eased the ploughman's task. At the same time, ploughs with two handles instead of one appeared, though even in areas of high farming with light soils – such as Norfolk, England – one-handed ploughs were common enough up to the end of the 19th century. Although such ploughs could tackle heavy work they were, by modern standards, exceedingly cumbersome.

Ploughs had to cope with a wide range of soil conditions and, while similar in principle, they varied a good deal in design from place to place. By 1523, the various adjustments had become so complex that Sir Anthony Fitzherbert wrote: 'It is harde to make a man to understand it by Wrytynge, without he were at the operation therof, to teach the practyce.'

clipper ships 172, 173 **early ploughs** 33 **merino sheep** 201 **refrigeration** 197, 199

From the 17th century increasing attention was paid to the mouldboard, which began to be curved so that the sod cut by the share was continuously inverted, burying the surface weeds. By 1784, in his *Treatise on Plough and Wheel Carriages*, James Small was giving a detailed geometrical analysis of the proper curvature of mouldboards to achieve this effect as smoothly as possible. A slight twist was possibly given to the mouldboard as early as the 14th century, in Flanders, but strong curvature was difficult until iron replaced wood. At about the same time, disc coulters, probably invented in Holland, began to replace the knife.

By the 18th century, the standard plough for northern Europe was the Rotherham plough – so called after an important manufactory in Yorkshire, but introduced from the Netherlands. It was light, had no wheels and the mouldboard was a continuation of the share. Although the Rotherham plough came to Britain from the Netherlands, its ultimate origin may well have been East Asian. As elsewhere, the traditional plough in China was the ard, but by Han times, some 2000 years ago, a turn-plough was in use which had a curved metal mouldboard. By the 17th century, such ploughs were widely used in South East Asia and the Dutch would certainly have seen them in use in the wet-rice areas. Their suitability for the water-logged soils of their homeland would have been at once apparent and they may even have brought back examples. There is thus a plausible, but not proven, connection between East and West.

Around the end of the 18th century the all-iron plough began to be introduced in northern Europe and in 1789 Robert Ransome patented a ploughshare which was made self-sharpening by being cooled in the foundry more quickly on one side than the other. Ransome set up a factory at Ipswich, England, where he made ploughs largely from interchangeable iron parts, so that worn ones could be quickly replaced. A further advantage of this system was that parts, especially shares, of different designs could be provided to meet local conditions and fitted to a standard frame: thus the possibility of a national market was opened for the first time. Indeed, an international market began to develop: from the mid-1850s James Oliver of Indiana began to market a light plough which not only sold well in America but was imported into Britain and other northern European countries.

Until about the middle of the 19th century all ploughing was done with oxen or, increasingly, horses. In the 1850s, however, steam power began to be used. The engine used was a stationary one, though it could be driven into position, and the plough was hauled across the field by a cable: generally it was made reversible so that each traverse was a productive one. While draught animals might be able to pull a two- or even three-shared plough on light soils, the steam engine could haul a multi-shared plough even on heavy soils. Up to 5 hectares (12 acres) a day could be broken up. The machinery was heavy and expensive and, except on the biggest estates, steam ploughing was done by travelling contractors. Towards the end of the century, mobile steam tractors began to be used, but the future of mechanical ploughing in the 20th century lay with the internal combustion engined tractor.

Steam ploughing by travelling contractors using stationary engines began to appear in the middle of the 19th century, and a few mobile steam tractors appeared at the end of it.

To provide a fine tilth, clods left after ploughing must be broken up, and medieval illustrations show this being laboriously done with wooden mallets. An alternative was a heavy wooden or stone roller, but early in the 19th century iron rollers began to appear, at first smooth but later serrated, as in the Crosskill clod-crusher of 1841. The Romans had used simple harrows from at least 300 BC to break down the soil still further, but an important advance was the zigzag iron harrow which came into use in the 1830s. This was followed by the disc harrow, which had the great advantage of rarely clogging.

After preparation of the ground, the next operation in the agricultural calendar is sowing, traditionally done by broadcasting the seed in handfuls and trampling it in. However, there are advantages in sowing seed systematically in rows: this is less wasteful of seed, the young plants suffer less from overcrowding, and weeding and singling with the hoe is easier. This was certainly practised in the Middle East in the third millennium BC – though to what extent we do not know – and widely in China in pre-Christian times. In the West, however, it was a comparatively recent development. Although some experiments with seed-drilling machines were carried out in Italy in the 16th century, it was Jethro Tull who effectively introduced the technique into Europe, with the seed-drill he invented in 1701. It was only slowly adopted though, mostly on large farms. His *Horse-Hoing Husbandry* of 1731 advocated the drilling of seed in regular rows and subsequent weeding of the crop with a horse-drawn hoe with the blades spaced to match the rows. Again, there is an undoubted Chinese connection, but it is a relationship of systems rather than of implements. Possibly the Italian seed-drills, the precise design of which is not known, were based on Chinese models. Italy then had strong links with the Orient, but the mechanism of all later European machines was quite different from those developed in China. Tull was a Cambridge graduate and read for the bar at Gray's Inn in London, and had his own estate on which to test his ideas. He was a well read man and would have been familiar with the wealth of information about Asian agriculture then coming back to Europe. We can suppose that he became aware of the advantages of row culture but developed his own devices to put it into practice.

Until the 19th century harvesting was still universally done with sickles, but in the first half of the century horse-drawn reaping machines appeared. The example shown here is that made in the USA by Cyrus McCormick in 1831. It had limited popularity until shown in London at the Great Exhibition of 1851.

Harvesting

Harvesting is one of the most labour-intensive of all farm operations and speed can be vital in uncertain weather. It was an operation in which almost everybody in the farming community took part and casual labourers could be sure of work. It is surprising, therefore, that reaping machines did not appear before 1800 and were not successful until 1826. In that year Patrick Bell, a Scottish farmer, devised a reaper in which the ripe corn was cut at the roots with a cutter-bar and laid alongside the machine in swathes. Bell's machine was pushed by two horses: in 1831 Cyrus McCormick, in America, invented a machine with a similar cutting mechanism, but drawn by horses rather than pushed. However, neither kind of machine attracted much attention until exhibited in London at the Great Exhibition of 1851.

A good hay crop was important for seeing beasts through the winter and this, too, was a labour-intensive process. Nevertheless, even such a simple machine as the tedder, in which rotating tines turned over the cut hay to speed up drying, did not begin to appear until 1800, and grass-mowing machines came into use in agriculture even later. Lawn-mowers similar to modern machines appeared in the 1830s.

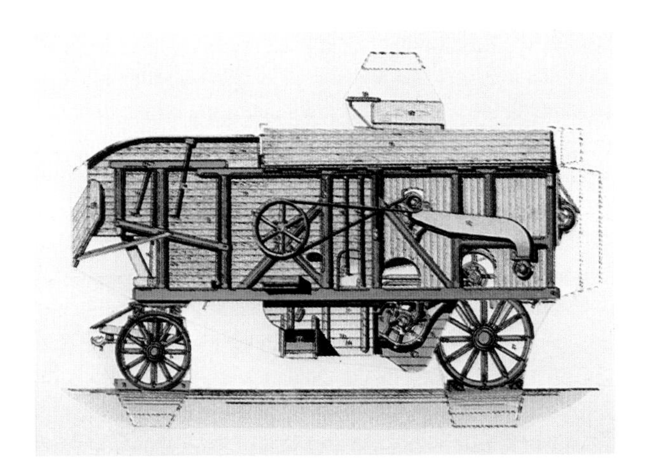

Also in the 19th century, flails and winnowing fans began to give way to threshing machines. This steam-powered example is representative of many in common use at the turn of the century.

Threshing

After harvesting corn, the next stage is threshing. The age-old method was to beat the ears with flails to separate the grain and chaff and then to throw the two together in the air in a light breeze, when the chaff blew away and the heavy grain fell straight to the ground in a heap. The straw was used for thatching or winter bedding for animals. Not until the early 18th century was any serious attempt made to mechanize this basic process, initially simply by imitating the beating action of flails. In 1786, Andrew Meikle devised the type of machine that was soon widely adopted. In this, the corn was fed into the narrow space between a rotating drum and a circular shield. Such a machine discharged grain and chaff together and the two had to be separated in a winnowing machine, in which the grain was shaken through a riddle into a sack and the chaff was blown away by a circular fan. From the middle of the 19th century, these two operations were combined in a single steam-powered threshing machine. Only very large estates would own their own machine and, like steam ploughing, the equipment was generally hauled round from farm to farm by contractors.

Milking machines

As a generalization, it can be said that up to about 1850 the main developments in agricultural mechanization were in Europe, and particularly in Britain, where large landowners had the necessary capital and incentive. From that time onwards, however, the initiative passed, not surprisingly, to the other side of the Atlantic, with its small population and vast tracts of arable land.

In either case, the emphasis was on crop production, but an important exception was in milking. All milking machines work on the suction principle and the first was devised in 1860 by an American engineer, L. O. Colvin. His machine was used in both America and Britain but had the defect that

modern agriculture 291–301 **primitive ploughs and seed drills** 33 **steam engines** 159–165

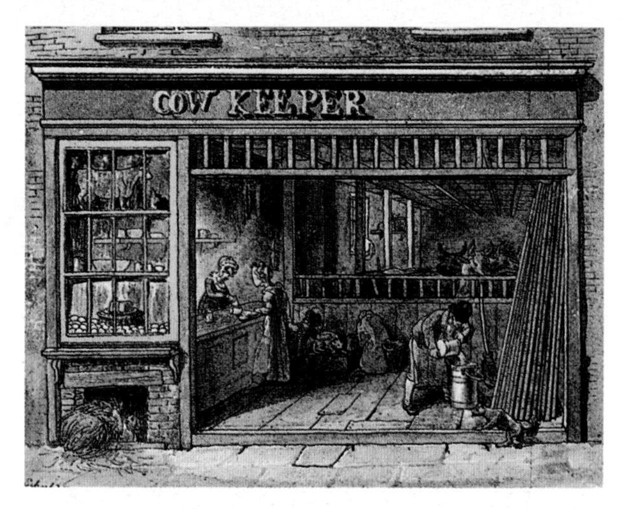

Until the availability of faster transport provided by the expansion of the railway network, urban households brought their milk from local cow keepers. This typical dairy traded in Golden Lane, London, in 1825.

continuous suction was painful to the cow and could cause inflammation and bleeding. A much better system was the 'Thistle' machine devised by Alexander Shields, of Glasgow, in 1895. This included a pulsating device which was more restful to the cow and approximated to hand milking. It was, however, too expensive to be quickly adopted.

An indirect effect of mechanization on dairy farming arose from the growth of the railways. Dairy cows were commonly kept in London and other large cities up to the last half of the 19th century to provide fresh milk for the table. From mid-century onwards, however, fast railway transport made it possible for the evening milk from country farms to be delivered to the city the next morning. This, together with increasing imports, led to a decrease in the farm production of butter and cheese, the traditional ways of stabilizing dairy products.

Fertilizers

That crop yields could be enhanced by application of animal manure, including human waste, was well known from the earliest times, and the basis of mixed farming was that arable land partly fed the stock which, in turn, produced the manure needed to keep the land in good heart. However, the balance between stock and crops was not easily maintained. Until the introduction of new fodder crops such as turnips, only a fraction of the stock could be brought through the winter, while more intensive cultivation made heavier application of manure necessary. In the 19th century, animal manure began to be supplemented by cheap organic waste such as dried blood, seaweed and fish waste in coastal areas, hoof and horn, and wool shoddy. Occasionally a light catch crop, such as mustard, might be grown and ploughed in. In 1839 guano, the accumulated droppings of seabirds, began to be imported into Europe from South America.

However, all these methods of soil enrichment were empirical. Not until the end of the 18th century was any serious study made of the scientific principles of agriculture. One of the pioneers was Daniel Thaer, who established a carefully managed experimental estate at Celle in Saxony, in 1802. Others were set up at about the same time elsewhere in Germany and France. In Britain, Humphry Davy gave lectures at the Royal Institution and his *Elements of Agricultural Chemistry* appeared in 1813. In 1834, John Bennet Lawes established the now famous agricultural research station at Rothamsted. Justus von Liebig's *Chemistry in its Application to Agriculture and Physiology* of 1840 was an important work. From this activity a clear picture began to emerge of the nutritional requirements of crops, with nitrogen, phosphorus and potassium emerging – particularly as a result of Liebig's work – as the most important elements. This paved the way for the artificial fertilizers which were to play a dominant role in agriculture in the 20th century.

Bones are rich in phosphorus and their value as a fertilizer had long been noted, but in the early 19th century it was discovered that they could be made soluble, and far more effective, by treating them with sulphuric acid. About 1817, James Murray, an Irish physician, began to manufacture this product, superphosphate, in Dublin and pointed out that mineral phosphate rock could be used in place of bones. Murray had limited success, but in 1843 Lawes established a superphosphate factory near London and 30 years later his output as 40,000 tonnes annually.

Throughout the 19th century, guano remained an important source of nitrogen, supplemented by naturally occurring sodium nitrate, mined as caliche in Chile. The value of wood ashes as a fertilizer had long been recognized on an empirical basis long before potassium was identified as its active ingredient. In the 19th century, huge quantities of ash, produced as a by-product of forestry, were imported into Britain from Canada, but from 1870 enormous deposits of a potassium mineral, carnallite, began to be worked in Germany, with disastrous consequences for the Canadian ash industry.

By the end of the century, the new artificial fertilizer industry had become of great economic importance: Europe was consuming annually around a million tonnes of sodium nitrate alone. In America, where the vast areas of arable land were not matched by accessible areas raising cattle providing natural manure, this was a particularly important development. Mixed fertilizers began to be manufactured in Baltimore in 1849 and by the end of the century annual production had risen to 2 million tonnes. Rock phosphate began to be mined in South Carolina in 1868.

Natural reserves of mineral phosphate and potassium were enormous. Nitrates, however, were a different matter and by the end of the century very serious depletion had occurred. In 1898, Sir William Crookes warned that the world faced starvation unless new forms of nitrogenous fertilizers could be found. In the event, salvation came from the Haber-Bosch processing for fixing atmospheric nitrogen, but this was a 20th century development to which we will come later.

agrochemicals 291–295 **cream separator** 165, 198 **Great Exhibition of 1851** 156, 157 **harvesting in antiquity** 26–27 **synthetic fertilizers** 264, 291

PART

4 The Transatlantic Surge

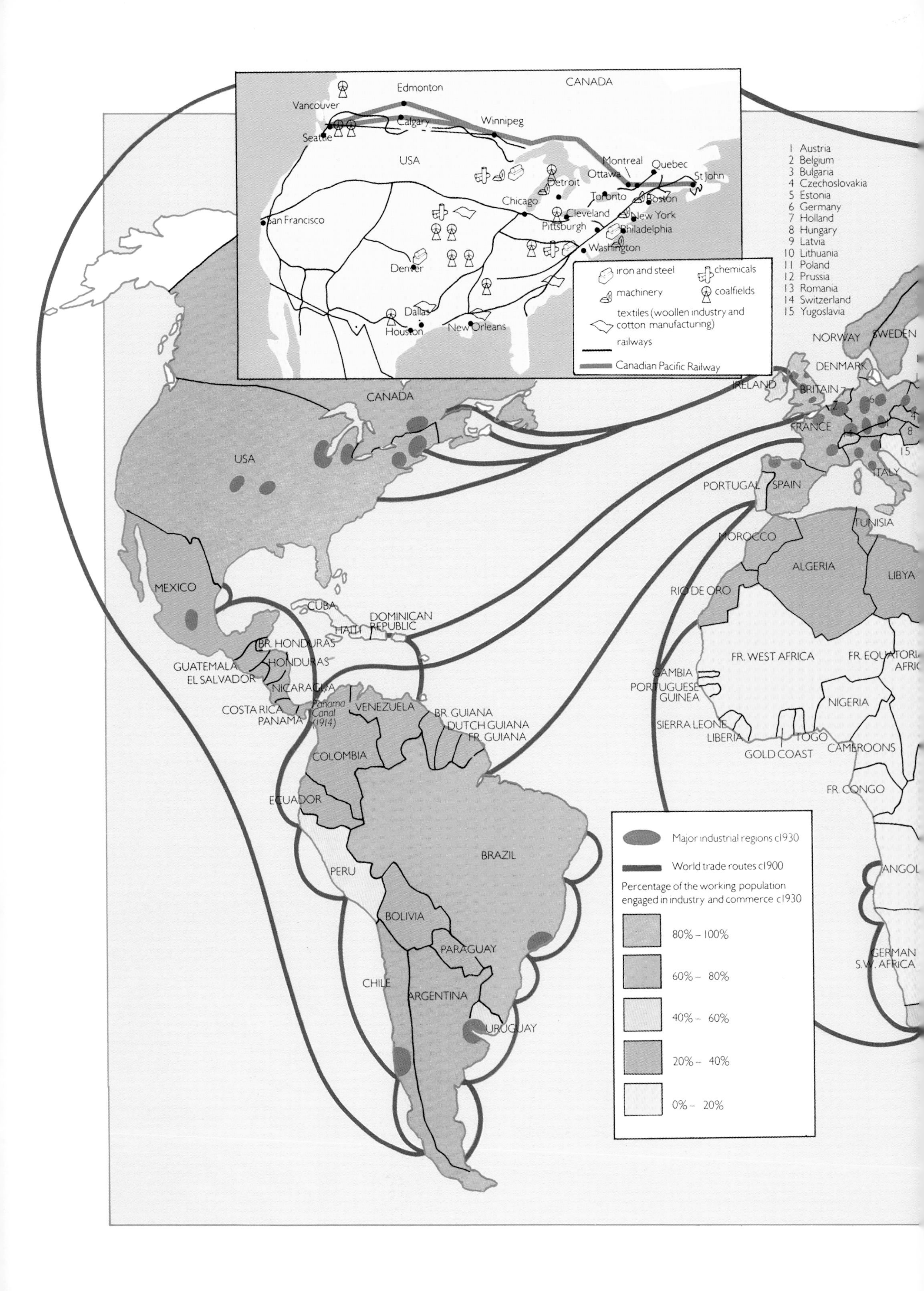

CANADA
Edmonton
Vancouver
Calgary
Winnipeg
Seattle
USA
Montreal
Quebec
Ottawa
St John
Detroit
Toronto
Boston
Chicago
Cleveland
New York
San Francisco
Pittsburgh
Philadelphia
Washington
Denver
Dallas
Houston
New Orleans
iron and steel
chemicals
machinery
coalfields
textiles (woollen industry and cotton manufacturing)
railways
Canadian Pacific Railway
1 Austria
2 Belgium
3 Bulgaria
4 Czechoslovakia
5 Estonia
6 Germany
7 Holland
8 Hungary
9 Latvia
10 Lithuania
11 Poland
12 Prussia
13 Romania
14 Switzerland
15 Yugoslavia
NORWAY
SWEDEN
DENMARK
IRELAND
BRITAIN
FRANCE
ITALY
PORTUGAL
SPAIN
TUNISIA
MOROCCO
ALGERIA
LIBYA
RIO DE ORO
CANADA
USA
MEXICO
CUBA
HAITI
DOMINICAN REPUBLIC
BR. HONDURAS
HONDURAS
GUATEMALA
EL SALVADOR
NICARAGUA
COSTA RICA
PANAMA
Panama Canal (1914)
VENEZUELA
BR. GUIANA
DUTCH GUIANA
FR. GUIANA
COLOMBIA
ECUADOR
PERU
BRAZIL
BOLIVIA
PARAGUAY
CHILE
ARGENTINA
URUGUAY
FR. WEST AFRICA
GAMBIA
PORTUGUESE GUINEA
SIERRA LEONE
LIBERIA
GOLD COAST
TOGO
NIGERIA
CAMEROONS
FR. CONGO
GERMAN S.W. AFRICA
Major industrial regions c1930
World trade routes c1900
Percentage of the working population engaged in industry and commerce c1930
80% – 100%
60% – 80%
40% – 60%
20% – 40%
0% – 20%

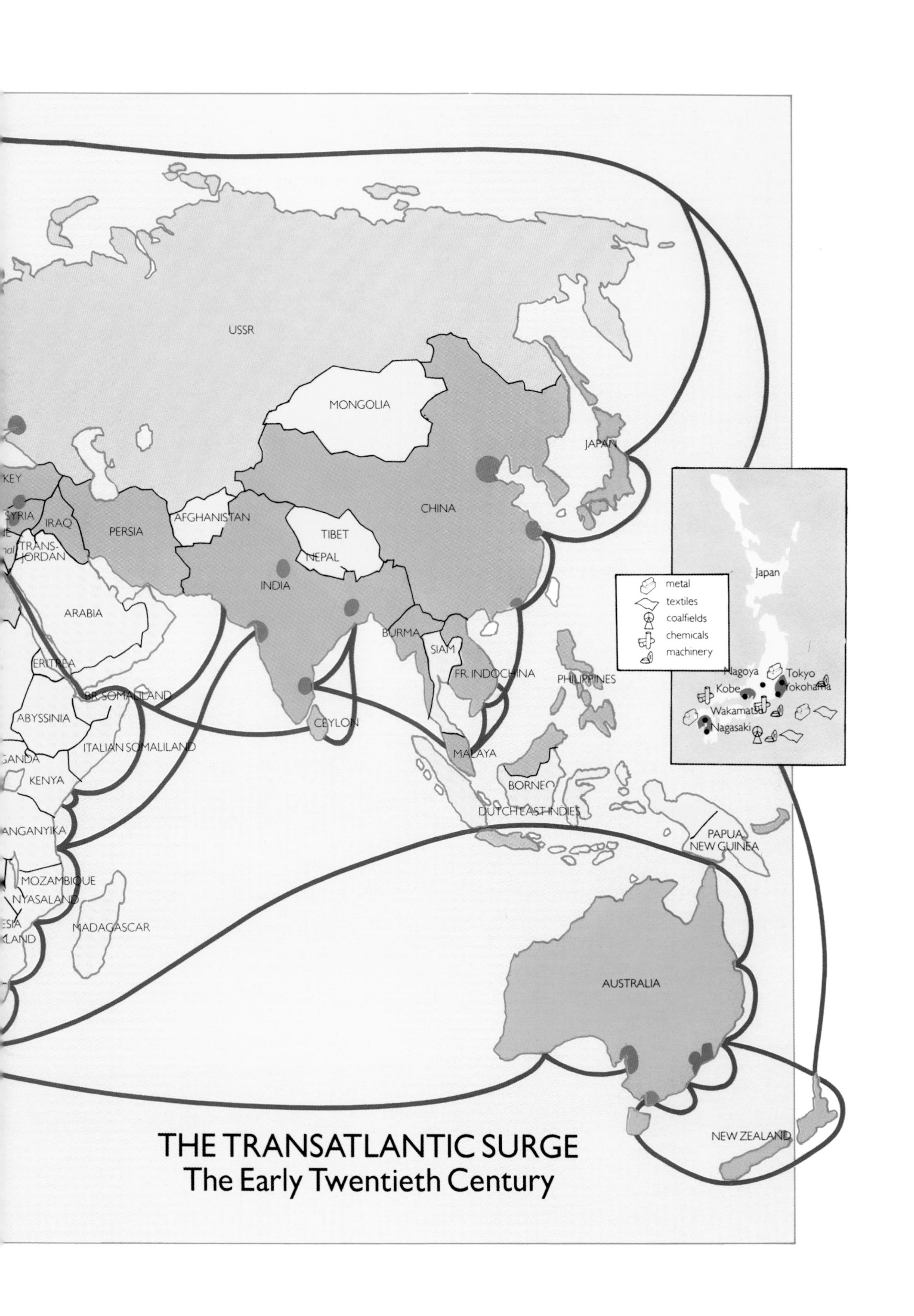

THE TRANSATLANTIC SURGE
The Early Twentieth Century

CHAPTER TWENTY-TWO

The Early 20th Century

The end of a century is a convenient punctuation point in history but it is not in itself normally of any intrinsic significance. In the history of technology, however, 1900 was an exception, as it roughly marked the first social impact of such diverse developments as the automobile and aeroplane, the telephone and wireless, sound recording and the electricity supply industry. These great developments took place against a changing international scene.

Britain, with its worldwide empire, still occupied the centre of the stage but was coming under pressure both militarily and industrially. The aggressive intentions of Germany were already apparent and a naval arms race began in 1898. Across the Atlantic, a thriving and innovative American industry was increasingly aggressive in the commercial field, not only towards Britain but also the rest of the world. In the East, Japan was becoming a world power, following the decision in 1868 to adopt a policy of Westernization: this began with the expansion of the textile industry in the 1880s, followed by heavy industry. In 1894–5, Japan had defeated an internally divided China and taken Formosa. Ten years later, to the consternation of the West, the Japanese convincingly defeated Russia and gained domination over Korea and south Manchuria. Russia was then slowly increasing the industrial content of an essentially agricultural economy, although it was not until after the Bolshevik Revolution that an intensive and rapid programme of industrialization – with emphasis on electricity as a source of power – was embarked on.

The uneasy politics of Europe reached a climax in 1914, when the assassination of the Archduke Franz Ferdinand sparked the First World War. This, like all major wars, was a powerful but uneven stimulant to technology, especially in the field of aviation. It was followed by a severe economic depression, originating in the USA and temporarily halting its onward march. From this, the world had scarcely recovered before the rise of the Nazis in Germany – with their keen interest in synthetic (*ersatz*) materials as a means of German self-sufficiency and in the mechanization of warfare – presaged the outbreak of yet another world conflict.

	1900	1905	1910	1915
WARFARE		first effective use of gyro-controlled torpedo; *Dreadnought* battleships	aerial bomb	First World War (1914-1918); gas warfare; tar
ENERGY AND NUCLEAR PHYSICS	offshore oil drilling; quantum theory; alternating current adopted for general use	geothermal energy; Special Theory of Relativity	atomic nucleus discovered	isotopes postulated; atomic structure recognized
COMMUNICATIONS	transatlantic radio signal; Caruso records for gramophone	diode valve; triode valve; first radio broadcast	colour photo; offset litho printing	
TRANSPORT	rigid airship	first aeroplane flight; experimental helicopter	Model T Ford; first cross-Channel flight	diesel-electric rail-car; Panama Canal opened
BUILDING	escalators	reinforced concrete bridge		
CHEMICAL INDUSTRY	rayon; indigo		salvarsan; stainless steel; Bakelite; helium liquefied; Haber-Bosch ammonia synthesis	hydrogenation of coal; thermal oil-cracking process
MEDICINE	veronal synthesized; hormones recognized	electro-cardiograph	vitamins recognized	

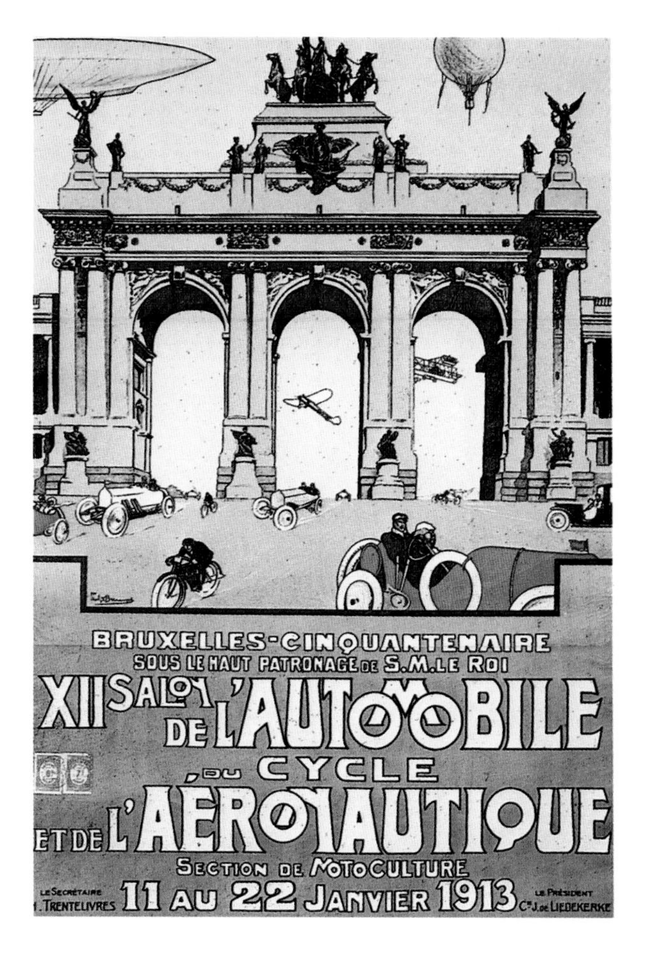

The First World War left lasting changes in the structure of society, not least in attitudes to the role of women. A shortage of manpower resulted in women coming out of their relative seclusion to engage in war work of many kinds. In Britain, this was a major factor in the first granting of women's suffrage in 1918. But all these changes were keenly affected by the increasing application of new technologies.

In 1900, although the motor-car was already more than 10 years old, it had still made little social impact, but Henry Ford's famous Model T was soon to bring motoring within the reach of the masses – 15 million were built between 1908 and 1928 – as well as establishing the assembly-line technique as a new feature of industrial life. In 1903, the Wright brothers for the first time achieved sustained flight in a heavier-than-air machine. Thus, in a very real sense, a new era of transport began with the new century.

Much the same may be said of communication. Telegraphy was well established on an international basis, but the telephone was still a novelty and only just beginning to make a real social impact. Radio communication, too, was still in its infancy: Marconi first spanned the English Channel in

By the beginning of the First World War the automobile and the aeroplane had become recognized as important new forms of transport.

This poster of 1913, advertising a transport rally in Brussels under royal patronage, epitomizes the new era.

1920 | 1925 | 1930 | 1935 | 1940

aircraft carrier · radar · Second World War (1939-1945) · jet fighter · magnetic mine

artificial ear · sformation · wave mechanics · mercury-arc rectifier · particle accelerator · neutron discovered · nuclear fission · natural gas in W. Europe

public radio broadcasting begins · Technicolor · IBM · sound film · transatlantic radio telephone · first public television service · quartz crystal clock · TWX and telex service · electronic television · xerography · ball-point pen · photo composition printing

transatlantic · airlines formed · autogyro · liquid-fuel rocket · synchromesh gearbox · front-wheel drive for cars · pressurized airliner · practical helicopter

Bauhaus, Dessau · prestressed and poststressed concrete · Empire State Building · concrete roads · cat's-eye reflectors

eum used · mical raw material · urea-formaldehyde plastics · pvc · synthetic rubber manufactured · selective herbicide · mercurial seed dressings · Houdry oil-cracking process · nylon · DDT · polythene

insulin · electroencephalograph · iron lung · penicillin discovered · sulphonamide drugs

Berliner's gramophone 241 **Edison's phonograph** 241 **Marconi and radio** 238–239 **Model T Ford** 248, 249 **Wright brothers** 219–221

1899 and the Atlantic in 1901. Initially conceived simply as a means of transmitting messages, radio was destined to become a worldwide source of entertainment for the masses. In sound reproduction, Thomas Edison had invented the phonograph in 1877 and by the turn of the century Emile Berliner's multiple recordings on flat discs marked the start of another great new home entertainment industry. The turn of the century was also a turning-point in the history of cinematography. The first public showing of a moving picture was staged in Paris in 1895: by 1900 sound had been linked with vision – though talkies were not commercially successful until the 1920s – and by 1905 the nickelodeon was an established feature of American life.

The main rival to the cinema was, eventually, television. Although this did not begin to be a major social force until after the Second World War, we can see the beginnings of television in Paul Nipkow's scanning disc, patented in 1884, and later adopted by John Logie Baird for his photo-mechanical transmissions in the 1920s. This was not, however, on the main line of development and the future lay with electronic systems, at the heart of which lay the cathode ray tube, introduced in 1897. Again, the seminal period, though not an immediately productive one, was the turn of the century.

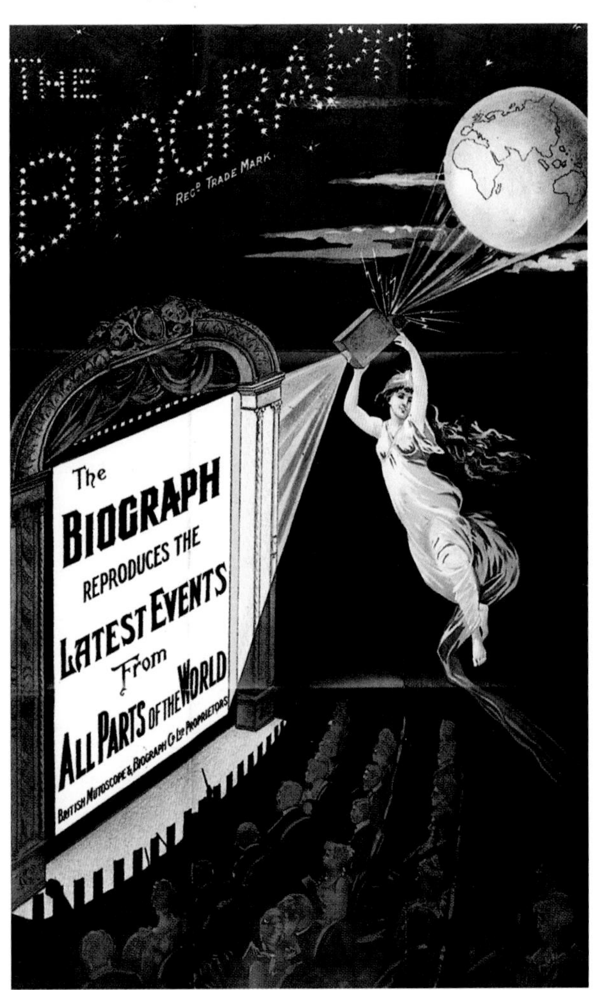

With the exception of the phonograph, which initially was purely mechanical, all these new developments were dependent in some way on electricity and reflect the rapid rise of the electrical industry. Even though many were dependent on small storage batteries for their operation, a mains supply was necessary for their convenient recharging. While it would be stretching the argument somewhat to identify 1900 as a particularly significant date in the history of the electrical industry – public supply companies had been established in London, New York, and elsewhere in the early 1880s – it is certainly true to say that it was in the first decade of this century that the electrical industry really took off. In Britain, for example, electricity consumption increased eightfold in the period 1900–10. Electricity was thus rapidly becoming a new source of energy for lighting and, later, heating.

Electricity also gave rise to one of the most important of all methods of medical diagnosis and treatment. In 1895, Wilhelm Konrad Röntgen discovered X-rays and by 1900 their use was firmly established in medical practice. X-rays were a key discovery in atomic physics, fundamental to the research that half a century later was to result in the release of atomic energy.

These developments are significant in that they were technologically novel – they were not an extrapolation of the methods of the past but something new and unpredictable. They were significant, too, in that they were as dependent on the application of new discoveries in science, especially in electrical science, as on improvements in mechanical engineering. It was, however, an interdependence: for example, the design of electrical generators depended on clear understanding of the basic principles, but the generators themselves were complex examples of precision mechanical

By the turn of the century the newly developed cinema was beginning to make a social impact. This advertisement (left) by the British Mutoscope and Biograph Company, for the equivalent of the later news theatre, appeared in 1900.

Marie Curie, (above) depicted in her laboratory with her daughter, Irene, was famous for her work on radioactivity. With her husband, Pierre, she discovered polonium and radium in 1898. Irene married Frédéric Joliot: they jointly discovered artificial radioactivity.

engineering, as much as the great steam engines or water turbines which turned them. Increasingly progress depended upon the availability of trained scientists, though the success of men like Edison, who had virtually no formal education of any kind, is a reminder that inventive genius can arise from the most unpromising backgrounds.

While novel technologies based mainly on electrical phenomena were the foundation of wholly new industries, many of the traditional technologies underwent important developments. The chemical industry was a case in point. While the large-scale mechanical generation of electricity favoured the development of new electrochemical processes – for the manufacture of caustic soda, for example – important new processes were developed on conventional lines. Thus the Haber-Bosch process for converting atmospheric nitrogen to ammonia, first worked in Germany in 1913, relieved anxiety about the adequate supply of nitrogenous fertilizers for farming. Indeed, the availability of enormous quantities of cheap nitrogenous fertilizers altered the whole economic basis of farming practice, with results that in the long run were not wholly beneficial.

The early years of the 20th century saw the beginning of the now enormous plastic industry, originally exemplified by phenol-formaldehyde resins such as Bakelite. Other important plastics developed before the Second World War were polyvinyl chloride (PVC), polystyrene, polymethylmethacrylate (Perspex) and polythene (polyethylene): the latter was to play an important part in the development of radar. Germany, for strategic reasons, paid much attention to the development of synthetic rubber. In the United States, the first nylon was made commercially in 1938, causing a revolution in fashion: 64 million pairs of nylon stockings were sold in 1939. The manipulative skills of industrial chemists bore other important fruits. In 1932, for example, the first of the very successful sulphonamide drugs was discovered in Germany.

As we have seen, the petroleum industry was well established by 1900, mainly supplying liquid fuel for heating and lighting and wax for candles. In the 20th century, however, the great demand on it was to supply fuel for the rapidly multiplying number of petrol-driven road vehicles and aircraft, as well as heavier grades for use as fuel in ships and industrial boilers. Petroleum is often associated with natural gas (mostly methane) and in America this was used as an alternative to coal-gas from as early as 1816. In the 1920s, the American chemical industry began to use petroleum as an alternative raw material to coal-tar. These developments were not matched in Europe, for no indigenous sources of gas or oil were exploited there until after the Second World War and there was a reluctance to depend unnecessarily on imports. At the beginning of the century, the United States was the world's largest producer of oil, but her share of the market slowly declined as the Middle East, Mexico and South America increased their output: much of their overseas development was, however, American controlled. This changing pattern of production was reflected in the transport world. The first oil tanker was launched in 1886, but up to the First World War such ships were quite small, rarely exceeding 800 tonnes; afterwards much larger vessels appeared until eventually tankers were the largest vessels afloat, far surpassing the giant transatlantic liners. On land, oil and petrol tankers became a familiar sight on railways and roads, and increasing use was made of pipelines to transfer fuel. By the 1950s gas trunklines thousands of kilometres long had been established in Russia and the USA.

New sources also had to be sought and developed for minerals other than petroleum. Although aluminium was the only new metal to be produced in substantial quantities, the traditional metals were required in greatly increased quantities for both old and new purposes. New techniques made it feasible to exploit ores formerly regarded as being of too low a grade. At the Bingham copper mine in Utah, for example, open-pit working on a very large scale – 4500 tons per day – made it profitable to mine a 2 per cent ore.

The electrical industry was a big new customer for copper, and the canning industry needed constantly increasing quantities of tin for tinplate. Ptolemy's map of the world shows a land of tin (Temala) which is probably identical with modern Malaysia. There was certainly some sort of tin industry there in the ninth century AD, but not until the turn of the 19th century was large-scale tin-mining established there, originally by French interests: in the 20th century it produced about one-tenth of the world's tin. When aluminium finally got into its stride, mainly as a consequence of the rise of the aircraft industry, new sources of bauxite had to be developed as far afield as Jamaica and Suriname.

Many minor metals assumed a new importance too. Chromium steels, first produced commercially in France in 1877, were of particular value for armour-plating as well as for machine tools. The vast chromite deposits on the Great Dyke in Zimbabwe began to be exploited from 1904, when the first claim was staked nearby. Tungsten, too, was an important alloying metal for hard steels and was increasingly in demand also for making the filaments of electric light bulbs: here China emerged as the largest single producer, followed by Burma. Although the quantity of such metals required was relatively small, they gradually became virtually indispensable.

In earlier chapters we have noted that although the Industrial Revolution originated and was sustained in Britain, the Great Exhibition of 1851 was a turning-point: thereafter, the initiative increasingly passed to Europe, partly because of the emphasis placed on technical education and partly because, coming late into the field, the Continental countries were not encumbered by old machinery and attitudes which in Britain were becoming outdated. By the beginning of the

Great Exhibition 156, 157 **petroleum industry** 223–227 **plastics** 267, 303–305 **Ptolemy's world map** 110 **X-rays** 283, 284

20th century, however, the centre of gravity was shifting once again, this time to the other side of the Atlantic.

There, the rise of American industry was phenomenal. The reasons for this are complex, but two were of particular importance. The first was that the Americans had available for exploitation not only vast tracts of agricultural land but also a great range of mineral resources. But they lacked manpower to develop these in the style of Europe – where a quite opposite situation prevailed – and this favoured mechanization of all labour-intensive processes. Secondly, there was a different attitude of mind. Europe was conservative and rather reluctant to try something new, intent on preserving the existing order: America, in contrast, was positively eager to try anything new, and was not unduly discouraged if it proved unsuccessful.

By 1900, there were, of course, many Americans who had been such through many generations, but others had more recently settled and others again were constantly arriving. There was, therefore, a general awareness of European ideas either through published accounts or through the firsthand knowledge of immigrants. Initially, American industry tended to thrive not so much on native innovation as on the vigorous and rapid exploitation of imported ideas and skills. But as all the infrastructure of an advanced industrial nation was established, especially in the fields of transport and education, this situation changed.

This may be exemplified by awards of Nobel Prizes, which in this century have become recognized as an index of original, creative work in the scientific field. The first of these were awarded in 1901, but not until 1907 was an American successful, when Albert Abraham Michelson gained a Prize for his novel method of measuring the velocity of light. Thereafter Americans figured increasingly, until the *annus mirabilis* 1976 when every Prize – in physics, in chemistry and in physiology or medicine – went to American scientists. For good measure, the Prize for literature also went to the United States.

In the Far East, China maintained a high, but rather static, level of civilization side by side with primitive conditions in the remoter areas. Japan, however, had embarked on a new course. For a brief period, from the middle of the 16th century to the middle of the 17th, European traders – largely Portuguese, Dutch and British – had received a guarded welcome, but under the shoguns a strict policy of exclusion had been followed. However, with the victory of the imperial party after a brief civil war in 1868, it was accepted that a policy of Westernization was essential for survival. From the 1880s, a competitive textile industry developed and by the turn of the century heavy industry was growing. All, however, was strictly on the Western model, adapted to local Japanese conditions: there was nothing comparable to the policy of high technological expansion that Japan has pursued so successfully since the Second World War.

After the Revolution in 1917, Russia's policy of industrial expansion relied heavily on a massive programme of electrification, extolled in this typical propaganda poster.

Geographically and technologically, Russia occupied something of a half-way house. Under the Tsars some degree of industrialization was encouraged, but it remained basically an agricultural economy, and after the 1917 Revolution the declared policy was 'peace, land and reform'. Soon, however, a new policy emerged, based on the collectivization of agriculture and a concentration on growth of heavy industry, which Lenin put forward in his famous equation for future success: electrification plus Soviet power equals Communism.

Proverbially, necessity is the mother of invention, but it is fathered by a favourable economic and social milieu. However, the normal pattern, if such there is amidst very complex interactions, is disturbed in time of war, when survival is the overriding factor. Under wartime conditions major enterprises may be embarked upon without regard to normal constraints, and never more so than in the 20th century, when 10 long years have already been wasted in global conflict and many others in major wars of more limited geographical extent. The following chapter is devoted to consideration of military technology as such, and several other later chapters will reflect the way in which peaceful purposes were served by developments of primarily military significance.

CHAPTER TWENTY-THREE

Military Technology & the First World War

Strategic preparations for war and the need to surpass the defensive and offensive weaponry of potential adversaries have always had a powerful influence on the development of technology. Apart from other considerations, substantial government contracts underwrite the risks of expensive experiments in innovation. Some developments in military technology make rather little impact on civilian activities: the use of the submarine, for example, is still almost exclusively as a weapon. By contrast, the aeroplane – initially developed almost entirely for military purposes – has completely transformed the world's transport systems.

The history of military technology does not lend itself to global review, for the major powers armed themselves differently according to their individual circumstances. Thus Britain and France, with powerful maritime interests, naturally paid particular attention to the development of their fleets, whereas Russia, with limited access to the open seas, concentrated on large standing armies. It is, therefore, easiest to consider military technology under the three separate heads of warfare on land, sea and air, though bearing in mind that these were necessarily parts of a total strategy. Nevertheless, certain technological developments were so fundamental to warfare generally that they can conveniently be considered first.

High explosives

By far the most important new factor in warfare in the 20th century was the introduction of totally new kinds of explosive to replace gunpowder which – no less revolutionary in its day – had been unchallenged for nearly 500 years. The invention and manufacture of these, and their use in mining, for which they were originally designed, is told elsewhere: here we will concentrate on their military use.

The new breed of explosives were made by the nitration of organic materials such as glycerine or cellulose and, later, toluene and phenol. They were many times more powerful than gunpowder, so that firearms generally had to be of more robust construction, but initially these materials were so unstable as to be very dangerous to use. However, by 1875 Alfred Nobel had largely overcome this problem and was beginning to supply his new product, dynamite, to the mining and civil engineering industries. Inevitably, thoughts turned to its military use but, as Nobel succinctly explained in a lecture to the Society of Arts in London in 1875, this was no simple matter: 'It is difficult, even with more powerful explosives at command, to supersede gunpowder. That old mixture possesses a truly admirable elasticity which permits its adaptation to purposes of the most varied nature. Thus, in a mine, it is wanted to blast without propelling; in a shell it serves both purposes combined; in a fuse, as in fireworks, it burns quite slowly without exploding. Its pressure, exercised in these numerous operations, varies between one ounce (more or less) to the square inch, in a fuse and 85,000 pounds to the square inch in a shell.' (That is, from 4 grams to 6 tonnes per square centimetre.)

These were demanding specifications to equal, but the military advantages ensured that they were vigorously pursued: a major demand was for a smokeless powder which would not give away the position of artillery or riflemen and reduce fouling of the barrel. To some extent this had already been achieved in Prussia, where J. F. E. Schultze about 1865 devised a propellant based on nitrated wood (cellulose) mixed with saltpetre. This aroused some interest, but although suitable for artillery Schultze powder was too violent for rifles: so, too, was EC Powder, a variant prepared in England in 1882 by the Explosives Company of Stowmarket. The first slow-burning smokeless powder suitable for a rifled firearm was Poudre B, invented in France in 1884 by Paul M. E. Vieille for use with the Lebel rifle: the 'B' was a tribute to General E. J. M. Boulanger, then Minister of War. This was followed in 1888 by Nobel's ballistite, a mixture of nitroglycerine and nitrocellulose. It made an excellent propellant because it burned fiercely rather than exploded. In Britain, Frederick Abel and James Dewar in 1889 patented, under the name cordite, a similar product – so similar indeed, that Nobel considered it a blatant infringement of his own patent and proceeded to expensive litigation, which he eventually lost.

By the end of the 19th century all the great powers had adopted military high explosives based on variants of Poudre B or cordite. A defect of the latter was that it causes excessive corrosion of the gun barrel: during the Boer War (1899–1902), one of the first major field tests of the new class of weapons, this proved so severe that a revised formulation, cordite MD, had to be introduced by Britain. In 1914, the protagonists faced each other fully equipped with explosives of a power totally different from that used in any other war in history.

Weapons

Although rifles had been used for sporting purposes from the 16th century, and were tested in several Continental armies in the 17th, they were only slowly adopted for military use, despite their greater accuracy. A major reason was that the spiral grooving made loading slow compared with the

American industry 208–209 **gunpowder** 136–137, 186 **Japanese industry** 208–209 **medieval warfare** 133–140 **warfare in antiquity** 82–88

The advent of the machine-gun in the latter part of the 19th century profoundly affected the role of the infantryman. The main contestants were the Hotchkiss and the Maxim guns, here being subjected to comparative trials by French officers during the First World War.

smooth-bore musket, and also the barrel fouled quickly. The turning-point was the War of Independence, in which the Americans used many rifles simply because, as sporting weapons, those were the firearms most ready to hand. In recognition of their accurate fire, the British recruited Jaegers (German huntsmen), and in 1800 the Rifle Brigade was founded. By the end of the century, the standard infantry weapon in the Western world was a breech-loading magazine rifle – this, first adopted by the Prussians, could be reloaded while the soldier lay prone – firing cartridges loaded with smokeless powder. For this, the old smooth-bore muzzle-loaders, still widely used in parts of Africa and Asia, were no match in accuracy or speed of fire. With his fast rate of fire a well-trained rifleman was as effective as, and more accurate than, a platoon of musketeers. This superiority in small arms was even more pronounced with the introduction of automatic rifles, when one man might equal the musket-fire of a battalion.

The pioneer of machine-guns was the American inventor Richard J. Gatling, whose gun appeared in 1862. It was accepted by the US Ordinance, but after trials in Britain it was rejected as being inferior to a field gun firing shrapnel. This was a type of explosive shell invented by Henry Shrapnel of the Royal Artillery in 1784: it had proved conspicuously successful in 1815 at the Battle of Waterloo. The gun invented in 1872 by Benjamin B. Hotchkiss, another American, was an improvement on Gatling's, firing 33 rounds a minute, but the first really efficient weapon in this class, and the one most widely used in 1914, was the Maxim gun, invented by Hiram S. Maxim in 1883. This was capable of firing an astonishing 666 rounds per minute: in an attack against the British in the Matabele War of 1893 four Maxim guns kept at bay 5000 picked native troops under Lobengula.

The advent of the machine-gun radically altered the power and role of the infantry, but for heavier and more destructive fire artillery was essential. With the introduction of the new smokeless powders it was possible to throw a shell weighing a tonne a distance of more than 30 kilometres (18½ miles), but for most field warfare smaller and more mobile guns were required: up to 1914 horses or mules had not been substantially displaced as gun-haulers by motor traction. In the 19th century, the rate of fire was limited because a field gun had to be re-laid after every shot fired. The famous French 75s overcame this by a hydraulic mechanism to absorb the recoil, with the result that a good gun crew could accurately fire one shell every three seconds.

The tank

The war of 1914–18 is remembered with bitterness for the long years of trench warfare in which trifling territorial gains were made from time to time at enormous costs in casualties. While it can be argued that the ultimate defeat of Germany was economic rather than military, the introduction of tanks by the Allies in 1916 was a powerful new factor and would probably have been decisive had the war continued into 1919.

The concept was not new, for we have already noted that in ancient times men had advanced against fortifications under the protection of some kind of armoured shield. In India, Alexander had encountered the armoured war elephants of Porus, and subsequently the Macedonians used them as a screen against cavalry, though occasionally in the vanguard of the advance. When the Seleucids forbade the export of Indian elephants, the Ptolemies and the Carthaginians obtained them from Africa. The battle wagons described in medieval and later literature was impracticable through lack of suitable motive power. Steam-power and the advent of caterpillar tracks to cover rough ground revived the idea of armoured battle wagons in the 19th century, but with no practical result. In the event, it was the adoption of the internal combustion engine as a source of motive power that made the idea feasible.

In the First World War the tank was one of the few new weapons to appear, though too late to affect the course of events until the closing stages of the conflict. Its ability to cross rough, roadless country gave it a mobility comparable with that of ships at sea. This picture shows some of the original British tanks being built in 1916 in the works of William Foster at Lincoln.

The man most largely responsible for the modern tank was Ernest Dunlop Swinton, who as early as October 1914 submitted to the War Office in London a scheme for a heavily armoured car, fitted with caterpillar tracks, to penetrate and crush barbed wire entanglement. Barbed wire which, with mud, had come to epitomize the misery of trench warfare, was introduced in 1874 in Illinois to fence off the cattle ranges. The prototype tank was completed in September 1915 by William Foster and Company of Lincoln and they went into action exactly a year later. At first too few were deployed to be effective – though they alerted the Germans to the new threat – but at the Battle of Cambrai in November 1917 nearly 400 massed tanks penetrated the Hindenburg Line to a depth of 2½ kilometres (4 miles). It was, however, on 8 August 1918 that a concentrated attack by nearly 600 tanks really demonstrated the power of the new weapon on what Ludendorff called 'the blackest day of the German Army'. The French, too, took up the tank and put 500 into the field at Soissons in the summer of 1918. These were lighter and faster than the British and had a fully rotatable gun turret. In the event, the war ended in November 1918, but the success of massed tank attacks had made such an impression that had the war continued tens of thousands of tanks would have been deployed by the Allies in 1919.

Perhaps because their industries were by then past responding, the Germans made virtually no attempt to use tanks: their total force probably did not exceed 50. But, as we shall see in a later chapter, they had fully realized the implications. The whole concept of the *Blitzkrieg*, fundamental to Nazi military strategy, was based on the swift and overwhelming advance of motorized troops strongly supported by tanks, of which they had 3000 in 1939.

Chemical warfare

Although particularly identified with the First World War, the military use of choking fumes goes back at least to the fifth century BC. During the Crimean War (1853–6), plans were put forward to reduce Sebastopol with the help of fumes of burning sulphur (sulphur dioxide gas) but rejected on the ground of inhumanity; this did not deter Chinese pirates using it at about the same time in the Hong Kong area.

The Germans resorted to poison gas in 1915, using chlorine released from cylinders in the Ypres salient. The Allies retaliated in kind the following month at the Battle of Loos, but the relative ease with which chlorine could be combated by masks, and the uncertainty of the effect due to the vagaries of the wind, discouraged further use. Instead, a variety of liquid gases – notably mustard gas and phosgene – were introduced and used by both sides: these not only vaporized and attacked the lungs but also produced severe and dangerous blistering.

Gas of all kinds is rightly regarded with abhorrence, but it is a matter of fact that in the First World War the ratio of deaths to casualties was much lower for gas than for all other causes. In 1936, the Italians used mustard gas in Abyssinia, but neither side used it in the Second World War, though both were equipped to do so.

After the Second World War various attempts were made to ban chemical weapons of all kinds by international agreement. In 1984 a formal protest was lodged with the United Nations charging Iraq with the use of chemical weapon against Iranian troops. In 1997 an agreement reached by the Chemical Weapons Convention prohibiting the development, stockpiling and use of chemical weapons, and calling for the destruction of existing stocks within 10 years, was signed by 167 nations; Iraq, Libya and Syria were among the countries with chemical weapons that did not ratify it.

Naval warfare

It is perhaps not without significance that responsibility for the production of British tanks was assigned to the Admiralty, for they enjoyed the same kind of mobility as do ships at sea. From the turn of the century, the design of surface vessels was strongly affected by two powerful influences, one political and the other technological. Up to the end of the 19th century, Britain's naval policy had been to maintain superiority over the combined fleets of the Franco-Russian alliance, while taking note of a strong Italian presence in the Mediterranean. The other major naval powers, but more remote, were Japan and the USA. Then, in 1898, Alfred von Tirpitz introduced the first navy bill in Germany, signalling her intention to become a major naval power too within a decade. Inevitably, this started a naval armaments race in Europe.

Technologically, there was a fundamental change in warship design. In the first half of the 19th century, they had been constructed as floating gun carriers, designed to pound each other at almost point-blank range, but with more

Britain's response to Germany's naval expansion programme of 1898 was HMS *Dreadnought*, completed in 1906. Carrying ten 305-millimetre (12-inch) guns, she outclassed every other battleship then afloat.

aircraft carriers 222 **early firearms** 139–140 **medieval warships** 105–107, 140 **siege towers** 87–88 **warships in antiquity** 50–52

The US navy's first submarine, USS *Holland*, designed by John P. Holland. It was built by the Crescent Shipyard, Elizabeth, New Jersey, in 1898.

powerful propellants and larger guns – by 1900 a calibre of 305 millimetres (12 inches) was normal – it was accepted that long-range engagements would be inevitable: by 1900 distances of 5 kilometres (3 miles) were accepted, and by 1914 this had been trebled. In 1906, Britain launched a new type of warship, the *Dreadnought*, which temporarily made obsolete all existing battleships. Very heavily armoured, she depended – except for a small number of small quick-firing guns – entirely on guns of a single calibre, ten 305-millimetre (12-inch) guns. The first to be fitted with steam turbines, she was for a time the fastest battleship afloat. By 1914, much bigger guns, of 380-millimetre (15-inch) calibre, firing a shell nearly twice as heavy as *Dreadnought*'s, had been introduced and powerful secondary armament had been brought back to counter the new threat of fast destroyers.

In the meantime, two novel underwater weapons had been developed – the submarine and the torpedo – and all warships had to be designed to resist the threat posed by these. The torpedo was invented by a British engineer, Robert Whitehead, who in 1856 set up a small business in Fiume, making silk-weaving and marsh-draining machinery. There, in 1866, he designed his first torpedo: driven by compressed air, it had a range of 700 metres (2300 feet) at 7 knots. By 1889, the range had been doubled and the speed quadrupled, but the accuracy was still unsatisfactory. This defect was overcome in 1896 by the introduction of gyroscopic stabilizers and, by then, a torpedo could carry a charge of nearly 100 kilograms (220 pounds) of high explosives.

The effectiveness of the Whitehead torpedo – soon licensed to all the major powers – was demonstrated in the Japanese attack on the Russian fleet off Port Arthur in 1904, and even now no wholly satisfactory defence has been developed. The ineffectiveness of the idea of lining the hull internally with coal bunkers was demonstrated when the obsolete British battleship *Belleisle*, so equipped, was sunk experimentally by a torpedo in 1904. Thereafter more reliance was placed on additional watertight divisions to limit flooding of the hull and extra speed to enable the ship to turn away when the revealing stream of air bubbles was sighted. When moored, anti-torpedo nets could be rigged.

The torpedo was the principal weapon of the submarine, the other main innovation in naval warfare. The concept was not novel, for as early as 1620 the Dutch engineer Cornelius Drebbel built some kind of submarine, propelled by oars, which travelled submerged in the Thames from Westminster to Greenwich. The Americans experimented with submarines in both the War of Independence and the Civil War, but with very limited success, though in 1864 a submarine sank the Federal sloop *Housatonic*.

Effectively, however, the submarine is a 20th-century development. In 1888, the French built an electrically driven prototype, the *Gymnote*, and by 1901 they had a fleet of about a dozen 30-tonne vessels. Among the pioneers was an Irish inventor, John P. Holland, who emigrated to the USA in 1873: he is said to have been inspired by the belief that such a weapon might contribute to Irish independence. The American Fenian Society financed the building of the *Fenian Ram* in 1881: this embodied most of the basic control systems of modern submarines, including compensation for the sudden loss of weight as a torpedo is fired. He designed various submarines for the US Navy Department, but none was successful. In 1898, he built his own craft, the 120-tonne *Holland*; this the Navy Department purchased in 1900, immediately laying down five more. The British Navy then commissioned five Holland submarines, and Germany and Italy two each. At the beginning of the century the world's entire submarine fleet amounted to no more than about 30 small vessels, but by 1918 the submarine threatened to decide the outcome of the war. Allied shipping lost, much of it destroyed by submarines, exceeded 12 million tonnes.

A major factor in this development was the availability of the large internal combustion engines. Under water, electrical propulsion was essential because of the problem of taking in and emitting air. The earliest French submarines had a very limited range because the batteries had to be recharged on shore. Later vessels, however, used their engines both to drive them on the surface and to recharge their batteries.

Both surface vessels and submarines had to contend with a further underwater danger, the mine. This, too, was not new, for the Dutch had used floating mines at the siege of Antwerp in 1585, and in the middle of the 19th century Immanuel Nobel, father of Alfred, had made mines for the Russian navy for use in the Crimea. Again, however, the mine did not become a serious menace until the First World War, when thousands were laid in the North Sea and elsewhere by both sides. They were anchored to the seabed and floated just below the surface: a glancing blow from a passing ship was sufficient to detonate them. This led to the development of a new naval vessel, the minesweeper, equipped with cutters to sever the mooring cables: when the mines floated to the surface they were destroyed by gunfire.

Aerial warfare

The 20th century literally added a new dimension to warfare. While armies and warships were confined to the surface of the globe, operating in two dimensions, submarines and aircraft could move not only forwards and sideways, but also up and down, giving them a three-dimensional range.

The ease with which birds fly and glide has from the days of antiquity encouraged inventors to imitate them, often through some kind of ornithopter simulating the beating of wings. Among the many who designed such a machine, in 1716, was the Swedish mining engineer and theologian Emanuel Swedenborg, but there is no evidence that his design, involving a complex system of springs, was ever put to the test. Undoubtedly the real father of heavier-than-air flight was Sir George Cayley, even though he never achieved it himself.

Cayley quickly lost interest in ornithopters, and rightly believed that the future lay with fixed-wing designs. He not only established the main design features necessary for stability and manoeuvrability, but also the power necessary to achieve a given speed and load. From the latter, it was apparent to him that no available power unit came anywhere near the threshold required. Accordingly, he restricted himself to flights with gliders and in 1853 dispatched his coachman in a 500-metre (1640-foot) flight across a valley. In the 1890s in Germany, Otto Lilienthal also experimented with gliders and made more than 2000 flights before being killed in an accident in 1896: he helped to create a vogue for gliding as a sport.

Before the end of the century, flight had been achieved in non-manned models powered by such devices as twisted rubber and clockwork and even – in the case of Samuel Pierpont Langley's large monoplane of 1896 – a steam engine. It was, however, the internal combustion engine that first made flight practicable in passenger-carrying machines. The decisive day was 17 December 1903, when Orville Wright made a 37-metre (120-foot) flight in North Carolina in an aeroplane built by himself and his brother Wilbur after long experiments with models and gliders. It was powered by a 10-horsepower petrol engine, also built by themselves in their bicycle workshop. This historic event attracted little public notice, but they persevered and by 1905 had built a larger and more powerful machine in which they flew a 40-kilometre (25-mile) circuit. This they patented and offered to the US Government but it was not until 1907 that they obtained a contract for their design – a tail-less biplane with the propeller at the rear, pushing the aircraft forward – for $30,000.

Nevertheless, their achievement still attracted little notice and in 1908 they went to France to give a number of demonstration flights, and incidentally to refute the claims of Henri Forman – who in fact had flown for the first time only in January of that year – to have anticipated them. In France, home of ballooning, there was much greater enthusiasm for aviation than in the USA. Louis Blériot, an established car manufacturer, introduced a different design – which eventually was widely adopted – for a monoplane with a tractor engine in the front. In this, he flew across the English Channel on 25 July 1909.

By this time aviation was beginning to be regarded as not merely an exciting sport for enthusiasts but potentially a new form of transport. This change of attitude was demonstrated by the staging of a week-long air show at Rheims at the end of August 1909. It was followed by similar gatherings elsewhere in Europe and, in 1910, at Los Angeles. Recognition of the industrial possibilities was reflected in the foundation of new companies, such as the Bristol Aeroplane Company in Britain: other existing companies, such as the big German electrical firm AEG, extended their interests into aircraft. By this time, all governments were becoming keenly interested

One of the great pioneers of heavier-than-air flight was Sir George Cayley. He confined himself to gliders because he realized that none of the motors available in his time had a sufficiently low weight:power ratio. When suitable engines became available at the turn of the century, his ideas were fully vindicated and their importance was acknowledged by the Wright brothers. This is the design of a machine he built in 1849.

first jet aircraft 274, 328–330 **nuclear submarines** 323–324 **propeller aircraft** 257 **submarine detection** 324 **submersibles** 341

Below: Aviation was only just getting into its stride when the First World War broke out, resulting in attention being directed almost entirely to military machines. This picture shows British DH4 bombers being attacked by Fokker triplanes in 1917.

Right: After the Wright brothers' epic flight in 1903 aviation developed rapidly. This picture commemorates Blériot's historic crossing of the Channel on 25 July 1909: this prompted the comment that 'Britain is no longer an island'.

Le Petit Journal

SUPPLÉMENT ILLUSTRÉ

ABONNEMENTS

DIMANCHE 8 AOUT 1909

LA TRAVERSÉE DU PAS-DE-CALAIS EN AÉROPLANE
Blériot atterrit sur la falaise de Douvres

Right: The first to achieve powered flight were the Wright brothers in 1903. This dramatic picture – Orville at the controls and Wilbur on foot – records their first flight on 17 December 1903.

Right: In the last decade of the 19th century Otto Lilienthal popularized gliding as a sport in Germany: he was killed in an accident in 1896 after making more than 2000 flights.

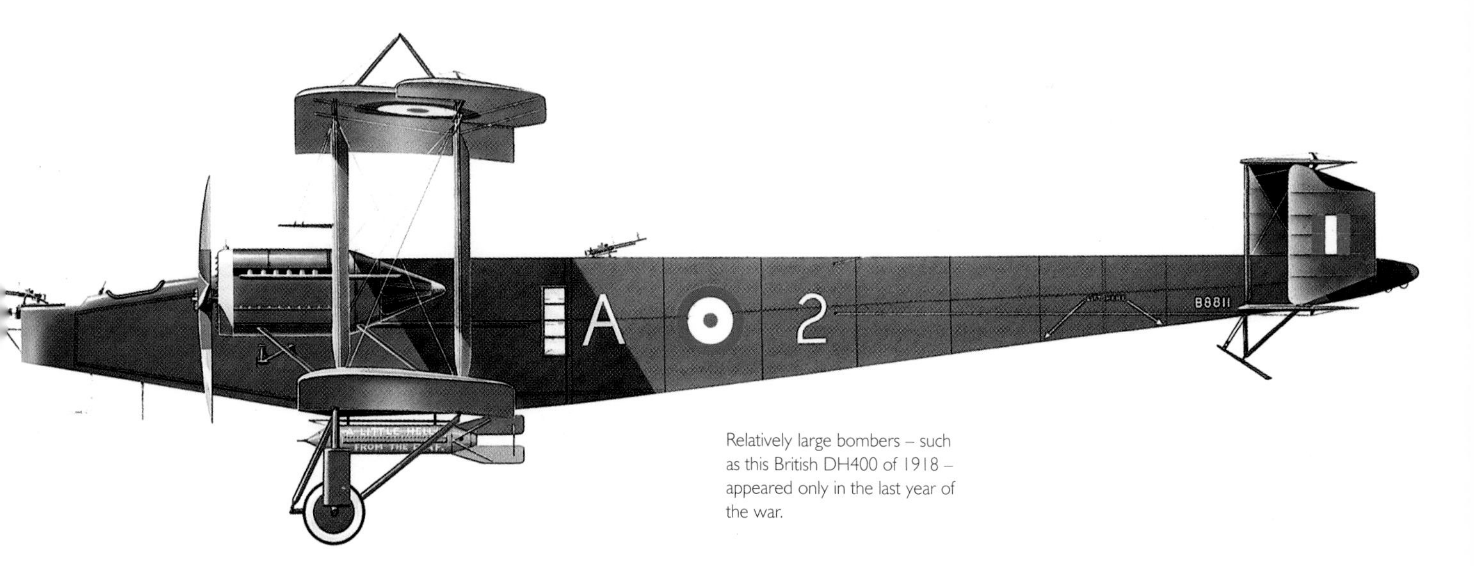

Relatively large bombers – such as this British DH400 of 1918 – appeared only in the last year of the war.

in the military possibilities of flying: after Blériot's cross-Channel flight in 1909 it had been shrewdly remarked that 'Britain is no longer an island'.

Again the French were in the lead, holding a *Concours Militaire* in the autumn of 1911 to try to identify types of aircraft with military potential. At that very time, however, the Italians were already making military history in their war against the Turks in Libya. It was a very small operation, using a few aeroplanes obtained from the Italian Aero Club, but it demonstrated the four main uses of aircraft that were to emerge during the war of 1914–18. These were observation, artillery spotting, bombing and photographic reconnaissance.

The advent of the aeroplane was not viewed with any great enthusiasm by most military leaders, who were generally distrustful of innovation. Nevertheless, with political persuasion, by the summer of 1914 the armed forces of all the great powers had air forces of some kind, even if they were by no means clear what their role might be. For the next four years, the history of aviation was dictated largely by military needs, and when civil aviation resumed its development in the 1920s it was inevitably very much dominated by the powerful military influence of its nascent years.

In the event, it was for reconnaissance and spotting shell bursts for artillery that aircraft were most used in the First World War, and this was partly a consequence of the shape the war took in its early months. Largely static trench warfare was dominated by artillery and machine-gun fire, and to locate enemy fireposts, to map trenches by photographic survey and to report the fall of shells was a task of great importance. To prevent such sorties was a major concern of both sides and, as gunfire from the ground proved largely ineffectual, the role of fighter aircraft became dominant. At first, combat was largely between individual pilots, but by 1917 actions involving a hundred or more aircraft were not unusual.

In these aerial combats, the main weapon was a light machine-gun similar to that used by infantrymen on the ground. In single-seater aircraft this involved firing through the path of the propeller; initially this technical problem was overcome by fitting the blades with metal plates to deflect any bullets which hit them. Later, however, Fokkers devised a satisfactory interrupter mechanism which prevented firing when the gun barrel was in line with a propeller blade: temporarily, this gave the Germans a considerable advantage. By 1917, two-seater aircraft were well established. The pilot fired forward as before, but a reargunner/observer protected the rear with a gun that could be fired in any direction.

Strategic bombing played a major role in the Second World War, though opinion is still divided on its effectiveness. It began to be practised in the latter part of the First World War, but in this case its material impact was undoubtedly marginal, though it was psychologically important. For this delay, there were various reasons, among them the fact that aeroplanes capable of carrying a sufficient bomb-load appeared only late in the war and bomb-sights were so primitive that most bombs fell wide of their target. In the first month of the war, a German bomb attack on Paris had a propaganda effect out of all proportion to the damage done, and later Zeppelin raids on London were grim reminders that England was indeed no longer an island. When Zeppelin losses became too high in 1917, the large Gotha IV bombers were used for daylight raids that caused consternation and prompted the withdrawal of urgently needed fighters from France. These were then followed by bomber planes built at the Zeppelin works, huge by the standards of the day, capable of delivering a bomb-load of up to 2 tonnes. Britain began to retaliate in 1916 with the comparable Handley Page 0/100 followed by the larger 0/400 version in 1918. In September of that year, in the last weeks of the war, 40 such bombers raided the Saar from Nancy. By that time a very small number of much bigger four-engined bombers, the Handley Page V/1500, had been built, but the Armistice was concluded before they could be evaluated. In the field, the favoured aircraft was the light bomber and, of these, perhaps the most

ballooning 171–172 **development of propeller aircraft** 257 **machine-guns** 216 **Zeppelins** 172, 256

successful was the de Havilland DH4, introduced in 1917. With modifications, this was widely adopted for civil use after the war.

The DH4, introduced in 1917, was the most successful light bomber of the First World War.

It was a two-seater, equipped with a machine-gun for the observer.

The aircraft carrier

At the start of the First World War every major power had two distinct armed services, the army and the navy: at its end a third service, the air force, had emerged as a powerful addition in its own right. The integration of these three to form a single strategic fighting force – in the face of intense inter-service rivalries – has nowhere proved easy as it is a major political problem. Here, we can consider only certain links based on technological factors. On land, as we have seen, the aeroplane was a valuable complement to the army for locating enemy positions and spotting the fall of shells. In such a situation its short range was no great disadvantage, for large distances were not in question. In the vast expanse of the oceans, however, the situation was very different. Land-based aircraft could effectively patrol only home waters and this led to the first experiments with ship-borne aeroplanes.

As early as 1910 an American pilot, Eugene B. Ely, successfully took off from a ramp constructed on the deck of the USS *Birmingham*. The next year he made history by landing on a similar platform on the deck of the USS *Pennsylvania*. Tensioned lines were stretched across the ramp to arrest his speed, a device subsequently widely used. During the First World War a few British cruisers were converted to make provision for launching seaplanes, which had later to be recovered from the sea. Apart from the inconvenience, the stationary vessel was very vulnerable during the retrieval process. In 1918, but too late to see action, HMS *Argus* was commissioned; converted from a liner, she had an uninterrupted flight deck and carried 20 aircraft.

The first aircraft carriers wholly designed as such were the Japanese *Hosho* and the USS *Langley*, both commissioned in 1922. These were closely followed by the much larger 11,000-tonne British carrier, HMS *Hermes*, completed in the summer of 1923. In the late 1920s, both Japan and the USA began to built much larger carriers – in the 35,000-tonne range – capable of carrying correspondingly large numbers of aircraft, 70 in the case of the USS *Saratoga* and *Lexington*. These were all built according to the so-called island design, in which the sweep of the flight deck was interrupted only by the navigating bridge at one side.

In the 1920s, there were divided views about the future role of the carrier in naval warfare: was it to be essentially a base for reconnaissance or was it to be an aggressive vessel in its own right? Britain inclined to the first view, but the USA and Japan conceived of carrier-borne aircraft as being capable of not only damaging, but sinking, heavily armoured capital ships at ranges far beyond that of the heaviest naval guns. This led both Japan and the USA to build very large carriers, capable of carrying over 80 aircraft. The matter was put to the test on 7 December 1941, when over 300 carrier-based Japanese aircraft attacked Pearl Harbor and sank four American battleships and severely damaged four others. Thereafter carriers played a major role in the Pacific area, culminating in the Battle of Leyte Gulf in 1944, when American carrier-borne aircraft sank the 60,000-tonne Japanese battleship *Musashi*, two other battleships and four carriers.

The post-war years saw the advent of very large carriers, such as the USS *Enterprise*, a nuclear-powered 86,000-tonne carrier commissioned in 1961 and carrying 80 combat aircraft. In this context nuclear power is particularly attractive – apart from the ability to stay at sea for long periods without refuelling – for the exhaust gases of conventional engines can create dangerous turbulence in the landing area.

To gain take-off speed on the relative short deck space available to them, heavier aircraft are launched by steam catapults, a British invention. An alternative to the general problem of taking off from a small area – apart from the helicopter, whose speed is slow – is the jump-jet, in which the jets can be swivelled for near-vertical take-off and landing or for fast forward flight. This is typified by the British Harrier: in the 1980s, the Sea Harrier was introduced into the Royal Navy.

CHAPTER TWENTY-FOUR

Sources of Energy

In the ancient world, as we have seen, there were only two major sources of power – manpower and animal power. The use of water power was minimal, and although wind power was very important for shipping the windmill as a stationary engine was a comparatively late arrival. But the main problem was not the source of power but how it would most effectively be deployed.

With the advent of the steam engine – and later the internal combustion engine – a rather different situation arose. Whereas men and animals derived their energy from their food, such engines depend on a combustible fuel – generally coal, but sometimes wood or peat – for the steam engine and some grade of petrol for internal combustion engines. For mine-owners, mainly concerned with pumping out floodwater or local steam traction around the colliery, the cost of fuel might be notional, in the sense that virtually unsaleable coal could be used, but for other industrialists generally the cost of fuel, and its assured availability, was an increasingly important consideration. In Britain, this was reflected in the foundation in 1807 of the Coal Exchange in London – whose later building (1847) was notable for its early use of cast-iron construction – to regulate the trade.

Concern over the cost of fuel led to increasing attention being directed to the efficiency of steam engines and, as we have already mentioned, some early measurements were made of their 'duty'. Such measurements were, however, necessarily empirical, for there was then no clear awareness of the true nature of heat – still regarded as some sort of intangible fluid known as caloric – let alone of any specific relationship between heat and mechanical work. There was certainly no understanding that there was any real limit to the amount of work that might be done in terms of the quantity of fuel burned.

Towards the end of the 18th century, however, clearer ideas began to emerge. In Munich, the flamboyant Count Rumford – who later, in 1800, established the Royal Institution in London as an outstanding centre of scientific teaching and research – was in charge of the royal arsenal. In that capacity, he was impressed by the great amount of heat generated when cannon were bored. By experiments involving measurement of heat in terms of the time taken to bring a given volume of water to the boil, he discovered – but could not precisely measure – a definite relationship between the mechanical work done and the heat generated. This line of thought was developed in Manchester by James Prescott Joule, who in 1849 published a definitive paper, *On the Mechanical Equivalent of Heat*, in which he stated: 'The quantity of heat capable of raising the temperature of a pound of water one degree on Fahrenheit's scale is equal to and may be converted into a mechanical force capable of raising 838 pounds to a perpendicular height of one foot.'

This was of profound importance, for two reasons. First, it clearly demonstrated that there was an absolute upper limit to the amount of mechanical power obtainable from a given quantity of fuel, though obviously this was unlikely to be attained in practice. Taken in conjunction with the theoretical work of Carnot, a yardstick of potential efficiency was established. Secondly, it paved the way for recognition that all forms of energy – not only heat and mechanical energy, but light, electrical and chemical energy – are interchangeable according to strict rules. We may compare this to international monetary exchange, in which different currencies can be converted at definite rates: an important difference, however, is that while money exchange rates may fluctuate, energy conversion rates are fixed for all time.

Fossil fuels

During the first two centuries of the age of steam, coal was by far the most important fuel, and, as we have seen, gas made from it not only provided a new source of light but also

After Edwin Drake first struck oil in 1859, the industry grew rapidly. This dance music published by a petroleum company in 1865 is indicative of popular interest.

early power production 64–70 **electricity production** 228–231 **nuclear reactors** 321–322 **steam & internal combustion engines** 159–165, 166–169

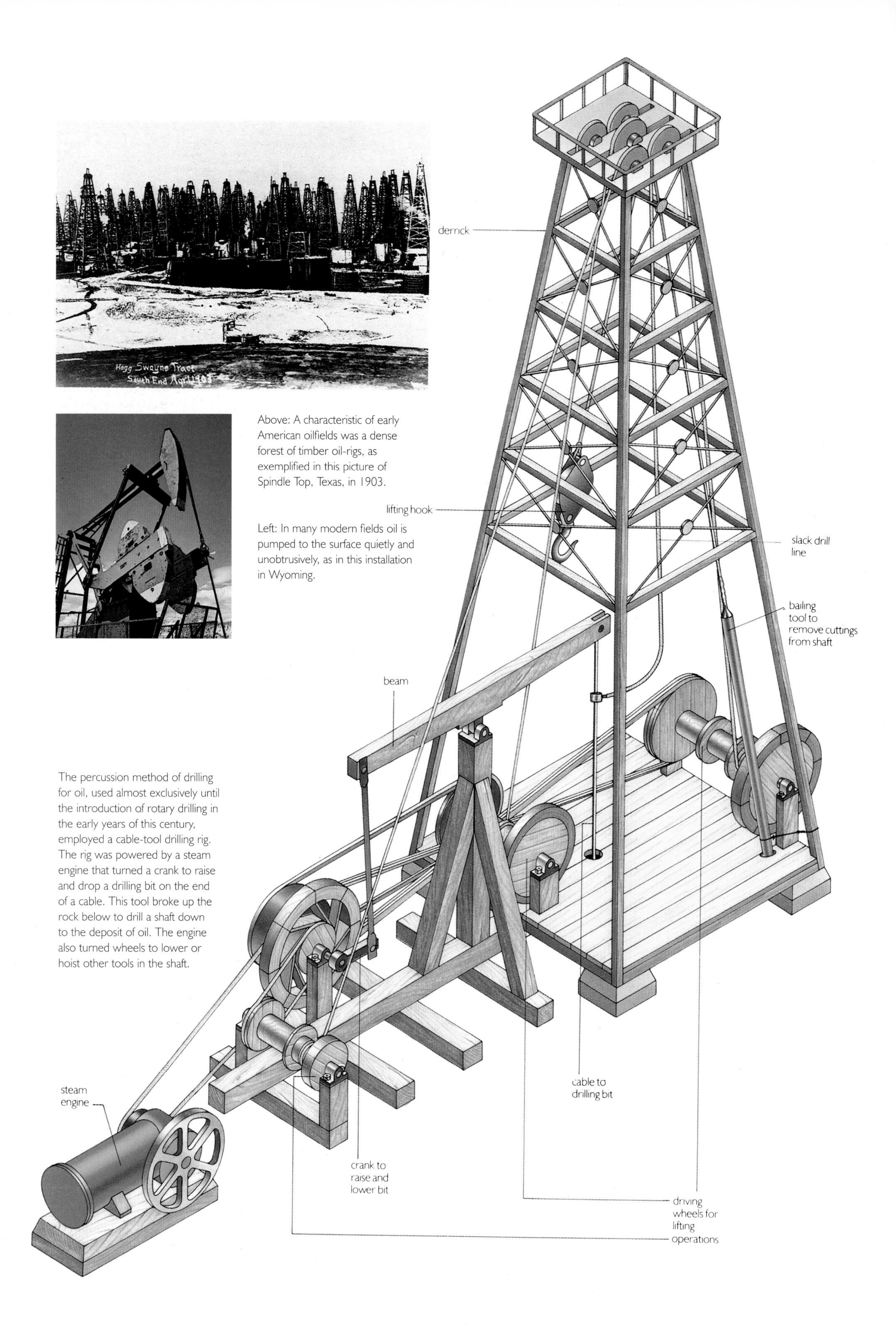

Above: A characteristic of early American oilfields was a dense forest of timber oil-rigs, as exemplified in this picture of Spindle Top, Texas, in 1903.

Left: In many modern fields oil is pumped to the surface quietly and unobtrusively, as in this installation in Wyoming.

The percussion method of drilling for oil, used almost exclusively until the introduction of rotary drilling in the early years of this century, employed a cable-tool drilling rig. The rig was powered by a steam engine that turned a crank to raise and drop a drilling bit on the end of a cable. This tool broke up the rock below to drill a shaft down to the deposit of oil. The engine also turned wheels to lower or hoist other tools in the shaft.

fuelled the first internal combustion engines. In the latter part of the 19th century coal also assumed a new importance as the main primary source of a totally new form of energy, namely electricity. Apart from a few hydroelectric installations, the generators supplying electricity were driven by steam engines. Before this, however, another fossil fuel, petroleum, was gaining rapidly in importance. This we have already discussed briefly in relation to road transport, but we have so far said nothing about its production and refining.

Like coal, petroleum is a fossil fuel, found largely in vast underground deposits, often associated with natural methane gas. Like coal, too, it sometimes appears at the surface of the earth, and for this reason was quite familiar to the ancient world: where available, it found uses for jointing masonry, caulking ships and as a lubricant. When Marco Polo visited Baku in 1272 he described the burning of petroleum gas at a fire-temple there. However, although these early uses are interesting, they are not significant: the world would have gone on much the same in their absence. Not until around the middle of the 19th century did petroleum become of any real industrial importance; by the middle of the 20th century it was the basis of any industry of vast international importance, as was forcibly demonstrated by the OPEC crisis of 1973.

Initially, interest in mineral oil was not as a source of heat but of lighting. In Scotland in the 1840s, James Young worked deposits of shale impregnated with oil, and discovered that an excellent illuminating oil could be distilled from it. Similar processes were worked at about the same time in America and the size of the potential market there encouraged George H. Bissell to seek liquid petroleum by drilling. The feasibility of this had long been known, for oil had often been encountered – but as an embarrassing by-product – when drilling wells for salt and water. Bissell directed his attention to Pennsylvania, where petroleum could be expected at no great depth. In the event, his contractor, Edwin Laurentine Drake, struck oil at a depth of only 21 metres (69 feet) on 27 August 1859: thus began the now enormous American oil industry, which, apart from domestic production, now has vast overseas interests.

However, this new industry was by no means exclusive to America. At about the same time oil wells were in production in Romania and soon afterwards the Nobel brothers built up a big oil industry on the Caspian. Long before the end of the century, they had developed various new techniques, including pipelines and oil tankers, the latter built in Sweden and brought to the Caspian via inland waterways. For their workers, they built what we would now call a 'garden city', complete with parks, schools and buildings for recreation. By the end of the century, Russia's annual oil production was 12 million tonnes, roughly equal to that of America.

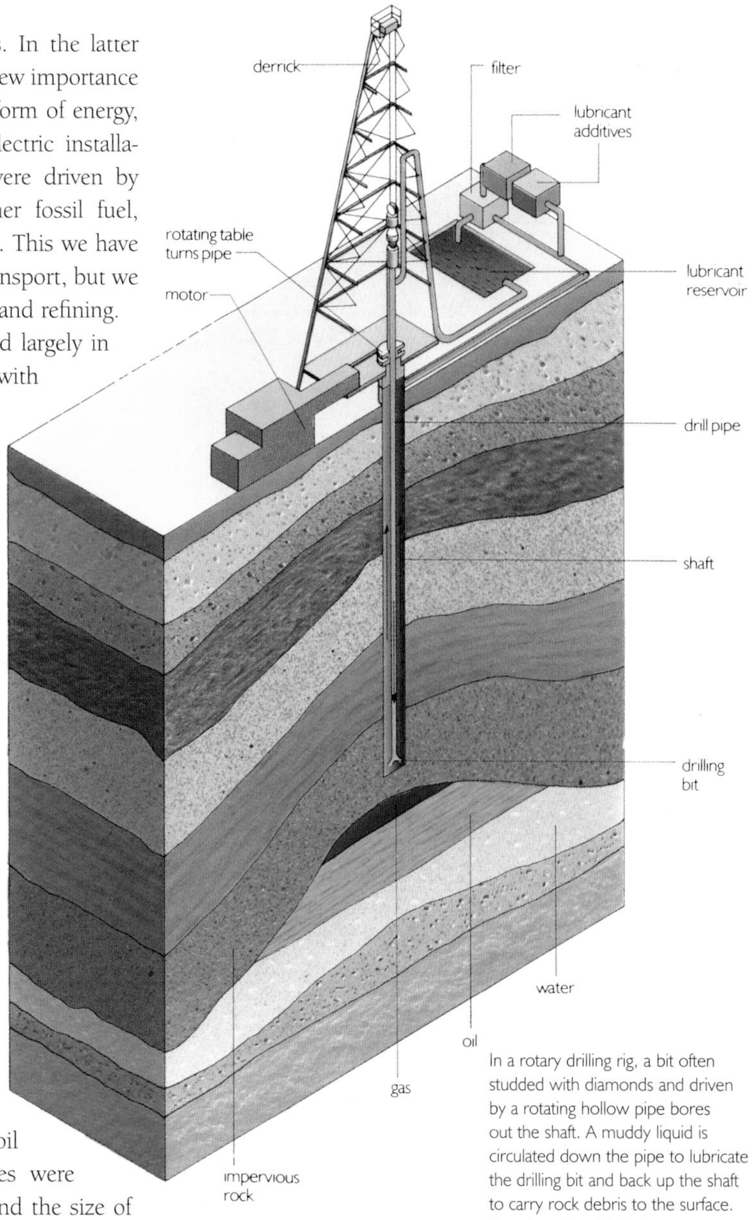

Drilling techniques were all-important in the quest for oil. Up to the end of the 19th century, this was done almost entirely by percussion, a method the Chinese had used for many centuries in drilling for brine. In percussion drilling, a heavy steel tool is repeatedly raised and dropped, shattering and penetrating rock and soil on impact: as the hole slowly deepened, it was lined with piping to prevent its collapse and the debris was flushed out with water. Traditionally the monotonous work was done by gangs of labourers, but by the middle of the 19th century steam-power was beginning to be used. By such means depths of 2500 metres (8200 feet)

In a rotary drilling rig, a bit often studded with diamonds and driven by a rotating hollow pipe bores out the shaft. A muddy liquid is circulated down the pipe to lubricate the drilling bit and back up the shaft to carry rock debris to the surface. Extra lengths of drill pipe are added as the shaft gets deeper.

early use of natural gas 120 **Marco Polo** 90–91, 96

This picture, taken on the world's first oilfield, shows Edwin L. Drake (right) in front of his second rig and engine house at Titusville (*circa* 1861).

had been reached by the 1920s. In Europe, an alternative method had been developed to reach underground water: a rotary drill was used and it worked well when only comparatively soft ground had to be penetrated. In America, however, where much drilling for oil was through hard rock, it was less satisfactory until the introduction of hard alloy steels, and, later, of very tough abrasives such as tungsten carbide or industrial diamond.

The characteristic feature of an oilfield is the great forest of derricks, from the tops of which are suspended the sections of the drill and of the lining. The taller the derrick, the longer the section that can be accommodated and the fewer the joints necessary. At first derricks were made of wood, but by the 1930s steel ones, rising as high as 45 metres (150 feet), were normal: by that time, the exhaustion of sources near the surface had led oilmen to sink wells up to 3500 metres (11,500 feet) deep. In some oilfields, the oil is under high pressure – like water in an artesian well – and gushes out as soon as it is encountered. More often, and when gushers have exhausted themselves, the oil has to be pumped to the surface.

At this sort of depth, the sheer weight of the drilling equipment, and correspondingly the power needed to manipulate it, began to be a major problem. In an attempt to overcome this, Russian engineers in the 1920s devised a turbo-drill, a self-contained unit which was gradually lowered down the hole. Early models were not very satisfactory, but they became widely used after the Second World War.

Oilfields, like coalfields, are not constrained by natural surface boundaries and many extend far under the sea. To reach this oil, new techniques had to be devised. In 1894, the Santa Barbara oilfield in California began to be exploited from jetties extending out to sea, and by the 1920s fixed offshore rigs were built in the Caspian. At about the same time, mobile drilling barges began to be used, at first off the Gulf Coast of America and later elsewhere.

On land or at sea, different parts of a field can be tapped from a single rig by the technique of oblique drilling: in this, the drill tube is deflected by a series of wedges, known as whipstocks. With deep wells a deflection of as much as 1½ kilometres (1 mile) can be achieved in this way.

Crude petroleum is a dark viscous liquid and, like coal, its composition varies considerably according to its origin. It is virtually useless in its natural state and from the earliest days of the industry some kind of refining was necessary. This was generally done by distillation, when three main fractions were successively obtained. The first and most volatile was petrol (gasoline) which, as we have seen, was initially a dangerous embarrassment but soon found a use as an almost ideal fuel for cars. The second, and initially the one most particularly desired, was paraffin (kerosene), in great and increasing demand as a lamp oil. Thirdly, there was a heavy fraction which also was for a time an embarrassment, but gradually came into use as a fuel oil, especially for ships. Left behind in the distillation was a thick black pitch.

Cracking processes

As the industry grew it became increasingly difficult to keep the different markets in step. The demands of the motor industry grew rapidly, but in competition with gas and electricity the demand for lamp oil decreased; on the other hand, paraffin found increased use for heating and cooking, especially with the advent of efficient pressure-burners of the Primus type. In the early days of the industry, the great objective was to make as much paraffin as possible, and as early as the 1860s increased yields were obtained by 'cracking' fuel oil. In effect, this process consists of subjecting the heavier fraction to strong heat, when the long heavy molecules break down into shorter, and more volatile, ones. From then until 1900 dozens of patents were sought covering variants of this basic process, but then interest diminished as the emphasis changed again from paraffin to petrol. Petrol, too, could be made from fuel oil by cracking, but up to the First World War most producers could make all the petrol they needed by simple distillation.

To this situation there was one major exception – Standard Oil of Indiana. In 1911, as a result of various legal judgments, the company found itself in the curious position of owning the world's largest refinery, having a large Midwest

Modern oil refineries combine a wide range of functions over a large area. This aerial view of a complex near Le Havre, France, shows docking facilities for tankers, with direct pipelines to the central plant, storage tanks, cracking plant, road access, etc.

sales organization, but not owning any oilfields. It therefore had to buy its raw materials and took a particular interest in fuel oil: the latter was locally available at very low cost as the demand was small, because coal was cheap locally. William M. Burton, head of the manufacturing side of the business, closely investigated the possibility of making petrol from fuel oil, and in 1913 patented a process involving high temperatures and high pressures which was not merely commercially viable but could actually slightly undercut the price of petrol suitable for lorries and stationary engines. The company promptly invested $1 million in 60 Burton-type cracking units and 10 years later had 800 in operation.

Encouraged by this, and by the greatly increased demand in petrol resulting from the First World War, other companies became keenly interested. This led to the patenting in 1915 of an improved process by Jesse Dubbs (and his rather incredibly named son, Carbon Petroleum Dubbs), and in 1919 Shell became the first licensee. Thereafter, cracking became a major activity of the petroleum industry and a vast number of new variants of the process appeared.

Not until 1937, however, was anything radically new devised. In that year Eugène Houdry in France described a process in which the cracking of vaporized oil was promoted by a catalyst. The great advantage of this was that it was quicker and could operate at a lower temperature, and in consequence was cheaper. No less to the point, it favoured production of the high-octane fuels with a low sulphur content increasingly demanded by the cars of the day. By that time, cracking was big business: the development of the Houdry process had cost $11 million. Strategically, as we shall see, this development was of the first importance. Hundred-octane petrol was necessary for the new generation of warplanes, and American demand alone rose from 40,000 barrels daily in December 1941, at the time of Pearl Harbor, to 600,000 barrels daily at the end of the war.

Coal and petroleum are similar in that their energy is released by combustion and for all practical purposes they are non-renewable, since they form so slowly in nature. An important difference is that, although various grades of coal exist – some most suitable, say, for making metallurgical coke and others for raising steam – relatively little treatment is needed between mining and sale. In the case of petroleum, however, not only are there important variations in composition, but the refining process is an essential and highly complex one. Yet another important difference is that while the production of coal is labour-intensive, that of petroleum is capital-intensive.

The petroleum industry is an interesting example of the way in which, under favourable conditions, a new technology can assume a dominant position in a remarkably short span of time. Coal, which had long been the basis of industrialization, quite suddenly found a serious rival. World production of petroleum around the end of the 19th century was no more than 25 million tonnes annually: by 1939 it had risen to 266 million tonnes and by 1960 it was around a billion tonnes. By contrast, world production of coal rose only from 750 to 1200 million tonnes annually between 1900 and 1939. These figures are striking, even when it is recognized that the two fuels are not wholly interchangeable – not all the new uses being found for oil were appropriate for coal – and that the overall demand for energy was increasing rapidly.

oil tankers 254, 255 **percussion drilling** 224, 225 **petrol engines** 167–168 **petroleum as a raw material for chemicals** 268

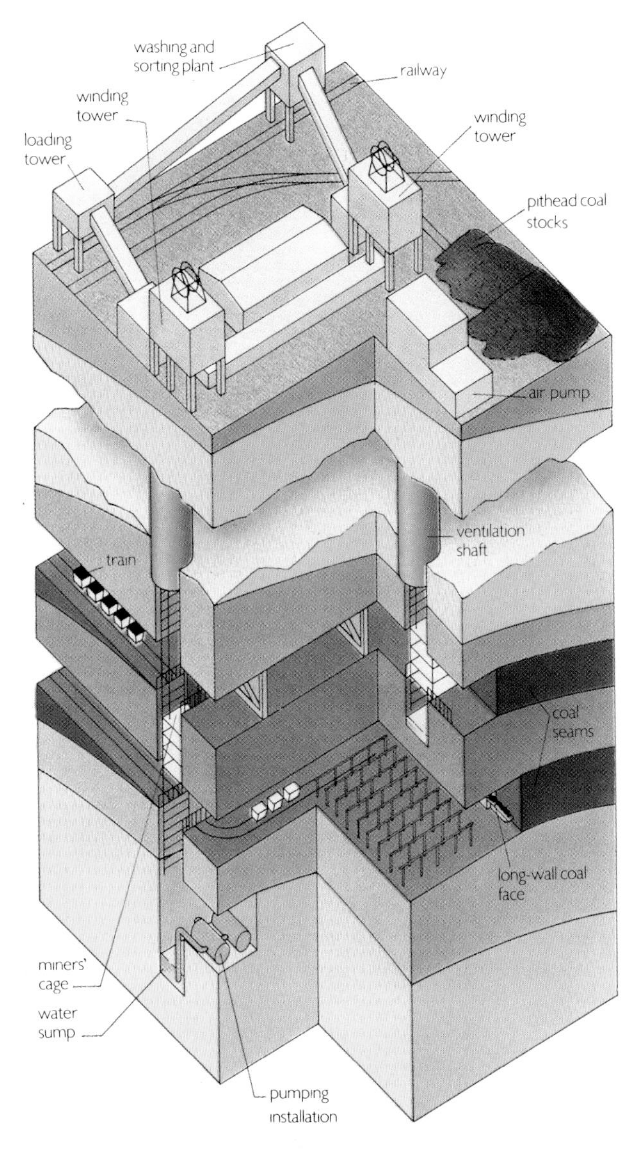

In a coal mine deep shafts and underground trains carry miners to the coal-face and bring coal to the surface. A ventilation shaft takes fresh air from the surface to the workings below ground. At the pithead, the coal is washed and sorted, then stored in heaps or transported by rail or road.

The electrical industry

Although lightning is a natural phenomenon as old as time and enormous quantities of energy are dissipated in electrical storms, usable electricity has in practice to be made by transforming some other form of energy. So far as the electrical supply industry is concerned, the immediate source is mechanical energy, which drives the generators. Apart from a few hydroelectric schemes, considered separately, the source of motive power for generating stations up to the middle of this century was almost entirely the steam engine, fuelled with coal or oil. In effect, therefore, electricity is simply another way of using fossil fuels. As we have seen, one of the important discoveries of 19th-century science was that when one kind of energy disappears an exactly equivalent amount of a different kind appears in its place. In practice, however, the utilization of the theoretically available energy can never be 100 per cent. Boilers can never be completely insulated, and heat is lost in bearings. As the amount of electrical energy ultimately obtained is, therefore, bound to be much less than the heat equivalent of the coal or oil burned in the first place, it is pertinent to ask why it is worth generating electricity at all. This very question was in fact argued widely in the early days of the electrical industry. *The Telegraphic Journal* wrote in 1874: 'No one would be found so foolish as to employ a steam engine to produce electricity for the machinery. It would evidently be vastly more economical to drive the machinery by means of the engine itself without the intervention of any complicated apparatus.' As with the gas industry, the basic answer is convenience: the consumer has at his command a source of energy which he can turn on or off at will and which involves him in no maintenance whatsoever. This outweighs the apparently adverse economics of generation.

Pioneers of the gas industry thought in terms of individual plants serving large consumers, such as cotton mills or big country houses. Major development occurred, however, only with the realization by Frederic Winsor that the future lay with central manufacture and a distribution network serving many customers. The early history of the electrical industry was similar: the first plants, in the 1870s and early 1880s, were situated in railway stations, theatres, factories and large stores. The electricity produced was used solely for lighting – initially by arc-lamps but soon by the new incandescent filament lamps as well.

With these early installations it was difficult to balance supply and demand, and proprietors were often ready to supply neighbours. A typical example was the Grosvenor Gallery in London: here a small power-station was installed in 1883, but it also supplied some adjacent premises. By 1887, however, a station designed solely to supply a small district was built in Brighton, England. That same year also saw a much more important development, the foundation of the London Electricity Supply Corporation. The guiding spirit was their electrical engineer, Sebastian de Ferranti, then only 23 years old. He conceived the idea of a large power-station at Deptford – 11 kilometres (7 miles) outside London, where land was cheap and there was easy access for delivery of fuel – which would supply a large, densely populated central area. His proposal was an ambitious one, involving generation at 10,000 volts, subsequently stepped down through transformers for local use. It was certainly ahead of its time, in that no available cables would work at such a high voltage. After a major failure, Ferranti designed a new type of cable, made in 6-metre (20-foot) lengths, in which the copper conductors were separated by waxed-paper insulators.

Left: Early power-stations supplied private customers: the first designed to provide a public supply to a district was that built at Brighton, England, in 1887. By 1910, as shown here, it had expanded enormously.

Below: Around the turn of the century battle raged between the advocates of direct and alternating current supply. In the event, a.c. prevailed, but as late as 1910 large d.c. generators were being built, as in this Siemens factory at Nuremberg.

Although the Deptford station is historically significant, in that it embodies most of the basic features of modern ones, it was not an unqualified success, and for many years the industry developed everywhere on the basis of power-stations providing local supply: in Britain, for example, in 1900 there were 250 suppliers and in America more than 3000. During this rather chaotic period a major battle was fought out between the supporters of alternating current (a.c.) and direct current (d.c.). The latter started in a strong position, in that batteries, the first regular sources of electricity, supplied direct current and this was what existing users – for example, in telegraphy – were accustomed to. Moreover, one of the intractable problems of the electrical supply industry is that the demand fluctuates enormously: with small d.c. stations surplus current could be stored in large accumulators to help meet peak demands, but this was not possible with a.c. installations.

In the long run, however, the advantage lay with a.c. generation. Implicit in the concept of stations supplying considerable areas was the question of transmission: this inevitably involved power losses in the main cables, but the higher the voltage the lower the loss. From the engineering point of view, it was much easier to construct high-voltage a.c. generators than d.c. ones; additionally, high-voltage a.c. can be readily reduced for local use with the help of transformers which involve no moving parts, whereas d.c. requires machinery. On the other hand, it was difficult to keep the cyclic output of different alternators in step with each other, and the consumer risked voltage surges which might damage his equipment. Eventually this design problem was overcome.

However, the situation in the industry remained chaotic for decades: where a.c. was supplied there were wide differences in both frequency and voltage, and even in London, as late as 1950, there were still small pockets of d.c. supply. This presented difficulties for appliance manufacturer and user alike: the former could not offer items for universal use and the latter might well find that on moving house his equipment was useless. Equally, fittings for electric lamps and power-points varied from country to country, and even within countries; travellers went equipped with a variety of adaptors. Not until after the Second World War was any effective degree of standardization reached: Europe then generally favoured 110 volts (Britain 240 volts) at 50 hertz (cycles per second), whereas North America favoured 120 volts – partly for its supposedly greater safety – at 60 hertz.

By the beginning of the century the basic character of the electrical supply industry had been established: namely high-voltage a.c. generated centrally and stepped down at transformer substations for local use. The size and efficiency of units was to increase enormously, however. In 1900, the biggest was one of 1500 kilowatts at Elberfeld, Germany. In 1933, the biggest in Europe was a 105-megawatt unit at Battersea, London. Larger units and improved design gave greater efficiency. Whereas the Elberfeld unit consumed 8 kilograms (18 pounds) of coal for every kilowatt-hour of electrical power, by the mid-1930s this had been reduced to 2 kilograms (4½ pounds), and in 1964 a consumption as low as 0.5 kilograms (1 pound) per kilowatt-hour was being achieved.

The increase in size posed serious engineering problems, but a limit was imposed by transport considerations. Certain units of a generator – notably the stator and rotor – must be

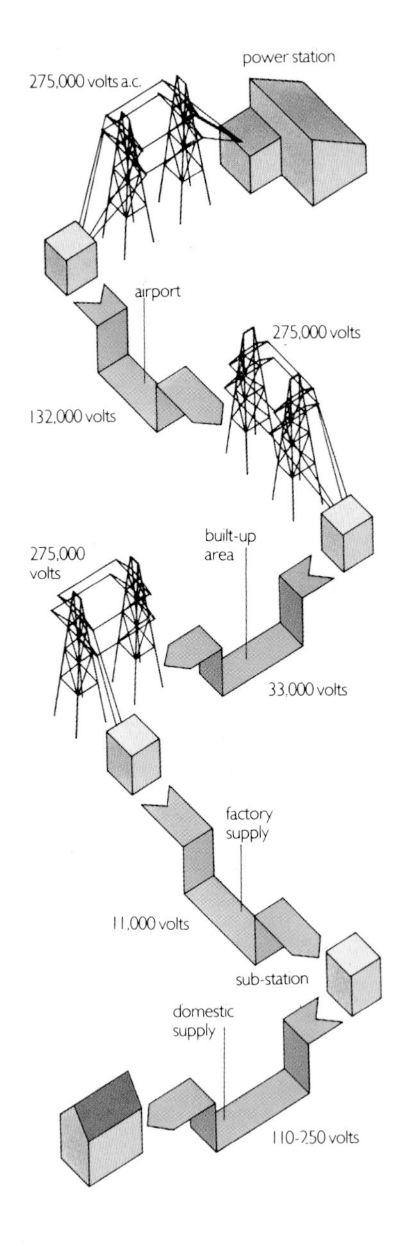

Typically, electric power is transmitted as alternating current at very high voltages by overhead power lines wherever possible to avoid loss. It is stepped down at substations to pass underground at various points, for example beneath industrial and other built-up areas, before being further lowered to supply factories and homes via underground cables.

assembled and tested where they are manufactured. When these reach a certain size there are difficulties in transporting them to their final destination: intervening bridges may not bear the load or roads and tunnels may be too narrow.

Central production for local use was characteristic of both the old gas industry and the new electrical one. In the case of coal-gas, however, it was not economic to pump it more than a few kilometres and this precluded anything in the nature of national networks: these were feasible only with the availability of high-pressure natural gas. The case was different with electricity, however: with the use of overhead cables – made of aluminium for lightness, reinforced with steel for strength – strung from pylons, great distances could be spanned without undue power loss. As the loss was smaller the greater the voltage, there was a strong incentive to use higher and higher voltages. As early as 1908, 110,000 volts was in use in America, and in 1961 some parts of Russian supply exceeded 500,000 volts.

In towns, local authorities often insisted on cables being underground, and this raised two problems. First, they joined a bewildering array of existing services – drainage, water, gas and telephone cables. Inductance in telephone cables caused by electrical conductors seriously interfered with the service; this was overcome by the use of coaxial cables, which Ferranti had used for the Deptford station. The second problem was dissipation of heat as loads became heavier. This was of little account with overhead cables freely exposed to the air but became serious with buried ones: from the 1930s oil-cooled cables were introduced.

Although the design and construction of generators favours a.c., power losses in transmission are in fact less for d.c. Initially, the cost of equipment for conversion and reconversion was too great to be economic, but the invention in 1928 of the mercury-arc rectifier, capable of handling very large currents, changed this situation. Since the Second World War a number of long-distance links have used d.c.: important examples are links between Volgograd and the Donbas Basin, and between North and South Islands, New Zealand.

The fortunes of the gas industry initially relied on domestic consumers, with industry a comparative latecomer as a major customer: in Britain, for example, industrial sales did not exceed domestic until 1973. But the electrical industry followed the reverse course. Industry has always been the largest consumer, with the domestic user well behind.

Electric lamps

By far the most important single piece of electrical equipment is the incandescent filament lamp or light bulb, in which a fine wire is heated to white heat in a vacuum or (after 1913) an inert gas. These have been made literally in billions: in London alone there were 2½ million even in 1900. Originally, the wire was made of carbon but this was fragile and soon blackened the glass by evaporation. High melting-point metals such as osmium, tantalum and tungsten began to be used around the turn of the century. Whereas the carbon filament lamp had an efficiency of only 3 lumens per watt – the lumen being the unit of light energy – that of a gas-filled tungsten lamp was nearly three times greater. A further improvement was effected with the coiled-coil lamp, in 1934, in which the coiled filament is itself an exceedingly fine coil. By 1920, the normal lamp had a working life of at least 1000 hours.

However, even the best of such lamps have a low efficiency: only about 6 per cent of the electricity they consume

Below: From the early days of steam, power was transmitted from a central engine through noisy and clumsy systems to overhead shafts and belting. These began to disappear with the advent of electric motors able to drive individual machines. The old system persisted until well after the First World War, however, as shown in this picture of the main lathe shop of BASF, Ludwigshafen, 1921.

Right: Although its overall contribution to world energy is small, hydroelectric power can be very effectively utilized in favourable situations. This installation is at the Sobradinho Dam, north-east Brazil.

appears as light and the rest is dissipated as unwanted heat. Although they are extremely cheap, there was a strong incentive to develop more efficient lamps, especially for large-scale users.

From this two alternatives emerged. One was the discharge lamp, in which an electric discharge passes through a gas or vapour – commonly neon gas or sodium or mercury vapour. While the colour of the light – blue or yellow – is unattractive, this does not greatly matter for applications such as street lighting and is more than offset by the high efficiency. This can be as much as six times greater than that of the best incandescent filament lamp of equal luminosity.

For domestic users, however, such colours were quite unacceptable at any price, and for them the fluorescent lamp was developed. This is based on the fact that the mercury vapour lamp emits ultraviolet as well as visible light: the inside of the tube is coated with substances known as phosphors, which absorb ultraviolet light and re-emit its energy as visible light. This can be a close approximation to daylight when the phosphors are appropriately blended. The efficiency of such mercury vapour lamps is high, but the relatively high cost of installation and replacement has limited their market: they are favoured more by commercial and industrial than domestic users.

Other sources of power

Until mid-century, the world's basic power sources were combustible fuels, notably coal and oil. While other sources had local importance, their overall contribution was very small. The traditional water-wheel was of only historical interest. Some large hydroelectric schemes were developed – as at Niagara, one of the first big a.c. installations – but many were less attractive than they appeared at first sight. The civil engineering work was far more expensive than the machinery: they might have to be located far from the consumers, involving power losses in transmission; and rivers are liable to seasonal variations in their flow. Overall, the proportion of power generated by hydroelectric schemes has diminished steadily. Old-fashioned windmills in working order still ornamented the landscape, but again their overall contribution was slight. In the main, they took on a new form: a simple, all-metal construction, used mainly on farms for pumping water from underground sources or for electricity generation for local use. Although they were built in enormous numbers – it is estimated that in 1950 there were nearly half a million in use in the United States alone – their average output was no more than ⅕ kilowatt-hour (¼ horsepower-hour) per day. Not until the advent of atomic power after the Second World War, a totally unexpected development, was the dominance of the fossil fuels to be challenged.

atomic power 321–322 **invention of electric lamp** 195 **Niagara power plant** 165 **renewable sources of power** 276–277 **solar energy** 277

CHAPTER TWENTY-FIVE

New Channels of Communication

At the very beginning of this history we remarked that, far from being a recent innovation, information technology is of great antiquity: some means of communication and record-keeping are fundamental to organized government. The advent of printing from movable type, first in the Far East and then in Europe, was an event of immense importance for the widespread dissemination of information and ideas. Although other important forms of communication have since been developed – telegraphy, telephony, television, photography and sound recording – printing still occupies a dominant position, and there have been many technological improvements, relating in the main to achieving greater speed and output. But there were surprisingly few developments of consequence between Gutenberg setting up his first works and the beginning of the 19th century, a period of some 350 years.

Printing processes

In 1719, Jacques-Christoph Le Blond of Frankfurt produced fine colour prints by carefully superimposing prints from separate blocks inked respectively in blue, yellow, and red: nevertheless, printing in colour on any considerable scale was not achieved until the 20th century. A more important development was the invention of the stereotype by William Ged, a Scottish goldsmith, in 1727. Until then type had to be reset if a second press was used or a later printing was needed: it was not economic to keep type standing for any length of time. Ged took a plaster mould of the forme of type, and then cast the whole page in metal: in a sense he was reverting to the old Chinese practice. Later, in 1829, papier-mâché was used instead of plaster.

In printing, the raised type gives finally a flat impression, but to the blind this is valueless. In 1824, Louis Braille devised the form of printing named after him: in this the letters are represented by patterns of raised dots which can be read with the fingertips.

Like most early machines, printing presses were made largely of wood, and an important advance was the introduction of the iron-framed press by Lord Stanhope, an amateur scientist, in 1800. This was stronger and more rigid, allowing a faster rate of printing and a sharper impression. The Stanhope press, which was widely used for many years, still used a hand-operated screw to press print and paper together, but it could print up to 250 sheets an hour. A considerable improvement was the Colombian press, introduced in America about 10 years later, in which for the first time the screw was eliminated in favour of powerful hand levers.

All these presses, and variants of them, had two features in common: they were manually operated and the flat surfaces of print and paper were pressed together, by either screw or lever, to produce the finished sheet. In 1810, a machine appeared which was different in both respects. This was Frederick Koenig's steam press, in which the paper was carried on a cylinder which rolled over the inked type: very shortly thereafter he introduced a second paper-carrying cylinder so that the forme made a printing on both the forward and backward passage. The importance of Koenig's machine was first realized by John Walter, the enterprising proprietor of *The Times* in London. He installed two of the double-cylinder machines in 1814, to replace his Stanhopes. Each printed 1100 sheets an hour, and he described it as the greatest advance in printing since Gutenberg.

As a general engineering principle, continuous rotary action is more efficient than constantly reversing reciprocating action, and the next development was to shift the type

Colour supplements to newspapers are no modern innovation. This Marinoric rotary press was used to print the *Petit Journal* colour supplement in 1901. Mass production of colour work was common by the turn of the century.

also from the flat bed to a cylinder. The first really successful machine of this type was patented by Richard Hoe in New York in 1847 and used to print the *Philadelphia Public Ledger*. Initially, metal type secured to the cylinder with wedges was used, but from about 1860 curved stereotypes cast from papier-mâché were used. Such machines were widely adopted by the publishers of newspapers and magazines who needed to deal very quickly with very large print orders. In 1857, *The Times* installed a multiple 10-feeder Hoe press capable of 20,000 impressions an hour – though it required 25 men to operate it. By the early 1860s, the labour of feeding sheets of paper into the machines was reduced by the use of long continuous rolls.

Until the early 19th century, type was still cast and set by hand. Not until 1838, again in America, did the first typecasting machine appear and by mid-century it was widely adopted in Europe, despite the restrictive practices typical of the trade. When *The Times* installed its first Hoe machine, it had a typecaster producing 1000 characters a minute. With this sort of output, it was unnecessary to break up the type after each edition: it was simply melted and recast.

To cast the type quickly was an important advance but this advantage would not be fully realized unless the setting also could be speeded up. The first mechanical typesetter was produced in 1842 by Henry Bessemer, the versatile inventor much better known for his converter for the manufacture of steel. It had a keyboard rather like that of a piano – hence the name Pianotype machine – to control the release of the characters and spaces from the magazines in which they were stored, and to arrange them in line; a second operator was needed to 'justify' the lines of type – that is, make them all of equal length. This machine and a number of successors had only qualified success. In 1886, however, Ottmar Mergenthaler introduced a machine which combined the casting and setting of type in whole lines, hence the name Linotype (line o' type). This was first used for setting the *New York Tribune* in 1886 and could set 6000 characters an hour. It was an immediate success, and was soon in use throughout the world, particularly for the production of newspapers and magazines. Only after the Second World War, with the

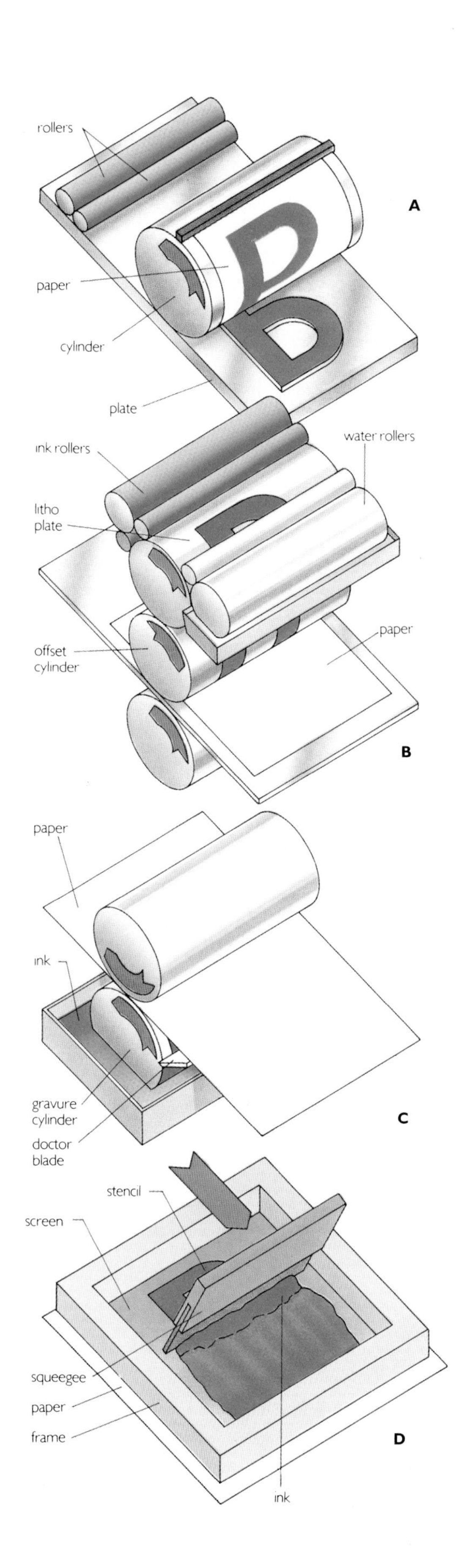

Principal Printing Processes

A. In letterpress printing, the type is raised above the surface of the printing plate and inked by rollers. The paper passes around a cylinder that rolls across the plate.

B. In lithography, the image is formed photographically on the cylindrical litho plate, which is treated so that only the parts corresponding to the type take ink. The plate is inked, and the image formed transfers to the offset cylinder and thence to the paper.

C. In gravure printing, the photographic image is etched or engraved on the surface of the gravure cylinder. Ink fills the incisions and the doctor blade then clears the surface of the cylinder before it prints by transferring the ink from the incisions to the paper.

D. In screen printing, a stencil in which the image is cut out is laid over a screen. Ink is then applied with a squeegee so that it is forced through the screen on to the paper beneath.

Henry Bessemer 188 **information technology** 310–312 **invention of printing** 100–104 **photocopying** 313

advent of photocomposing techniques, was it seriously challenged. During that time its only rival was the Monotype machine, introduced in 1887. This, as its name implies, cast, set and justified individual characters, which made it easier than with the Linotype to make corrections: it was, therefore, favoured for high quality work, such as book production. It, too, was widely used for many years.

These machines were operated through a keyboard and there was no necessity for the latter to be adjacent to the casting/setting equipment. A satisfactory teletypesetter was first demonstrated in America in 1928 and shortly afterwards one was installed in the Houses of Parliament in London to transmit editorial material to *The Times*.

During the first half of the 20th century, the printing industry was dominated by hot-metal casting and setting processes, but nevertheless there were alternatives. Among the most important of these was that known as lithography – literally 'stone writing' – first described in 1798. If a design is drawn on a smooth stone surface with a greasy pencil, that part of the stone will repel water, while the untreated surface will still absorb it. If a greasy ink is then applied with a roller, it will adhere only to the parts covered by the original design, from which it can be transferred to a sheet of paper. By using a succession of stones and differently coloured inks, fine colour prints could be made. In this simple form, the process was far too slow and elaborate for any sort of general printing, but in the latter part of the 19th century it was adapted for flat-bed printing machines, the original stone being replaced by zinc, or later aluminium, plates. A more important development was the offset litho process, in which the design was transferred from the plate to the paper not directly but via an impression on a rubber roller. This process became popular for small office printing machines, though as early as 1912 Vomag in Germany offered an industrial machine capable of printing 7500 sheets an hour.

The first successful machine for casting and setting type was the Monotype, introduced in 1897.

In the long term, however, the future of printing lay with photographic processes and computer typesetting, doing away with metal type altogether. Although the first photocomposing machine was invented by W. C. Huebner in the United States in 1939, it came into widespread use only after the Second World War.

The enormous expansion of the printing industry which started in the 19th century had far-reaching consequences. Not least of these was a corresponding growth in the paper industry, especially in the cheaper grades used for newspapers, magazines and popular books. From the 1970s, this was increasingly made from chemically treated wood pulp. By 1939, annual production of this was 22 million tonnes, of which one-third went to make newsprint. So severe were the inroads on the world's forests that climatic changes threatened, but only very recently have serious attempts been made to counter these.

The typewriter

The printing press is designed to produce very large numbers of copies of a single original. For mechanical writing aimed at producing only one, or very few, copies the typewriter was introduced. The first satisfactory machine was that devised by Christopher Latham Sholes in 1867, and produced from 1874 by the Remington Company. Although clumsy by today's standards, this incorporated most of the features of machines manufactured during the next 50 years or so. These included the inked ribbon and characters carried on the end of strike bars. The familiar QWERTY keyboard was devised to minimize the risk of these bars clashing. Carbon paper had been used for manuscript copying since the early 19th century and was thus immediately available for making a few extra copies of typewritten material.

A major defect of this design is that only one style of type can be used but, in 1933, the VariTyper appeared. This adopted a feature of William Austin Burt's typographer of 1829, in which the characters were mounted on a metal band which could be easily replaced to give a different typeface. The same result was later achieved with the 'golf-ball' head which appeared in the IBM Selectric electric typewriter of 1961.

With all early typewriters, and many still in use, the line of type was not justified: that is, the lines ended unevenly on the right of the page. A line-justifying device was introduced in 1937 and this paved the way for crisp typewritten copy that could be directly used in photoprinting processes.

The typewriter was an invention of great social consequence, for it provided a new form of employment for women – like the telephone exchange a generation later – that was, at that time, socially acceptable. It can fairly be said to have been the means by which the doors of the business world were first opened to them.

Electrical communication

The first concept of an electric telegraph seems to have been formulated in 1753, when an unidentified correspondent in the *Scots Magazine* suggested that words could be spelled out letter by letter along a 26-wire system activated by static electricity. The receiver was to consist of 26 light pith balls, each carrying a different letter of the alphabet, which would be attracted to the activated wire. The system was not feasible with the apparatus of the day, but became more so with Alessandro Volta's invention of the battery (voltaic pile) in 1800. This led Francisco Salvá, in 1804, to use a similar system of transmission in which the electrically activated wire was identified by the stream of hydrogen bubbles it produced when immersed in weak acid: he succeeded in sending messages over a distance of 1 kilometre (⅝ mile). Eight years later Samuel T. von Sommering increased the range to 3 kilometres (2 miles). Among those to whom von Sommering demonstrated his electrochemical telegraph was Baron Schilling, an attaché in the Russian embassy in Munich. He developed the idea, and by 1832 he had made an electromagnetic device using fewer wires, in which the individual letters transmitted were identified by the movement of needles suspended over wire coils connected to the transmitter.

This type of detector was the basis of the effective electric telegraph patented by William Cooke and Charles Wheatstone in 1837, which was the basis of all later systems. It required five wires, activating five needles: the movement of any two of these indicated the letter tapped out by the transmitting operator. By 1842, a two-needle receiver had been devised.

Cooke and Wheatstone's telegraphic transmitter of 1837 was the prototype for many later devices. This model with punched tape input and pen-tape output was made in 1858.

Telegraphic transmitter, with magnetic induction, manufactured by Siemens in 1856.

One of the earliest installations of the Cooke–Wheatstone telegraphs was on the railway from Paddington to Slough and, in 1845, this was put to dramatic use which attracted much public attention to the possibility of the invention. A suspected murderer was seen to board a train at Paddington and a message telegraphed to Slough resulted in his arrest there. As we shall see later, a similar episode, involving the notorious murderer Doctor Crippen, gave tremendous publicity to Marconi's new system of wireless telegraphy.

Several technical developments assisted the rapid development of the telegraph. Using the famous code invented by Samuel Morse in 1835, messages could be received as long and short stokes on a paper strip, subsequently decoded by the operator at the receiving end. Less than 20 years later, in 1855, David Hughes devised a printing telegraph: in this the message was typed out in full on the keyboard of a transmitting instrument and at the receiving end it was printed. In the 1870s, multiplex working allowed several messages to be sent simultaneously over the same wire.

By 1900, the principal countries of the industrialized world not only had their own extensive internal systems but were also linked through an international network. In 1851 Britain – which by that time already had 6400 kilometres (4000 miles) of internal line – was joined with France and in 1858 with the USA. It was a system that affected all levels of

invention of paper 99–100 **pictorial communications** 312–313 **photocopying** 313 **radio telegraphy** 238–239

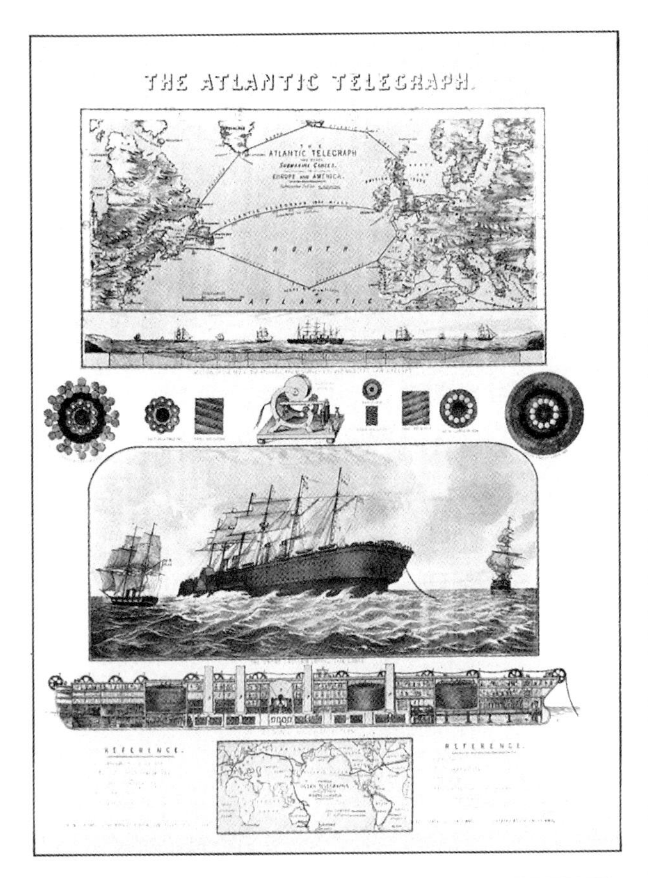

Top: After various proposals for northern and southern routes involving intermediate land stations, a 2960-kilometre (1850-mile) transatlantic submarine telegraph cable was first laid in 1858. The huge *Great Eastern*, a failure as a liner, was well suited to carry the huge coils of cable and ancillary equipment.

Above: The first commercial picture transmitted by wire, Cleveland to New York, 9 June 1924.

society. In 1849, Paul Julius von Reuter in Aachen formed an organization for transmitting commercial intelligence and from this grew – in line with the growth of newspapers – a vast international news agency. Governments for the first time could keep in constant touch with their embassies abroad: from 1865 Queen Victoria could send telegrams direct to her Viceroy in India. The military consequences were quickly perceived. In the Crimean War, a submarine cable was laid across the Black Sea to Varna in Bulgaria, enabling the French and British commanders to keep in touch with their governments via the existing European network. The ordinary citizen could use it for urgent messages: in Britain, nearly 400 million telegrams were being sent annually at the end of the century, at a cost of only one shilling (5p) for 20 words. By that time there were 25,000 kilometres (15,500 miles) of telegraph wire in Britain, 130,000 kilometres (81,000 miles) in Europe and 80,000 kilometres (50,000 miles) in the USA.

Utilization of the photoelectric properties of selenium led to the introduction in the 1920s of an effective system of facsimile or phototelegraphy, by which pictures could be transmitted through the telegraph network: this had profound consequences for journalism. This was quickly followed by the adaptation of the printing telegraph to provide a service between individual subscribers. Originally on a national basis – TWX, in 1931, in the United States, and Telex, in 1932, in Britain – it was beginning to be developed as an international service at the outbreak of the Second World War.

The consequences of the advent of the telegraph as a fast means of international communication cheap enough to be used in time of need by all classes, can fairly be described as revolutionary. Cheap though it was, however, users had to exercise a strict economy in words and this led to the concise and elliptical phraseology – omitting punctuation and minor words – known from the 1880s as telegraphese. Commercial users developed codes so that frequently used phrases could be transmitted as single words. Of necessity the telegram was brief, impersonal and to the point. It was thus in marked contrast to the telephone, which enabled people to speak directly to each other, responding immediately to questions and reacting to inflections of voice.

The telephone

The principle of the telephone is very simple. A reed or diaphragm vibrates in tune with the voice of the caller: the vibrations are transmitted along a wire as a correspondingly fluctuating electric signal and at the receiving end the electric signals cause a second reed or diaphragm to vibrate in step with the first. The first to achieve success, though by no means the first to experiment, was Alexander Graham Bell, a Scotsman who had emigrated to the United States and became a citizen in 1882: his special interest in speech arose

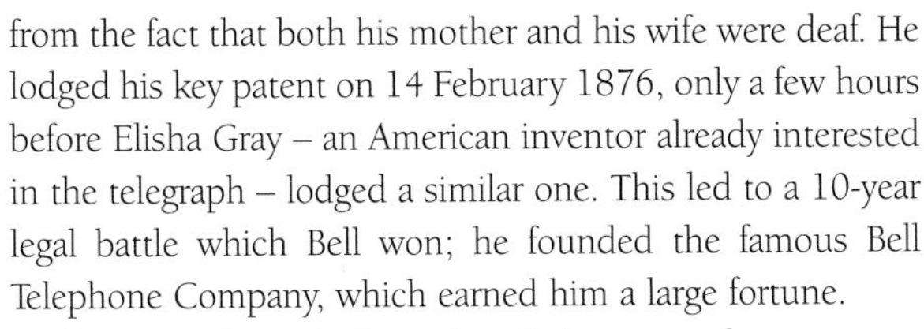

Left: This simple manual telephone switchboard was installed in Cincinnati in 1879.

Above: Employment for women was a consequence of the new telephone systems. This photograph shows a manual exchange in London, 1904.

from the fact that both his mother and his wife were deaf. He lodged his key patent on 14 February 1876, only a few hours before Elisha Gray – an American inventor already interested in the telegraph – lodged a similar one. This led to a 10-year legal battle which Bell won; he founded the famous Bell Telephone Company, which earned him a large fortune.

The very first telephone installations were between two fixed points but, in January 1878, the first exchange, with 21 subscribers, was opened in New Haven, Connecticut. By January 1884, Boston had been linked with New York, a distance of 480 kilometres (300 miles), at a cost of nearly $75,000. In Europe, Bell promoted interest in the telephone by a series of talks and demonstrations – including one to Queen Victoria in 1878, the year in which the first telephone company was formed in Britain. In Germany, in 1877, the telephone was from the first adopted as a state monopoly, as it was in Britain 20 years later. By 1900, there were over one million telephones in the USA alone.

Initially exchanges were worked manually by an operator, the caller giving the number to which he wanted to be connected. As early as 1889, however, Almon B. Strowger of Kansas City devised an automatic system, by which numbers could be called by push-buttons or, later, a numbered dial. The engagement of operators posed no great problem, for the work was one of the rather few kinds that women could then acceptably undertake. As a result, automatic exchanges only slowly came into use: Bell Telephone did not adopt them until after the First World War.

Like gas and electricity, the telephone began as a local service. One reason was that the iron wires originally used had rather poor transmission qualities – though Australian farmers found they could get tolerable results using their miles of iron fencing wire. Copper alloys proved much more satisfactory. By 1900, not only were local systems being linked nationally, but a number of international networks were being built up. This involved laying a number of submarine cables: England was first linked with the Continent in 1891, by which time the European network extended to 20,000 kilometres (12,000 miles), much of it making use of the existing telegraph system. This important development stemmed from the introduction of anti-interference 'chokes' by a Belgian engineer, F. van Rysselberghe, in 1882. However, long-distance telephony demanded signal boosters at intervals and even on overland wires this presented a difficulty. New York was not linked with San Francisco until 1915 and the first transatlantic telephone cable was not laid until 1956: it was 3600 kilometres (2200 miles) long and included more than 50 boosters. Meanwhile, the Atlantic had been bridged from England by introducing a radio link into the telephone service in 1927: this had, however, limited capacity and was subject to much atmospheric interference, sometimes causing total closure. Similar radiotelephone links were established with Australia and South America, and from the early 1930s the telephone was truly a global service, though rather an expensive one.

For the individual subscriber the most important developments were in the strength and quality of the signal and in the instrumentation. The very earliest telephones involved a single microphone, used alternately for speaking and listening, but separate ear and mouthpieces were introduced by

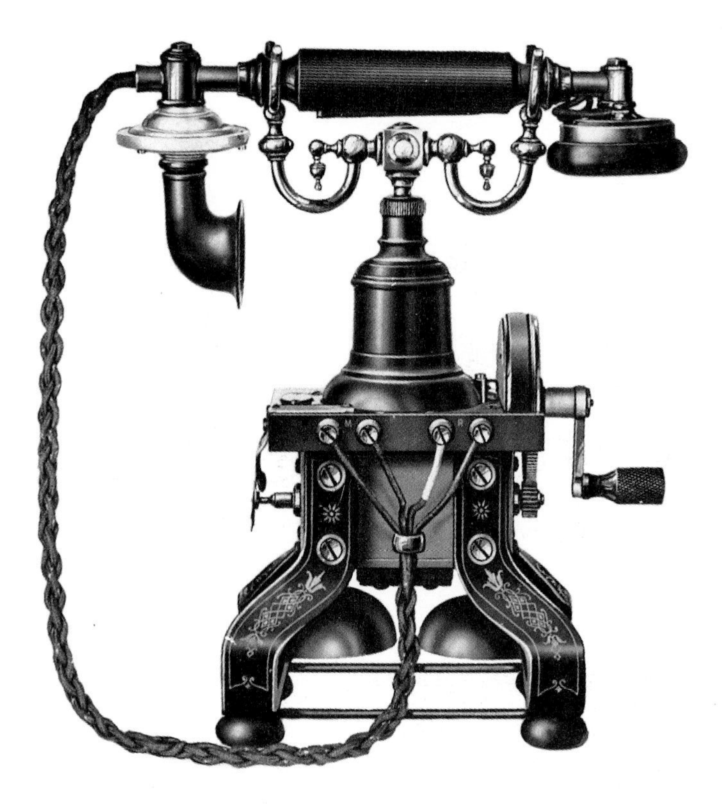

With early telephone instruments, like this Swedish Ericsson model of 1892, the caller alerted the exchange operator by cranking a magneto-generator. Later this was done automatically as the handpiece was lifted off its rest.

Edison, in 1877, and very quickly became general. Up to the 1920s the candlestick design was the most popular: the subscriber was connected to the exchange when he lifted the earpiece from its hook. From the 1930s, however, the modern type of instrument – in which the mouth and earpiece form part of a single unit held in the hand – became standard. While most local calls could be made automatically by the 1920s, most long-distance calls had to be made through an operator until after the Second World War.

To these important advances we will return later. As far as the first half of this century is concerned, the general pattern of the service was established during its first decades, though it was not until the 1920s that the telephone became a common piece of household equipment. In 1934, the number of telephones in the world was about 33 million, slightly more than half of them in the USA. Until the First World War the telegram was still the popular means of quick communication. Indeed, in many countries the availability of the telephone has always tended to follow a social pattern. While in Britain, for example, it has long been a normal part of upper- and middle-class homes, it initially attracted relatively few subscribers among the working classes, who preferred to use public pay phones involving no permanent liability by way of rental.

Wireless telegraphy

The importance of the telephone and telegraph as a cheap and rapid form of international communication needs no emphasis. Nevertheless, both had their limitations. In particular, they required the installation of enormous networks of wire and associated exchanges, both needing maintenance and subject to breakdown. Furthermore, it was not possible to communicate with places – such as ships – not connected to the network. There was, therefore, much interest in any system of communication that made all this equipment unnecessary.

Wireless telegraphy was just such a system, but its origins lay in some very theoretical science. In Cambridge, James Clerk Maxwell – developing Faraday's work on electromagnetic induction – had concluded that there must exist in nature a range of electric radiation of which light is only one manifestation. Seeking evidence for the existence of such radiation, Heinrich Hertz in Kiel showed in 1887 that such waves, travelling with the speed of light, were generated by an electric spark, and succeeded in detecting them at a distance of 20 metres (66 feet) from the source. Similar experiments to test Maxwell's theory were carried out at Liverpool by Oliver Lodge, who demonstrated them to the British Association for the Advancement of Science in 1894. In Russia, in 1895, Alexander S. Popov detected electromagnetic waves 80 metres (262 feet) away from their origin.

Such demonstrations, in the face of scepticism, of the validity of Maxwell's revolutionary ideas opened completely new fields of physical science, but their intention was not to develop a completely new system of electrical communication. The first seriously to realize the practical implications was a young Italian aristocrat, Guglielmo Marconi, who had little scientific knowledge but much inventive genius and business acumen. Within a short time of reading an account of Hertz's work in 1894, when he was only 20 years old, he had succeeded not only in transmitting signals over a distance of 2 kilometres (1¼ miles) but had pulsed them so that messages could be conveyed in Morse code. Gaining no support in his homeland, he went to London and secured the interest of William Preece, then engineer-in-chief to the Post Office, which was just about to expand its interest in electrical communication by taking over most of the telephone system in Britain. He formed his own company in 1897, which in 1900 became the internationally organized Marconi's Wireless Telegraph Company.

Marconi went from strength to strength. In 1899, he sent a signal across the English Channel using tall masts for transmission and, by 1901, he had bridged the Atlantic from Poldhu in Cornwall to St John's, Newfoundland. This last achievement was something of a surprise, but a very welcome one. If radio waves were electromagnetic in nature, they ought to travel in straight lines, like light, and should therefore be blocked by the curvature of the earth. An explanation was not found until 1924, with the discovery of an electrified layer in the upper atmosphere capable of reflecting such radiation.

From this time on, Marconi devoted himself to his business interests, but in 10 years he had done sufficient to be awarded a Nobel Prize in 1909: this he shared with the German physicist Karl Braun, another pioneer of radio telegraphy but now better remembered as the inventor in 1897 of the cathode ray oscilloscope, which played a key role in television. Marconi's Nobel states that 298 merchant ships and the principal warships of the British and Italian navies had been equipped and there was already a public transatlantic wireless telegraph service. Five years previously, the Cunard liner *Campania* had begun to publish the first of the *Cunard Daily Bulletins* for the benefit of travellers. But perhaps the event that most impressed the public with the power of the new system was the arrest of the notorious murderer Doctor Crippen and his mistress, after the suspicions of the captain of the *Campania*, *en route* from Antwerp to Canada, had been aroused by news received by radio. Except for the fugitives, virtually the whole world knew what was afoot. Equally, the sinking in 1912 of the *Titanic*, with the loss of 1500 lives, stressed the importance of ships' wireless.

The next three decades saw important technical advances. An essential component of a wireless receiver is a valve to allow the electric current to flow in only one direction, exactly like a valve in a water main. Originally, these were of the cat's whisker type – which, after the Second World War, appeared in a new guise as the transistor – but in 1904 John Ambrose Fleming, professor of electrical engineering at London University, invented the diode (two-electrode) valve. In 1907 Lee De Forest, in the USA, patented the triode (three-electrode) valve which could be used as an amplifier and detector. Fleming successfully challenged De Forest's patent, claiming that it was no more than an extension of his own for the diode – though 40 years later this ruling was overturned. The triode was immensely useful and was made in tens of millions: when coupled with an oscillator a very powerful signal could be generated.

The earliest radio transmitters incorporated a simple spark-gap, and this gave a broad spectrum of radiation, so that nearby transmitters seriously interfered with each other. Nevertheless, in an emergency at sea this was a positive advantage, for a message was more likely to be picked up by other vessels. Valve transmitters, which could be tuned to a narrow waveband, were widely adopted but not until 1927 was it internationally agreed that spark transmitters above a certain power should be phased out.

The development of valve transmitters and receivers meant that more powerful signals could be transmitted and these would be more clearly and loudly received: as a result, the early headphones could be replaced by loudspeakers. By using a microphone to modulate the amplitude of the transmitted wave, sound – as opposed to a simple signal – could be transmitted and received. As early as 1906 Reginald Aubrey Fessenden – who later, in 1902, invented the heterodyne circuit in which the weak incoming signal modulated a much stronger locally generated wave – broadcast a spoken Christmas message which was picked up by ships' operators off the American coast. In 1915, speech was transmitted across the Atlantic, from Virginia to Paris, for the first time. Public broadcasting systems were, however, essentially a post-war development. For nine months, commencing on 23 February 1920, Marconi broadcast a regular news service from his transmitter at Chelmsford in England: on 2 November in the same year the Westinghouse Company began regular radio transmissions in Pittsburgh, USA.

Top: In 1902 Marconi began to build this permanent transmitting station at Poldhu, Cornwall.

Above: In 1920 Marconi's company began regular news broadcasts from their transmitting station at Chelmsford, Essex.

Thus began not only a new form of public entertainment but a vast new industry. Progress was very rapid indeed. By 1922, there were 600 commercial broadcasting stations and a million listeners in the USA alone. In London, the famous 2LO station came on the air in the same year and the British

Faraday and induction 170 **radio astronomy** 336 **telecommunications** 314–315 **transistor radio** 315

Broadcasting Corporation was founded as a public corporation in 1927: by that time there were already two million receiving sets in Britain.

Television

Although experiments with picture transmission by telegraph began in the 1850s it was, as we have noted, the 1920s before a satisfactory system was developed for pictures possessing tone, as opposed to black-and-white line drawings. From the end of the 19th century, the cinema had effectively demonstrated the possibility of reproducing motion. In principle, therefore, there was nothing very novel about the idea of transmitting moving pictures by wireless. Nevertheless, reliable and acceptable television programmes are essentially a feature of the years since the Second World War, even though the basic principles had been established much earlier.

In the early years, inventors pursued two separate lines. although it was not in the end the one that proved successful, it is perhaps appropriate to consider first that of John Logie Baird, for it was the one that first attracted public notice. His was a photomechanical system, in which the picture to be transmitted was systematically scanned by a rotating disc carrying a series of small holes arranged in a helix. This method of scanning a picture had been devised long before, in 1884, by Paul Nipkow, a German inventor. Bright parts of the picture transmitted a strong light beam through the corresponding hole, and dark parts a weaker beam. The light signals were transformed by a photoelectric cell into electric pulses. At the receiving end the reverse procedure took place: the electric signals were turned into a fluctuating light beam which fell on a corresponding Nipkow disc and then on to a screen. Later a scanning system based on a rotating mirror-drum was introduced. On the other side of the Atlantic, Charles Francis Jenkins also experimented at about the same time with photomechanical television.

Baird's photomechanical system – here illustrated in apparatus he built in 1925 – was not in the main line of development, but nevertheless served a very useful purpose in arousing public interest in television.

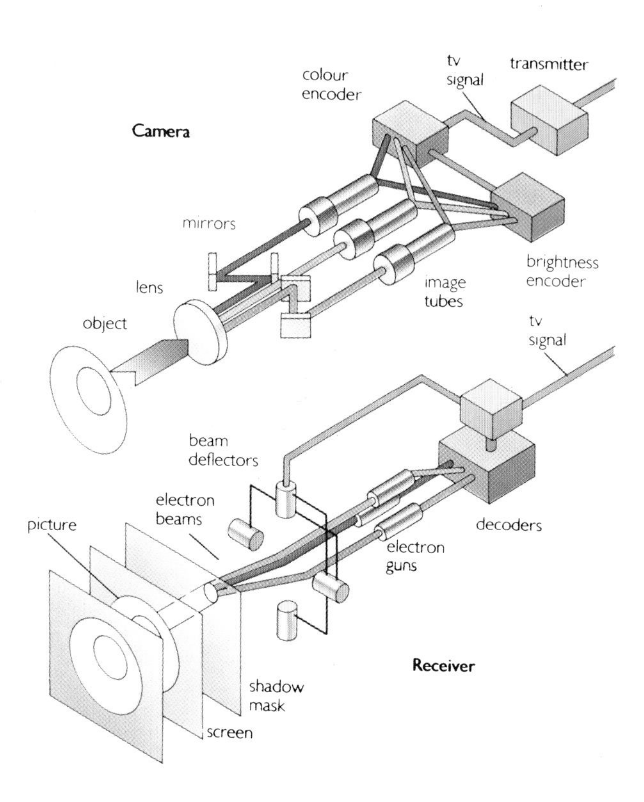

Colour television uses three superimposed images in red, green and blue. In the camera, prisms or mirrors separate the three coloured images, which are then transmitted to the receiver. Three electron beams activate red, green and blue phosphors in the screen to form a full-colour picture.

It was a crude and unsatisfactory system, but it worked after a fashion and in 1929 the newly founded British Broadcasting Company licensed Baird to start a regular public television service: this they took over themselves in 1932 and continued until 1937. It was never very satisfactory, partly because the receivers were clumsy and expensive – no more than 2000 were ever sold – and the quality of the 30-line picture was poor. By 1935, the BBC was disillusioned and turned to the alternative all-electronic system, which finally carried the day. As early as 1908 Alan Archibald Campbell-Swinton, a Scottish electrical engineer, proposed – but never developed – a scheme for using the flying light spot of the cathode ray tube invented by Braun both to scan the picture and form the image. This idea was put into practice by Vladimir Zworykin, a Russian electrical engineer who joined Westinghouse in the USA in 1920. In 1933, he produced his iconoscope: soon afterwards, the British company EMI produced a similar television tube, the emitron. The way was thus open to a much improved service and on 2 November 1936 the BBC was able to offer a regular 405-line transmission service: the Radio Corporation of America followed suit on 30 April 1939, just before the outbreak of the Second World War, showing President Franklin D. Roosevelt opening the New York World's Fair. However, on both sides of the Atlantic the impact was small. Although receiving sets were much improved, they were still expensive and, by modern standards, the picture was blurred and flickery: in 1942, there were no more than 20,000 sets in Britain and only half that number in the USA. The full burgeoning of public television was strictly a post-war development.

Sound recording

The telephone and radio provide a means of transmitting sound – whether it be speech or music – but not to record it so that it can be referred to in the future like the printed word. Such devices as the musical box, invented in Switzerland in the middle of the 18th century, do not record sound but create it mechanically. The first to achieve true sound recording was the versatile Edison, whose first phonograph was built in 1877. In this the vibrations caused by sound falling on a thin metal diaphragm were transmitted by a steel stylus to a sheet of waxed paper, later tinfoil, wrapped round a rotating cylinder. As the cylinder rotated, it was gradually pushed forward, as in a screw-cutting lathe, so that the stylus cut a spiral track. The vibration of the stylus led to a 'hill-and-dale' track being cut, and when the procedure was reversed the original sound was reproduced. Although the phonograph brought Edison fame, it brought him little profit: the reproduction was poor, the recording was brief and the cylinders survived only a few playings. In 1886 Chichester Bell – a cousin of Alexander Graham Bell – produced an improved model, the graphophone, in which a hard wax cylinder was used: its success, and a belief that his own patents were being infringed, brought Edison back into business. But the big advance was that of Emile Berliner, who introduced the modern disc record in 1888, to be played on what he called a gramophone. The vibrations of the recording stylus were reproduced by lateral – as opposed to hill-and-dale – movements in the helical soundtrack. In his system recording and reproduction were for the first time separated. The original recording was transferred to a copper plate, from which large numbers of replicas could be made in hard shellac resin. Previously, only single recordings could be made, the singer repeating his performance according to the total number required. The public then owned machines that only played and did not record, purchasing the flat, easily stored discs according to their taste. The cylindrical record died hard, however: as late as 1908 the mail order firm Sears Roebuck offered their customers a choice of both machines, which they still called graphophones, at approximately the same price of $15. The first recordings were trivial, but by the turn of the century great artists like Enrico Caruso

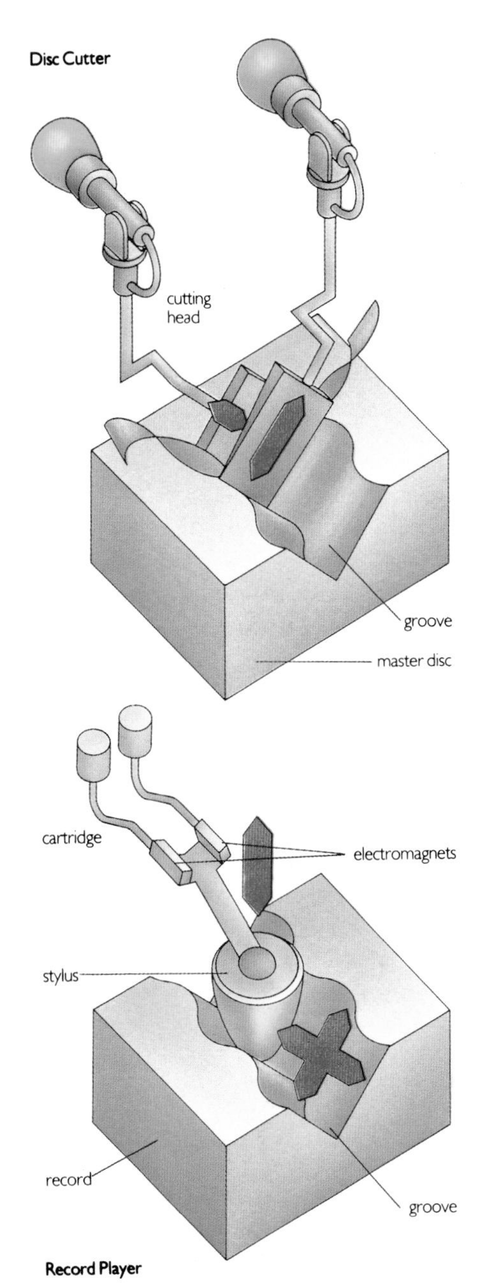

Below: In Edison's standard phonograph the soundtrack was cut spirally on a cylinder, and players and records of this design were available up to the First World War. By then, Emile Berliner's double-sided disc records were dominating the market.

Above: In the manufacture of stereophonic records, a master disc is produced by a delicate cutting machine that cuts the two walls of the groove at 90 degrees in response to the two stereo signals. The commercial record is a copy of this master disc. The stylus of the player is vibrated by both walls when placed in the groove for playing. It vibrates the cartridge, which contains two electromagnets also at 90 degrees, to separate the vibrations and reproduce the two stereo signals.

colour television 274–275 **compact disc players** 310, 314 **film soundtrack** 246 **future of television** 315, 340, 341 **video recording** 275

realized that a new public had become open to them: his first recording from *Pagliacci* sold more than a million copies. Yet another great new branch of the entertainment industry had been founded. Whereas the sale of a million copies of a Caruso recording in 1902 was a sensation, the 'golden discs' for sales of over a million records, launched by Radio Corporation of America in 1942, find many claimants today.

Early recording machines and gramophones were entirely mechanical – first turned by a hand crank but later by clockwork – and not until 1925 was an electrical system introduced. In this a microphone replaced the huge horns used hitherto in recording studios, and a valve amplification system served as an intermediary between that and the cutting tool on the master disc. Corresponding changes were made in record players, a name that began to supersede that of gramophone.

Although stereophonic records were not available to the public until 1958, the pioneer work was done in the 1930s, roughly simultaneously by EMI in Britain and Bell Telephone in the USA: the latter rather pleasingly referred to it as the reproduction of music 'in auditory perspective'. To get this three-dimensional effect, two separate microphones are necessary, and two separate tracks must be cut in the disc. The latter was achieved by combining lateral recordings with the old hill-and-dale kind originally used. An even more sophisticated technique, quadrophonic sound – using four speakers – was introduced in 1971.

The availability of new plastics in place of shellac made it possible to make narrower and more closely spaced soundtracks and this led in 1948 to the LP (Long Playing) record, typically playing 23 minutes per side. But by then, however, all conventional records were finding a new commercial rival. This was the tape, in which variations in sound are recorded as variations in the magnetism of a tape coated with iron oxide or chromium oxide. This technique, using a wire rather than a tape, had been patented by Valdemar Poulsen, a Dane, as long ago as 1898. Although tape recorders found limited use in Britain and Germany in the 1930s, they were not widely available to the public until after the Second World War. In 1985 tapes outsold records for the first time, but were subsequently superseded in 1991 by the compact disc, which was first marketed in the early 1980s. Later, the same process was used to make videotape to record television programmes.

The 1960s saw the advent of the synthesizer, developed by RCA, for creating music electronically without any human intervention. The concept was not new – pipe organs had long included facilities to mimic the principal musical instruments – but the means of achieving it was. Synthesized music is generated with the aid of a multiple keyboard and can be recorded on magnetic tape. Obviating the need for an orchestra, it is naturally popular for providing background sound for radio and television programmes.

Photography and the cinema

In the late 17th century, artists were using the lens of a camera obscura to project a picture on to paper, and in the early 18th century Johann Heinrich Schulze had observed that silver salts darkened when exposed to light. Surprisingly, these two observations were not combined to produce the modern photograph for more than a century. Indeed it was a quite different kind of photochemical reaction – the hardening of bitumen when exposed to light – that in 1826 enabled Joseph Nicéphore Niepce to produce the world's first photograph, as distinct from shadographs of leaves, insect wings and similar objects. It was a crude picture taken with an eight-hour exposure. Nevertheless, it embodied all the essential features: the image of the scene was focused by a lens on a light-sensitive plate in a dark box, the camera.

In 1839, a much improved process was demonstrated in Paris by Louis-Jacques-Mandé Daguerre, who had collaborated with Niepce. His light-sensitive material was a copper plate coated with silver iodide: the image was developed by exposing the plate to mercury vapour. In bright sunlight, the exposure was only 30 minutes, but at the end of the

Daguerre's photographic process was first demonstrated in Paris in 1839. The plates he used had very low light sensitivity – an exposure of up to half an hour was necessary even in bright sunlight – and a stand-mounted camera was essential.

Left: Fox Talbot's *Pencil of Nature* (1844) – now a very rare and valuable collector's piece – demonstrated many applications of photography. It is of historic significance in being the first book to be illustrated with photographs, each individually stuck in: this picture from the book is of a boulevard in Paris.

Below: The Kodak camera of 1888 was the first hand-held camera to incorporate roll film, each yielding 100 circular pictures. Weighing just over 1 kilogram (2 pounds), it had a fixed aperture and a single-speed shutter: film processing was done by the manufacturer.

operation one photograph, known as a daguerreotype, was obtained. In Britain, William Henry Fox Talbot described, also in 1839, a process called calotype in which a negative picture was formed on transparent paper coated with silver salts: the immense advantage of this was that from the one original any number of positive copies could subsequently be made. It was a development comparable with Emile Berliner's system of making a master recording of sound on metal, from which an unlimited number of copies could be made. Exposure times were still fairly long – not less than a minute under the best conditions – but portraiture was just possible, though the subjects looked a little strained. In 1851, Frederick Scott Archer introduced his wet-plate process: in this the photographic plate – now made of glass to give a sharper image than was possible with paper – had to be dipped in silver nitrate solution immediately before exposure, and it then had to be developed immediately. The great advantage was the far faster speed: for brilliantly lit subjects half-a-second would suffice.

However, Archer's and similar 'wet' processes demanded that the photographer carried a vast load of paraphernalia with him. Dry plates were introduced in 1853: these had the advantage of not needing immediate development but were relatively very slow. One manufacturer of dry plates was George Eastman, who about 1874 conceived the idea of putting the light-sensitive emulsions not on glass plates but on long strips of paper, later celluloid, which could be tightly rolled on spools. This at once opened a mass market for amateur photographers; promoting his famous Kodak camera in 1888, Eastman boasted: 'You press the button, we do the rest.' Each camera contained sufficient film for 100 photographs. Initially the camera itself had to be returned for reloading, but soon only the roll of exposed film was required. Although the 20th century saw many improvements in the performance of lenses, shutters and filmstock, the evolution of the hand-held camera was virtually complete with Eastman's model.

In lenses, one of the most important developments was that of wide-aperture lenses for use in poor light. In 1908, a good working aperture was f6.8, but by the 1930s f2.8 was not uncommon in the more expensive cameras. Whereas, at the end of the 19th century the average exposure time was around 1/25th second, by 1935 even amateurs were using 1/500th second. Filmstock was improved by increasing sensitivity and reducing 'grain'. While early film was sensitive mainly to visible light of short wavelength, at the blue/violet end of the spectrum, the later orthochromatic and panchromatic films were sensitive to a wide range of colours.

Early photographers relied on bright sunlight, but with the availability of powerful electric lighting and faster films, indoor work in studios was increasingly feasible. For intense light, but one apt to cast harsh shadows, flashpowder – a mixture of magnesium powder and sodium chlorate – came into use in the 1880s. A very much more convenient form of flash was the electric flashbulb, introduced in 1929.

The Polaroid camera (above right) introduced in 1946, combined negative and positive film as well as developing chemicals, and gave an instant photograph (the first is shown above). In 1963 the system was developed to give colour prints.

Right: The Autochrome process – launched by the Lumière brothers in 1907 – was the first to bring colour photography within the range of the amateur. This Cibachrome print is of an original Autochrome in the Lumière collection of Ciba-Geigy, Switzerland.

Very large cameras continued to be used for studio work, and still are, but the keen amateur was offered increasingly compact models. One way of achieving compactness was to mount the lens on a bellows which folded away when the case was closed. More expensive cameras were reduced in size by precision engineering, pioneered by the famous German Leica of 1925.

Colour photography

As we have seen earlier in the context of lithographic printing, the principles of colour printing by superimposing red, blue and yellow images were understood and practised early in the 18th century. Although professional photographers used similar techniques in the 19th century, and produced some very fine results, no film suitable for popular use was available until 1907. In that year the Lumière brothers – Auguste and Louis – introduced a process in which the sensitive emulsion was underlaid by a transparent film containing a mixture of starch particles dyed in the three primary colours – red, yellow and blue. A variety of such additive processes – so called because the differently coloured lights are added to each other – appeared over the next 25 years. In the mid-1930s, Kodak and Agfa offered colour film based on a subtraction process: in this three layers of sensitive film, each dyed in one of the three primary colours, were superimposed. All such processes gave transparencies which had to be looked at in a hand-held viewer or projected on to a screen. Not until 1942 did Agfa introduce a process that gave coloured prints, but these were not widely available until the 1950s.

Another interesting post-war development was the Polaroid camera introduced by Edwin H. Land in 1946. In this ingenious device negative and positive film, and the chemicals necessary for development, are contained in a single pack, thus an instant photograph can be obtained. From 1963 the process – confounding the considered views of sceptics – was available also for colour pictures.

Cinematography

The phenomenon known as persistence of vision was familiar to Ptolemy in the second century, for he describes how a spinning disc appears uniformly coloured even if only a sector of it actually is. In the 19th century, various devices were used for home entertainment in which the illusion of animation was given by viewing a number of pictures – of a dancing clown, say – in such rapid succession that the image of one had not faded on the retina before the next appeared. One such device was the zoopraxiscope devised in the 1870s by Eadweard Muybridge, an English professional photographer. Basically, this is still the technique of the cartoon film maker.

In 1877–8, Muybridge investigated the action of a galloping horse by taking, with the help of an ingenious mechanical contrivance involving two dozen cameras, a succession of pictures at short intervals: even at that time he was able to use exposure times as short as 1/1000th second. These photographs conclusively showed that at certain points the animal had all four feet off the ground – hitherto a subject of much argument. Muybridge's work led a French physiologist, Etienne-Jules Marey, to develop a 'photographic gun' by which a sequence of pictures could be recorded in rapid succession round the circumference of a circular photographic plate. Later Marey made his recordings on roll film and, very much to the point, projected the successive images on to a screen: by making the speed of the projection less than that of photography, a slow-motion effect could be obtained. Marey had evolved all the essential features of modern cinematography by 1890, but his interest was in the scientific investigation of natural phenomena, especially the movement of animals, and he did not pursue the possibilities of his technique in wider fields.

About 1890, Muybridge conceived the idea of combining his zoopraxiscope with Edison's phonograph, thus producing, in effect, a 'talkie'. For once Edison missed an opportunity, for when approached he dismissed Muybridge's device as a mere toy – which, indeed, it was: it was the concept it embodied that was important. However, Edison clearly did not forget the idea, for in 1891 he produced his Kinetoscope, a viewing device in which one person could see about 15 seconds of action on a succession of film frames exposed at the rate of 15 per second. It was popular in American entertainment halls – especially after he added music and vocal effects in 1896 – but he did not think it worthwhile to patent it abroad. As a result, there was nothing to stop the Lumière brothers from adapting it to project the moving picture on to a screen so that it could be seen by several people. Their first public showing to a paying audience was given in Paris on 28 December 1895 in the Grand Café, 14 Boulevard des

Above: Muybridge's zoopraxiscope disc of 1880 did not record movement but created the illusion of it by showing the viewer a succession of slightly differing pictures – in this case of a galloping horse.

colour television 240, 274–275 **Edison's phonograph** 241 **first colour cinema film** 246 **lithographic printing** 233, 234

Left: The roaring lion that prefaced all MGM films from 1929 became familiar to hundreds of millions of cinemagoers throughout the world. The cameraman and sound technician required much patience – and nerve – to obtain the desired result. This picture of one of the recording sessions illustrated equipment typical of the day.

Below: In a cine camera, the film is transported by a sprocket wheel and claw mechanism through the gate so that it halts as the rotating shutter opens and moves as the shutter closes. In this way, a series of static images is produced by the lens on the film. A cine projector works in much the same way to project the images on to a screen.

Capucines. This event certainly marked the beginning of the cinema as we know it today.

It was the start of what was to become a vast new industry which grew with surprising speed: it is estimated that when the Second World War broke out, the world investment in the film industry exceeded $2.5 billion: the figure is appropriately expressed in dollars because after early dominance by Europe, especially Britain and France, the centre of the industry moved decisively to the USA, and especially Hollywood, about 1915. This growth was due partly to the immense appeal of the new industry – bringing to the public not only live news from all parts of the world, especially identified with Charles Pathé of Paris, but full-length dramas such as *The Battleship Potemkin* of 1905, portraying revolution in the Russian navy. The eagerness of the public to pay made possible more and more elaborate and costly productions. But success was also a result of major technical improvements, of which the most important were sound and colour.

Sound as an accompaniment to film was recognized as important from early days, and pianists were employed to play music appropriate to the changing mood and tempo of the film. Later, gramophone records were used, but there were problems in synchronizing sound and vision: to get them out of step was a recipe for disaster. The real advance was to record sound – not only background music but also the voices of the characters – on a soundtrack. The sound was made by a modulated light beam that formed a soundtrack along the edge of the film. The beam formed either a dark band with a serrated edge, the varying area corresponding to the variations of the sound, or it formed a band of varying opacity, the denseness corresponding to the sound variation. When the film was projected, the sound was simultaneously reconstituted through a valve-amplifier system of the type already well established for radio. After various false starts, the phonofilm system – invented by Lee De Forest, pioneer of the triode valve in radio – was introduced in the USA in 1926: it was the predecessor of the better-known Movietone system. Almost overnight sound dominated the cinema: by 1930 95 per cent of major new films were talkies.

The principles of colour photography were already well established, and Lumière colour film was available as early as 1907. Although some short films had been laboriously coloured by hand even before 1900, colour cine film was not really successful until the 1930s. The first full-length colour film was *Becky Sharp* in 1935. As we have noted, it was just at that time that television was beginning to emerge as a rival, albeit unsuspected, to the cinema as a great medium of popular entertainment: the film industry had virtually reached its peak.

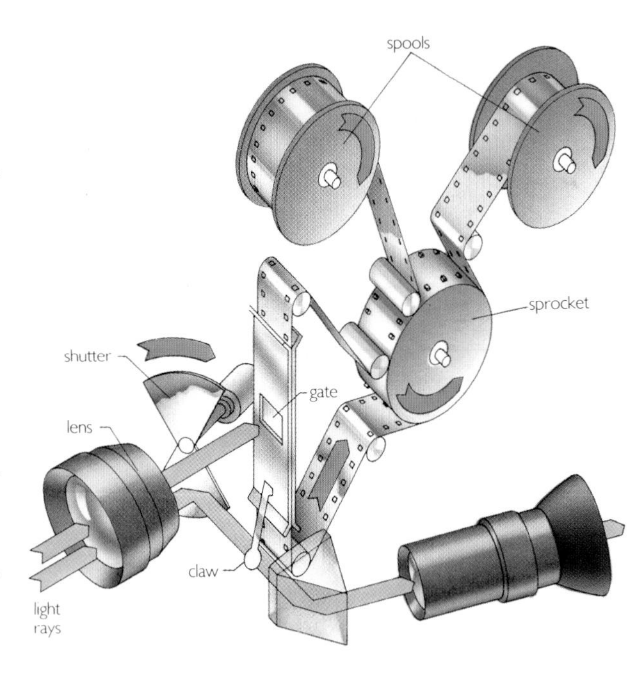

CHAPTER TWENTY-SIX

Transport: The Rise of Road and Air

Until the early 19th century, transport had altered little for 2000 years. Then, quite suddenly, the whole situation changed with the advent of railways. Not only were speeds far in excess of anything previously thought possible – or even survivable – easily attainable but long-distance travel was brought within the means of ordinary citizens. At mid-century British railways carried 100 million passengers annually, but by 1900 this had risen to well over a billion; the goods traffic, too, was heavy by the end of the century, some 700,000 wagons being in regular use. This was indeed a transport revolution, but the 20th century was to see two further revolutions of comparable magnitude, first on the roads and, later, in the air.

The bicycle

The revolution of road transport was, of course, the result of utilizing the internal combustion engine, but before considering the early history of the automobile something should be said of the bicycle, especially as many of the early automobile manufacturers came from the bicycle industry. This, too, was important as a means of mechanical transport within the reach of everybody, including women. The role of the bicycle in the emancipation of women should not be underestimated: Amelia Bloomer's daring 'bloomers' of the 1860s were designed with lady cyclists in mind.

In 1818, Karl von Drais von Sauerbronn invented his two-wheeled dandy-horse, or Draisine, on which the rider sat astride and propelled himself along by kicking with his feet. About 1839, a Scottish blacksmith, Kirkpatrick Macmillan, improved this by adding pedals which drove the rear wheel through a system of cranks, and in 1861 a French coachbuilder working in Paris, Pierre Michaux, attached pedals and cranks directly to the front wheel. Such velocipedes, as they were called, began to be manufactured in Britain at Coventry and there, in 1870, James Starley built the first 'ordinary' bicycle. This, too, had pedals attached directly to the front wheel, which was very large, nearly 2 metres (6½ feet) in diameter, with a smaller rear wheel – hence the popular name 'penny-farthing' (a penny and a farthing). This arrangement was no freak of design but a simple way of ensuring, without gearing, that the rider's legs were comfortably adjusted to the rotation of the wheel. This was solved in a different way in the Rover 'safety' bicycle of 1885: this went back to the original system of placing the rider comfortably and safely between two wheels of equal size, and the drive from the pedals to the rear wheel was effected by a chain. The necessary gearing was obtained by making the sprocket-wheel on the pedal shaft considerably bigger than that on the wheel. Two subsequent British innovations of importance were John Dunlop's pneumatic tyre of 1888 and W. T. Shaw's encapsulated crypto-dynamic gearing of 1885, which was followed by the more widely used Sturmey-Archer three-speed hub of 1902. On the Continent and in the USA, the dérailleur type of gear was favoured, in which the chain is transferred from sprocket to sprocket of different sizes. The free-wheel came into general use about 1903.

With the advent of the safety model, bicycling became immensely popular not only as a pastime but for everyday transport. Enthusiasts quickly showed that the bicycle could take its rider long distances. In 1884, Thomas Stevens began a 19,000-kilometre (12,000-mile) journey, completed in 1887, which took him across America and Europe and then on to Asia; in the autumn of 1888, the Reverend Hugh Callan cycled from Glasgow to Jerusalem and back. Racing cyclists attained high speeds: J. Michael reached nearly 75 kilometres per hour (47 miles per hour) in Paris in 1902.

By 1885 most of the features of the modern bicycle had appeared and bicycling had become a popular pastime – as well as a useful means of transport – throughout the world. This picture shows a road rally at Buffalo, NY, in 1894.

development of railways 177–182, 252–253 **Lee De Forest** 239 **sound recording** 241–242 **steam tricycle** 177

From 1900 onwards, there were virtually no fundamental changes in the bicycle. Weight was reduced by better design and use of light aluminium alloys – though wooden rims could be had up to the First World War – and multiple rear-wheel gears, giving 10 ratios or even more, were introduced. By the outbreak of the Second World War, world annual production of bicycles was around seven million. After the war, the pedal bicycle industry went into a decline with the introduction of light motorcycles with extremely small and economical petrol engines, but the 1970s saw a great revival.

The automobile

Rather surprisingly, by 1900, virtually all the essential features of the modern car had also been developed: had an automobile engineer of that time been reborn in the 1990s there would have been little to surprise him. An important difference, however, is that the number of cars in existence at that time was small. In 1900, there were probably no more than 9000 cars in the world, and half of these were in the USA: by contrast, in 1892, more than 1000 cyclists could be seen on the London–Brighton road alone in a single day.

The main feature of the car is, of course, the petrol engine and the early history of this has already been discussed. The very earliest vehicles were, in effect, little more than light horse carriages that had been motorized, much as the first railway locomotives were colliery wagons fitted with steam engines. Only gradually did the automobile emerge as a systematically planned entity, a development not assisted by the fact that, until very recently, few manufacturers were vertically integrated: the automobile industry is traditionally an assembly one, buying in its main components from a variety of specialist suppliers.

Many of the earliest cars were built for individual customers and the separate making of engine, chassis and bodywork gave a desirable flexibility. Later, however, as production came to be numbered in millions, standardization became essential. The first to realize the possibility of a mass market engendered by highly efficient methods of assembly was Henry Ford, who began making automobiles in Detroit in the 1890s. In 1908, he launched his famous Model T or Tin Lizzie – of which 15 million were eventually built – and, in 1913, he introduced the assembly-line principle. This was not new, however, as it was already in use in the great Sears Roebuck mail-order company founded in Chicago in 1893, for assembling the contents of packages for customers.

Between the engine and the wheels three pieces of equipment are necessary. First, a clutch so that the engine can be kept running when the car is stationary. Second, a gearbox to enable the engine to adjust to varying road conditions, such as steep gradients, and to allow the car to reverse. Third, a mechanical link between the gearbox and the wheels.

The standard clutch was of the friction-plate type, in which two circular plates – one attached to the crankshaft and the other to the mainshaft of the gearbox – are engaged and disengaged by a pedal. In the early days, various kinds of gearboxes were experimented with, but the one generally adopted was of the sliding-pinion type, manually operated by a lever. To effect a smooth change, the driver had to master the technique of double declutching in order to get the two shafts rotating at roughly the same speed: failure to do so resulted in a hideous grating noise. This problem was removed, at least for the more expensive cars, when General Motors introduced the synchromesh gearbox in 1929.

By 1905, standard practice was for the engine to be mounted at the front of the car and power to be transmitted to the rear wheels. To bridge this gap, some early cars had belt or chain drives – and for a long time this was used on motorcycles – but, by 1910, the propeller shaft passing underneath the chassis was almost universal, though in Britain Trojan continued with the chain drive until the 1920s. The alternative was to have front-wheel drive, but this presented engineering problems, as these wheels had also to be steerable. However, Citroën in France introduced this system in 1934 (together with integral chassis and body), as did some other European manufacturers, but it was not generally adopted in the USA until after the Second World War.

The Panhard-Levassor motor-car of 1890 was essentially a dogcart fitted with a petrol engine: it was not incongruous, therefore, that braking should be effected by wooden shoes applied to the solid rubber tyre, exactly as in the horse-drawn vehicles of the day. Very soon, however, more sophisticated brakes appeared, in which either a belt tightened on a drum or shoes were applied via a foot pedal to the inner surfaces of drums attached to the wheels. Until the 1920s, two-wheel brakes, applied at the rear, were normal, but four-wheel brakes then became general – though the Dutch firm of Spyker used them in 1903. This introduced problems of even application to all four wheels, without which skidding was a hazard. The wire brake cables in general use tended to stretch unevenly and needed constant adjustment; for the more expensive cars this system was replaced by a hydraulic one, ensuring that the same force was transmitted to all four brakes. Although Frederick Lanchester experimented with disc brakes as early as 1902, they were not common until the 1950s: in this type, two pads grip a rotating disc.

In 1900, metalled roads were rare outside urban areas and the early cars had to face rough surfaces. Some form of springing was, therefore, essential and initially this consisted simply of leaf springs of the kind used for carriages and railway locomotives. Later, coiled springs were used and, on a few models even before the Second World War, torsion bars. In the British 'Mini', one of the most successful of all post-war cars, produced in millions from 1959, shocks were absorbed by the twisting of rubber.

Again in the tradition of the carriage, very early cars had wooden wheels with steel tyres, but solid rubber tyres were

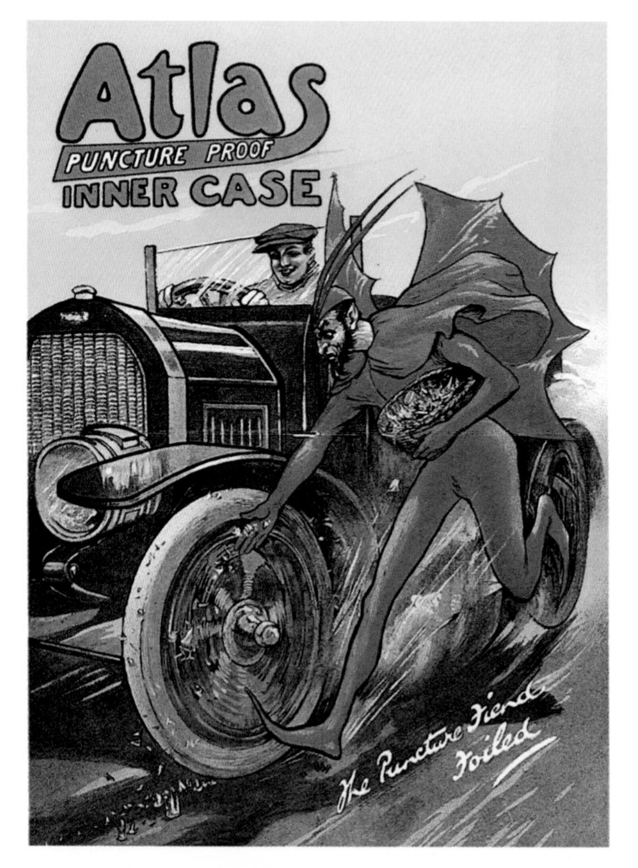

Left: Almost from the outset automobiles had pneumatic tyres. Punctures were a constant hazard, and manufacturers vied with each other in their claims.

Above: Citroën cars have always been noted for the elegance of their design. This 7CV of 1936 was noted also for being the first to incorporate front-wheel drive.

The Model T Ford was perhaps the most famous of all cars. 15 million were eventually made. This 1913 model (left) was built for the British market. A major factor in Ford's success was his introduction of the assembly-line system of manufacture (above).

development of petrol engine 167–168 **diesel engine** 168, 169 **invention of pneumatic tyre** 247 **workings of petrol engine** 251

soon introduced and continued in use until the late 1920s on heavy goods vehicles. As early as 1895, however, Peugeot began to fit pneumatic tyres, which by then were in regular use on bicycles. These gave a smoother ride but were very prone to punctures. In 1905, tougher rubber was made by formulating it with carbon black and, from 1910, they were reinforced with cord. In Germany, striving for a self-sufficient economy, synthetic rubber appeared in the 1930s.

Commercial vehicles

What has been said so far applies largely to private cars, but commercial vehicles have almost as long a history. For the carriage of passengers, the motor omnibus appeared at the turn of the century and soon drove the horse-drawn buses from the road: in London, for example, the motor-bus had completely replaced horses by 1914. For carrying goods, motor lorries could not at first compete with the railways over any considerable distances, but they were soon adopted for local use – including use by the railway companies for distribution from their freight termini. But in the inter-war years, with better roads and more efficient vehicles, big lorries were making long hauls. In the 1930s, 15-tonners were common, and in these additional axles were introduced to spread the load. After the Second World War, with the development of an extensive system of motorways, much better vehicles appeared, ranging up to 32.5 tonnes in the 1980s and some having as many as five axles.

The choice of fuel for commercial vehicles depends on several factors. Diesel engines are cheaper, and more economical in fuel consumption, than an equivalent petrol engine. Moreover, being slow-running they require less frequent overhaul. Against this, they are relatively much heavier in proportion to their power, but this weight factor naturally becomes less important the greater the overall weight of the vehicle and its load. Where these factors balance each other depends on local circumstances. The cost to the consumer of petrol and fuel oil depends not so much on production costs as on the tax levied on them, and this is subject to arbitrary fiscal control. Small commercial vehicles usually run on petrol, but for heavy lorries diesel engines became widely used in the 1920s.

Virtually all internal combustion engines used on cars are of the reciprocating type: that is, the pistons work up and down in a cylinder. In principle, continuous rotary motion is preferable, and this is achieved in the type of engine devised by Felix Wankel, a German engineer, in 1956. Although developed by NSU in Germany and Mazda in Japan, it presents technical problems which have not yet been fully overcome.

Steam deserves passing mention here in that, as we have already noted in an earlier chapter, the first mechanical road vehicles were steam driven. They could not, however, survive the competition of the internal combustion engine, even though – like steam locomotives – they avoided the need for a gearbox. A few sophisticated steam cars were made in the USA up to 1932 and steam traction engines were to be seen even later drawing very heavy trailers. For practical purposes, however, steam had no place in the road transport of the 20th century.

At mid-century, the petrol-fuelled road vehicle had not only created a vast new industry but effected a social revolution, giving the individual a freedom of movement hitherto undreamt of. The centre of this revolution was the USA. In 1950, world production of motor vehicles was almost 10 million: of those on the road 70 per cent were in the USA, which then had no more than 7 per cent of the world's population.

Above: By the 1960s commercial vehicles had become far larger, with as many as five axles to distribute an overall weight of over 30 tonnes.

Left: The motor-bus quickly ousted the horse-drawn v[illegible]cle in the first decade of this century. This example, with solid rubber tyres, was photographed at Hindhead, Surrey, in 1906.

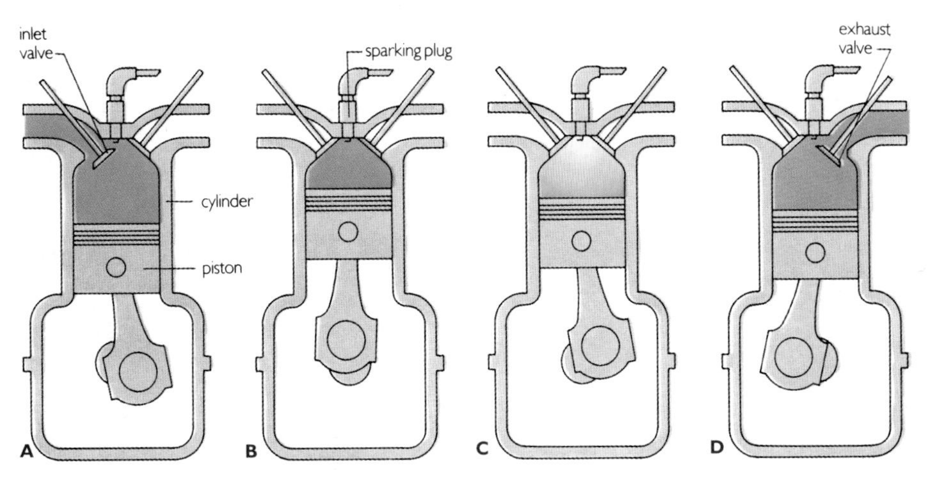

The four-stroke cycle in a petrol engine begins with the induction stroke (A), in which the inlet valve opens as the piston descends, admitting the fuel/air mixture to the cylinder. The compression stroke (B) occurs as the piston rises, closing the valve and compressing the mixture. Ignition follows, producing the power stroke (C), in which the piston again descends. On the exhaust stroke (D), the exhaust valve opens and the piston rises to expel the exhaust gases.

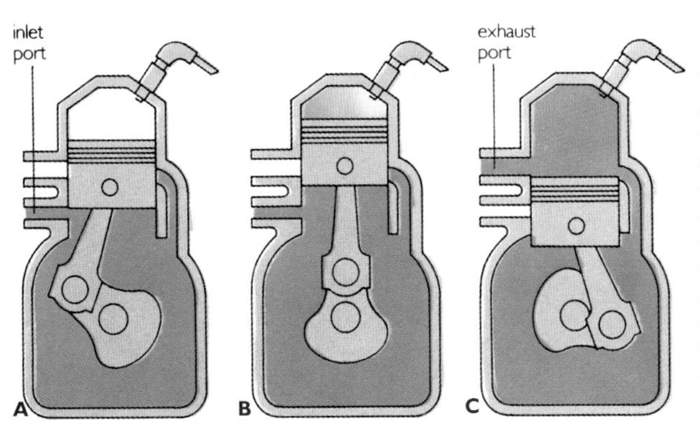

The two-stroke cycle in a petrol engine (left) commences with the upstroke (A), in which the fuel/air mixture in the cylinder is compressed as the piston rises and more mixture enters through the inlet port below the piston. Ignition (B) then occurs, forcing the piston down. On the downstroke (C), the exhaust gases leave through the exhaust port while the fresh fuel/air mixture enters the cylinder from below the piston.

The Wankel engine (below) has a triangular rotor instead of a piston. It has gas-tight seals at its edges and rotates inside a chamber so that three separate compartments are produced between the rotor and chamber walls. In each compartment, induction (A), compression (B), ignition by two sparking plugs (C) and exhaust (D) occur in a four-stage sequence. The rotor drives directly without a crankshaft.

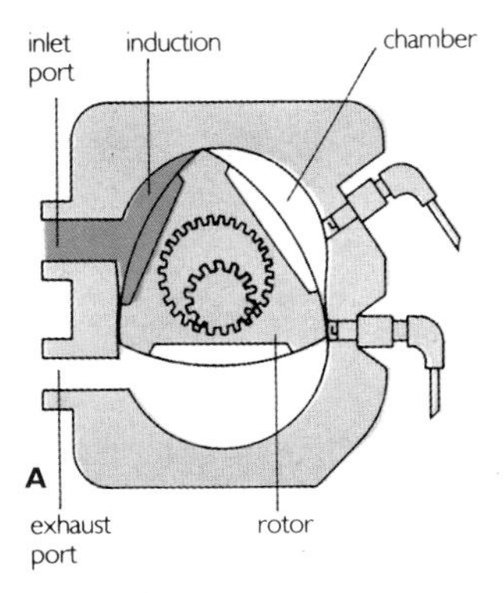

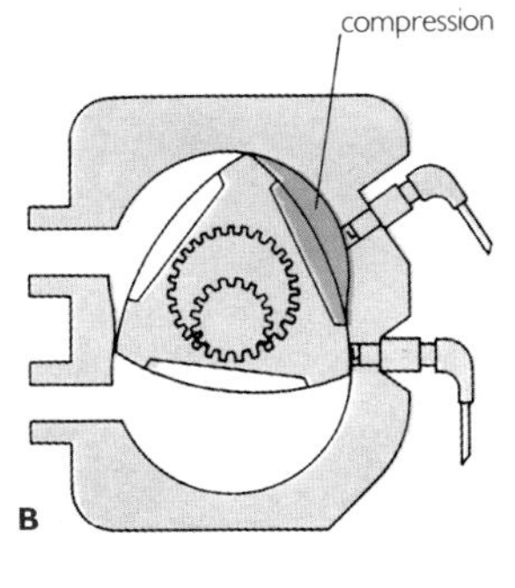

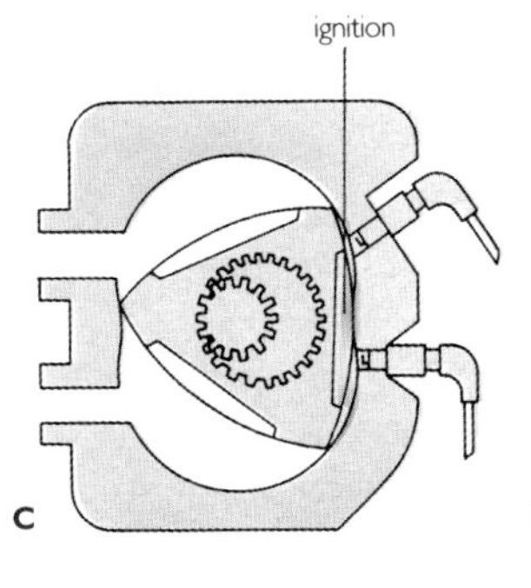

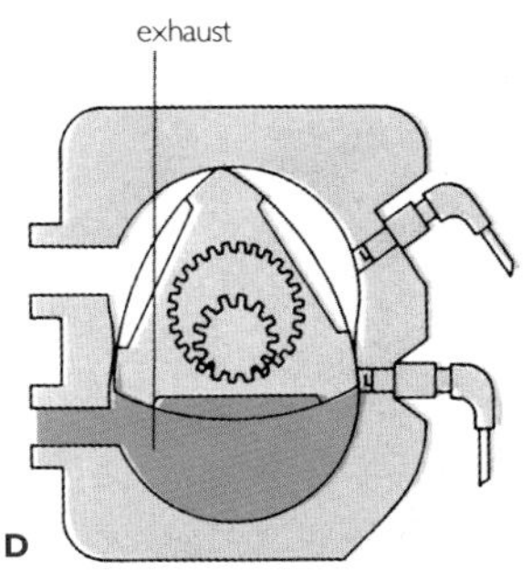

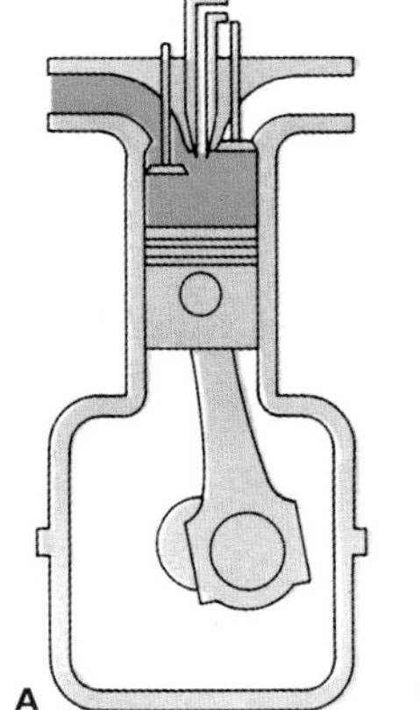

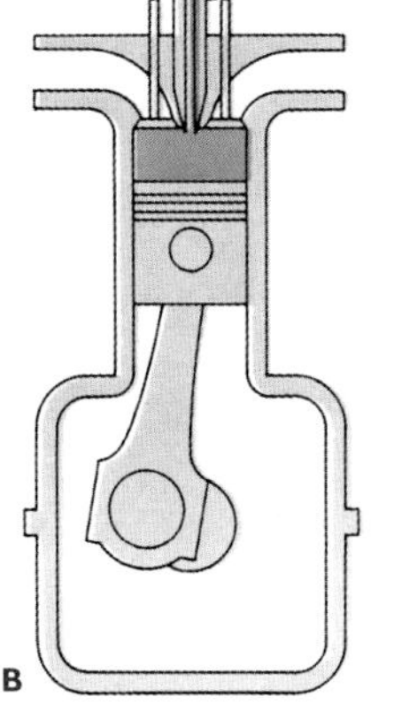

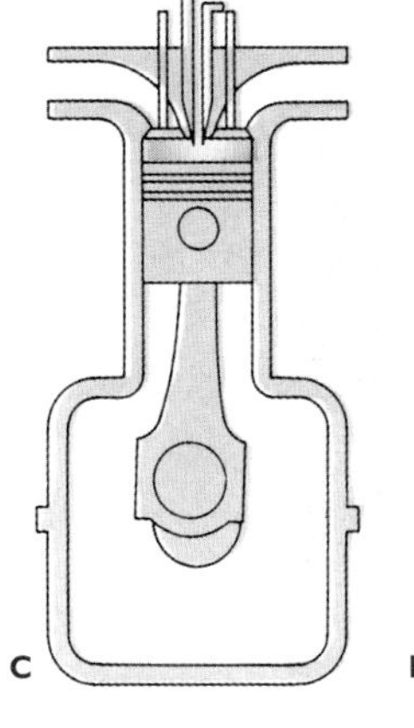

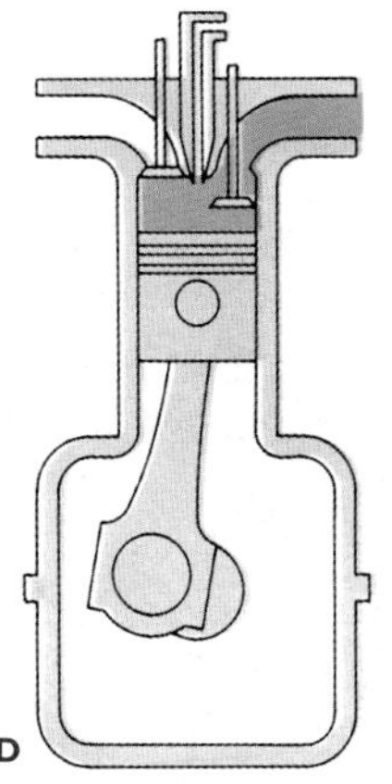

In a four-stroke diesel engine, air enters the cylinder through the inlet valve on the induction stroke (A). The piston rises and both valves close on the compression stroke (B), compressing the air and heating it as fuel is injected into the cylinder. The fuel ignites (C) without the need of a sparking plug, forcing the piston down on the power stroke. The piston then rises on the exhaust stroke (D) to expel the exhaust gases through the exhaust valve.

Although electrically propelled road vehicles have been in use for nearly a century for commercial purposes, they virtually disappeared from private motoring after the First World War. This picture shows a fashionable Baker electric runabout in the USA in 1916.

Electric vehicles

Although also very much in the minority, electric vehicles have a continuous history from the early 1890s. They are of two kinds: those that draw their power from batteries and those that are supplied with electricity from an external source. Up to the First World War, battery-powered electric vehicles were quite popular as local runabouts and taxis, but the operative word is local: only relatively short runs were possible between recharges. From the 1920s until the present time electric delivery vehicles – easy to stop and start in the absence of a gearbox – have found a use on short runs with very frequent stops, as in milk delivery to private households or small local buses.

As we have seen, steam-powered railways evolved from colliery tramways using horses, and horse-drawn trams appeared as public service vehicles. They were introduced by Enoch Train in New York in the 1830s and in 1860–1 he extended their activities across the Atlantic to Birkenhead and London. Electric traction in place of horses had an obvious appeal; for such large vehicles batteries were not practicable, but an external supply – via centre rail or overhead cable – was, however, quite feasible. From the 1890s, electric trams became an important part of the world's public transport systems in urban areas. Although some are still in use, as in Hong Kong, many were discontinued after the Second World War as they were incompatible with heavy motor traffic. An alternative to the tram was the trolley-bus: this, too, derived its power from an overhead cable, but was steered by the driver and not confined to rails. They appeared first in Germany in 1882 and were gradually introduced elsewhere in Europe and in other parts of the world. By mid-century many thousands were in use, but now they have virtually disappeared, victims – like trams – of urban traffic congestion.

In the context of electric vehicles the cable-car deserves passing mention. Under this heading come both surface-bound vehicles, such as the famous cable-cars of San Francisco and the aerial cars familiar in mountainous countries. The popularity of the latter dates from 1889, when a cable-car lift was built to the top of Vesuvius.

Railways

Tramways are in effect light railways and thus can be regarded as an extension of the railway system, in the aggregate amounting to many thousands of kilometres. For the railways proper, however, the great period of growth was past by the turn of the century. Thereafter, in the developed parts of the world, expansion tended to be a matter of business organization – a multiplicity of small companies being merged within a relatively few large ones – rather than physical extension of the network. Thus Britain, France, Germany, Italy and the USA had substantially the same railway mileage in 1960 as they had in 1900: collectively 770,000 kilometres (480,000 miles) in 1960 compared with 710,000 kilometres (440,000 miles) at the beginning of the century. In developing countries, however, the situation was very different. China, for example, had no railways at all until 1883: in the 1980s, the Chinese railway network extended to 50,000 kilometres (31,000 miles), more than half constructed since 1949 – and it is still growing. Russia embarked on a great railway expansion programme after the Revolution, culminating in the 3000-kilometre (1964-mile) Baikal-Amur Mainline opened in 1984 as a strategic alternative to the Trans-Siberian Railway of 1903, which passes uncomfortably close to the Chinese frontier. Today, Russia is unusual among industrialized nations for its high dependence on railways for the transport of both passengers and freight: elsewhere road transport has made heavy inroads into railway traffic.

The 19th century epitomized the Age of Steam, but in the 20th there was increasing competition from electric and diesel traction. Steam, however, continued to evolve, and faster and more powerful locomotives appeared. One such was the British *Mallard*, which set a world record for steam of 202 kilometres per hour (126 miles per hour) in 1938: in power, the ultimate was reached in the USA, with the mammoth 600-tonne Union Pacific 4000s of 1941, each with sixteen 1.7-metre (6½-foot) driving wheels. While there are still plenty of steam locomotives around the world, rather few are now being built.

For underground railways the advantages of electric traction were obvious, as they were in countries like Switzerland and Sweden, with cheap hydroelectric power. Elsewhere, however, the technical problems of main-line electrification and the continuing abundance of cheap coal ensured that progress was slow. Technically, the choice was between low voltage d.c. picked up by a shoe from a third rail and high-voltage a.c. obtained via a pantograph from an overhead cable. The latter eventually carried the day, especially after French engineers demonstrated in the 1950s that the ordinary commercial frequency of 50 hertz, supplied at 25,000 volts, could be satisfactorily used.

In spite of difficulties, electrification had established a foothold even before the First World War. By then most of the European countries had some electrified track: in the USA, the 112 kilometres (71 miles) from New York to New Haven began to be electrified in 1907. Between the wars most countries expanded their systems, especially France and Italy.

By comparison, the internal combustion engine entered the railway domain as a rival to steam quite late in the day. As with electrification, there were economic and technical difficulties. With their immense investment in steam and the continuing availability of cheap coal, in most countries, the incentive to explore any alternative was not great. Additionally, both steam and electricity had the advantage of requiring no gearbox. With the diesel engine, however, this was essential and the construction of gearboxes capable of handling the heavy power loads involved was not easy. The kind of gearbox so satisfactorily developed for cars could not simply be scaled up, and the most widely used alternative was an electric system avoiding a gearbox. In effect, the diesel was used to generate electricity for traction.

Although Sweden had a diesel–electric railcar as early as 1913, the diesel engine made little impact until the 1930s. Then, in 1935, General Motors made a major advance in the USA with their 567 diesel locomotives, which had a far better power:weight ratio than any previously available. Rated at 1800 horsepower, they were much less powerful than many of the steam locomotives in general use, but this limitation was deliberate. An intrinsic quality of the diesel engine – as opposed to the steam engine – is that it is at its most efficient when working at maximum load. The General Motors' units were designed for multiple use: they could be coupled together to provide exactly the amount of power required for the task in hand.

Above: After the 1917 Revolution, Russia's industrial programme included an expansion of the railway system. This poster by Vladimir Mayakowski was part of a national promotion campaign.

Left: This 160 kilometres per hour (100 miles per hour) M 10,000 three-car articulated unit for the Union Pacific was completed in 1934. After a publicity tour of 68 cities – including President Roosevelt as a passenger – it ended up at the Century of Progress Exhibition at Chicago.

Left: Turbine propulsion was adopted for many of the luxury transatlantic liners built before the First World War. Here the Cunarder *Mauretania* (1906) is seen alongside the diminutive *Turbinia*, the first turbine-driven ship, which caused a sensation at the Spithead Review in 1897.

Below: After the Second World War, the great liners were far exceeded in size by giant oil-tankers – such as *Tactic*, illustrated here – the largest of which exceeded 500,000 tonnes.

Shipping

In shipping too, until after the turn of the century, steam reigned supreme, though it developed along two separate paths. The triple or quadruple expansion piston engine, working with superheated high-pressure steam was the ultimate achievement along traditional lines. Its main rival, though still a steam engine, was the turbine. The latter was more expensive to build and run, but its greater smoothness and flexibility and its favourable power:weight ratio made it acceptable for warships, where economy was not the first consideration. For similar reasons, it was attractive also for the big passenger liners. Cunard made a first trial in 1904 and subsequently adopted the turbine for such famous pre-war ships as the *Mauretania* and *Aquitania*. Until the Second World War, however, the real rival to steam for smaller vessels was the diesel engine. Although installed in a French canal-boat in 1902 and a Russian tanker in 1904, the first large vessel to be so equipped was the Danish *Selandia* of 1912. Nevertheless, the diesel engine was not widely used at sea until the 1920s, an important example then being the 18,000-tonne Swedish liner *Gripsholm* built in 1925. However, well before this, oil was increasingly being used not as a fuel for diesel engines but as a substitute for coal for raising steam, its advantage being its cleanliness and greater ease of handling.

By 1900, steel had almost completely replaced iron in shipbuilding, but in either case units were assembled by riveting. In the early 20th century, however, welding was increasingly used: this was not only quicker but also more economical of material, as butt joints could be used as opposed to the overlaps necessary for riveting. For warships even small savings in weight – such as could be achieved in this way – became significant after the London Naval Convention of 1936. Two methods were used: in arc welding the heat of a powerful electric arc is used to fuse the metal, whereas in oxy-acetylene welding it is an intensely hot flame.

Another important change in methods of construction was the increasing use of prefabricated parts: by 1939, units weighing up to 200 tonnes were common. Assembling these required more space than the traditional method of attaching individual sheets to the frame of the ship, and such space was not always available at shipyards in Britain, which in 1939 was still building more than 30 per cent of the world's shipping. This was one reason for the post-war success of the more modern and spacious shipyards of Japan and Scandinavia.

Another major development was the size of ships. In 1900, the typical seagoing vessel was the general-purpose tramp steamer – so called because it did not have a fixed itinerary but picked up cargoes as and when it could – with a deadweight of around 10,000 tonnes. Although the ill-fated *Grand Eastern* had been surpassed in tonnage in the 1880s, her length of 211 metres (692 feet) was not exceeded until the German *Kaiser Wilhelm der Grosse* of 1897. This was built for a duel-purpose passenger trade: a high standard of

Train ferries were a specialist link between transport by land and sea. The *Drottning Victoria*, built in 1909, connected Denmark and Germany through the ports of Trelleborg and Sassnitz.

comfort for the growing numbers of well-to-do transatlantic passengers and an austere steerage class for poor European emigrants who congregated at her home port of Hamburg. Similar British liners sailed from Liverpool.

The *Kaiser Wilhelm* was powered by enormous quadruple expansion steam engines, but soon the giants of the Atlantic, in fierce competition for the lucrative trade, were equipped with turbines. The largest ship built before the Second World War was the 84,000-tonne Cunard liner *Queen Elizabeth*, completed in 1939: she went straight into service as a troopship and not until 1946 was she commissioned for her original purpose. But by then, she and the other Atlantic giants were already obsolescent: from 1957, more passengers flew across the Atlantic than went by sea and, by 1970, the airlines had virtually a monopoly of the trade.

The association between ships and oil was not limited to propulsion. By 1939, world production of petroleum had reached 250 million tonnes annually and much of this had to be moved by sea. This led to the development of tankers built specially for this trade, though some could carry other cargoes as well, such as iron ore. Up to the Second World War they rarely exceeded 10,000 tonnes, but later they were far larger, several times greater than the largest of the liners. In 1981, the *Seawise Giant*, 485 metres (1590 feet) long, was the first ship to exceed half-a-billion tonnes deadweight. However, it seems possible that such vessels will prove the dinosaurs of the shipping world: with a changing pattern of supply and demand in the wake of the OPEC crisis, there is likely to be a return to smaller, though still large, tankers capable of traversing the Suez and Panama canals.

Another specialist craft, of considerable importance in some areas, was the train ferry, by which a whole train could be transported across water from one railway system to another. The first was used on Lake Constance in the latter part of the 19th century – and others plied on the Great Lakes of America – but the construction of many later vessels and their port facilities was complicated by the fact that they were intended for tidal waters and some means – usually a hinged ramp – had to be found to compensate for the rise and fall of the ship. In the early 20th century, a number of ferries were built to connect the Danish mainland with offshore islands. In Britain, a Dover–Dunkirk service was launched in 1934, giving passengers a through service between London and Paris.

Train ferries often also carried cars, and after the Second World War the explosive growth of car ownership, a corresponding demand for holidays abroad and the growth of fast motorway systems made this a big business. The slow and clumsy technique of hoisting cars individually aboard on pallets or in slings was replaced by a drive-on drive-off system modelled on the tank landing craft of the Second World War: by this means it was possible also to accommodate large lorries. Most car ferries, such as those that cross the English Channel and the North Sea, are quite small, carrying perhaps 300 cars, 50 lorries and 1000 passengers. For long runs a few larger ferries have been built, such as the 25,000-tonne *Finnejet* linking Helsinki and Travemünde in the Federal German Republic, an overall distance of 1200 kilometres (745 miles). After the Second World War hovercraft, too, were adopted as car ferries for short crossings, for which they are very suitable because of the ease with which they can be loaded and unloaded.

Until aircraft established a dominant position in the passenger transport industry, rival shipping companies competed fiercely to provide the greatest comfort, the best cuisine and the fastest service. But all this was little consolation to the many passengers stricken with sea sickness, and it is not surprising that much attention was paid to devices which might counter the pitching and rolling which causes it. The problem of pitching is virtually insoluble, in that if a ship does not lift to an oncoming wave it would sweep over her. Rolling, however, is a different matter. In the First World War experiments were made with huge stabilizing gyroscopes, but the most satisfactory solution to date is the underwater fins introduced by the Japanese: these are automatically extended or retracted as needed at the command of small gyroscopes responding to the ship's motion.

development of shipping 50–53, 105–113, 172–176 ***Great Eastern*** 147, 173, 175 **hovercraft** 331 **steam turbine** 164–165

The Junkers F13 – which made the first east–west crossing of the Atlantic by air in April 1928 – was the workhorse of the Lufthansa air fleet between 1926 and 1932. It was a strut-free, all-metal, low-wing monoplane carrying four passengers in a closed, heated cabin.

Aeronautics

The development of aircraft has always been so closely identified with their military use that their early history has already been discussed in this context. We must, therefore, now pick up the threads and consider how civil aviation developed in the aftermath of war. How enormous were the effects of four years of hostilities is indicated by the bare statistics. In 1914, there were in existence probably no more than 5000 aeroplanes, many of them very primitive. By the end of 1918, the major combatants had between them built well over 200,000, many of the later ones being of quite a sophisticated nature. Understandably, the first move was to adopt some of these surplus military aircraft, available quite cheaply, for civilian use: not for 10 years were new aircraft designed and built specifically for this purpose. In this field, a strong lead was, almost inevitably, given by Germany. She had much experience and skill in aircraft design, but was forbidden by the Treaty of Versailles to build military machines – a restriction which, incidentally, encouraged her to develop gliders as a good means of training future pilots.

Airships

From this point onwards the history of aviation is concerned almost entirely with heavier-than-air machines, but in the inter-war years there were still those who had faith in the future of airships. The most enthusiastic were the Germans, who built in all about 160 craft of the Zeppelin type. After the war they established a few commercial services with 'flying hotels' to South America and the United States, but some major disasters in the 1930s – notably the loss of the British R101 in 1930 and of the huge 250-metre (820-feet) Hindenberg in 1937 – severely shook public confidence. As a means of public transport the airship was finished, though a few are still used for special purposes, such as coastguard patrol.

Aircraft construction

Until the end of the First World War, the typical aeroplane was a biplane with a wooden frame covered in 'doped' (varnished) fabric, the cockpit was closed and there was a single engine. A few of the larger bombers – with wing spans up to 40 metres (130 feet) – had two engines. One important development was to make the frame of welded steel tube and the cladding of duralumin. The latter was a light alloy developed in Germany before the war by Alfred Wilm, who accidentally discovered that aluminium containing a little magnesium and copper gradually becomes very strong if it is heated and then quenched in water, a process known as age-hardening. It was used in the building of Zeppelins, but only in the last year of the war did Hugo Junkers' company use it for aircraft. The Junkers F13 of 1919 was the first civil aircraft to be built in this way, but the duralumin was still only a cladding. A more significant development, early in the 1920s, was to stress the cladding so that, as an integral part of the machine, it contributed to the strength of the wings and fuselage: this was comparable with the combination of chassis and body in the automobile industry. For commercial aircraft, this form of construction quickly became universal, but for the tens of thousands of one or two-seater light aircraft built for amateur fliers the fabric cover remained popular. This outlet was very important for the aircraft industry. One of the earliest was the British Moth introduced by de Havilland in 1925, selling for the remarkably low price of £650. Of these 2000 were built, as well as nearly 10,000 of its successor, the Tiger Moth.

By the 1930s, civilian aircraft had assumed something like their modern form, though far smaller: in particular there

The Douglas DC-3 was the leading American transport aircraft of the 1930s. Its control panel, shown here, was very sophisticated by the standards of the day.

was a comfortably appointed cabin for passengers. Very typical was the highly successful two-engined American DC-3 – made in thousands from 1935 – which could seat up to 30 passengers. By that time flying was still a rare experience, but the needs of passengers began to be seriously studied. One problem was analogous to the pitching and tossing of ships: in bad weather flights could be not only uncomfortable, and even frightening, but also cause sickness. This could be mitigated by flying higher – which also gave better fuel economy – but then passengers suffered the equivalent of mountain sickness through lack of oxygen. The answer to this was the pressurized cabin, first introduced on the Lockheed XC-35 in 1937, closely followed by the Boeing Stratoliner in 1939. These aircraft were designed to fly at 6000 metres (20,000 feet) or higher, twice the height of their predecessors.

The French Gnome rotary engine was widely used in Allied fighter aircraft during the First World War.

Power units

For all flying machines – whether lighter or heavier than air – weight is of paramount importance, and the engine is the heaviest unit involved in construction. Indeed, as George Cayley perceived, powered flight could not be seriously contemplated until a suitable lightweight engine was available. Until the advent of jet-propelled aircraft – the first of which was airborne in Germany only one week before the outbreak of the Second World War – the standard engine was a piston engine working on the Otto cycle.

The first Otto engines of the 1880s were rated at about 200 kilograms (440 pounds) per unit of horsepower, but for the Wrights' first flight in 1903 this had been reduced to about 6 kilograms (13 pounds) per unit horsepower. In the years immediately before the First World War, the centre of interest in flying moved to France and there the Gnome, designed by Laurent Seguin and rated at 1.5 kilograms (3.3 pounds) per unit horsepower, was the most successful. Thereafter, better rating was increasingly hard to achieve. The V-12 Liberty engines, built in thousands in the USA during the First World War, reduced the rating to around 1 kilogram (2 pounds) per unit horsepower, and 25 years later the Wright Cyclone developed for B29 bombers in 1944 cut this to 0.5 kilograms (1 pound) per unit horsepower. However, for commercial aircraft the future lay with jet propulsion, though the internal combustion engine remained unchallenged for small private aircraft.

The last generation of aircraft piston engines were of three kinds. The first were water-cooled linear engines modelled on the contemporary automobile engine. Secondly, there were two kinds of air-cooled engine of quite different design – these were respectively the rotary and the radial. In the rotary type, such as the Gnome, the whole engine carrying the propeller, rotated about a fixed crankshaft. The radial engine was nearer to the linear type, but the cylinders were arranged in a circle around the crankshaft. For larger units the latter prevailed and reached its ultimate development in the Wright Cyclone. This had 28 cylinders arranged in four circles of seven; it developed 2200 horsepower.

The air transport industry

By the end of the First World War, commercial air transport was technologically feasible, and in 1919 the first crossing of the Atlantic by John Alcock and Arthur Whitten Brown emphasized that more than local services were possible. Although technological problems abounded – not only in relation to aircraft but also to all their related facilities, such as airports and navigation – the real difficulty was economic. Operating costs were high and the nascent air transport industry had little to offer, save speed, in competition with the firmly established international railway and shipping networks.

However, there were more than strictly economic factors to be considered. The military importance of the aeroplane had been forcibly demonstrated in the First World War and no government was in any doubt that it would be far more important in any future conflict – the likelihood of which became more apparent as the years went by. There was, therefore, a strong incentive to encourage both the aircraft manufacturing industry and commercial flying as incidental to military programmes, and the principal European governments were prepared to subsidize their own national airlines to get them established. One of the first was the Dutch KLM founded in 1919; in Britain, Imperial Airways was created in 1924.

But by then the lead had once again been taken by the USA, where flying had started 20 years earlier. There four major companies had emerged as domestic operators and Pan American as an overseas one operating in all parts of the world in competition with Europe.

balloons and airships 171–172 **first aerial warfare** 219–222 **invention of aircraft** 219–221 **modern airlines** 328–329 **Otto cycle** 166–167

CHAPTER TWENTY-SEVEN

New Building Techniques

Throughout history the building industry has in general been highly conservative and in many ways 20th-century building has proved no exception. If we think in terms of various structures – houses, shops, other small commercial buildings and the like – the materials and principles were very much the same as in the 19th century or, indeed, a great deal earlier. There was some mechanization in methods, with motorized cement mixers, pile-drivers, hoists to take materials to upper storeys, and, from the 1930s, bulldozers to shift soil quickly; some degree of prefabrication of basic units such as roof trusses, windows and stairs; and bolted steel scaffolding replaced lashed wooden poles. Nevertheless, even in the industrialized countries much everyday building was carried out by methods which differed very little from those of medieval craftsmen, and in the remoter parts of the world these prevailed unchanged.

Yet to say that building changed very little is by no means to imply that there were no important developments at all. As in the past, however, these found expression mainly in large public buildings, within which we may now include very large commercial buildings, such as the prestigious headquarters erected by the growing number of great national and international corporations. Here, there were not only changes in design, reflected in the appearance, but also new materials and new techniques for distributing stresses to give the necessary strength and safety. There is, of course, nothing surprising in this, for it is only for major building projects that money is available to take full advantage of architectural knowledge and ingenuity and of the practical skill of contractors to translate design into reality.

Unlike some aspects of technology – such as wireless communication, electrical supply, the car and aviation – the turn of the century was of no particular significance in the history of building. In the main, the first half of the 20th century was a period of evolution: virtually the only innovation – but an exceedingly important one – was the introduction of pre-stressed concrete in the 1920s.

Growing use of iron and steel

The use of iron as a reinforcing agent dates from classical times; as it became cheaper and more readily available from the latter part of the 18th century this use naturally increased. We have seen, for example, how the timberwork of heavy machinery was increasingly replaced by iron. Civil as well as mechanical engineers made use of cast iron. The first iron bridge in the world spanned the Severn at Coalbrookdale in 1779; the main structural units were 20 metres (66 feet) long and the total weight of the castings was nearly 400 tonnes. Elsewhere in Britain, iron appealed at the same time to the owners of textile mills, where naked flames were used for illumination and the risk of fire was high. However, although iron does not burn, it loses its mechanical strength when strongly heated. It followed, therefore, that in iron-framed buildings masonry replaced timber as far as possible for floors and partitions: if timber had to be used, it was as far as possible enclosed in nonflammable materials.

Factories are, of course, strictly utilitarian buildings and appearance is a secondary consideration – though this did not prevent some early 19th-century industrialists modelling their factories on great country mansions. The use of iron as a deliberately visible feature of the architectural design really dates from the second quarter of the 19th century. Important early examples are the Coal Exchange in London (1847); the domed Reading Room of the British Museum (1857); and a number built by James Bogardus at about the same time in the USA. Even church architects experimented with iron. Thomas Rickman designed three churches in Liverpool which were built between 1813 and 1818: the framework

Methods in the building industry changed little in the 19th century, though some mechanization was introduced – as shown in the use of a steam crane during the building of bridge caissons on the Whitby–Scarborough Railway in 1882.

was of iron, and the infilling of grey Welsh slate. The effect was sombre and the architect is said to have repented of his design; however, this did not deter St John's College, Cambridge, from commissioning him to design their New Court in traditional style, in 1826. A far more successful venture into the ecclesiastical field is the elegant cast-iron spire of Rouen Cathedral, completed in 1876 after more than 50 years of work. But perhaps the world's most famous iron building is the great 300-metre (984-foot) tower in Paris designed by Gustave Eiffel for the Paris Exhibition of 1889.

The Eiffel Tower marked the end of constructional work in iron, for by that time steel had become generally available as an alternative. From 1890 in the USA, and from about five years later in Europe, steel-framed buildings replaced iron ones. The availability of large steel ingots meant that heavy flanged girders could be rolled, in contrast to the original casting of iron members. It was the steel frame that gave structural strength: such buildings were therefore not dissimilar to the old medieval timber-frame buildings. The walls were no longer load-bearing but simply a weatherproof curtain – sometimes made even of glass – so that they could be built simultaneously at all levels. It is interesting to note, however, that technological advance was often impeded by bureaucratic regulation, just as the adoption of the car was hindered by the Red Flag Act restricting speed to walking pace. Not until 1909 did London's building regulations reduce the statutory minimum thickness of the outer walls, regardless of whether they were load-bearing or not.

In the great new cities of the USA and in some central city sites in Europe, where land values were particularly high, there was a strong incentive towards multi-storey steel-framed buildings. One of the earliest was the 21-storey Masonic Building in Chicago, completed in 1892. The best-known steel-framed skyscraper – though no longer the tallest – is perhaps the 102-storey Empire State Building in New York: completed in 1931, it towers to 385 metres (1263 feet), extended to 450 metres (1476 feet) by the addition of a television mast in 1951.

Such skyscrapers, as they came to be called, created two complementary problems. One was that of water supply, for the local mains pressure could not service the upper storeys: it was therefore necessary to provide pump-fed cisterns at various levels. The second was that the occupants could not be made dependent on stairs to reach any but the lowest levels: it was, therefore, essential to provide lifts, originally of the hydraulic type devised by Elisha Otis in 1854. An important feature of this was the system of self-locking pawls which would automatically stop the lift in the event of the cable breaking. In 1889 electric motors were introduced and in 1903 the system of counterweights now in common use. As the heights of the buildings grew, express lifts operating at up to 500 metres (1640 feet) per minute were introduced.

Top: The Eiffel Tower, built for the Paris Exhibition of 1889 commemorating the French Revolution, was the last great building constructed of iron; thereafter steel was increasingly used.

Above: Elevators for goods, often hydraulically operated, appeared about 1850, but passenger elevators were rare before the advent of reliable electric motors. This lift at Salzburg was built by Siemens in 1890.

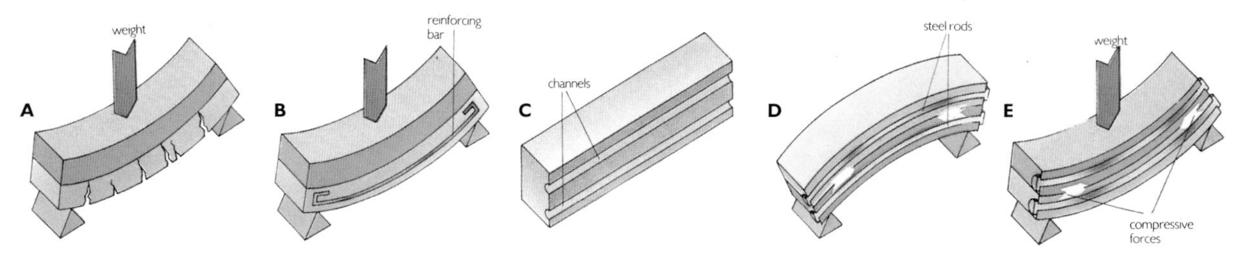

Ordinary concrete fails under load with tension caused by bending (A). Reinforcing with steel bars (B) gives strength in tension. Post-stressed concrete (C) has channels for steel rods that are then placed under tension, usually by tightening nuts at either end, compressing the concrete and giving it strength when bent in any direction (D, E).

A major use for reinforced concrete was in the building of dams for hydroelectric and irrigation schemes, as in this installation at Itaituba on the borders of Brazil and Paraguay (right). It is one of the largest in the world, as indicated by this view (below right) from the Paraguayan side.

For short rises, generally not more than 30 metres (100 feet) – as in department stores or railway stations, moving stairways, or escalators, were introduced at the very end of the 19th century. They became popular after being demonstrated at the Paris Exhibition in 1900.

New uses for concrete

In the ancient world, the Romans made much use of a form of concrete known as pozzolana, made by mixing lime and volcanic ash. It was very similar to the kind of cement made by calcining chalk and clay, invented in 1824 by Joseph Aspidin, a builder in Wakefield, England: he named it Portland cement in the hope that it would be an effective substitute for the popular Portland stone, a form of limestone. This proved optimistic, for its appearance was not attractive; nevertheless it was widely adopted by builders and civil engineers because of its cheapness and ease of working.

Mixed with sand or shingle, cement can be easily cast to shape in moulds, but suffers from the same defect as natural stone: while it resists compression, it fractures easily under bending forces. In an attempt to overcome this, various trials were made, from 1849, of embedding iron rods in the concrete. The first to achieve a real measure of success was Joseph Monier, a French professional gardener, who took out a series of patents from 1867. In the 1890s, François Hennibique developed the technique further by incorporating not only longitudinal reinforcement but also iron hoops to resist shear. Reinforced concrete could be in precast units, and the iron reinforcement could be introduced as concrete was poured on the site.

The conferring of strength by using metals under tension is very old. Thus the wheelwright would fit an iron tyre as tightly as possible to a wooden wheel while it was red-hot and then quench it quickly with cold water. As the iron contracted, it gripped the wheel tightly as a result of being in a state of permanent stress. In 1928, the French engineer Eugène Freyssinet successfully applied this principle of prestressing to concrete. The reinforcing rods of high-tensile steel were kept under tension as the concrete was poured round them. When the concrete was set, and the rods immovably embedded, the ends of the rods were cut off: the resulting structural unit was thus strengthened by being kept permanently under compression. This made possible a great reduction in the amounts of both concrete and steel required to achieve a given strength and thus enabled architects to design lighter and more graceful buildings. Freyssinet also post-stressed concrete: in this technique the structural unit is cast with channels running through it. Steel rods are passed through these and then kept in permanent tension, usually secured in place with nuts.

Concrete in various forms was used for a number of constructional purposes, from farm buildings to the dams of major irrigation schemes. For the latter, the enormous quantities required were far beyond the scope of hand-mixing: small steam-powered mixers appeared in France in the 1850s, but by the 1930s major projects such as the Boulder and Grande Coulee Dams in the USA were each calling for many millions of tonnes of concrete.

While steel-framed buildings were essentially rectilinear, the strength and variability of reinforced concrete lent itself equally to graceful curves. Thus Freyssinet's aeroplane hangars at Orly, built in 1916, had an unprecedented semi-circular span of 75 metres (246 feet). The German-American architect Walter Gropius was a pioneer of elegant building in glass and concrete, such as in his school of design, the famous Bauhaus at Dessau (1926). Such buildings were revolutionary and by no means generally acclaimed in their day, but in the current era they are an accepted part of the architectural scene.

Possibly part of the prejudice was as much against the material as against the controversial new designs it permitted, for many strictly utilitarian concrete buildings – such as farm buildings and small factories – were remarkably unattractive. Their stark appearance was not helped by the widespread use of corrugated sheeting as a cheap roofing material. If sheeting is corrugated, it becomes very much less easy to bend across the ridges. In 1844 John Spencer made use of this fact in manufacturing the first corrugated iron roofing sheets. Originally, the corrugations were made individually, but with the availability of heavy rolling mills it became possible to corrugate large sheets in a single operation. From the beginning of this century, increasing use was made also of corrugated sheets of concrete reinforced with asbestos – a mineral fibre which had been prized since classical times because of its fireproof properties, but now regarded as a health hazard and banned in many countries.

Building for the needs of transport

Much of the building needs of transport in the first half of the 20th century were satisfied by the further development of techniques used much earlier for canals and railways. Such changes as there were, were dictated mainly by the greatly increased scale of operations and the appearance of some novel needs. The increased scale of operations had two main consequences. First, although pick-and-shovel techniques persist to this day, even for major works, in developing countries where labour is cheap, there was increasing use of power-driven machinery for the heaviest tasks. Secondly, the wealth generated by the new transport systems, and their national importance, made feasible civil engineering projects on a scale not previously contemplated.

The change to mechanization is exemplified by two great canal projects. The 160-kilometre (100-mile) Suez Canal begun in 1859 and opened 10 years later, was originally planned on the basis of forced labour, but in 1864 the Egyptian government put a stop to it. This obliged the construction company to use machinery, mainly mechanical dredgers, which deepened an originally shallow canal. In contrast, the Panama Canal – finally opened in 1915 and then the biggest single civil engineering work ever undertaken – was from the outset planned on the basis of intensive mechanization. In all, some 170 million cubic metres (6000 million cubic feet) of rock and soil had to be excavated using huge 100-tonne steam shovels unloading directly on to flat wagons on a specially constructed railway system. When the St Lawrence Seaway, linking Montreal with Lake Ontario, came to be built in 1954–8, civil engineering machinery had reached such a scale that 500 houses were moved bodily.

Walter Gropius was one of the pioneers of building in glass and concrete. This early example of his work (1911) is the Fagus factory in Alfeld, Germany.

The method used for constructing docks in the 20th century differed little from those already established, except that provision had to be made for much larger vessels. The most important development was, again, the transition from masonry to reinforced concrete around the turn of the century. An important port facility is the dry dock, in which work can be freely carried out on the exposed hull. Mostly these were land-bound, the water being pumped out after the vessel had been floated in and the lock gates closed. In addition, some very large floating dry docks were built which could be used away from ordinary port facilities: by 1939 several exceeding 350 metres (1150 feet) in length had been built and immediately after the Second World War floating docks capable of accommodating 50,000 tonne vessels were available.

Roads

The importance of maintaining a good network of metalled roads for both military and civil purposes was well recognized by the Romans, but not until the 17th century was road-building again taken seriously. In 1747, the *Ecole des Ponts et Chaussées* was set up to control some 40,000 kilometres (25,000 miles) of state-owned roads. Under Pierre Trésaguet

canals 63, 96, 152–153 **modern glass manufacture** 302 **road building** 153, 176 **Roman concrete** 55 **Roman roads** 63

As late as the 1930s much road-making was still extremely primitive, even in the technologically advanced countries of western Europe. This picture shows workmen laying asphalt in Germany in 1936.

good roads were built with a lower layer of large stones and an upper one of smaller ones well packed down. In Britain improved roads are particularly identified with John Loudon McAdam, who realized that ultimately it is the soil on which a road is laid that bears the load: if this is made firm and well drained, and the road surface is compacted to make it waterproof and cambered to throw off water, then a long-lasting highway can be achieved. From the 1830s such 'macadamized' roads were widely adopted in Europe, where they were almost universal for main roads by the end of the century. Although a short length was laid in the USA as early as 1832, that country had virtually no metalled roads even in 1900.

Such roads sufficed for horse-drawn traffic, with rather few vehicles, averaging no more than 15 kilometres per hour (9 miles per hour), but would not stand up to the fast, heavy motor traffic that developed after 1900. The first satisfactory solution was to provide a top dressing of tar or asphalt, coated with stone chips to give a rough surface to prevent skids. This proved additional waterproofing and avoided dust in dry weather, though in hot sun the surface could melt. A later alternative, especially favoured for the very heavy traffic of motorways, was again reinforced concrete. Although this had been experimented with in Europe in the 1890s, it did not begin to be extensively used in road-making until the 1930s, with the introduction of fast-setting mixes. One advantage of concrete is that it is very rigid, so that the weight of heavy vehicles is evenly spread: tar, by contrast, is slightly plastic.

From 1934, roads throughout the world began to incorporate one simple but very important safety feature. This was the prismatic 'cat's eye' reflector for defining traffic lanes at night.

Bridges

Reinforced concrete was quickly pressed into service also by bridge builders. The first such bridge was built in France in 1907: the largest, built under the direction of Freyssinet himself, was the Gladesville Bridge over the Parramatta near Sydney, Australia. This massive structure, weighing over 25,000 tonnes, has a single span of 304 metres (997 feet).

If conditions are favourable, long distances can be bridged by joining a succession of piers, as widely used to carry canals and railway viaducts across valleys. If this is impracticable, however, recourse can be made to a suspension bridge. This is a very ancient device, long used by primitive people, in both the Old World and the New, for foot passengers. Strictly speaking, such primitive bridges are catenary bridges, in that the footway follows the natural curve assumed by the hanging footrope. By the 15th century, however, the Chinese had conceived the idea of using the catenary to suspend a level roadway and from that time iron chains were often used as an alternative to bamboo in the construction. In the West, however, such suspension bridges were of no importance before the 19th century, where they may well have been an independent invention.

In the first decade of that century James Finlay built a number of suspension bridges in the USA: the first, with a span of 25 metres (82 feet), was built in 1801 over Jacob's Creek, Pennsylvania. In Europe, the first major bridge of this kind was Telford's bridge over the Menai Straits in Wales, completed in 1826. This was a massive construction in which the suspension units were heavy iron bars and rods. In the 1830s French engineers, such as Marc Séguin, began to use multiple-wire suspension units which were much lighter. But the great pioneer of this kind of bridge was the American engineer John Roebling, who spanned the Niagara Gorge in 1854 and completed the Brooklyn Bridge, New York, in 1883. The latter had a then record span of 490 metres (1600 feet).

In the 20th century the main developments in suspension bridges were the achievement of greater spans and lighter and more graceful designs. In 1940, however, a disaster in the USA – fortunately involving no loss of life – directed new attention to what had long been recognized as a potential weakness, namely the destructive effects of high winds. In that year the 860-metre (2821-foot) Tacoma Narrows Bridge spectacularly collapsed over a period of some three hours. The immediate response was to revert to stronger construction along conventional lines, but a better solution, devised in Britain and first used for the Severn Bridge in 1966, was to design the deck on aerodynamic principles to minimize the effect of the wind on it. This was used, too, for the 1423-metre (4667-foot) Humber Bridge. Such spans are spectacular but theory suggests that suspension bridges of up to 2400 metres (8000 feet) are feasible.

CHAPTER TWENTY-EIGHT

The Rise of the Chemical Industry

In earlier chapters we developed the argument that the Industrial Revolution was largely initiated and maintained by ambitious, practical men rather than by trained scientists and engineers. To this an obvious exception is the chemical industry. Here, although the business control might still be in the hands of men with no technical knowledge, progress increasingly depended on understanding the fundamental processes, in the light of the best knowledge of the day. As a generalization it is fair to say that modern chemistry was born with the publication of Lavoisier's *Traité élémentaire de chimie* in 1789 and consolidated in Dalton's *New System of Chemical Philosophy* of 1808, evaluating the theory that elements are distinguished by the weights of their atoms.

Chemistry began to emerge as an organized profession, as opposed to an intellectual pursuit, somewhat before the middle of the 19th century. The oldest national chemical society is the Chemical Society of London, founded in 1841: in the USA the short-lived Chemical Society of Philadelphia, founded in 1792, was perhaps the earliest scientific society devoted to chemistry, but the national body, the American Chemical Society, was not founded until 1876.

By 1900, the chemical industry was a fully fledged science-based industry but one whose roots already lay a century or more in the past. In an evolutionary process, it is hard to identify a single event as being of decisive importance, but in this case we can fairly single out Nicolas Leblanc's process for making alkalis – hitherto largely obtained as potash from wood ash – from salt and sulphuric acid. For this, he was awarded a prize offered for such a process by the French Academy of Sciences in 1775, but the Revolution intervened, his patron was executed and he never received the prize. His method was undeniably a nasty process in that clouds of hydrochloric acid gas were evolved and for every tonne of sodas made 2 tonnes of galligu were produced, a noxious waste not only difficult to dispose of but a cause of loss because it contained a considerable amount of valuable sulphur. Nevertheless, with various improvements both technical and economic, the Leblanc process was the most important source of soda, of particular importance to the textile industry, for more than a century. Not until the beginning of the 20th century was it finally supplanted by a process, based on salt and ammonia, developed in Belgium by the Solvay brothers: although patented in 1861, it took more than 10 years to overcome all the technical difficulties. In 1900, world production of soda was nearly two million tonnes annually, of which nine-tenths was manufactured under licence by the Solvay process.

For certain purposes, such as soap-making, a stronger alkali – caustic soda – was required. In 1894, Hamilton Y. Castner, the American chemist whose process for making aluminium had been made obsolete by an electrochemical one, devised an electrochemical process for making very pure caustic soda from brine. This was first worked at Runcorn, England in 1897.

Soap manufacture requires not only caustic soda but also fats and oils, and shortage of the latter led Germany in the First World War to experiment with synthetic detergents as an alternative. In the 1930s, these found many applications and since the Second World War they have also established a huge domestic market, especially for use in washing and washing-up machines.

The other chemical required in quantity – including that needed for the first stage of the Leblanc process – was sulphuric acid. In the 17th century, this was made at Nordhausen in Saxony, by distilling green vitriol (iron sulphate), but by the middle of the 18th century John Roebuck – first of Birmingham and later of Prestonpans in Scotland – had devised the lead chamber process, so called because of the nature of the reaction vessels, chosen for their resistance to the highly corrosive acid. In this, oxides of sulphur – derived from brimstone or pyrites – reacted with water in the presence of nitrous oxide, made from saltpetre (nitre). Like the Leblanc process, this continued well into the 20th century, but before then it was being supplanted by another, the contract process, developed about 1870 by the German chemist Rudolf Messel. In this, the sulphur dioxide was directly converted to sulphur trioxide – which in turn reacts with water to form sulphuric acid – by making it combine with oxygen in the presence of a catalyst, the latter being a substance which accelerates a chemical reaction without being consumed by it. At first expensive platinum-based catalysts were used, but after the First World War vanadium was widely adopted as a catalyst instead.

Sulphuric acid soon became of central importance in the chemical industry not only for the manufacture of soda but also of a host of other important chemicals and for a variety of processes in the metallurgical and other industries: it was Roebuck's interest in metal refining that first led him to sulphuric acid manufacture. It came to be known as 'the barometer of industry', the demand for it rising and falling in tune with the general state of the industrial economy. In favourable times the amount manufactured was very large indeed. In 1925, just before the Great Depression, world demand was 15 million tonnes: by mid-century it had nearly doubled. However, it is now less reliable as an index of

birth of chemical industry 119–123 **bridges** 63, 125, 187, 258 **Hamilton Y. Castner** 189 **sulphuric acid production** 191

industrial activity as the picture is distorted by the unduly large proportion of acid – nearly half – that goes into the making of superphosphate fertilizer. By 1939, more than 10 million tonnes of rock sulphate was being used for this purpose every year. As we shall see, however, energy consumption is now a more realistic index of industrial activity.

Synthetic fertilizers

Plant growth depends not only on adequate supplies of chemically combined phosphorus but also on the availability of combined potassium and nitrogen. By 1900, natural sources of nitrogenous fertilizers were severely depleted and the chemical industry increasingly turned its attention to the possibility of 'fixing' the nitrogen that forms nearly three-quarters of the air around us into a form that plants can utilize. In Norway, where there was an abundance of cheap hydroelectric power, Olaf Kristian Birkeland and Samuel Eyde developed a process for making nitrogen combine with oxygen by blowing the mixture through an electric spark spread out into a large disc by a magnetic field.

In Germany, quite a different approach was made. Fritz Haber, who was awarded a Nobel Prize in 1918 for his work, devised a process by which nitrogen and hydrogen were made to combine, in the presence of a catalyst, to form ammonia. The chemistry was simple but the technology was exceedingly difficult, for satisfactory yields could be obtained only at high temperatures and at far higher pressures – some 200 times greater than atmospheric – than had ever previously been achieved in an industrial process. The first processing plant came into use in Germany in 1915 and by the end of the war total output was around 100,000 tonnes annually. Apart from providing nitrogenous fertilizers, the ammonia could be converted into nitric acid, essential for the manufacture of high explosives in a blockaded Germany. After the war, the Haber process was widely worked throughout the world.

Yet another route to obtain the all-important chemically combined nitrogen was via calcium carbide, made by fusing lime and coke in an electric furnace. If water is allowed to drip on to this, acetylene gas is evolved. In the 1890s, this was used as the fuel for lights – from bicycle lamps to lighthouses – and later the intense heat of an oxy-acetylene flame was used for cutting and welding iron and steel. Additionally, however, the reactivity of acetylene made it a useful intermediary in the chemical industry. In the 1930s the Germans, at Badische Anilin und Soda Fabrik (BASF), used it extensively for this purpose: the great advantage was that it could be made from indigenous raw materials. Calcium carbide also found another important use. Around the turn of the century, it was discovered that if heated in an atmosphere of nitrogen it formed a new compound, calcium cyanamide: if this is treated with superheated steam, ammonia is formed. Cyanamide can also be used directly as a nitrogenous fertilizer, and was widely applied in Europe for this purpose.

The fixing of atmospheric nitrogen relieved the shortage of nitrogenous fertilizers that was causing such anxiety at the end of the 19th century, but in the long term this relief had some unforeseen consequences. The abundance of cheap fertilizers from the 1920s had a powerful influence on the agricultural economy worldwide: farmers found they could increase yields quite out of proportion to the additional costs. In the 1970s, however, this was to prove a mixed blessing. By then, petroleum was the main source of hydrogen for making ammonia and the sharp increase in its cost after the OPEC crisis of 1973 was immediately reflected in the cost of fertilizers: farmers found it no longer cheap but expensive. Secondly, the prodigal use of nitrogenous fertilizers led to excess being leached out into drinking water supplies, creating a health hazard.

The kinds of chemical we have considered so far have two common characteristics. First, they are what are called 'heavy' chemicals: that is to say, they are produced in huge quantities, measured in millions of tonnes annually. Secondly, they are inorganic – or mineral – chemicals, as distinct from organic (carbon-containing) chemicals such as are derived from plants or animals. The distinction is now of only historic significance, however, going back to the old doctrine of vitalism, which held that the constituents of living matter could never be synthesized. This belief has long been exploded, for not only have many constituents of living matter been synthesized but so too have a vast number of carbon-containing compounds not found at all in nature. Nevertheless, chemicals are still broadly categorized into two families in this way: organic chemicals contain carbon, inorganic chemicals do not.

In the first half of the 20th century, it was the organic chemical industry that developed most strongly, producing a whole range of novel materials – explosives, dyes, drugs, dry-cleaning fluids, plastics and many others – and it is to these products that we must now turn our attention.

Explosives

As we have seen in an earlier chapter, the introduction of gunpowder completely revolutionized the conduct of war and, later, of mining and civil engineering. For 500 years, however, no other explosive was invented. Then, about 1845, the German chemist Christian Schönbein, working at the University of Basel, discovered that if cellulose, in the form of cotton, is treated with nitric acid an exceedingly powerful explosion is formed. Unfortunately, it was very unstable and attempts to manufacture it in England, and later elsewhere in Europe, led to some disastrous explosions and it was soon abandoned, except in Austria. Some 20 years later Frederick Abel, adviser on explosives to the British Government, showed that guncotton, as it was called, could be made safe by repeated washing to remove the last traces of acid.

Meanwhile, however, a different line of progress had begun in Italy. There Ascanio Sobrero, professor of chemistry in Turin, had discovered in 1846 that treating glycerine – a by-product of soap manufacture – with nitric acid also yielded a powerful explosive. This, too, proved dangerously unstable but, in Sweden, Alfred Nobel in 1867 showed that nitroglycerine could be tamed by absorbing it in a form of clay known as kieselguhr. This product he marketed for mining as dynamite: in 1875 world production was already 3000 tonnes.

To some extent, however, Nobel's improvement was self-defeating, for the dilution with kieselguhr naturally weakened the force of the explosive, although it was still five times stronger than gunpowder. In 1876 he marketed a new product – blasting gelatine – in which the nitroglycerine was mixed with a few per cent of guncotton. Surprisingly, these two fiery individuals seemed to tame each other.

Initially, these new explosives were developed for civil use, but it was inevitable that they should quickly attract the attention of the military. Here, as we have seen in an earlier chapter, they had a revolutionary effect almost as great as that of gunpowder centuries earlier.

Above: Dynamite being packed in one of Nobel's factories, *circa* 1900.

Nitration of glycerine had to be carried out with extreme care to avoid overheating. In this Scottish factory (left), the operator sat on a one-legged stool: if by chance he dozed off, he would be rudely awakened.

Dyes

The earliest textiles of which we have knowledge were dyed to make them more attractive. Until the 19th century, however, these dyes were all of vegetable or animal origin but, in 1856, a chemical discovery was made which was to create yet another great new industry. In that year William Henry Perkin, a student at the Royal College of Chemistry in London, conceived the idea that the important natural anti-malarial drug quinine might be synthesized, in a series of chemical reactions, from benzene. Even in the light of the still limited chemical knowledge of the day, his ideas lacked substance and all he succeeded in obtaining was a depressing black deposit in his testtube. However, he did not throw this away, and from it he obtained not quinine but a brilliant mauve dye which, on the advice of Pullars of Perth, well-known Scottish dyers, he patented and began to manufacture at Greenford, near London. It quickly became fashionable, especially among French couturiers, through whom it found its way back to Britain.

Mauve was the first of a whole range of synthetic dyes, many of them originating in Germany. Their manufacture was facilitated by growing understanding of the chemical principles involved. This led to a second line of development – synthesis of some of the more important natural dyes. Here Perkins himself, having been defeated by quinine, scored a spectacular success with the synthesis of alizarin, the red constituent of madder, in 1869: he lodged a patent, but to his dismay found that a German chemist, Heinrich Caro, a director of the newly founded BASF company, had beaten him by a single day. In the end, there was a satisfactory outcome, in that BASF licensed Perkin to manufacture alizarin in Britain. In 1897, indigo was successfully synthesized.

By 1900, some scores of new synthetic dyestuffs were available to the textile industry, but an important change of emphasis had occurred. While the new industry had been created in Britain, the leadership had decisively moved to Germany, thanks in large measure to its encouragement of chemistry as an academic study and a more enthusiastic support of industrial enterprise. By 1914 Britain was importing three-quarters of its dyes from Germany and when war broke out it was difficult even to dye the khaki uniform the army required: recourse even had to be made to re-establishing the cultivation of indigo in India, a crop which – like madder – had been ruined almost overnight by the advent of the synthetic product.

By improvisation, the British dyestuffs industry emerged from the war considerably strengthened, but the eclipse of Germany was only brief. By 1924, world output of dyestuffs – and by that time this meant almost entirely synthetic dyestuffs – was about 150,000 tonnes annually. Half of this was again being made in Germany, but the USA had moved up to second place, with about 20 per cent of the market, roughly equal to the contribution of Britain and Switzerland combined. Nevertheless, Britain did evolve a few important new products such as Caledon Jade Green, in 1920, and Monastral Blue, in 1935.

dyes 37, 120–122, 156, 157, 196 **explosives** 185–186, 215 **fertilizers** 206, 213, 291 **gunpowder** 136–137, 186

The dyeing of fabrics is almost as old as civilization. Until the mid-19th century, however, only natural dyes were available. Then, in 1856, the textile industry was transformed with the preparation of the first synthetic dyestuff, 'Mauveine', by William Perkin.

By that time, the nature of the dyeing process was much better understood, and it was possible logically to develop dyes suitable for the new synthetic textile fibres exemplified by rayon and nylon.

Drugs

Until the latter part of the 19th century, most drugs used in medicine were of vegetable origin. The old 'herbals' – such as those of John Gerard and Nicholas Culpeper – listed many remedies, some very valuable but others useless or even harmful. Useful ones included quinine, a specific against malaria; digitalis, a foxglove preparation used to treat cardiac disease as manifested by dropsy; opium, a soporific and painkiller; atropine and hyoscyamine, from deadly nightshade and some other plants, to facilitate examination of the eye by dilating the pupil.

Many new drugs derived from plants have been adopted in recent years, of which the mould product penicillin is by far the most important. In 1942, curare – derived from a deadly South American arrow poison – was introduced as a muscle relaxant. As recently as 1952, chemists in the laboratories of the Swiss chemical company Ciba identified reserpine, present in the plant *Rauwolfia serpentina*, as a valuable drug for the treatment of hypertension.

With the growth of chemical understanding, two new lines of progress were open. One was to treat the natural product and extract the pure active principle from it: in this way dosage could be accurately controlled, a very important consideration in view of the great variability of natural material. In some cases, the active principle could then be made synthetically: aspirin, perhaps the most widely used drug in the world, originally derived from salicin, present in willow (Salix) bark.

The other line was to test the increasing number of synthetic compounds available for indications of particular kinds of therapeutic activity: it was this that in the 1930s revealed the sulphonamides and some related substances to be very effective drugs against certain bacterial infections. At about the same time, atebrin was introduced as a synthetic alternative to quinine for the treatment of malaria, and in 1945 ICI chemists in Britain invented Paludrine (proguanil) as another alternative. The quinine-containing Indian Tonic Water which Jacob Schweppe's mineral water company – founded in London in 1792 – had been supplying to troops throughout the British Empire found a rival.

Polymers

Man-made fibres and plastics represent an important new class of industrial chemicals: more properly known as polymers, they are almost entirely a product of the 20th century. Virtually all are organic in nature and they result from the carbon atom's unique ability to form itself into very long chains by joining together large numbers of quite simple units. Their name derives from the Greek *poly* (many) *meros* (parts). In polythene, for example, the basic molecule is ethylene gas, the molecules of which contain two atoms of carbon and four of hydrogen: the molecule of the polymer, a hard, waxy-looking solid, may contain as many as 50,000 such units. Polymers abound in nature and include the traditional textile fibres – cotton, wool, linen and silk – as well as proteins and carbohydrates. To these, the chemical industry quite suddenly added a variety of new substances cheap enough to be widely used.

In the second half of the 19th century, some attempts were made to modify natural polymers for special purposes. Thus John Wesley Hyatt in the USA, where billiard halls were very popular, in 1869 patented celluloid – made by treating cellulose with nitric acid and blending the product with camphor – as a substitute for the natural ivory that was becoming increasingly scarce and expensive for billiard balls. It also found a use in making dental plates, and later as a base for photographic film, until replaced by cellulose acetate which is far less flammable.

While celluloid, which is still manufactured, can fairly be regarded as the parent of the modern plastics industry, it was never produced in bulk, partly because of the fire hazard and partly because it does not lend itself to the modern techniques of injection moulding. The same can be said of Galalith, a semi-synthetic polymer derived from the casein of milk. This was invented in Germany in 1897 and was popular for small articles, especially buttons.

The huge modern plastics industry really dates from 1909, with the introduction of Bakelite, technically the invention of Leo Hendrik Baekeland – who had already made a fortune from his invention of Velox gaslight papers for photographers in the USA – but discovered independently in Britain by James Swinburne, who lodged his patent only one day later. The organic chemists of the day were primarily interested in pure crystalline chemicals which they could readily identify and manipulate. When the great German chemist Adolf von Baeyer, in 1872, mixed phenol and formaldehyde, he observed a reaction between them, but as the product was a dark sludge he investigated no further. Yet in the hands of Baekeland and Swinburne this unpromising material became the first industrially important thermosetting plastic – so called because once the mixture had been heated in a mould its shape was permanently fixed. Thermoplastic resins, in contrast, can be softened and hardened repeatedly by heating and cooling, like wax.

By mid-century, what von Baeyer had contemptuously dismissed as a mere *Schmiere* was being manufactured at the rate of 200,000 tonnes annually. This growth was greatly assisted by the concurrent growth of the automobile and electrical supply industries, which both required vast numbers of small items with good insulating qualities – such as switches, rotor arms and lamp holders. In this context, its unattractive dark brown colour was of no consequence. For other purposes, however, this was a disadvantage: this led to the development of colourless urea-formaldehyde in 1928, and melamine-formaldehyde in 1935, resins which could be attractively coloured by adding pigments. This opened up new markets, especially for domestic articles.

A selection of domestic articles moulded in celluloid, xylonite and other early plastics.

Meanwhile, other quite different chemical configurations were being explored as a basis for new plastics. The 1930s saw the introduction of a tough transparent material which in many applications was a valuable alternative to glass because it was light and did not easily shatter. This was based on acrylic acid, and marketed in Britain as Perspex and in the USA as Plexiglass: initially, a major use was in the manufacture of windscreens and cockpit covers for aircraft, but after the war – when this priority demand suddenly fell – vast quantities were used for various kinds of sheeting, including corrugated sheeting for roofs, and for mouldings.

The introduction of the Haber-Bosch synthetic ammonia process stimulated interest in the possibility of other industrially useful reactions occurring only at high pressures and temperatures. This led in 1932 to the discovery by ICI in Britain of polythene. This proved to have exceptional qualities as an electrical insulator, and when war broke out in 1939 its use in submarine cables was being explored: 1½ kilometres (1 mile) of polythene-coated cable was laid between the Isle of Wight and the mainland in 1938. During the war, it proved immensely valuable in the development of radar equipment, but in the event these highly specialized uses, originally thought to limit the market, proved of minimal overall importance in terms of quantity. After the war polythene, in various forms, found quite new uses in the manufacture of all kinds of mouldings, from kitchenware to toys. By 1960, against all expectations, production of what seemed to be a product of extremely limited use had soared to over one million tonnes annually: in the next 10 years this multiplied fivefold.

In 1938, Du Pont in America made a very interesting variant of polythene. In this, the hydrogen atoms in the basic ethylene molecule are substituted by atoms of fluorine. The polymer, PTFE or Teflon, is expensive and difficult to work – a modified form of powder metallurgy is commonly used – but its special properties outweigh this in certain applications. These include high-resistance electrical components, non-stick coatings for pots and pans and – because of its low coefficient of friction – coatings for skis to make them run faster.

During the First World War, blockaded Germany suffered from lack of many essential imports, among them rubber. To mitigate this, small amounts of a rather poor synthetic rubber were made. As part of this policy of self-sufficiency inaugurated in the 1930s, provision was made for manufacture of much improved synthetic 'Buna' rubbers derived ultimately from acetylene. By 1938, output had risen to 5000 tonnes annually and this was much increased during the war. The USA, with supposedly unrestricted access to natural rubber, did little in this field until the Japanese suddenly overran the Far East rubber plantations. This led to a frantic crash programme to produce synthetic rubber there – though of a rather different kind: by the end of the war annual output was nearly one million tonnes.

biotechnology 297–298 **chemotherapy** 280, 282–283 **penicillin** 283, 285 **plant remedies** 32 **synthetic materials** 303–305

Man-made fibres

As all natural fibres are polymers of one sort of another, it is not surprising that attempts were made to make them artificially. In 1892 two British chemists, Charles Frederick Cross and Edward J. Bevan, devised a chemical process for regenerating cellulose in fibre form. In 1904 Courtaulds brought the British rights in this viscose rayon and, in 1907, began to manufacture it at Coventry: by 1939 world production was in excess of 500,000 tons annually. By that time an alternative form, known as acetate rayon, had been developed. This had its origins in the need to find a use for surplus cellulose acetate after the First World War, when it was widely used for doping aeroplane fabric.

However, the really big development in this field was nylon, a polyamide, invented in 1934 in the USA by Wallace Hume Carothers of Du Pont: full-scale manufacture began in 1938. It first appeared before the public in the form of toothbrush bristles, but it immediately appealed to the stocking industry: 64 million pairs of nylon stockings were sold in 1939.

In a sense, the ultimate polymer is simply a long chain of carbon atoms with no side attachments at all. Such fibres were in fact achieved in Britain in 1963 by very carefully controlled heating of certain synthetic polymer fibres such as rayon. They are very flexible, conduct electricity well and resist high temperatures. However, they are costly to manufacture and thus can be used only for exceptional purposes – and then usually as composites, where their unique combination of properties confers a real advantage. Applications range from fishing rods to critical aircraft components, from pacemaker leads to golf clubs. World production – mostly in Britain, the USA and Japan – in the late 1980s was not much more than 1500 tonnes annually. In quantity, the aerospace industries are one of the biggest users: every modern jumbo jet contains about 2 per cent of the CFPR (carbon fibre reinforced plastic).

Raw materials

Organic chemicals are by definition compounds of carbon, so it follows that they cannot be made without a corresponding source of carbon atoms. In the early days of the industry, this source was coal-tar, a cheap waste product of the coal-gas industry – and later of the metallurgical coke industry when it found it worthwhile to collect its by-products instead of discharging them into the air. On distillation, this tar yielded a variety of basic chemical intermediaries such as benzene, naphthalene, anthracene and toluene. So universally was this raw material used that the first synthetic dyes were commonly referred to as coal-tar dyes: in Europe, coal-tar was all-important until well after the Second World War. In the USA, however, which by 1939 had superseded Germany as the world's greatest manufacturer of chemicals, events took a different course. There, petroleum became an increasingly important raw material from 1920, and by 1939 a million tonnes of oil a year was being turned into chemicals.

One of the first successful man-made fibres was viscose rayon, a semisynthetic product made by regenerating cellulose. This advertisement was issued in the 1920s by Courtaulds, one of the biggest manufacturers.

The raw material used is not unrelated to the nature of the final product. In coal-tar derivatives, the molecules of the main constituents have the carbon atoms arranged in rings, which happen to be very suitable for making dyes and drugs. Such ring compounds occur also in petroleum, especially that from Borneo, but most of the constituents of natural oil have short-chain molecules, very appropriate for making plastics.

Finally, we must mention the use of coal itself – nine-tenths of which consist of carbon – as a raw material. In 1913, Friedrich Bergius developed in Germany a process for hydrogenating coal at high pressures – akin to those of the Haber process for ammonia – to make petrol. It proved valuable in both World Wars.

In beginning our review of the new world of plastics, it was stated that they were all organic compounds, because carbon alone had the necessary power of forming long chains. In fact, this is not strictly true: the element silicon, too, has this quality, though to a far more limited extent. From the beginning of this century, F. S. Kipping, at the University of Nottingham, pioneered the study of silicon chemistry and this ultimately led, in the early 1940s, to the production of a range of plastics known as silicones, by the Dow Corning Corporation in the USA. Where cost was not a prime consideration, silicones found applications based on their exceptional resistance to heat, their power of repelling water and their anti-adhesive properties.

coal gas 154, 196 **Haber process** 264 **plastics** 303–304 **rubber plantations** 291, 292 **synthetic fibres** 304–305

PART

5
The Modern World

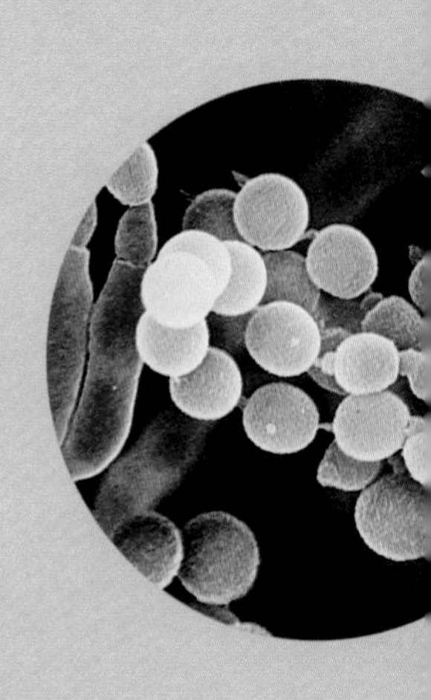

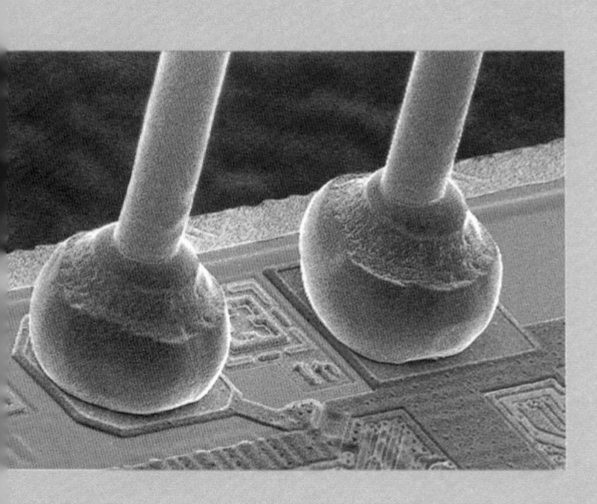

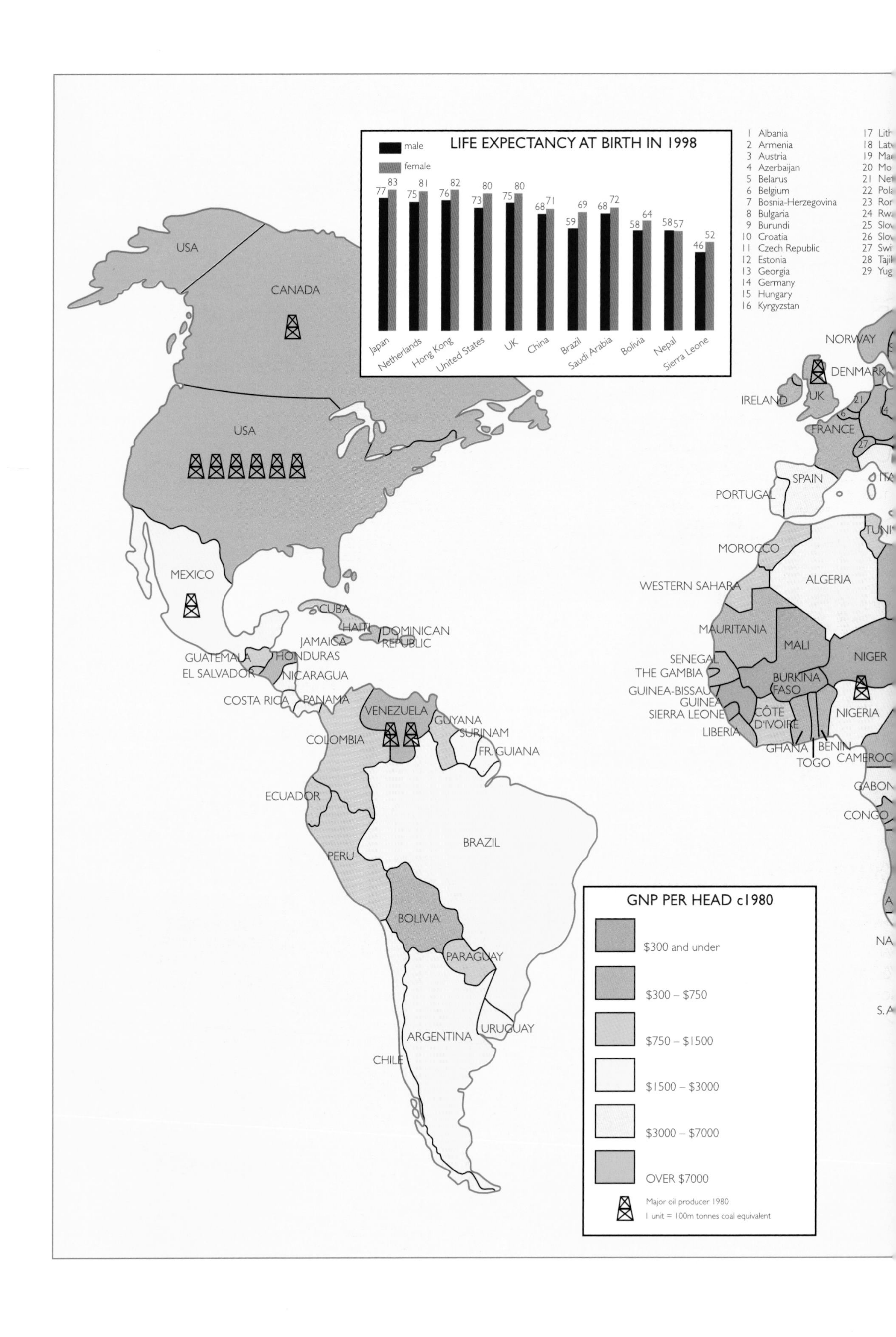

LIFE EXPECTANCY AT BIRTH IN 1998
male
female
Japan 77 83
Netherlands 75 81
Hong Kong 76 82
United States 73 80
UK 75 80
China 68 71
Brazil 59 69
Saudi Arabia 68 72
Bolivia 58 64
Nepal 58 57
Sierra Leone 46 52
1 Albania
2 Armenia
3 Austria
4 Azerbaijan
5 Belarus
6 Belgium
7 Bosnia-Herzegovina
8 Bulgaria
9 Burundi
10 Croatia
11 Czech Republic
12 Estonia
13 Georgia
14 Germany
15 Hungary
16 Kyrgyzstan
GNP PER HEAD c1980
$300 and under
$300 – $750
$750 – $1500
$1500 – $3000
$3000 – $7000
OVER $7000
Major oil producer 1980
1 unit = 100m tonnes coal equivalent
USA
CANADA
USA
MEXICO
CUBA
HAITI
DOMINICAN REPUBLIC
JAMAICA
GUATEMALA
HONDURAS
EL SALVADOR
NICARAGUA
COSTA RICA
PANAMA
VENEZUELA
GUYANA
SURINAM
FR. GUIANA
COLOMBIA
ECUADOR
PERU
BRAZIL
BOLIVIA
PARAGUAY
URUGUAY
ARGENTINA
CHILE
NORWAY
DENMARK
IRELAND
UK
FRANCE
SPAIN
PORTUGAL
MOROCCO
ALGERIA
WESTERN SAHARA
MAURITANIA
MALI
NIGER
SENEGAL
THE GAMBIA
BURKINA FASO
GUINEA-BISSAU
GUINEA
SIERRA LEONE
LIBERIA
CÔTE D'IVOIRE
NIGERIA
GHANA
TOGO
BENIN
GABON
CONGO

THE MODERN WORLD

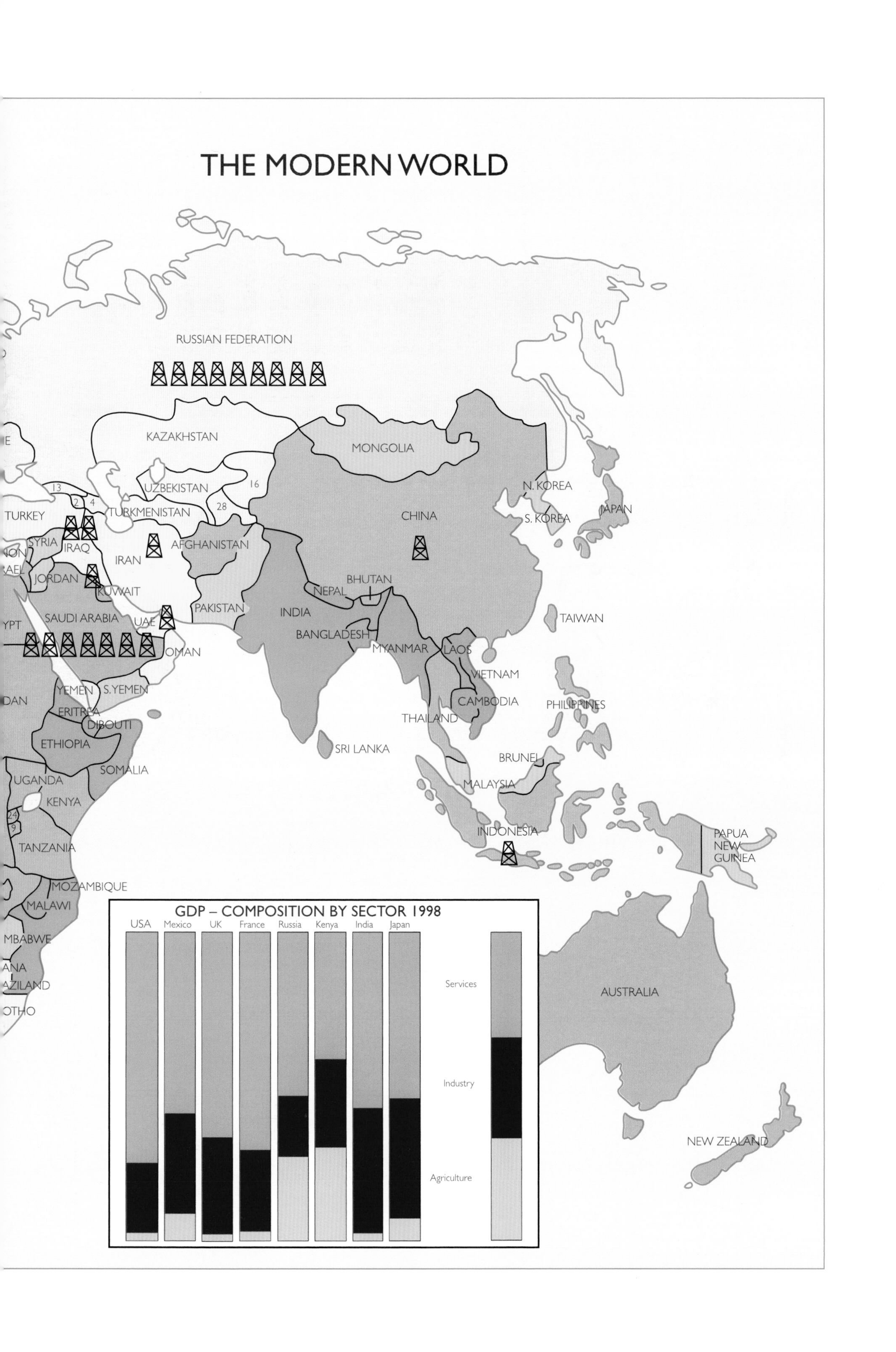

CHAPTER TWENTY-NINE

The Post-War World

The world that slowly emerged from the cataclysm of the Second World War was politically, socially and economically very different from that which embarked upon it. Not only had the whole balance of power changed, with the USA undisputedly the most powerful nation, but the course of events had again forcibly demonstrated that in war superior technology was the best guarantee of success. This meant not only superiority in weaponry and all the immediate auxiliaries of the military machine, which we will come to later, but also in the industries which supported the national economies and civilian populations in wholly abnormal circumstances.

In the abnormal circumstances of war, where national survival is the over-riding consideration, the distinction between military and civil technology is inevitably blurred, but in neither case do normal economic constraints apply. It is for this reason that wars tend to be forcing grounds for certain kinds of innovations, although, equally, developments not immediately relevant may be slowed down. This proposition can be exemplified from half a dozen different fields.

The technical possibility of atomic energy was implicit in experiments published just two days before war broke out. The military possibilities led to its extraordinarily rapid exploitation in the American Manhattan Project. Although the ultimate development of atomic power was inevitable even without the war, its history would surely have been very different. Much of the current antipathy towards nuclear power stems from its military connotations, without which its adoption for peaceful purposes might have been much more generally acceptable.

A second example of technological advances accelerated by war can be taken from the medical field. It was, again, literally on the eve of war that research workers at Oxford embarked on the project that was eventually to introduce penicillin – arguably the most important single discovery made in medicine – to medical practice throughout the world. Undoubtedly this, too, would have happened in any case in the fullness of time, but it was the potential value of penicillin for treating military casualties that led the USA – especially after Pearl Harbor in December 1941 – to give a favourable reception to emissaries from Oxford and launch an immediate crash programme for penicillin production.

In the 1930s, various chance observations had suggested the possibility that the position of aircraft might be determined from the ground by means of reflected radio waves.

	1940 – 1945 – 1950 – 1955 – 1960 – 1965
MEDICINE	cocaine in eye surgery; penicillin manufactured; artificial kidney; fluoridation; radioactive isotopes; heart-lung machine; synthetic steroids; polio vaccine; kidney transplant; pacemaker; contraceptive pill deve
AGRICULTURE AND FOOD	battery farming; artificial insemination techniques improved; microwave oven; malathion incesticide; frogs cloned; *Silent Spring*; test-tube plant prop
MATERIALS AND ENERGY	Terylene (Dacron); nuclear reactor; silicones; beryllium/titanium; basic oxygen steel process; float glass; catalytic polymerization; nuclear power station; breeder reactor; laser; shape memory alloys; tokamaks; North Sea oil and gas exploited
COMPUTERS AND COMMUNICATIONS	electronic computer; principles of holography; transistor; first commercial computers; colour television service; transatlantic telephone calls; stereo recording; video tape recording; integrated circuit; electronic telephone exchange; communica satellite (Telstar)
WARFARE	V-2 rocket; schnorkel; atomic bomb; air-to-air missile; hydrogen bomb; nuclear submarine; sea-to-a missile
AEROSPACE	radar systems improved; supersonic aircraft; jet airliner in service; artificial satellite; radio telescope; lunar probe; hovercraft; weather satellite; first manned spaceflight; Venus fly-by; Mar fly-b

Above left: Space research owes much to Robert Goddard, who began to experiment with rockets for investigating the upper atmosphere while professor of physics at Clark University, USA, 1919–43. He is seen here (second from the left) with a four-stage rocket launched in 1936.

Above right: The space shuttle *Challenger* deploys satellite communications systems as it orbits the earth. The Shuttle programme revolutionized the space industry by providing spacecraft that could be relaunched.

Although civil aircraft were already navigating by means of radio beacons, it was the military possibilities of this discovery that first attracted strong support for it. This led to almost all development being carried out in strict secrecy. Most of the major powers evolved some kind of radar, though that of Britain was perhaps the most successful, mainly because British scientists developed not merely a detecting device but a detecting system, which is a very different matter. The availability of highly developed radar was a contributing factor in the burgeoning of the civil aviation industry after the war.

The aviation industry provides another example of technological development being greatly accelerated by the needs of war. In this case it is jet propulsion, which was being experimented with for military reasons immediately before

970 1975 1980 1985 1990 1995 2000

CAT scanner
synthetic organs
MRI scanning
test-tube baby
Human Genome Project
GE internal organs

single cell protein
artificial twinning
life patented
mouse patented
GM foods
sheep cloned

solar power station
controlled fusion
'smart' materials

microprocessor
email
Multi-User Domains
public database
fibre-optic communications developed
personal computers
fifth-generation computer development begun
optical computer
WWW
brainwave controller

Exocet missile
Strategic Defence Initiative research
digital battlefield
'smart' bombs
telemedicine MASH

nar landing
airliner
Venus lander
space station
Mercury fly-by
Mars lander
Jupiter fly-by
Saturn fly-by
space shuttle
Uranus fly-by
comet probe
Venus orbiter
Mars rover
Titan probe

atomic energy 316–322 **Pearl Harbor** 222 **penicillin** 283, 285 **radar** 324–326

In the 1950s television was still a novelty in Germany. One of the earliest major outside broadcasts there was the football match between Germany and Italy in 1955 (below). Today, hundreds of spectators who are unable to obtain seats for the big tennis matches on Centre Court at Wimbledon can watch them on large screen displays (right).

the war. In the event, this was a race won by the Germans, whose Heinkel He178 first flew in the week before the outbreak of war. By the end of the war, Britain, the USA and Russia were all involved with jet aircraft, which soon also became widely used in the civil field, pioneered by the de Havilland Comet in 1952.

Finally, there was the conquest of space. In the inter-war years, rocketry had been closely studied in Russia, Germany and the USA. During the war, Germany developed long-range rockets for the bombardment of southern England and it was from these that the USA developed its successful moon-landing programme with the Apollo series of manned spacecraft. Although the first landing on the moon was achieved by the USA in 1969 it was the Russians who, in 1961, had first successfully launched a man into space.

The rise of television

If we look at the other side of the coin to seek some very important aspects of technology which were hampered rather than promoted by the war, television provides a very good example. As we have seen, photo-mechanical systems such as Baird's had served a useful purpose in stimulating public interest, but it was the all-electronic system that triumphed. By the summer of 1939, the main technical problems had been overcome and public transmissions had begun on both sides of the Atlantic. These services were, however, curtailed or abandoned in most countries for the duration of the war. When progress was resumed its speed was phenomenal, both socially and technologically. By 1960 there were an estimated 10 million television sets in Britain and 85 million in the USA. On the Continent, progress was slower. A decade later there were a quarter of a billion sets in the world, and the broadcasts they received powerfully shaped public taste and opinion. By 1998 there were television sets in 227.5 million households in China (equivalent to one for every six people) – almost twice as many as in the USA. This huge expansion of television from almost a standing start in 1945 created a correspondingly large number of new jobs, both in the production and transmission of programmes and in the manufacture of receiving sets and related equipment.

This worldwide expansion went hand-in-hand with important technological developments. One major achievement was transmission in colour. In 1940, Peter Goldmark, of the Columbia Broadcasting System (CBS), developed a colour system that reverted to Baird's photo-mechanical scanning disc in part but, as with black-and-white, the future lay with all-electronic systems successfully introduced by RCA in 1953. Apart from their cost, the first colour sets were not very satisfactory, the rendering – especially of flesh tones – being irregular and distorted. By 1970, however, these technical problems had been overcome and black-and-white sets – though still used today – were beginning to look old-fashioned. By that time, too, TV sets had become much more compact. Up to 1960, all receivers had operated with thermionic valves, but the Japanese firm Sony then began to replace these with transistors: today, valves are obsolete. This made possible the manufacture of small portable sets, such as flat-screen battery-operated sets which are literally pocket-size.

Two other developments in television were too important to ignore. One was the development of an international network. Generally speaking, TV signals have a short range and booster stations are necessary if long distances are to be covered. The coronation of Elizabeth II in 1953 received wide coverage in Europe, but truly international link-ups had to await the availability of relaying satellites, first achieved with Telstar in 1962. In 1985, a Live Aid concert, staged in aid of famine relief in Ethiopia, was beamed to 169 countries and watched by an estimated 1.5 billion viewers, but it was the funeral of Diana, Princess of Wales, in September 1997 that received the largest global audience of an estimated 2.5 billion people.

The second major development was the introduction of recording devices. These were first conceived for use within broadcasting systems, so that major programmes, or extracts from them, could be reshown at a later date, but it was soon perceived that there was a huge potential market among individual viewers who might want both to see a particular programme again or record it to watch at a more convenient time. Additionally, entrepreneurs entered the field and began to offer specially prepared material for home use. The pioneer of videotape recorders was the Ampex Corporation in the USA, and a range of devices suitable for domestic use came on the market in 1956.

So far we have discussed television only in the general context of public service transmissions. It must be remembered, however, that television also serves many purposes as a remote eye. Closed-circuit television can be used, for example, to investigate the ocean depths – both for scientific research and to assist the submarine plumbing carried out by big oil companies; to monitor the flow of road traffic; to detect shoplifters; to make observations in industrial environments unfavourable to human operators; and to direct surgical procedures from a distance or robots on Mars. Through the medium of television, hundreds of millions of viewers throughout the world witnessed 'live' the historic moon landing of Neil Armstrong and Edwin ('Buzz') Aldrin on 21 July 1969.

Energy

In the last chapter we remarked that for many years the consumption of sulphuric acid was a reliable barometer of industry. Now, however, it is the consumption of power that most realistically reflects the level of industrial activity. In the past 50 years, two major events profoundly affected the general pattern of power production. One was technological: the development of atomic energy in the 1950s gave the world its first completely new source of energy since the earliest civilizations.

The second development was economic in origin. The action of the OPEC countries in dramatically forcing up the price of oil in 1973 inevitably provoked a strong reaction in the hard-hit industrialized countries of the West. On the one hand, national policies were adopted to provide incentives

In the early 1950s Britain built two gas-cooled reactors at Sellafield, Cumbria for plutonium production. The world's first large-scale nuclear power station designed for peaceful purposes, Calder Hall, was completed nearby in 1956.

for fuel economy. On the other, many research projects were launched to develop alternative sources of power, particularly renewable sources such as solar energy and heat stored deep in the earth.

The development of nuclear power followed a surprisingly nationalistic pattern. The predominantly Anglo-Saxon nations – such as the UK, New Zealand, USA and Germany – adopted an extremely cautious approach in the face of a strong environmentalist lobby. But the Latin countries, together with Taiwan and Japan, have been far more relaxed. France, in particular, went ahead with an ambitious programme, in the 1980s deriving more than half its electricity from 60 nuclear power stations, at the cheapest rate in Europe. However, in the aftermath of accidents at Three Mile Island in the USA in 1979 and Chernobyl in Russia in 1986, the environmental lobby was considerably strengthened, especially as this roughly coincided with a sudden collapse in oil prices. Since 1979 no new nuclear reactors for the production of electric power have been commissioned in the USA and several older ones have been decommissioned.

Thermonuclear power

Although practical fusion reactors have yet to be built, several research facilities have managed to produce some power by the brief fusion of hydrogen plasma. Should fusion reactors become possible, they will provide an almost limitless source of power. Nuclear fusion is the process by which the sun generates energy and occurs when two isotopes such as deuterium (one proton and one neutron) and tritium (one proton and two neutrons) collide at high speeds and fuse to form helium (two protons and two neutrons) and release energy as heat and an extra neutron. Fusion can occur only under particular conditions of extreme temperature and density, but if these conditions could be matched, then nuclear fusion could take place with no external supply of energy.

Because of the extreme pressures and temperatures necessary (100 million°K), special generators called tokamaks are necessary to suspend the plasma in a magnetic field. The first tokamak (abbreviated from the Russian *Toroidal Kamera Magnetic*) was invented by Lev Andreevitch Artsimovitch of the former Soviet Union and put to use in 1963. In 1991, the Joint European Torus (JET) in Culham, England, achieved controlled fusion for two seconds from plasma heated to 220 million°K. The flash produced only 2 megawatts of energy for the 15 megawatts supplied, but it represented a significant breakthrough.

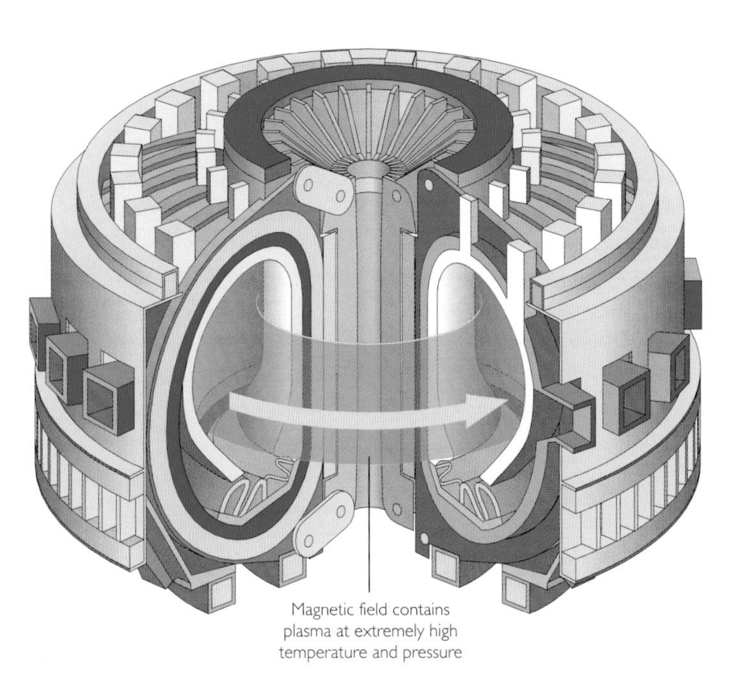

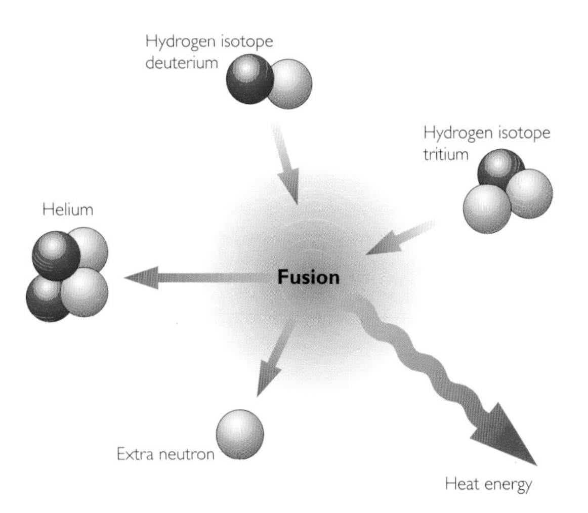

Although not yet a feasible means of generating energy, tokamak fusion generators offer possibilities for the future. They operate by suspending plasma in a magnetic field until the extremely high temperature and pressure cause hydrogen isotopes to fuse into helium molecules, releasing energy in the form of a stray neutron.

Renewable sources of power

One of the great unsolved problems of technology is that of storing very large amounts of energy. This is one reason why fossil fuels are attractive: their energy remains firmly locked up until they are burned, as it has been for millions of years. In coal- and oil-fired electric power stations, for example, large quantities of fuel can be stored and transferred into power according to demand, which fluctuates widely. Daytime needs are far greater than those in the early hours of the morning; winter demands are greater than summer.

By contrast, even the most promising alternative sources suffer from the defect that they are variable. Solar energy not only disappears as the sun goes down, but is also subject to changes in weather. Hydroelectric schemes are vulnerable to long periods of drought. Tidal schemes are a problem because not only does the time of the tide vary from day to day but its height varies from season to season. These variations could be evened out if energy generated during periods of slack demand could be economically stored for use at peak periods. Unfortunately, this is not at present feasible for really large quantities of energy, though for relatively small amounts, or individual use, various satisfactory solutions have been found. In nuclear power schemes, for example, surplus power can be used to pump water up to reservoirs, from which it can be drawn back through generating turbines when demand rises. Another possibility that has been explored is to store surplus energy in huge rotating flywheels.

The discovery in 1986 of certain superconducting ceramics by Georg Bednorz and Alex Müller, two IBM scientists working in Switzerland, opened up yet another possibility for long-term storage. Superconductive metals – those conferring high conductivity and low resistance at extremely low temperatures – had been discovered as early as 1911 by Heike Kamerlingh Onnes, but the cost of cooling them with liquid helium proved too great to foster widespread use. The ceramics, however, could achieve superconductivity at temperatures as high as -140°C (-220°F) rather than the -269°C (-452°F) necessary for metal alloys such as niobium-titanium. Because a superconducting coil has extremely low resistance, a current in a closed loop of such a coil could flow indefinitely, offering a means to store vast amounts of energy. The loop is simply broken with a circuit when the stored energy needs to be retrieved.

But the very real practical difficulties of storing large amounts of energy tend to be ignored by those who think simply in terms of total energy resources. Of course, it is true that a tiny fraction of the solar energy falling every day on the earth could fully meet all mankind's needs, and the same can be said of the overall power of the wind or the tides. But to harness these to produce a reliable supply in all circumstances is a very different matter. A possible solution is to use these variable resources to supplement conventional ones. One could, for example, imagine a large wind-powered

The OPEC oil crisis of 1973 directed increasing attention to the possibilities of renewable energy sources such as solar power – as in this complex of reflecting mirrors at Barstow, California.

source linked with an oil-fired power station: when a steady wind blew, the consumption of oil would fall; when the wind dropped, oil burning would correspondingly increase. The difficulty here is that every power system must provide a guaranteed supply and must therefore have sufficient stand-by capacity to meet peak demand even in periods of calm, which might last for some time. There would, therefore, be no capital saving on equipment but only a saving on fuel: in some circumstances, however, the latter might be the over-riding consideration.

Solar energy

There is nothing new in the utilization of solar energy. Archimedes is said, though very improbably, to have set fire to enemy ships with a burning glass, and Antoine Lavoisier and others certainly used such a glass to achieve high temperatures in chemical experiments in the late 18th century. After the Second World War the French government set up an experimental solar energy plant on Mont Louis in the Pyrenees, and in Israel and India solar cooking stoves were developed. But the world's first operative helioelectric power plant was EURELIOS in Sicily, initiated by the EEC, which – although essentially a pilot installation – began to feed electricity into the grid of the Italian Electricity Board in 1981. In this, suntracking mirrors – with a total area of 6200 square metres (66,737 square feet) – reflect solar energy on to a central boiler, which raises steam to drive a turbine. Simple calculation shows that to supply all Europe's energy needs in this way would require about one per cent of the total land area, roughly equal to that of the existing road system. A different kind of approach has, therefore, been proposed. This would locate the solar generator in Third World countries which have strong and assured sunshine, and plenty of open space. There they could lock up the energy for export to energy-hungry countries in fuels such as hydrogen or alcohol.

Apart from such ambitious schemes, the future of solar energy seems to lie in domestic heating panels which, again, could not be expected to meet all demands but could complement conventional supplies. By 1995, 88 per cent of solar thermal collectors shipped in the USA were used for heating swimming pools, while another 10 per cent were used for domestic hot water systems. In Israel, more than 30 per cent of buildings use solar water heating systems, which are now required in all new houses.

In addition to solar thermal collectors – those which receive solar energy and convert it directly into thermal energy – various photovoltaic methods are adopted for providing electricity. In 1954, D. M. Chapin, Calvin Souther Fuller and G. L. Pearson of Bell Laboratories in New Jersey produced a silicon photovoltaic cell which transferred solar energy into an electric current, following an idea proposed earlier that year by Paul Rappaport. The photovoltaic effect was discovered by Edmund Bacquerel in 1839, but it has only been in recent years that effective use of this knowledge has been made. The silicon cells produced in the 1950s only achieved a 6–10 per cent efficiency, yet even at an early stage of development they were used to power satellites — the first, Vanguard I, was launched in March 1958 and operated for eight years, powering a 5-megawatt backup transmitter. By stacking gallium arsenide cells, which capture blue (shorter) wavelengths of light, on top of gallium antimonide cells, which capture red (longer) wavelengths, scientists at Boeing were able to increase the efficiency to 37 per cent. The most commonly used photovoltaic cells are still based on silicon, although there are a wide range of options, including single crystal thick, ribbon and thin films. Transparent solar panels which could replace ordinary windows were patented by Michael Gratzel in 1991.

Tidal power

Medieval tidemills were mentioned in an earlier chapter. In these, the rising tide was trapped and the water then allowed to flow back through a sluice to turn a water-wheel. On the face of it, such systems could readily be scaled up to provide large sources of power with efficient modern water-turbines replacing the crude water-wheels. Careful study shows, however, that there are rather few places in the world where local circumstances – including a minimum tide of 4 metres (13 feet) – are favourable. One such is on the Rance in France, where Electricité de France installed a 750-metre (2460-foot) barrage in 1966.

Wind power

Wind power has long ranked with water power as one of the most important sources of mechanical power and it, too, is a promising candidate as an alternative source of energy today. In one sense, it is already very successful: hundreds of thousands of small windmills have been erected in the 20th

medieval tidemills 116 **OPEC oil crisis** 255, 264, 275, 278, 297, 340 **solar power station** 277 **windmills** 114–115, 166, 231

'Wind-farming' has become popular in California because of government grants and high local wind velocities in certain mountain passes. This 300-unit installation is at Altamont Pass.

century, many of them since the Second World War, for local use in remote areas – to pump water or generate electricity for domestic use. But so far attempts to scale such devices up to a size that, even in groups, might contribute to national needs have not been successful, despite some ingenious variations in design. A number of experimental installations have been built and some interesting results obtained. In the Orkney Islands of Scotland, for instance, two large windmills with blades of up to 60 metres (200 feet) across provide a significant proportion of the electricity required. However, it is difficult to see wind power as a viable option in the short to medium term. One major problem is that large windmills must work efficiently under a wide range of conditions, from winds of a few metres per second to the fiercest winter gales.

Geothermal power

The interior of the earth below its outer crust is very hot, as is evidenced from time to time by volcanic eruptions and the more permanent phenomena of hot springs. In some places this surplus energy has been successfully harnessed. Reykjavik, in Iceland, has long had a central heating system fed from hot springs some miles from the capital and the superheated steam that gushes from the ground at Larderello, in Italy, has been used to generate electricity since 1905.

But these are all very special cases, taking advantage of unusual local circumstances. Of more general interest is the possibility of utilizing the general temperature difference between the earth's surface and rocks at a depth of say 3000–4000 metres (10,000–13,000 feet). Here the general approach has been to circulate cold water through permeable rock, returning it hot to the surface, and some successful demonstration schemes have been launched to supply hot water to dwellings. A practical problem, however, is that rocks are not good conductors of heat and this severely limits the rate of flow. Also, unfortunately, by no means all rocks are permeable, especially the hottest ones at great depth, which have been heavily compacted by the huge weight above them. In such cases, it is possible that the rock might be explosively shattered deep underground to provide a suitable heat-exchange area. Experiments on these lines have been conducted at Los Alamos in the United States, at a depth of 3000 metres (10,000 feet).

The direct technological response to the 1973 crisis was to develop conventional fuel sources outside the OPEC area, to exercise fuel economy and improve the efficiency of machinery of all kinds, and to investigate alternative sources. No less important, however, was the advent of fuel accounting. There was at last a general recognition that almost all industrial operations contain an accumulated energy content. On the farm, for example, there is an energy content in the manufacture of fertilizers, in their distribution and their spreading. Awareness of the real cost of energy was one of the major changes in Western industry from 1873 onwards.

Heat locked up in the earth's crust, and the rise and fall of the tide, are also potential sources of energy. These pictures show a geothermal installation at Monitombo, Nicaragua (below right), and a tidal power station in the Rance, in France (below left).

CHAPTER THIRTY

Medicine and Public Health

Until around the middle of the 19th century medicine hovered somewhat uneasily between an art and a science – an art in the sense that physicians depended very much on judgments made empirically in the light of experience; a science in the sense that over the centuries clinical observations had been to some extent systematized. In the 16th century, for example, Vesalius's magnificently illustrated *De humani corporis fabrica* (1543) revealed an extraordinarily acute observation of the details of human anatomy; Thomas Sydenham, a century later, was an equally acute recorder of the clinical symptoms of diseases. The great 16th-century herbalists, such as Leonhard Fuchs and John Gerard, accumulated a wealth of empirical knowledge on medicinal properties of plants: from this emerged some very valuable and specific remedies, such as the use of foxglove (digitalis) in certain kinds of heart disease and salicin (derived from willow) for relief of rheumatism. Later, in the 18th century, Edward Jenner's observation that infection with cowpox conferred immunity to smallpox led to a dramatic decline in the incidence of the disease through widespread vaccination. These were but solid islands of truth in a sea of ignorance and, very often, superstition. The great hindrance to the advancement of medicine was a lack of understanding of the underlying principles. Even when the diagnosis was clear – as in, for example, tuberculosis or rickets – the physician was all too conscious that the best he could offer were palliatives and sympathy. With some diseases, this is still effectively the position today. Meanwhile, the whole nature of medical practice has been transformed by applying many new advances in technology in addition to the fruits of scientific research over a wide field. It is fair to say that while modern medicine has now achieved the status of an exact science it is now recognized as a branch of technology in its own right.

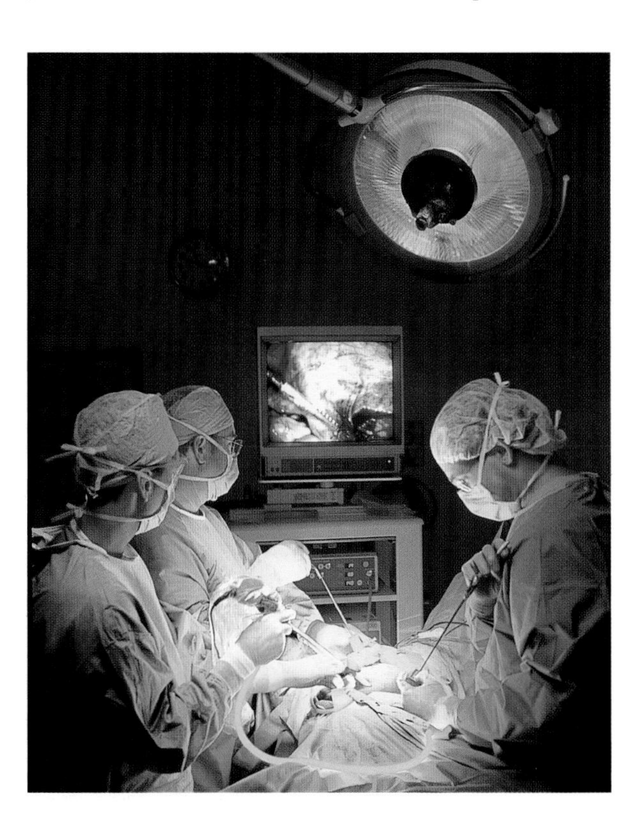

Surgery

To one major branch of medicine this generalization does not fully apply, however, for the surgeons had from the earliest times used a formidable array of instruments in pursuit of their profession. Many prehistoric skulls, for example, show evidence of trephination, the cutting away of bone to remove splinters, relieve pressure or, no doubt, provide an escape for evil spirits. Many of the patients clearly survived, as is evidenced by healing of the originally rough edges of the incision. Military surgeons used their sharp knives to clean up flesh wounds and adapted the saw and other tools of the carpenter to perform amputations. A set of surgeon's instruments uncovered in Pompeii is not very different from those used in the 19th century. Not all surgery was rough and

Left: Laparoscopic surgery enables surgeons to operate through very small incisions without direct contact with the patient. Surgical staff guide their instruments by looking at a monitor; a tiny camera sends an image from inside the patient.

Above: Until well into the 19th century surgeons' instruments differed little from those used in classical times. This Roman collection includes a box for drugs, bleeding cup, rectal speculum, scalpel, probes, forceps, hook and spoon.

anaesthesia 280–281 **antisepsis and asepsis** 281 **chemotherapy** 282–283 **early medicine** 32–33

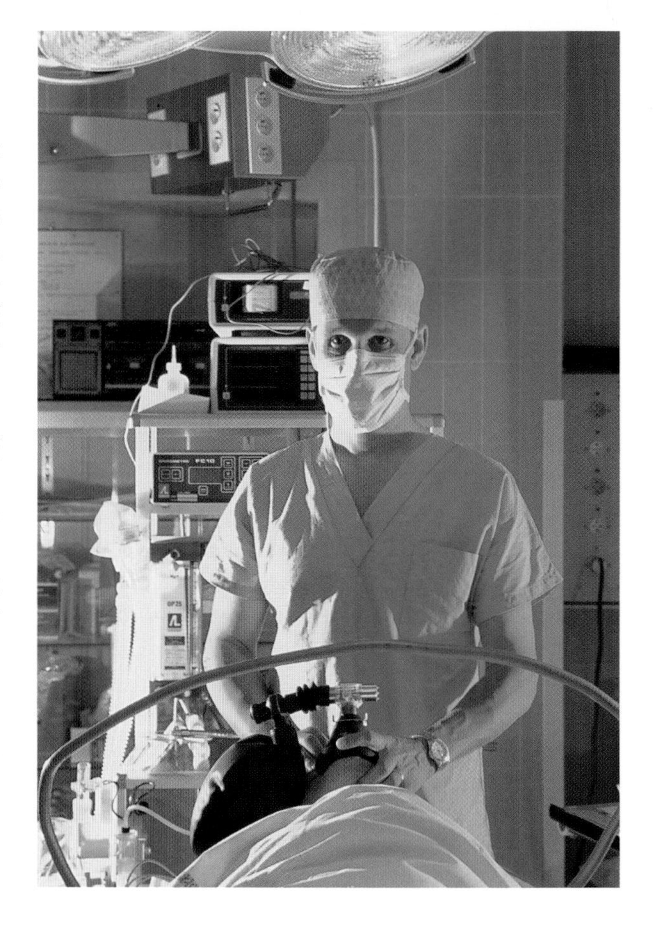

ready, however. For example, the *Sushruta Samhita*, a great Sanskrit medical encyclopaedia probably compiled about the beginning of the Christian era, already minutely describes the technique for removal of a cataract (opaque lens) from the eye. Indian surgeons also devised a skin-grafting operation to build new noses: apart from the normal hazards of life, in many regions amputation of the nose was the penalty for adultery.

Even where surgery was a success, many operations required a lengthy and painful recovery period as a result of the trauma incurred by the surgeon's scalpel. In efforts to speed recovery and lessen discomfort, many surgical procedures are now performed with minimally invasive techniques using laparoscopic or 'keyhole' surgery. Rather than requiring an incision in a patient large enough to accommodate a pair of hands, this type of surgery relies on openings of only 1–2 centimetres (½–¾ inches) into which long-handled instruments, fibre-optic light sources and tiny remote cameras are inserted. Where once surgeons responded to direct contact with their patients, surgery now takes place by remote feedback, with surgeons responding to visual images displayed on a television screen and to haptic (tactile) feedback from the resistance felt at the handles of their instruments.

The success of a surgeon depended very much on his manual dexterity – particularly the speed with which he could complete his operation – and his knowledge of anatomy, but even the most skilful had two major problems to overcome. One was the limited capacity of the patient – perhaps already in a state of debilitation or shock – to withstand the severe pain and trauma of a major operation. The other was that even if the operation was successful, the patient was very likely to succumb to subsequent infection. Not until the 19th century did these problems begin to be overcome, with the advent of anaesthesia, and the allied techniques of antisepsis, asepsis and chemotherapy. Today, mortality after surgery is still distressingly high; however, death is now rarely the consequence of the operation itself, but rather of its failure to arrest the course of the disease treated.

Anaesthesia

Although many primitive people are reputedly able to bear severe pain with great stoicism, one of the main difficulties of a surgeon was the struggles of a frightened, pain-stricken patient. While various painkillers were used – such as henbane, hemp, opium and alcohol – all too often the surgeon had to rely on the brute strength of a team of assistants, or some sort of harness, to hold his patient still. Not until the very end of the 18th century did any hint of relief appear. Then, in 1799, Humphry Davy, working at the Pneumatic Institute in Bristol, prepared nitrous oxide gas and discovered that it had anaesthetic properties – it also provided the essential ingredient of fashionable laughing gas parties – which he suggested might be used in surgery. Not until 1844, however, was this suggestion seriously followed up, when Horace Wells, in the USA, used it as a dental anaesthetic. Two years later, William Morton, another American dentist, used ether as an alternative; Robert Liston, a Scottish surgeon, introduced ether anaesthesia into Europe. At about the same time, chloroform also began to be used and, in 1880, William Macewen, another Scotsman, made the important advance of administering chloroform through a tube introduced into the trachea. This is particularly valuable when the lung is collapsed, as in operations involving opening the chest cavity.

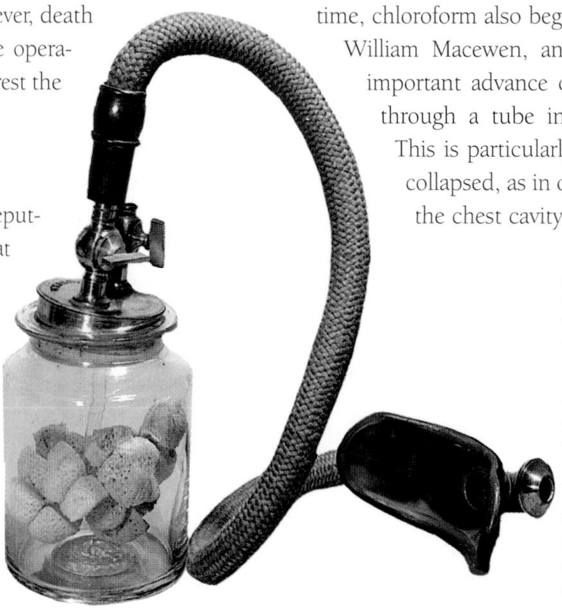

Advances in anaesthesia are perhaps the single most important factor in the success of modern surgery. Today's anaesthesiologist (above) has at his command a complex array of apparatus for administering anaesthetics and monitoring the patients.

Today's sophisticated equipment stands in sharp contrast to the primitive inhalation equipment (left) that sufficed at the turn of the century.

At the beginning of the 20th century anaesthesia was still primitive. Today, however, the situation is very different. The anaesthesiologist is a highly specialized and vital member of the surgical team, having at his or her command elaborate equipment to monitor the patient's heartbeat, brainwave activity, rate of respiration, temperature and blood pressure; to administer not one but several anaesthetics as changing circumstances demand; and to give oxygen or carbon dioxide as the patient's condition requires them. It is only this transformation of anaesthesia that has made possible the prolonged surgery – often lasting for many hours – that such modern techniques as organ transplants demand.

Inhalation anaesthesia still remains of great importance, but other techniques, including endotracheal anaesthesia, already mentioned, are now also widely used. Perhaps the greatest single advance was the introduction of intravenous anaesthesia. This seems to have been first used about 1874 by the French physician Pierre Ore, who used Chloral. Various barbiturates were tried after the discovery of barbital (veronal) by Emil Fischer in 1902, but it was not until the 1930s – when the German physicians Helmut Weese and W. Scharpff introduced evipan – that a really satisfactory intravenous anaesthetic for short operations was available. For obvious reasons, the technique was especially valuable for operations involving the head and neck. Another variant, also practised from the beginning of this century, was spinal anaesthesia (now commonly referred to as a 'spinal tap') – in which the patient remains conscious – involving injection into the spine between the lumbar vertebrae. This required great skill because of the ease with which the needle broke, but by the time a satisfactory needle had been developed, in 1950, relaxing drugs such as curare had come into use.

For many minor operations, as in dentistry, a purely local painkiller is all that is needed. At the end of the last century Carl Koller used cocaine in eye surgery, but this was quickly superseded by the much less toxic procaine, a synthetic variant. Today it is seldom used and lignocaine, which is very stable and acts rapidly, is probably the most widely used local anaesthetic. Local anaesthetics, and many drugs, are administered by a hypodermic syringe, invented in 1853 by a French physician, Charles Gabriel Pravaz.

It must be remembered that the modern pharmaceutical industry offers a multiplicity of different formulations of almost every basic drug. While decisions about the marketing of these formulations are naturally strongly influenced by commercial considerations, they are often based on sound medical reasoning.

Antisepsis and asepsis

The value of anaesthetics was quickly recognized by surgeons all over the world and the number and range of operations increased rapidly. Unhappily, so too did the incidence of post-operative infection – every large hospital was subject to

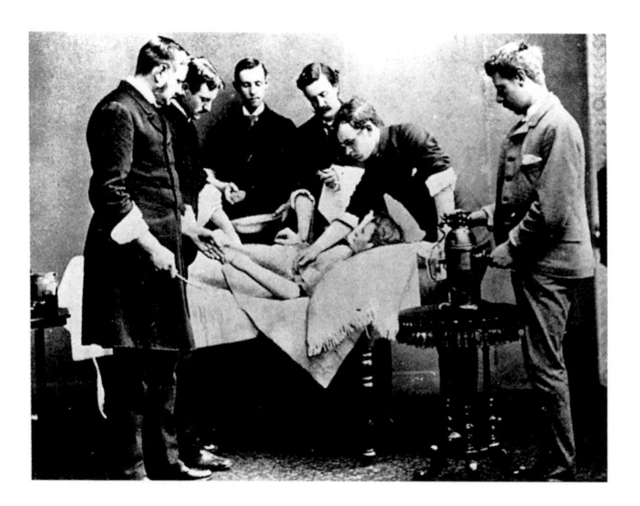

In former times, post-operative infection was as great a hazard as the surgical operation itself. The mortality rate was much reduced by the adoption of Lister's antiseptic spray technique, shown above in use in an Aberdeen hospital in 1869.

'hospital fever'. In many, the mortality after surgically successful amputations, for example, was more than 50 per cent. How this fever spread was not properly understood, though it had long been recognized that fever could be spread by contact from one person to another, as witness the social segregation of lepers and the victims of plague. There was some notion that infection was carried by impure air but then, in 1857, Louis Pasteur demonstrated that natural fermentations are caused by micro-organisms and he went on to show that many human and animal diseases, such as anthrax, are also caused by micro-organisms.

The first surgeon to see the practical significance of these results was Joseph Lister, who in 1865 was professor of surgery at Glasgow University. He discovered that carbolic acid was a powerful germ-killer and used this both to impregnate wound dressings and to spray the air in the operating theatre. The result was a dramatic drop in the post-operative mortality rate. After initial scepticism, Lister's technique was eventually widely adopted, thanks in considerable measure to the enthusiasm of his disciple, Richard von Volkmann of Halle. In paying tribute to Lister, who can fairly be said to have revolutionized surgery, we should not forget the much more obscure Hungarian doctor, Ignaz Semmelweis, who in the 1847, suspecting that infection was carried on the hands of doctors and midwives, drastically reduced the incidence of puerperal fever in his hospital by insisting that everybody in obstetric wards thoroughly wash their hands in bleach before attending patients. But his views, too, aroused much controversy and he lacked the forcefulness of a Lister to overcome this; ironically, he died in 1865 from an infected finger before his contribution was recognized.

Lister's work was important in showing what could be achieved by a systematic attack against lurking germs, but it was not the only method to this end. It was succeeded early in the 20th century by the technique of asepsis, which does

development of drugs 266 **laparoscopic surgery** 279 **organ transplants** 287 **sanitation and disease** 193–194, 289–290

not seek to destroy germs but rather to exclude them altogether before the start of the operation. The surgeon and his assistants wear sterilized clothing and face masks to avoid droplet infection; every instrument and dressing is sterilized before use; and the patient's skin is thoroughly cleaned and sterilized before any incision is made.

Asepsis radically changed medical practice. Most instruments can be satisfactorily sterilized by boiling them in water, but a few dangerous germs can resist this treatment. A very important piece of equipment in every modern hospital is the autoclave, a superheated steam sterilizer in which water is raised above its normal boiling point through the use of pressure: it is, in effect, no more than a scaled-up version of the familiar domestic pressure-cooker – which itself is none other than the digester invented by Denis Papin in 1679.

However, not everything can be heat-sterilized. The walls, floors and working surfaces of operating theatres, for example, still have to be washed down with strong disinfectants. For the patient's skin, milder products must be used, and in the 1930s the chemical industry developed a number of synthetic products which are strongly antiseptic yet non-irritant in action. Among the best known of these in Britain are TCP (tricresyl phosphate) and Dettol (chlorxylenol). Powerful post-war additions are cetrimide and Hibitane (chlorhexidine).

Immunization

From Pasteur's crucially important discovery of the true nature of infectious diseases evolved two important methods of treatment. One, developed by Pasteur himself, arose from his chance observation that inoculation with attenuated germ cultures can induce a mild attack of the corresponding disease, but thereafter confer lasting immunity. This method of immunization provided a logical basis to Edward Jenner's empirical technique of vaccinating against smallpox by means of a mild infection with cowpox. In 1885 Pasteur's fame was assured by his successful use of vaccination against the deadly disease rabies.

Today, vaccination has virtually eradicated smallpox throughout the world, and poliomyelitis, diphtheria and tetanus also have largely been defeated in the Western world. With the development of civil aviation, travellers can be exposed to a variety of infections which they would not encounter normally and immunization against a number of diseases is now common practice: in many countries it is mandatory as a condition of entry. These diseases include cholera, typhoid and yellow fever.

Chemotherapy

Immunization can protect the individual and can be very successful in containing epidemics, but combating an established infection is a very different matter. The identification of specific microbes as the causes of disease suggested another

In the 18th century Edward Jenner demonstrated that immunity to smallpox could be conferred by a relatively mild infection with cowpox. As late as 1905 Parisians (right) were being vaccinated in this primitive way: cowpox infection was transferred direct from the shaven skin of a heifer to the arm of the patient. This contrasts with the modern picture of yellow fever immunization in a developing country (below right).

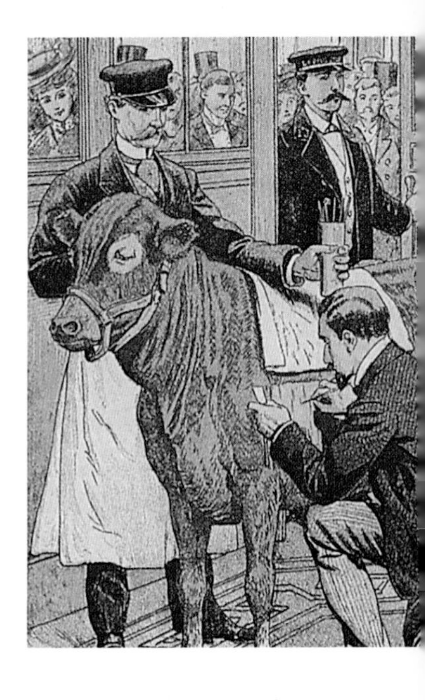

line of attack, namely to find substances lethal to germs but harmless to the tissues of the body. At first this seemed a vain hope, for it appeared that in this respect all antiseptics, such as those discussed above, were alike – they attacked all living tissue indiscriminately. Then, in 1906, Paul Ehrlich, in Frankfurt, discovered that a synthetic arsenical compound, which he named salvarsan, was active against the organism causing human syphilis and this was soon introduced into medical practice.

In his enthusiasm Ehrlich called salvarsan a 'magic bullet', a chemical which could seek out germs but leave the body unaffected, and for his new method of treatment he coined the word chemotherapy. In fact, salvarsan was a far from ideal drug. At best it had unpleasant side effects and at worst, in unskilled hands, it could be fatal. Moreover, it was not effective against other microbial infections. Nevertheless, salvarsan established a vital principle: it convinced the sceptics that some differential toxicity between human tissue and germs was attainable.

However, little progress was made until 1927, when Gerhard Domagk, of the great German chemical company IG Farbenindustrie, embarked on a systematic screening of all his company's great stock of chemical samples to see which, if any, were active against the germs responsible for such serious infections as streptococcal septicaemia, meningitis, and – that dreaded 'captain of the men of death' – pneumonia. Any proved to be active would then be tested for toxicity towards experimental animals and, if this hurdle were passed, submitted for clinical trial. For five years, no significant result was obtained but then, dramatically, a substance originally made as a red dye for leather, Prontosil Rubrum, proved highly effective in treating streptococcal infections in mice. At the Pasteur Institute in Paris, J. Trefouel showed that the active part of the dye molecule involved was a relatively simple substance known as sulphanilamide. Thus was born the extremely important and effective range of drugs known collectively as the sulphonamides. Very quickly thousands of

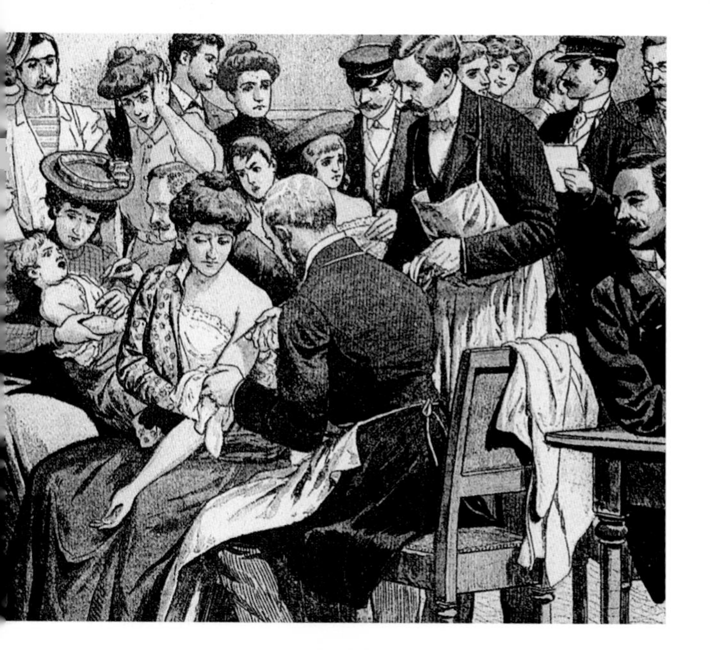

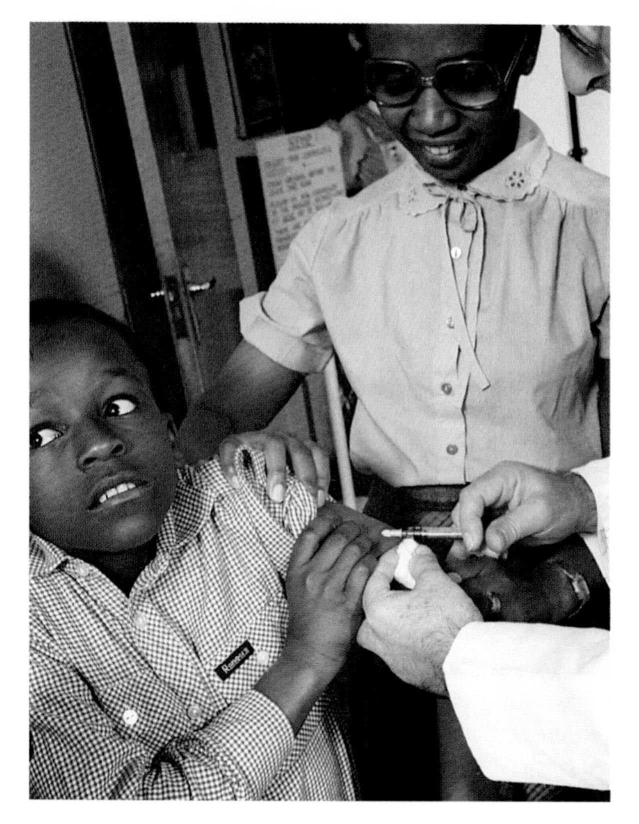

variants were synthesized, a few – such as sulphapyridine and sulphathiazole – proving active against other pathogenic organisms and acceptable for chemical use. Here, at least, was something approaching a magic bullet.

Meanwhile, an even more successful development was unfolding. In 1928 Alexander Fleming, a bacteriologist working at St Mary's Hospital, London, accidentally discovered that a mould, *Penicillium notatum*, produced a substance active against staphylococci and a range of other pathogenic organisms. He named this substance penicillin, but he was unable to isolate it and failed to appreciate its unique properties: as a result his interest lapsed and he turned to other research. Not until 1939 was there any further serious interest in penicillin and then not because of its potential medical value but because the antagonism between the mould and microbes was an interesting example of a very general phenomenon known as antibiosis.

In that year Howard Florey, an Australian by birth and a pathologist by profession, and Ernst Chain, a refugee biochemist from Germany, embarked on a general study of antibiosis at the Sir William Dunn School of Pathology at Oxford. By a fortunate chance, one of the examples they chose involved Fleming's penicillin. Quite quickly penicillin was purified sufficiently to show that it promised to be not only uniquely powerful as a chemotherapeutic agent but also virtually non-toxic to human tissue. By that time, however, Britain was at war and air raids had begun. Under the circumstances, the British chemical industry could not embark upon development of penicillin and in 1941 Florey and a colleague, N. G. Heatley, made a three-month tour of the United States to enlist the support of American pharmaceutical firms. This they did successfully; in view of the immense potential importance of penicillin for treating infected war wounds – as important to the USA as to Britain after the attack on Pearl Harbor on 7 December 1941 – a crash programme of penicillin manufacture was embarked on with high government priority. As a result, sufficient penicillin was available to treat all serious casualties after the invasion of Europe and in the Far East.

Penicillin was the first – and still overall perhaps the best – of a whole range of natural chemotherapeutic agents which, because of their origin, are known as antibiotics. Among the most important of the later ones are the cephalosporins, also developed at Oxford.

Diagnosis and treatment

Mathematics has been called the handmaiden of the sciences because it is of value to all of them. Equally, science could be called the handmaiden of medicine, for the progress of medicine has always depended very much on the application of advances in science not made in direct response to medical needs.

A classic example is X-rays, discovered in 1895 by Wilhelm Röntgen, a physicist working at Würzburg. Their penetrating power is quite different from that of light: they will pass readily through soft tissues but bone and other dense materials are almost opaque to them. Actinically, however, X-rays resemble light and they can be recorded on photographic film.

The diagnostic value of X-rays was quickly realized: they could be used not only to examine fractured bones clearly and painlessly, but also to locate and identify embedded foreign bodies, such as bullets and needles, or those accidentally swallowed. The doctor or surgeon therefore knew exactly what had to be done before he began any surgical

gene therapy 344 **IG Farbenindustrie** 294 **Papin's digester** 159 **Pearl Harbor** 222 **X-rays** 212

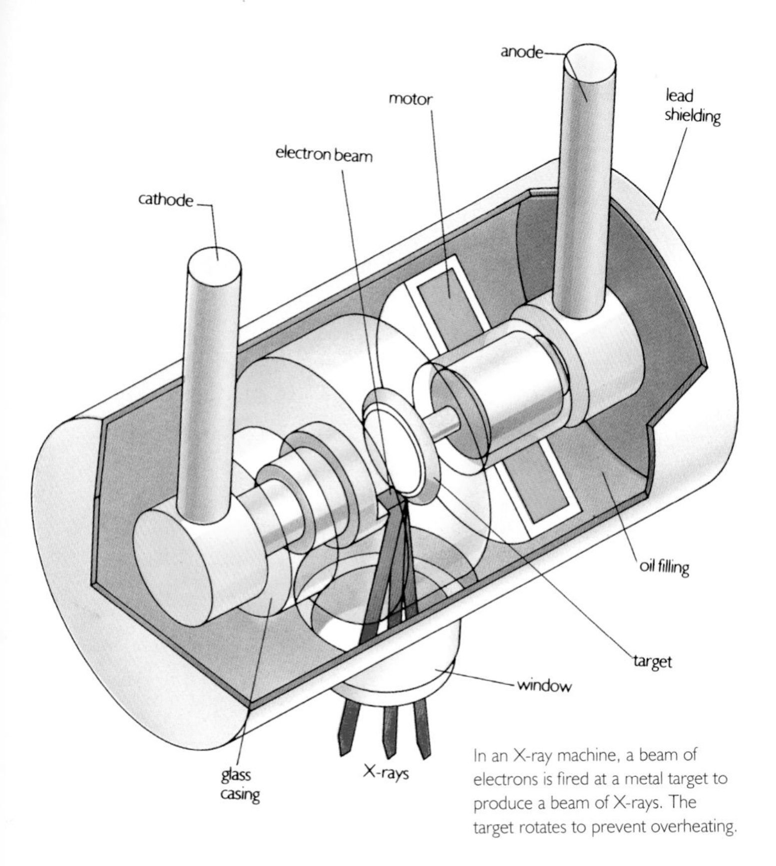

In an X-ray machine, a beam of electrons is fired at a metal target to produce a beam of X-rays. The target rotates to prevent overheating.

procedure. It was difficult to distinguish between different soft tissues, all being more or less transparent, but in 1897 an American physician, William Cannon, devised a dense bismuth (later barium) meal so that abnormalities of the gastrointestinal tract could be examined. By the 1930s, X-ray examination was routine medical practice, an important development being its use to detect tuberculosis of the lungs.

The major post-war development in X-rays was the advent of the CAT (computerized axial tomographic) scanner. This was invented in 1973 (although experimentally tested in 1972) by Godfrey Hounsfield, of the British firm EMI, and was first used to give a complete three-dimensional picture of the skull. In 1979, he and Allan Cormack of South Africa and the United States received the Nobel Prize for their development of computed axial tomography. It was soon developed as the body scanner which enables the same procedure to be applied to the whole body. Its diagnostic value is enormous but unfortunately it is also extremely expensive.

Magnetic Resonance Imaging (MRI) scanners have also been developed and are now the more commonly used equipment for full-body scanning. MRIs use radio waves, magnetism and a computer to examine the structure of bones and joints as well as soft tissue. In this procedure, the patient lies down inside a circular magnet which aligns the protons of hydrogen atoms inside their body. Radio waves then cause the atoms to give off a faint signal which is picked up by a receiver and developed by the computer into an extremely detailed image which can indicate any abnormalities in the structures within the body. It is a painless and safe method for attaining information and has no known side effects such as the radiation from X-rays.

X-rays also found a second use in medicine, for the treatment of cancer. This involves focusing an X-ray beam on tumour tissue, and at the same time exposing surrounding tissue to much less intense radiation. For this, various devices are used. In one, the patient's body is slowly rotated, while the beam is focused firmly on the tumour: in another, a number of rays, each relatively weak, are brought to a single focal point at the site to be treated. For this sort of treatment very powerful radiation is required, presenting a hazard of injury from excessive exposure, especially among those who regularly operate the machinery. Unfortunately, this possibility was not immediately recognized and some pioneers suffered fatal illnesses. Today, with careful monitoring and the use of lead screening, this is a thing of the past.

X-rays were so called because their nature was at first unknown. At about the same time another novel kind of radiation was discovered. In 1896, the French physicist Antoine Henri Becquerel discovered that uranium emitted a penetrating radiation: for this discovery of the phenomenon known as radioactivity he was awarded a Nobel Prize in 1903. Among the pioneers of this new science were Marie and Pierre Curie, who discovered radium in 1898. It was a long and difficult task: from 800 kilograms (17,600 pounds) of pitchblende they extracted only a single gram (1/28th of an ounce) of radium. Unfortunately, this powerful new radiation, too, was an unsuspected hazard, particularly as a cause of leukaemia, of which Marie Curie herself died in 1934.

Not realizing the risk, Becquerel casually carried some uranium in his waistcoat pocket and suffered a burn as a result. Ironically, this suggested that the new radioactive elements – and particularly radium – might be used therapeutically in the same way as X-rays, which proved to be the case. Because of the different nature and source of the rays, one method of applying them was to insert a small needle carrying a radioactive seed into the tissue to be treated. After the Second World War the scope of this treatment was much extended with the availability of a great range of artificial radioactive elements, such as radio cobalt, from atomic piles. The radioactive gas radon is particularly useful in treatments because it loses nearly all its radioactivity in three weeks and therefore the seed need not be removed.

The electrocardiograph

Heart diseases are both common and serious, and clear diagnosis of what is wrong is an essential first step to treatment. In this field one of the most valuable instruments is the electrocardiograph, which depends on the fact that a contracting muscle generates a small electric current that can be detected and measured through electrodes suitably placed on the body. In this way, it is possible to monitor and record the pulsation not only of the heart as a whole but of each of the four main cavities separately.

A

D

B

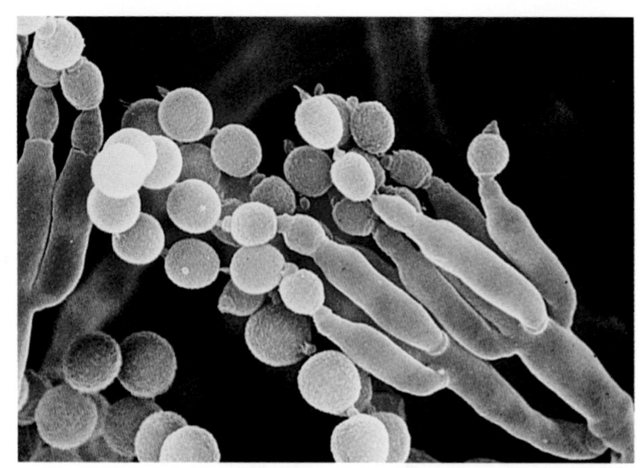

E

C

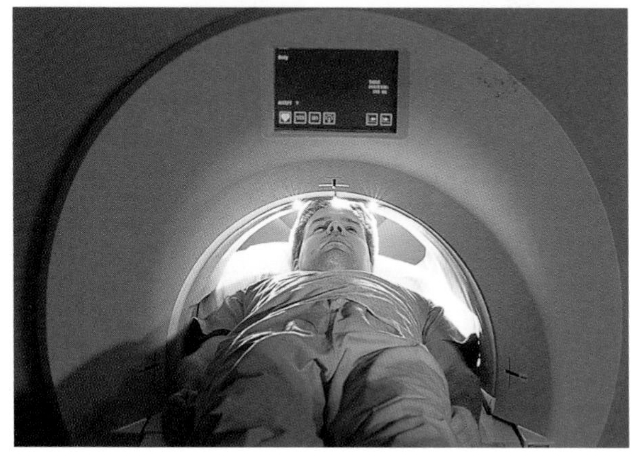

F

Penicillin was discovered in 1928 by Alexander Fleming, a bacteriologist at St Mary's Hospital, London: this memorial window is in the nearby church of St James (A). He failed to recognize its importance, however, and its identification as a uniquely powerful chemotherapeutic agent was due to the research of Howard Florey and Ernst Chain. Their early (1942) laboratory (B) contrasts sharply with the fermentation vats later used for mass production (C). The micro-organism first identified as a source of penicillin was the mould *Penicillium notatum* (D) but other species such as *Penicillium chrysogenum* also produce it: the fruiting body of this mould is shown at × 850 magnification (E). CAT scanners produce detailed images of bone as well as soft tissues by directing thin beams of X-rays in a circular tube at a patient who lies on a mobile bed (F).

Marie and Pierre Curie 212 **penicillin** 283, 285 **use of radiation in food preservation** 300–301 **X-rays** 212

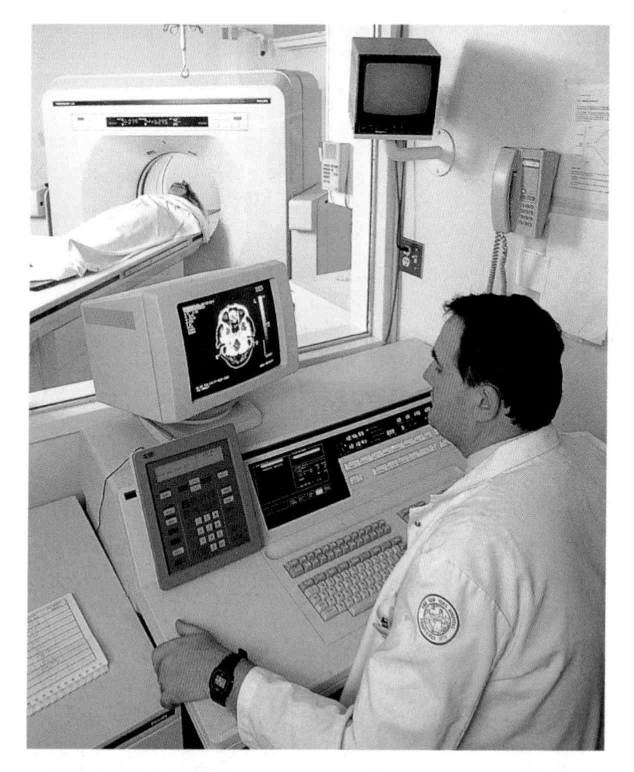

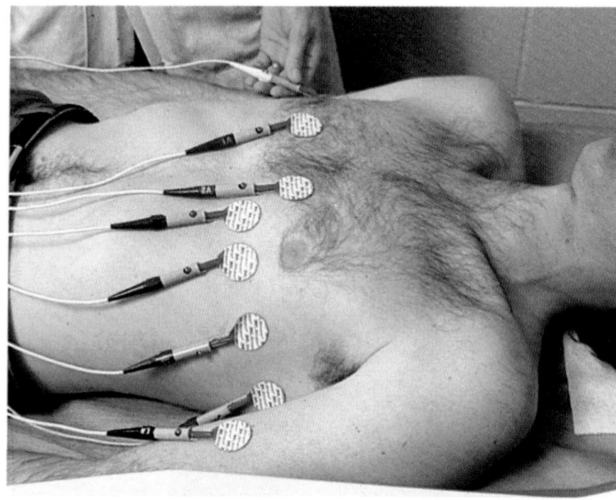

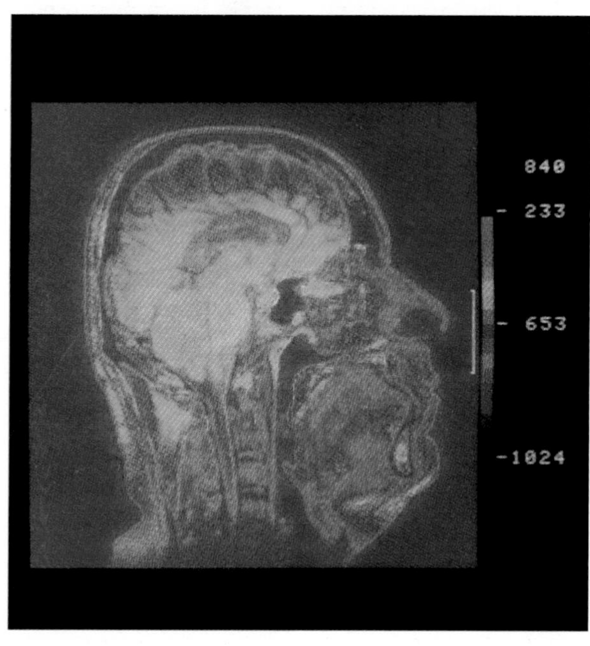

Above left: The CAT scan is directed by a laboratory technician from an instrument panel like the one shown.

Above right: In recent years physical methods of diagnosis have become of increasing medical importance. Here a patient's heart rhythm is being monitored with an electrocardiograph.

Far left: The electroencephalograph analyses electric currents in the cortex of the brain, and is a valuable diagnostic device in cases of head injury and diseases of the brain.

Left: The computer axial tomographic (CAT) scanner was devised in 1973 to scan the human head.

An early pioneer of electrocardiography was Augustus Désiré Waller, who recorded the first human electrocardiogram in 1887. It was not until 1901, however, that Willem Einthoven, professor of physiology at Leiden, developed the loop galvanometer to measure more accurately the changes of electrical potential during the heartbeat; he went on to build the first electrocardiograph (EKG) in 1903. It was very large and weighed nearly 300 kilograms (660 pounds), but over the years has been made very much more compact. By linking it with a monitor, the beat of the heart can be viewed over a prescribed period, enabling intermittent irregularities to be picked up: the electrocardiograph is among the instruments commonly used by anaesthesiologists to monitor patients.

Another valuable diagnostic aid in the treatment of heart disease was the cardiac catheter, devised in 1929 by Werner Forsmann, who was awarded a Nobel Prize 30 years later. This is a very fine flexible tube which can be gently pushed through a vein in the arm until it reaches the heart and can sample its contents.

The brain, too, generates electricity when it is active, especially in the cortex. Although this was known in the 19th century, few physiologists thought it worthwhile to investigate the phenomenon. The first to do so systematically was Hans Berger, professor of physiology at Jena in Germany. From 1924 he began to record the brain rhythms of dogs and later of patients in his psychiatric clinic. He constructed his first electroencephalograph (EEG) five years later, but when he published his results psychiatrists showed little interest. Fortunately, the local optical firm of Carl Zeiss supported him and enabled him to build a much improved instrument. Since the Second World War the electroencephalograph has become as important in the diagnosis of cerebral disease and injury and in the support of patient monitoring while under anaesthesia as the electrocardiograph in the study of the heart.

However, it was not only the relatively complex machines that improved diagnosis. Scipione Riva-Rocci's simple sphygmomanometer – introduced in 1896 and used throughout the world still – enabled doctors to measure blood pressure with no more than a stethoscope and an inflatable cuff.

Vitamins and hormones

By the beginning of this century the basic factors in human nutrition had been worked out: it was clear that a healthy diet required a suitable balance of carbohydrates, proteins and fats. It was beginning to be realized, thanks particularly to the work of Frederick Gowland Hopkins at Cambridge, that it was also essential to have very small amounts of what he called 'accessory food factors', now known as vitamins. This quickly opened up a whole new field of medicine, for many major diseases proved to be the consequence of vitamin deficiencies: scurvy (vitamin C), beri-beri (vitamin B), rickets (vitamin D) and pellagra (nicotinic acid). Such diseases can be prevented by suitably supplementing a deficient diet and the preparation of vitamins – some synthetic, such as vitamin C – is today an important activity of the pharmaceutical industry.

Although some are synthesized by micro-organisms normally present in the gut, the body's need for vitamins is largely met through the normal intake of food: vitamin C, for example, is present in fresh fruit and green vegetables. It also became apparent that the healthy functioning of the body depends too on the internal production of small quantities of other extremely active substances known as hormones: the source of these are the ductless glands, so called because their products are discharged directly into the bloodstream instead of through ducts, like bile. One of the most important of the hormones is insulin, normally produced by the pancreas; failure to do so causes the disease diabetes, which can be fatal.

This hormone was discovered in Canada in 1922, by Frederick G. Banting, Charles H. Best and John Macleod. Diabetes can now be readily controlled by regular intake of insulin, and most diabetes sufferers can lead relatively normal lives. Today, many hormones are in regular medical use, including hormones to treat a variety of sex-related complications, as in the menopause. Perhaps the most significant of all developments in this field, in terms of social consequences, has been the contraceptive pill.

Organ transplants

The regular use of skin grafting in ancient times has already been noted. The success of this depended upon the use of the patient's skin, often taken from the forehead but sometimes from the cheek. Attempts to use another person's skin – except possibly that of an identical twin – ended in failure because the graft was rejected. This is a general effect, made use of in immunization, when the body is made to produce antigens which will destroy the infectious micro-organisms.

In many ways the human body resembles an automobile: if one major component, or organ, fails it comes to a halt. An important difference is that whereas an automobile can easily be made to function again by fitting a new part, the human body will normally reject a new organ from another person if it is substituted for a diseased one. It is this rejection mechanism, as much as the delicate and prolonged surgery involved, which has hitherto limited the long-term success of organ transplants. Nevertheless, in the last 30 years or so a good deal of progress has been made in the identification of the factors most conducive to satisfactory transplantation which may enable suitable donors to be selected, in the development of immunosuppressive drugs which diminish the tendency towards rejection and in the development of synthetic organs and cell/tissue engineering techniques which permit new organs to be cultured in a laboratory and subsequently to develop fully inside the patient.

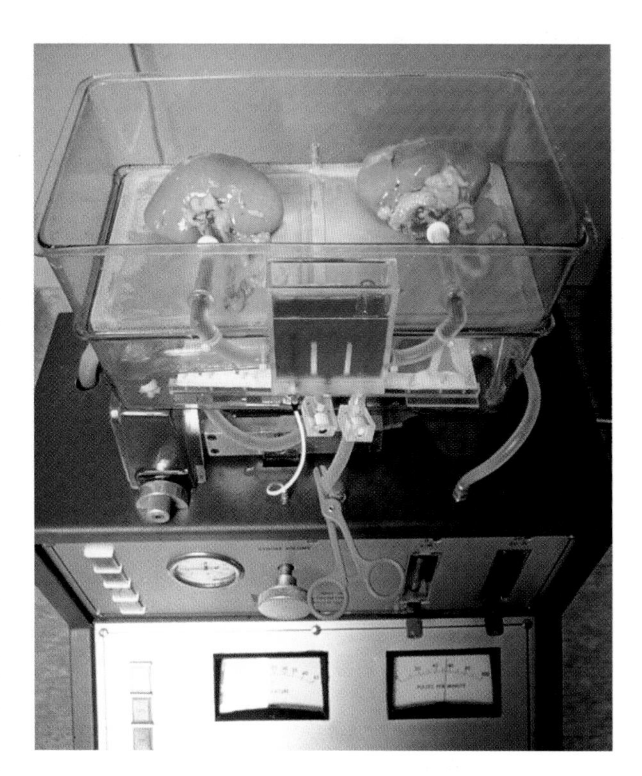

In organ transplant surgery the greatest success has been with kidneys. An essential requirement for success is to keep the donor kidneys in a healthy state until the graft is made. This can be done either by storage in ice-cold saline or perfusion with plasma, as here.

CAT scanner 284 **cell/tissue engineering** 288 **immunization** 282 **synthetic organs** 288–289, 343

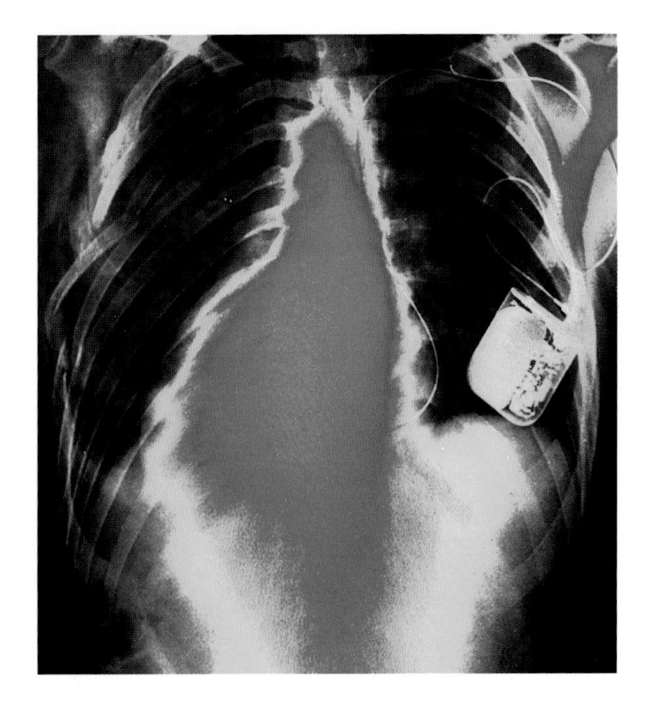

The success rate with heart transplants has been disappointing but millions of patients benefited from the insertion of a pacemaker.

The greatest success has been with kidney transplants, first effected by John Merrill in the USA in 1953. Since then tens of thousands of successful operations have been performed. A factor in this success is that the kidney is a paired organ, and it is perfectly possible to survive with only one. In December 1967, Christiaan Barnard of Cape Town University performed the first human heart transplant; three years earlier, James Hardy, at Jackson University Hospital, Mississippi, had transplanted a chimpanzee heart into a 58-year-old man. Neither was successful in the long term, with the recipient of the human heart surviving for 18 days, while the latter survived a scant three hours. The technique is very specialized, and the success rate is far lower than for kidneys, but surgical units have been set up in Britain, the USA and elsewhere. Liver and lung transplants have also been effected. Another field in which there has been a considerable degree of success is that of bone marrow transplants for patients suffering from aplastic anaemia or severe leukaemia.

At this stage, it is difficult to assess the eventual role of organ transplant surgery. Although the technique has been steadily improved, there is a lack of suitable donor organs and it is still far too demanding a specialist resource to be available to more than a tiny minority of patients who might benefit from it. However, this situation might change considerably if the problem of rejection could be fully solved. Significant advances have been made recently which have generated three feasible alternatives to human transplantation: xenotransplantation; the use of artificial organs or other synthetic aids; and tissue engineering.

Xenotransplantation is the transplantation of organs between species. For human organ replacement, the most likely donor species is the pig, whose organs bear a remarkable similarity to humans and consequently have been used in organ research for over half a century. Two possible means of decreasing the probability of rejection by the host are through genetic modification of the animals to create organs with greater compatibility and the formation of hybrid synthetic/biological organs through tissue engineering.

In the 1990s, chemical and production advances led to the creation of a range of immunosuppressive plastics for the medical industry – synthetic polymers which do not prompt rejection and which, through their structure and composition, can support the growth of human tissue. Hybrid organs are formed by separating the donor organ from the host with a thin layer of synthetic polymer which contains a layer of human cells. The cells are harvested from the host and grown *in vitro* in the immunosuppressive polymer 'scaffold' until they have firmly 'taken root'. Obviously, this technique requires time to generate a new organ and is, therefore, at present, not a viable option for trauma victims. In addition to hybrid organs, the development of artificial bone, cartilage and other connective tissue has improved greatly.

Tissue engineering is also enabling some organs to be grown entirely from a few cells of the host. The best method is to rely upon the body's natural biochemical processes by transferring a three-dimensional matrix of appropriate cells – grown *in vitro* – to the desired site where growth may proceed within the person. This approach was pioneered in the 1970s and 1980s by researchers such as Ioannis V. Yannas, Eugene Bell and Robert S. Langer at MIT and Joseph P. Vacanti at Harvard Medical School, and is in use with some skin wound patients and those with cartilage damage.

The complexities of engineering human internal organs stymied 20th-century medicine. In fact, it wasn't until 1999 that the first internal organs of any species were grown and transplanted successfully. A project headed by Anthony Atala at the Laboratory for Tissue Engineering at the Children's Hospital in Boston and the Harvard Medical School led to the growth and transplantation of six beagle bladders. Two types of cell, muscle and urothelial (bladder skin), were co-cultivated on a polymer scaffold before insertion into the dogs, where they performed normally for another 10 months. As Atala had already grown human bladders *in vitro* by the same techniques, the beagle operation provided critical data for comparison before attempting implantation in humans. It also encouraged the belief that human organ growth developments will proceed quickly.

Artificial organs and other synthetic aids are hardly new; however, their development may still be considered to be in its infancy. For example, since the first implantations in the early 1960s, the pacemaker has proved a simple and effective device for regulating the heartbeat in many patients – over 200,000 are inserted annually worldwide – and research in Japan and the USA suggests that mechanical hearts may eventually prove feasible substitutes for defective natural ones. Again, the artificial kidney, first introduced by Willem

Kolff in 1943, has proved an effective – if cumbersome and restrictive – form of long-term therapy for renal conditions. It was also Kolff who invented the artificial heart in 1957. In 1976, his colleague, Robert Jarvik, invented a pneumatic heart for use with patients for whom there is no alternative form of treatment. Although some 90 Jarviks were implanted, their benefits were widely disputed and eventually they were banned in the USA at the end of 1989.

Many of the difficulties of the Jarviks – prevalent blood clotting and a refrigerator-sized pneumatic pump – seem to have been eradicated with the left ventricular assist device (LAVD). Created by Thermo Cardiosystems of Woburn, Massachusetts, and approved for clinical use by the FDA in 1995, the device is comprised of a palm-sized titanium pump, implanted in the patient, and a compact controller with two batteries which is worn outside the body on a harness. Rather than replacing the diseased heart, the HeartMate augments its functions; in many cases, the old heart, surprisingly, regained almost all of its former pumping ability. The device is now designed to be a long-term therapy for heart failure, rather than simply a temporary measure until a transplant can be effected.

The most successful artificial implants have been tissues and structures: skin, bone and cartilage. Recently, research projects have generated artificial muscles created with a polyelectrolytic gel which expands and contracts under electrical stimulation. Bionic limbs, driven by sensors which respond to the electrical impulses created by contracting muscle, have also been successfully utilized.

Telemedicine

The practice of medicine benefited greatly from the growth in global telecommunications. Telemedicine emerged from the convergence of technologies and practices in telecommunications and medicine, and holds significant potential to increase access to underserved populations, to lower costs of consultations and to facilitate training and certification. Telemedicine can broadly be defined as the use of telecommunications technologies for the purpose of conducting research, making diagnoses, transferring patient data and improving medical treatment in remote areas. As such, it is not a new phenomenon, with telephone and radio employed in medicine as early as the 1950s. Telepsychiatry and distance education programmes with state mental hospitals were initiated by the Nebraska Psychiatric Institute shortly after an interstate demonstration at the 1951 World's Fair in New York, and a teleradiology programme began in Montreal about that time.

One of the most likely reasons that radiology was among the first medical specializations to adopt telemedicine is that radiologists had long relied on distance communication, by mail or fax machine, to deliver films from processing facilities or to send them to colleagues for consultation. Current digital imaging techniques, combined with high bandwidth transfer of data have considerably hastened the processing, transfer and sharing of radiology images.

Historically, the telephone, fax and radio have been used in telemedicine, but recently interactive video consultation has become more commonplace. Once the initial investment costs have been covered, video-conferencing facilities lower running costs significantly. Video-conferencing also reduces the patient's expenditure of money and time in travelling great distances for face-to-face consultations with a specialist. Such facilities could support several medical needs, such as home health care and follow-up visits to physicians, as well as training or professional consultation for physicians who live and work in rural areas.

Trauma cases have been improved through the use of portable communication systems which enable specialists in a hospital to diagnose more effectively patients who are still at the scene of an accident and to direct staff in the field.

Recent research projects have proven telesurgery to be possible, provided a network of sufficiently high-speed transfer of data is available. To date, a human gall bladder removal has been performed from across the room, and test operations on animals have been performed over a 5-kilometre (3-mile) distance. Earlier it was noted that minimally invasive procedures were transferring surgical interactions from direct to mediated interfaces where a surgeon no longer comes into direct contact with the patient. Telesurgery, therefore, requires a suitable infrastructure and safety measures rather than new expertise.

Preventative medicine

Preventative medicine is as old as civilization, in the sense that a clean water supply was recognized not only as a valuable amenity but as conducive to public health. Pasteur's identification of micro-organisms as the causes of many diseases led to increasing recognition of the important relationship between water supply and sewage disposal. This brought two results. First, the gradual reconstruction of water supply and sewage systems to ensure that they were kept quite separate. Second, water was not only purified by improved methods of filtration but also chlorinated to sterilize it. Despite the overwhelming evidence for its value as a public health measure, there was considerable opposition to this, which was revived with the introduction of fluoridation, initially in the USA in 1945, to reduce tooth decay.

Sewage disposal is no great problem for small towns on fast-flowing rivers, for the natural processes of decay quickly render it innocuous. In great conurbations such as those situated on sluggish rivers, however, the situation is very different and huge sewage farms have had to be built to treat the waste before discharge. Up to mid-century these relied mostly on filtration, but today a faster air activation process depending on oxygen to break down organic material is used.

genetic engineering 297–299, 344 **nanotechnology** 343 **sanitation and disease** 193–194, 281–282 **surgery** 279–280 **telemedicine** 342

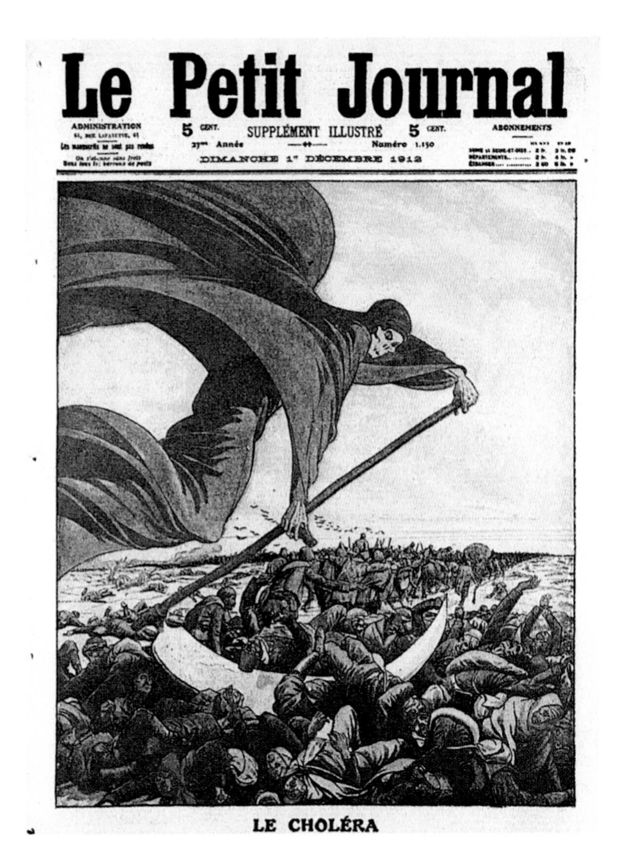

Cholera is one of the most serious of water-borne diseases. In the Balkans Wars of 1912–13 the Turks were losing up to 100 men a day from it, more than from casualties in the field (left). Sewage purification plants (above) are a vital part of the public health system.

Water supply and waste disposal are the two most important weapons in the public health armoury, but there are others. We have referred already to the widespread use of immunization. Other important public health measures include regulations affecting the preparation and sale of food, the control of dangerous drugs and the prevention of pollution, especially air pollution.

Although preventative medicine has long been practised, it has come to assume much increased importance in recent years. Governments have had to face the hard reality that while new medical techniques can reduce suffering and mortality, many can do so only at a cost that puts them far beyond the reach of most patients. Indeed, in the world at large four-fifths of the population have no regular access to any sort of health services, and even in the USA, the world's richest nation, the World Health Organization estimates that 50 million people live in areas formally recognized to be medically underserved and 30 million have no regular source of medical care. In the first year of life, mortality in the advanced Western countries is less than 2 per cent; in Third World countries, such as Afghanistan, it is 25 per cent.

Against this sort of background, and in the face of hard economic reality, there is a feeling that medical strategy has concentrated too much on disease and not enough on health. Overall, money spent on health care for the population as a whole can pay a much higher dividend than that spent on increasingly sophisticated specialist and hospital services. Certainly a few dollars per head annually can produce striking results in the Third World. In Latin America, for example, if the proportion of deaths from infectious and parasitic diseases were reduced from its present level of 22 per cent to 6 per cent – a perfectly feasible target with modern antibiotics and anthelmintic (worm-destroying) drugs – the expectation of life would increase from 45 to 68 years. In the rich and sophisticated countries much is being achieved by campaigns to discourage over-indulgence in tobacco, alcohol and foods known to present a health hazard.

In 1978, at a joint meeting of the World Health Organization and the United Nations Children's Fund, a campaign was formally launched to encourage all governments to revise their health policies in favour of promoting health rather than treating disease. In 1999, the United Kingdom imposed its strictest regulations against the tobacco industry by banning all tobacco advertising on television or billboards by the year 2000, and a near total ban on sports advertising. Only a few sports have a respite for one or two years in which to realign their financial structures so that they are not fully reliant upon tobacco sponsorship.

Below: As health care costs have escalated, many countries have begun to pursue active preventative campaigns, such as the one against smoking promoted in this poster.

CHAPTER THIRTY-ONE

New Aspects of Agriculture and Food

By far the greater part of the world's food is still produced by traditional methods of soil tillage, cultivation and harvesting or by animal husbandry. Indeed, in the underdeveloped parts of the world the agricultural scene is still little different from what it was in pre-Christian times, and likely to remain so for many years to come. At the other end of the spectrum, however, modern agriculture has seen radical changes which have come to be known as the 'green revolution'. Apart from an acceleration in the trend towards mechanization, already noted, and a related move towards larger – and sometimes enormous – holdings, there has been a revolutionary change in the use of chemicals. Apart from the extensive use of fertilizers to promote growth and yield, a whole battery of chemical weapons has emerged to combat the principal pests and diseases of crops. Other agrochemicals have reduced the amount of labour involved in preparing the land before sowing. Rather few new crops have been introduced – though the kiwi fruit of New Zealand has been a notable exception – but selective breeding has resulted in greatly improved varieties of established ones. In terms of animal husbandry the picture is similar – healthier stock and systematically improved breeds.

Food is the main product of agriculture but by no means the sole one: some crops, such as rubber and cotton are grown solely for industrial use, and everyday agriculture and animal husbandry produces industrial by-products such as wool and leather and glandular material for the preparation of drugs such as insulin. Some genetically engineered livestock – so-called 'pharm' animals – are raised solely to produce pharmaceuticals in their milk. New processes of food technology, such as freeze-drying and deep-freeze preservation, have done much to even out the differences between supply and demand, while the great ease of transport between northern and southern hemispheres gives, in effect, two seasons in one year.

Agrochemicals

In bulk, by far the greatest quantity of chemicals used in agriculture is applied to the land as synthetic fertilizers. Properly used, these can produce striking increases in crop yields, but over the years their very cheapness encouraged excessive use which not only produced no additional benefit but might actually be counter-productive: for example, over-fertilized cereals may have excessively long stems which are easily beaten down in storms, making the crop difficult to harvest, and overuse of herbicides can lead to increased immunity in pests, weeds and disease-causing agents. Additionally, excess fertilizer, especially nitrate, may ultimately find its way into drinking water, with potential danger to health. However, in recent years the sharp rise in fertilizer prices following the energy crisis of 1973 has encouraged more responsible use.

The more dramatic change has been in the development of high-value synthetic chemicals for specialist use. What have been produced are products similar to Ehrlich's 'magic

Top: Modern agriculture includes a number of purely industrial crops. Among them is rubber, used in enormous quantities for automobile tyres. This picture shows latex being tapped on a Singapore plantation. Above: Genetic engineering research and development escalated rapidly at the close of the 20th century, generating capital and controversy.

Ehrlich's 'magic bullet' 282 **fertilizers** 206, 213, 264 **food preservation** 43, 152, 299–301 **problems of medical care** 338

bullet' in medicine: chemicals which do not affect crops but destroy pests and diseases to which they are subject. Global estimates of damage by pests, weeds and diseases are always rather suspect, but the figure of £60 billion, produced in the 1970s, is at least indicative of the size of the problem.

Among the worst enemies of growing crops are weeds, especially in their early stages. Traditionally these have been controlled by hoeing either by hand or, later, as we have seen, by horse- or tractor-drawn implements. For clearing ground of vegetation, chemicals such as sulphuric acid or sodium chlorate could be used, but these were, of course, useless near growing crops and made the land unusable for some time. Then, in 1926, F. W. Went, in Holland, discovered the existence of growth-controlling substances: these were first known as auxins but later as phytohormones, because of their similarity to the hormones that, also in minute quantities, regulate the activities of animal bodies. Further research revealed that such substances could not only stimulate plant growth but also inhibit it. It was this that led to the concept of selective herbicides – chemicals which can destroy weeds but not associated crops. The first such chemical to be developed was dinitro-ortho-cresol (DNOC) patented in France in 1932. By modern standards its selectivity was small, being active against one of the two main groups of plants, the dicotyledons, but not the monocotyledons. It so happens, however, that all cereals are monocotyledons (producing one leaflet when the seed germinates) while most weeds are dicotyledons (producing two), so DNOC achieved a considerable measure of success. It was followed, in the USA, by the more active and selective 2,4-dichlorophenoxyacetic acid (2,4-D) produced in large quantities in the Second World War: by 1950 production there had risen to 10,000 tonnes a year, a very large amount indeed when one remembers its strength. Subsequently, a variety of chemically related, selective herbicides were developed partly to meet particular crop situations and partly to replace products to which certain weeds had become resistant.

Meanwhile, in the 1950s, a totally different class of herbicide was being developed by ICI in Britain: chemically, these are known as bipyridyls and the best known is paraquat. These were active against a variety of grass weeds and found a number of applications, for example in the control of aquatic weeds and of grass weeds in rubber plantations. But

Bipyridyl herbicides will kill surface weeds by desiccation (above), but are inactive in the soil. Crops can be grown on land thus treated without the need for the traditional preparative processes of ploughing and harrowing.

In the last 50 years the economic basis of agriculture has been radically changed by the use of enormous quantities of synthetic nitrogenous fertilizers. Demand is seasonal: here (right) supplies are being built up in a huge silo.

much more interesting was their initiation of a new technique of crop production, a rare event in agricultural history. This arose from the fact that, while paraquat will kill surface growth, it is very quickly inactivated in the surface layers of the soil. It is thus feasible first to prepare the land by spraying and then – without the age-old process of ploughing and harrowing – drill the new crop directly. The appearance of this new technique of minimum tillage or 'ploughless farming' proved to be very well timed. Not only was the cost of labour a major element in conventional soil preparation, and increasing steadily, but so, too, was the cost of fuel.

Weeds are not the only plants which cause very serious damage to crops: a variety of fungal diseases also do great harm. Among the earliest chemical fungicides was Bordeaux mixture, a mixture of copper sulphate and lime introduced in 1885 to combat downy mildew of grapevines. Because it is cheap and effective, it is still widely used for plantation crops such as tea, coffee and cocoa. Increasingly, however, copper fungicides have been supplemented by a wide range of synthetic organic fungicides. Some of these (dithiocarbamates and phthalimides) are applied to the foliage as sprays but now increasing use is being made of systemic fungicides. These are absorbed through the roots, travel up through the xylem – much as a drug travels through the bloodstream – and then attack fungi infesting the leaves and stem. It is a war that is never really won, for the fungi can acquire a resistance to certain fungicides: this can be countered only by, as it were, catching them unawares with another new product. In Japan, a promising start has been made in the use of antibiotics to control fungal disease, especially in rice. However, it is becoming increasingly difficult to keep up the supply of novel products, partly because of the expense of the research and development involved but even more so because of the complexity of regulations that have to be met worldwide before a new agrochemical can be offered for sale.

Fungal and other diseases – such as smut and bunt in cereals – can be spread through seeds, and from the 1930s organic mercurial compounds have been used as seed dressings. These are effective and, under properly controlled conditions, quite safe, but they are nonetheless toxic. There was, therefore, an incentive to develop non-mercurial dressings, pioneered by Du Pont's thiram in 1930, and a variety of these have become commercially available since the Second World War.

The third great enemy of growing crops is insects, ranging from clouds of big locusts advancing in tens of millions and devouring everything in their path to the small flea-beetles that infest root crops. Overall, their damage, too, must be measured in billions of dollars annually. The first major development in this field was DDT, the powerful insecticidal properties of which were discovered in 1939 by Paul H. Müller, working in Basle: the importance of this discovery earned him a Nobel Prize in 1948. Its manufacture

Growing crops of all kinds are a prey to a variety of pests and diseases. These pictures show mould infection (bunt or smut) of ears of wheat (top) and barley (above centre). A great variety of agrochemicals have been developed to combat such diseases: here (above) dry rice is being sprayed in the Philippines.

Pests can easily acquire resistance to insecticides and the farmer needs a succession of chemicals to keep them under control. Here cotton has responded to treatment with pyrethroids (above) but not to an organophosphorus insecticide (above right).

was developed as a matter of urgency during the war and it attracted public notice in 1943/4 when it was instrumental in controlling an outbreak of typhus, a louse-borne disease, in Naples. This was followed in Britain during the war by another organic compound of chlorine known as BHC: this quickly proved very effective against a variety of pests, such as wireworms in soil and flea-beetles on brassicas. It was also shown to be very active against locusts and paved the way for modern methods of locust control. These are based on attacking locusts in their limited breeding areas as soon as they show a tendency to swarm. In the event, BHC proved particularly useful in the public health field – for controlling vectors of disease such as mosquitoes and tsetse fly and in veterinary medicine – rather than in agriculture because it was found that many crops, such as root crops and black-currants, acquired an unpleasant taste.

Meanwhile, chemists had been directing their attention elsewhere. Immediately before the war IG Farben in Germany had been developing phosphorus-containing insecticides, including one known as parathion. While this was very effective, it was also rather toxic to mammals, including man, and for a time it seemed that this was an inherent defect of all insecticides of this type. However, after several disappointments, two useful new products emerged in the 1950s, menazon in Britain and malathion in the USA.

These new products, especially the chlorinated organic chemicals such as DDT, were both highly effective and cheap and correspondingly widely used throughout the world. With their capacity to destroy many of the most serious pests of important crops – of great importance in a world in which the population was growing explosively – and the insects that serve as vectors of widespread and deadly disease, it seemed that a considerable and timely blessing had been bestowed on mankind. Yet the chemical industry was in reality sitting upon a timebomb, for evidence was accumulating that chemical pesticides were by no means an unmixed blessing. These products have a significant animal toxicity, though one varying from species to species, but if properly used do not pose a danger to the operators. Unfortunately, in the field, especially in primitive conditions, they are not always properly used and can be a serious health hazard. They also pose a hazard to a broad variety of wildlife, especially fish and birds.

These problems first came to wide public notice with the publication of Rachel Carson's *Silent Spring* in 1962. Subsequently, most countries imposed much stricter controls on the use of insecticides and chlorine-based ones such as DDT were largely phased out, except for essential public health purposes, and replaced by others regarded as less hazardous. It also led to renewed interest in long-used natural insecticides such as derris root, the active principle of which is rotenone, or certain chrysanthemums which contain pyrethrin. Such products tend to allay public anxiety simply because they are natural, but this ignores the fact that some of the most deadly poisons known, such as strychnine or botulinus toxin, are also of natural origin. Uneasiness about synthetic insecticides also led to renewed interest in other, biological, methods of pest control. For example, in the 1920s millions of acres of prickly pear were eradicated in Australia by introducing highly specific predators from Texas. More recently red spider mites and white fly have similarly been controlled in glasshouses by another parasite, *Encarsion formosa*. Even with such methods there is a hazard, for in new surroundings a predator may change its habits and become a pest. However, some success has been achieved in luring insect pests to traps baited with their highly specific sex attractants, known as pheromones. From these different approaches has developed integrated pest control, a judicious combination of chemical and biological methods.

Most benefits can be abused and the powerful synthetic insecticides are no exception, but two important points should be noted. The first is that most large-scale users are aware of their responsibilities and sufficiently cost-conscious not to use excessive quantities of these very expensive products. The second is that literally millions of human lives have been saved by these much abused chemicals. In the early 1950s, a vast global attack was launched to eradicate malaria, an often fatal and always debilitating disease: this was based partly on antimalarial drugs and the drainage of swampy areas where mosquitoes breed, but also relied heavily on the use of synthetic insecticides. In 1971, the World Health Organization, monitoring the campaign, reported dramatic results. Of an estimated 1814 million people living in the originally malarious areas of the world, 1347 million then lived in areas where the disease has either been eradicated or was no longer a serious problem.

A single seedling is progagated in a test-tube. Genetic selection is one way in which existing plants can be modified to become disease-resistant and new plant varieties created.

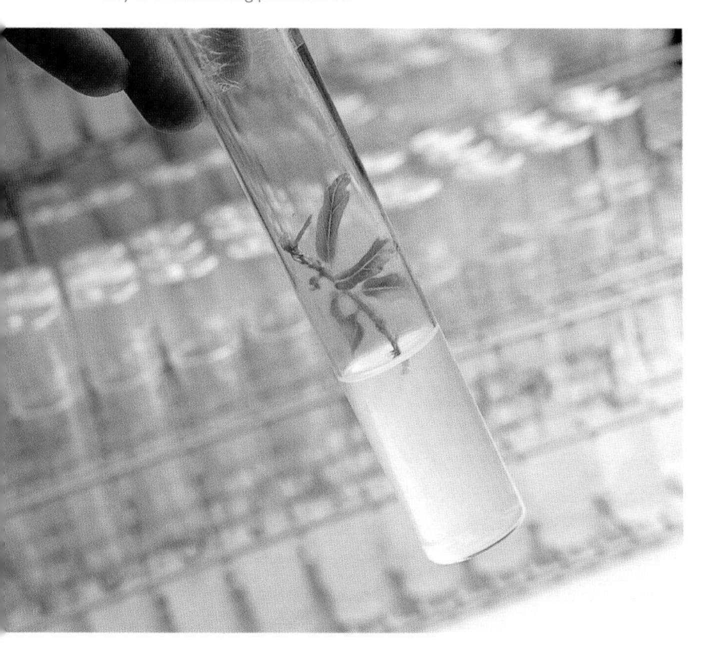

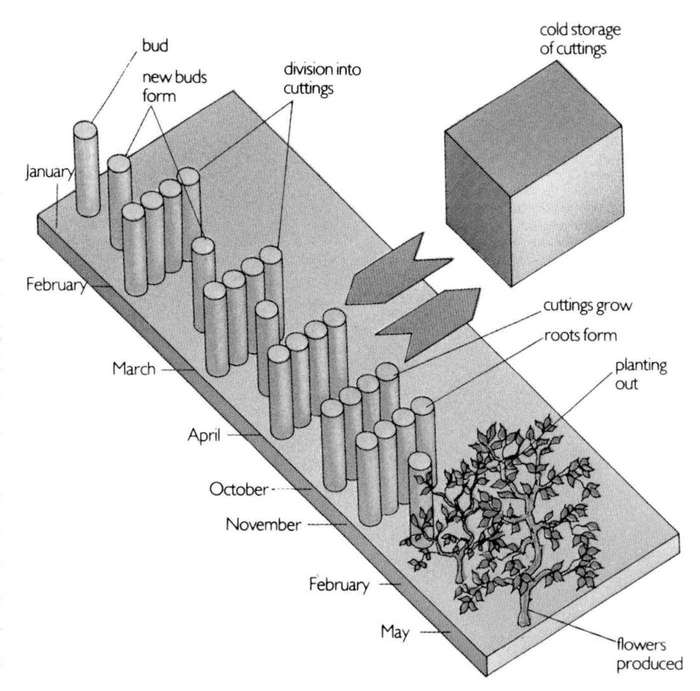

Test-tube propagation, here of roses, enables thousands of plants to be grown quickly from a single parent plant. A tiny piece of bud or meristem tissue cut from the parent is grown in a culture medium in a test-tube. This is then divided into four new cuttings, and each cutting further divided and so on. After a year, the resulting young plants are planted out and flower after six months.

Plant and animal breeding

A long practised method of combating plant diseases is to develop resistant varieties and over many years plant breeders have had much success in this field. In the past 20 years or so, the potential of this method has been much improved by a new technique of 'test-tube' propagation or cloning. In this, a tiny fragment of meristem tissue from the growing point of a plant is first grown on a nutrient medium in a test-tube until it has developed roots and shoots: it can then be grown on by conventional methods. In this way a pure strain can be propagated extremely quickly: thus in a single year 200,000 new rose plants can be generated from a single parent. This technique has two important consequences. First, whereas by conventional methods it takes at least 10 years to produce new varieties of a plant in sufficient quantities to begin marketing them, test-tube breeding cuts this to two years. Second, while virus diseases cannot be transmitted via seed, they can be through the normal vegetative propagation systems – such as layering, grafting and dividing – used for many major crops like potatoes, sugar-cane, bananas and strawberries. By starting with virus-free micro-cuttings, completely healthy plants can easily be propagated in great numbers, though they are still, of course, prone to reinfection if not planted in disease-free situations. The success of this technique can be shown by the excellent potato variety Belle de Fontenay, which in 1954 was almost extinct through viral infection: today, through test-tube propagation of disease

environmental damage 338 **genetic engineering** 297–299, 344 **IG Farbenindustrie** 282

free tissues, it is once again widely grown. This technique has also found increased applications in animal breeding.

Until recently, the livestock breeder, too, could make only slow progress – so slow indeed that it might not keep in step with the kind of beast most in demand. At the beginning of the century, for example, the demand in Britain was for heavy sheep, around 45 kilograms (100 pounds), with plenty of fat: 30 years later the most popular animal was less than half this weight and quite lean. To improve breeds by normal mating with the best sires was a slow business and even in the 19th century some horse-breeders resorted to artificial insemination. Not until the 1920s, however, was this technique at all widely practised in agriculture generally; it became popular largely through the work of I. I. Ivanov in Russia on sheep and cattle. It was adopted on a limited scale in Denmark and the USA immediately before the Second World War, but for practical purposes it was essentially a post-war development. In the USA, only 7500 cows were bred by artificial insemination in 1939, but this had risen to over a million in 1947 and to six million in 1958. This was due to a large extent to the discovery of methods of preserving semen for long periods by freezing: in this way the genetic qualities of good sires can be perpetuated long after their death. Conversely, to perpetuate desirable genetic qualities on the female side, fertilized ova from top-grade animals can be transplanted into other cows for onward development into full-grown healthy animals.

The position with regard to horses is rather different from that of other farm animals. While breeding for racing purposes remains a big business in the Western world, the number of horses employed in agriculture has – sadly, for sentimental reasons – declined sharply. The British position illustrates the trend. At the beginning of the century horses numbered over three million, and were virtually the sole source of traction power on the land; in 1950 there were only 347,000 and by 1960 so few that they were no longer recorded in the government's census of agriculture. The decline of the horse was matched by the rise of the tractor. These made a marginal contribution in 1900, but by 1950 the global total was around six million – two-thirds of them in the USA – and today it is near 20 million. Yet the pattern of change is very uneven. In many developing countries tractors are a rarity and the age-old buffaloes and oxen are still dominant: statistics are not very reliable, but probably the world population of these two draught animals today is around 70 million and possibly increasing rather than declining.

The new developments in breeding were based on growing understanding of the scientific principles of inheritance stemming from the pioneer research of the Silesian monk Gregor Mendel in the late 19th century, and developed in the 20th century by Hugo de Vries in Holland and T. H. Morgan in the USA. One consequence is the diminishing importance of purity of breed and the lessening dominance of the old herd books designed to perpetuate this: breeders had become obsessed with creating animals which fitted into a very precisely defined mould. Today, the approach is more empirical and based much more on performance than on breed. Increasingly, scientific advances are causing breeders to anticipate a future in the new biotechnologies: genetic engineering, cloning and transgenics.

Some countries, notably Brazil, have countered the oil crisis by producing alcohol by fermentation as a substitute for petrol. This Brazilian filling station is typical of many throughout the country.

Biotechnology

Modern biotechnological techniques have created means of producing plants and animals other than by propagation or breeding. Contrary to public perception, biotechnology is not a recent development, and does not mean simply genetic engineering by the manipulation of DNA, but the employment of biological processes to make useful products. Although the term 'biotechnology' was not coined until 1919 by Karl Ereky, many of the principles have, in fact, been in practice for millennia, from the first beer that was brewed and cheese that was cultured around 2000 BC in Egypt and Sumeria. As an historical example we may take citric acid, widely used in the food, soft drinks and pharmaceutical industries. In the second half of the 19th century, it was made – as some still is – from lemon juice, then virtually an Italian monopoly. American production was about 1000 tonnes annually but fluctuations in supply resulting from varying citrus harvests led to a search for more reliable sources. In 1923 Pfizer, in New York, began to make citric acid by fermenting sugar with a mould *Aspergillus niger*, and this is now the main source of supply. During the First World War a critical shortage of acetone, vital for the manufacture of high explosives, was relieved by a fermentation process devised by the Manchester chemist Chaim Weizmann, who later became the first president of Israel. Other important chemicals made by such biological processes include alcohol and, more recently, penicillin and other antibiotics.

These fermentation processes have lately been developed in two ways, the first of which has been a simple scaling up of production. Alcohol can be used as a fuel in suitably adapted internal combustion engines and in the face of the OPEC crisis Brazil adopted a policy of utilizing surplus vegetable material of all kinds to produce it on a very large scale by fermentation. While this has met with some success, the process results in the production of very large volumes of dilute waste liquid liable to become offensive by decay and the disposal of this has become a considerable problem. Kenya, too, has sought to follow a similar biofuel policy, but turning over ½ million hectares (1¼ million acres) of land to fuel-crops for fermentation has made necessary increased imports of food: whether there has been an overall economic gain is questionable.

The products considered above are all relatively simple chemicals, but since the 1970s the rising cost of animal foodstuffs has led to much interest in fermentation processes designed to produce single cell protein. These depend on the fermentation of a variety of materials which only a few years ago would have looked unpromising; among them are methane gas, gas oil and higher paraffins. One process, developed by ICI, utilizes fermentation of methyl alcohol by a species of bacteria known as *Methylophilus methylotrophus*: the processing plant has a capacity of 60,000 tonnes a year and the product is marketed as 'Pruteen'. The former Soviet Union is reputed to have plants with a total capacity of 200,000 tonnes a year utilizing paraffins as the raw material.

These chemical transformations are effected by natural catalysts, enzymes, contained in the micro-organisms; another interesting development has been to isolate the enzymes concerned and attach them to some sort of solid base, such as a membrane or fine particles. Suitably prepared, these immobilized biocatalysts have long lives before their activity falls to a level too low to be useful.

Modern biotechnology has developed rapidly from advances in chemistry, physics, biology and computing, and has dramatically affected our understanding and practice of agriculture and medicine. Three major branches have emerged: diagnostic techniques, cell/tissue techniques and genetic engineering. Through the identification and manipulation of biological matter at the cellular and atomic level, these techniques are leading to advances in crop production, disease control and resistence, medical research, organ transplants, organic/inorganic hybrids, bionics and neural nets.

Much of the research in genetics has involved the study of bacteria and other single-celled organisms, primarily because they are naturally cloning: they reproduce asexually by separating the halves of their DNA double-helix and recombining with appropriate nitrogen bases to form two identical strands of recombinant-DNA (rDNA) from one original. Testing various drugs and compounds on bacterial clones means that any differences in results are attributable to differences in the compounds, not the bacteria.

Diagnosing the effects of new drugs advanced further when, in 1971, Stanley Cohen of Stanford University and Herbert Boyer of the University of California developed the initial techniques for recombinant-DNA technology. Their research proved that genetic traits could be transferred from one organism to another by genetic engineering – splicing genes (segments of DNA) from one to the other. This discovery provided scientists wishing to study the effects on a particular gene with a means for mass-producing that gene by splicing it into a bacteria or yeast plasmid (a strand or piece of DNA that can exist and reproduce autonomously) which would then naturally replicate in a carefully controlled laboratory culture. Using these methods it is possible that common and readily propagated bacteria such as *Escherichia coli* can be given the capacity to synthesize insulin, currently extracted from the pancreases of pigs and cows. Another possibility is to confer on cereals the nitrogen-fixing capacity of leguminous plants such as peas and beans, thus diminishing the need to apply artificial fertilizers.

Ultimately, these advances also led to the creation of transgenic animals. Once it was understood that all life shares the same genetic makeup, and that DNA could be transferred between organisms, the concept of transgenic animals – animals engineered to carry genes from species other than their own – was a natural progression of thought.

The number of commercial biotechnology ventures multiplied following the 1980 Diamond v. Chakrabarty case, involving an engineered oil-decomposing bacterium, in which the US Supreme Court ruled that genetically engineered life forms could be patented.

Within two years the first transgenic plant was created and the first human gene was successfully transferred into a mouse. The first genetically engineered livestock were reported in the 1980s, a decade in which we also saw the first patenting of an animal – the 'oncomouse', a mouse engineered by Timothy Stewart, Paul Pattengal and Philip Leder of Harvard University. Recent research has focused, in particular, upon transgenics which could enable the mass production of various proteins which are otherwise cost-prohibitive to generate.

Genetic engineering and modification

In 1990, the US Food and Drug Administration (FDA) endorsed the first food product to be created by genetic

fermentation 43 **gene patenting** 344 **livestock breeding** 201 **OPEC oil crisis** 255, 264, 275, 278, 340

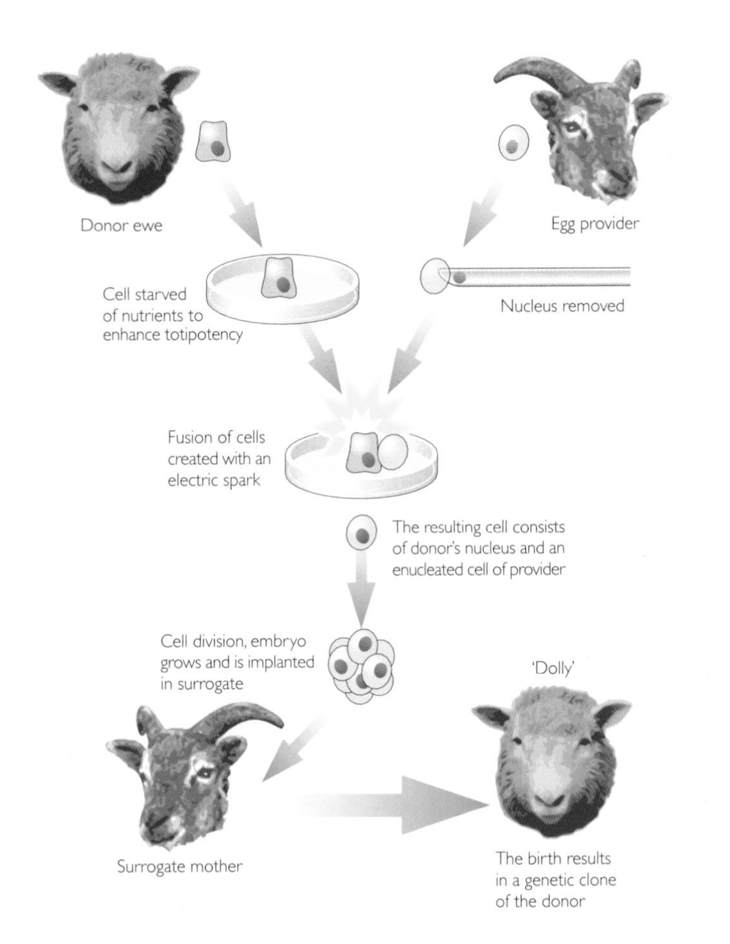

In 1996 Dolly, the sheep, became the first mammal to be cloned from an adult cell. The process involves taking the nucleus of a cell from one animal and fusing it with the shell of the egg of another animal.

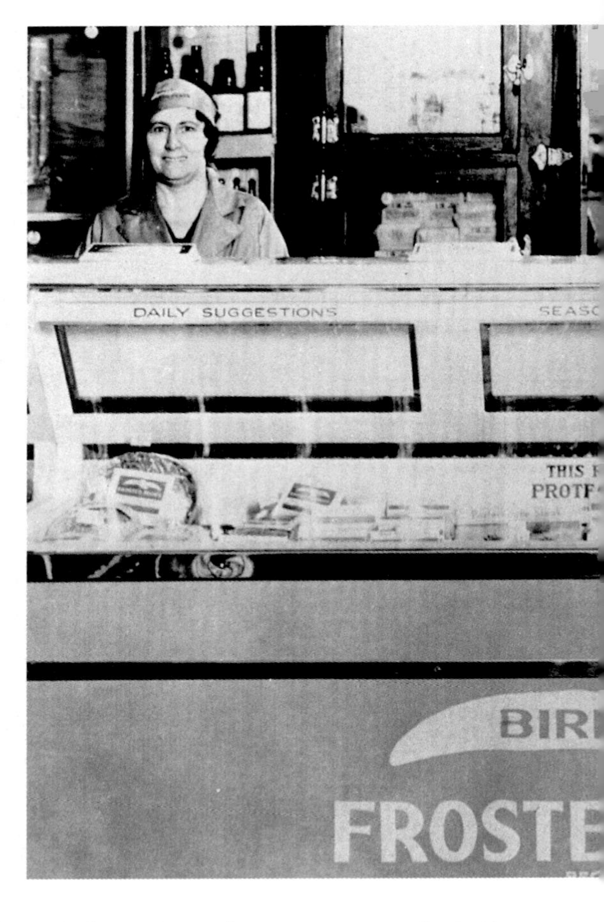

engineering. Interestingly, it was chymosin, an enzyme that occurs naturally in rennet in calves' stomachs and has been used for centuries and up to the present day in the production of cheese. The following year, Calgene Inc. requested a review of the Flavr Savr, a tomato modified to resist rotting and softening. In 1994, it became the first FDA-approved whole food produced by rDNA biotechnology; because the FDA had found it to be as safe as tomatoes bred by conventional means, it also became the first rDNA-created food to be sold for public consumption. By the end of the decade, several other genetically modified (GM) foods were available, including oilseed rape plants engineered to create high-yield hybrids and pesticide-resistant soya beans.

The process of cloning involves the creation of genetically identical cells or organisms. Although the German scientist Hans Spemann first proposed nuclear transplantation as a method for cloning in 1938, it wasn't until 1952 that Robert Briggs and Thomas King, developmental biologists at the Institute for Cancer Research (now the Fox Chase Cancer Center) in Philadelphia, were able to clone several tadpoles from an embryonic cell. In their experiments, they were able to remove the nucleus from one frog embryo (recipient) and replaced it with the nucleus – the part containing the DNA controlling growth and development – of another. They used cells from an embryo because these had yet to become specialized, meaning they still had the potential to create any other type of cell. In contrast, cells in an adult would have become increasingly specialized – a skin cell can normally only create other skin cells.

However, in 1966, John Gurdon, a molecular biologist at Oxford University, by producing adult frogs from intestinal cells of a tadpole, was able to prove that even after a great deal of specialization, some cells remain *totipotent* – capable of being 're-programmed' to create an entire organism. Following this discovery, scientists were destined for years of setbacks in which either the nuclear transfer was unsuccessful or the young never survived to adulthood. The world had to wait for 30 years before the next significant breakthrough.

In the interim years, an easier method for creating animal clones was developed in the 1980s and was quickly taken up by livestock breeders. Embryo splitting, or artificial twinning, involves the splitting of individual cells to achieve results similar to naturally occurring twins or multiple births.

In July 1996 a team of researchers, led by embryologist Ian Wilmut at the Roslin Institute in Edinburgh, announced the birth of Dolly, the first mammal to be cloned successfully from an adult cell. Their procedure involved removing

Above left: Deep-freeze food was introduced by Clarence Birdseye in the USA in 1929. The picture shows a typical locker in an American store in the early 1930s.

Above right: Public outcry against GM foods has fostered the growth of organic farms. This one uses mustard plants to attract pests away from the potato crop.

udder cells from adult sheep, placing them in a solution to starve them of nutrients until growth ceased for a few days, and then fusing each by means of an electric spark to the enucleated embryonic cells of another sheep. Once the resultant cells had grown into embryos, they were placed into surrogate ewes to carry to term. Grown from the cell of a six-year-old sheep, Dolly was the first of some 300 attempts to survive the process. When, in April 1998, she gave birth to a healthy lamb of a natural mating process, she dispelled some concerns over the possibility of latent genetic disorders.

In 1998, biologists at the University of Hawaii cloned more than 50 mice from an adult and improved the technique with two modifications. First, they used naturally dormant cumulus cells (those surrounding the ovaries) that did not require starvation to be re-programmed. Secondly, they used an extremely fine needle to inject the donor nuclei into the enucleated recipient eggs. Because this caused much less damage to the egg than electrical fusion, the chances of a healthy embryo were much greater. As mice bear striking similarities to humans in their genetic structure and development, this experiment – with procedures which were also less stressful to the developing embryo – was significant in the quest for human clones. Dolly had proven adult cloning to be possible; the Hawaiian mice strengthened the belief that human cloning was feasible.

The food industry

So far we have considered the production of food in bulk rather than its subsequent processing for the consumer. In the main, there has been rather little innovation in recent years and the main change has been in the scale of processing rather than in its kind. In the Western world, this can be seen as a general tendency towards larger industrial units. The making of butter and cheese moved from the farm to the factory; the big steam bakery replaced the small local one; shops supplying tea and coffee from their own bulk supplies lost their trade to firms selling national brands in packets. There was a change, too, in the geographical pattern of production. While most countries – with the notable exception of Britain in the 1930s – produced enough milk to meet their own needs for cheese and butter, New Zealand, Australia, Holland and Denmark became big exporters of dairy products.

However, in recent years there has been renewed interest in the produce of small farms, including free-range eggs and meat, unpasteurized cheeses and organic vegetables, as people become increasingly concerned about where their food

biotechnology 296–297 **cloning** 344 **Human Genome Project** 344

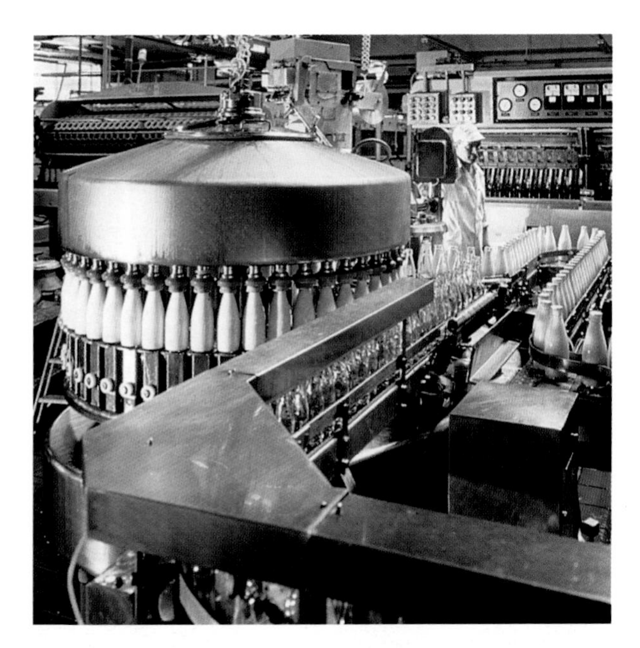

Dairy farming has undergone a high degree of mechanization with the introduction of milking machines and milk bottling plants.

comes from, who produces it and how. This is, in part, a response to widespread food scares about salmonella and bovine spongiform encephalitus (BSE) and adverse publicity about genetically modified crops.

Although some new techniques – such as the use of post-mortem electric shock treatment of meat in abattoirs to make it more tender – were designed to improve quality, the emphasis tended to be on preservation. This serves a double purpose: it enables seasonable peaks in production, as in peas and soft fruits, to be spread over the whole year and makes it possible for big producing areas to supply distant markets around the world.

Some of the oldest methods of food preservation, such as smoking and salting, are still widely practised. In the post-war world, even canning can be considered to have become an old industry – by 1950 the USA alone was using 10 billion cans annually – but though the canning process remained essentially unchanged there were changes in the materials used. Aluminium cans began to replace tinplate after the Second World War and lacquering was introduced to reduce corrosion and discoloration. Some products, such as beer and soft drinks and even wine, began for the first time to appear in cans rather than glass bottles: glass also increasingly found a competitor in plastic bottles.

We have seen already that refrigerated food, especially meat, was being transported long distances even before the beginning of this century, but refrigeration acquired a new significance in the Western world after the Second World War. One important change was the very rapid increase in the number of domestic refrigerators, which today must be regarded as virtually standard pieces of domestic equipment. This enables perishable foods to be stored safely for short lengths of time, but the advent of the domestic deep-freeze has radically extended the time-scale. With this, people can store quantities of bought-in food – meat, fish, vegetables and fruit – for months on end and can also preserve home-prepared dishes until required.

Long-term refrigeration in cold-stores, shop display units and domestic refrigerators is an energy-consuming process: a small domestic deep-freeze locker will consume about two kilowatts of electricity. This is not a negligible expense, nor is it a good thing in a world doing its best to save energy. There is, therefore, a considerable attraction in another refrigeration process, freeze-drying. This was developed in Sweden in the 1930s to preserve various biological products such as blood plasma. It depends upon the fact that at low temperatures ice can pass directly from the solid to the vapour state without melting in between: this is why snow slowly disappears even in the absence of a thaw. The technique depends on first freezing the material and then removing the water contained in it by subjecting it to a continuous vacuum. This not only leaves a dehydrated product – which can be kept for long periods in closed containers – but one in a light porous form which readily reconstitutes itself when water is added. Freeze-drying rapidly became popular after the war for a variety of products such as coffee, soup and vegetables.

The long-storage qualities of freeze-dried products depends on the elimination of moisture and, as this is done at a low temperature, the flavour is unimpaired. But water can, of course, also be driven out by heat and a variety of new heat-drying processes have been developed to produce stable products that can be kept for a long time. Milk, for example, can be dried by letting it run over a hot roller and scraping off the resulting crust. A more sophisticated method, used for preserving eggs and milk during and since the Second World War is to spray the material into an uprising column of hot air, when the droplets fall to the bottom as dry powder. Such processes are highly successful from the preservative point of view but the heat usually changes the taste and texture of the product and there are difficulties in reconstituting it later to anything like its original self by the addition of water.

Compared with refrigeration, canning and drying have the advantage of a once-for-all consumption of energy. Drying processes have the further advantage of saving on storage space and transport, for the water content of most foods is very high: milk, for example, contains around 87 per cent water and even potatoes around 80 per cent. However, many people do not want to be bothered with reconstituting dried milk and prefer it in a form that can be stored and used directly. This demand is met by various heat sterilization processes. In Ultra High Temperature (UHT) processes the milk is superheated to about 150°C (390°F) for a few seconds, giving a 'long-life' milk, sold in a carton, with a shelf-life of several months. The UHT process is used also for cream, fruit juice and other liquid products.

Left: Sterilization at Ultra High Temperature (UHT), using superheated steam from boilers, is today widely used to provide 'long-life' milk, fruit juice and other liquid products.

Above: Fish-farming is virtually as old as civilization: it was practised in early times in Egypt and stew ponds were common in medieval Europe. Lately, there has been much interest in intensive cultivation: this picture shows a large state fish farm in Colorado.

The essence of all food preservation treatments is to sterilize it with the least possible alteration in its appearance and taste. For this reason, irradiation is an attractive proposition, using radioactive isotopes or fast electron beams as the source. Although there are some restrictions on the use of this process, on safety grounds, some Spanish and other growers are now using it for strawberries and other soft fruits.

Aquaculture

Fishing is virtually the last major activity of man as a hunter-gatherer. During the 1970s, world catches totalled around 70 million tonnes annually and it is estimated that the maximum that can be achieved without serious depletion is around 90 million tonnes. Whaling, long an important sector of the fishing industry, seems destined for extinction as the nations of the world reach agreement on policies to protect the dwindling population of whales.

However, by no means all the world's fish is caught in the wild: some 7 million tonnes annually are produced in fish farms, and this is expected to rise to 10 million tonnes by the end of the century. In Asia, fish-farming is an age-old practice, and in Europe the stew pond was a feature of large medieval establishments such as monasteries. In recent years, there has been much interest in the development of large commercial fish farms, especially as a means of providing much-needed protein in developing countries. In Africa there is a particular interest in the tilapia, an important food fish in the Nile in ancient times. It lends itself particularly well to intensive cultivation, under strictly controlled conditions, which seems first to have been practised in Kenya in the 1920s. By 1965 tilapia culture was widespread in Africa and had been adopted also in the Far East, the southern states of the USA and Latin America. Today, some 500,000 tonnes of tilapia are produced annually.

canning 152, 183, 197–198 **early food preservation** 43 **refrigeration** 152, 199 **whaling** 194

CHAPTER THIRTY-TWO

New Materials

Surprisingly, considering how different it looks from that of even a generation ago, the modern world uses rather few materials that have appeared since the Second World War. In the main, what has happened is a profound change in the pattern of use – exemplified by the explosive growth of the plastics industry – improvements in manufacturing techniques, in the working properties of existing products and new uses resulting from combining several materials to achieve a given end.

Metals

First, we should mention briefly a somewhat anomalous metal, sodium. This soft, silvery, highly inflammable metal is never encountered as such in everyday life but nevertheless has very important uses. It is an intermediary in making tetraethyl lead as an antiknock agent for petrol – now generally phased out – and it is the source of the yellow light of some kinds of street lamp. In recent years, it has assumed a new importance as a cooling agent in certain kinds of nuclear reactors: it is an exceptionally good conductor of heat and liquefies just below the boiling point of water. A large reactor of this kind may require 1000 tonnes of sodium.

The metal to have a meteoric rise, and become familiar to everybody, was aluminium and its alloys: it has become the most widely used metal after iron and steel, when a combination of lightness and strength is desired. It is used in great quantities in all the transport industries and space craft; in high-tension cables; in window-frames and doors; in cooking utensils and domestic appliances; in tennis racquets and caravans. It is very much the 20th-century metal: in just 20 years from 1939 annual production increased fivefold.

Only two new metals, titanium and beryllium, have come into use since the Second World War on any substantial scale, and then in ways which the ordinary citizen would not often become aware of. Although titanium was isolated by

Aluminium was little used until the turn of the century. Among major outlets developed after the First World War was the rapidly growing aircraft industry (left), in which its combination of strength and lightness made it particularly valuable.

Optical fibres (above), made from tiny flexible glass rods, are used to carry high volumes of data over long distances by means of light pulses, and have become indispensable to modern telecommunications.

the German chemist Martin Klaproth in 1794, until the middle of this century it was almost entirely encountered in the form of its white oxide. This was widely adopted by the paint industry because, unlike lead-based pigments, it does not blacken if there are traces of sulphur compounds in the atmosphere. Present interest in the metal is due to the fact that some of its alloys are lighter than steel but, unlike the even lighter aluminium alloys, retain their strength at high temperatures: this makes them of particular interest to aeronautical and space engineers.

Beryllium was discovered by the French chemist Louis Vauquelin in 1798. From the 1920s, small quantities were used to alloy with copper, giving a metal with unimpaired electrical and thermal conductivity but much harder. Although alloys still absorb most of the metal manufactured, its present importance stems mainly from its use in the nuclear power industry. Its manufacture involves a complex electrochemical process, made more difficult by safety precautions necessitated by its high toxicity.

Like beryllium, other metals are rarely used in their pure state but rather as alloys: bronze and brass are familiar examples. The alloys in regular use are far too numerous even to catalogue, but recently alloys with unique properties have become of interest. These are the so-called 'shape memory alloys' (SMAs), which originally consisted of roughly equal parts of nickel and titanium, but are now seen in alternatives such as copper-aluminium-nickel, copper-zinc-aluminium and iron-manganese-silicon. If a component made of such an alloy is strongly heated it will 'remember' its shape after it has cooled. However it is manipulated when cold, it will resume its original shape when re-heated. Although William J. Buehler, a researcher at the Naval Ordnance Laboratory in Maryland, discovered the alloy, it was David S. Muzzey who accidently revealed its shape memory properties in 1961 by idly heating with his pipe lighter a strip which had been bent many times; to his surprise it resumed its original shape. Later, Frederick E. Wang located the structural changes which caused the effect – a rearrangement of particles within the crystalline structure.

Pottery and glass

Pottery and glass are, of course, amongst the oldest constructional materials, and pottery, in particular, because of its great durability, is always the most abundant relic of ancient civilizations. It is, therefore, not surprising that by the middle of the 20th century innovation was rare and in the main progress lay in improving existing products and processes. An increasingly important new need in modern industry is for highly refractory materials capable of withstanding very high temperatures, as in the new generation of jet and rocket engines and the protective panels on the space shuttle. Here a very interesting development was that of composites known as cermets which, as their name implies, combine metals fused with ceramics. These are very hard and can be machined like metals: one possible use is to make the cylinder blocks of automobiles. Recently, there has been an increased use of the piezoelectric effect – the variation in electrical resistance resulting from pressure changes – in some cermets as a sensing mechanism for vibration or shock.

In the manufacture of glass in bulk, the major innovation since the Second World War has been the float process for making sheet glass, introduced by the firm of Pilkington Brothers in Britain in 1952. In this a ribbon of molten glass is allowed to flow over the surface of molten tin, chosen because it imparts a particularly smooth glaze to the lower surface; the upper surface of the glass is continuously fire-polished with gas jets to achieve the same end. This gives a sheet of absolutely uniform thickness with a fine surface on both sides, eliminating the need for grinding and polishing.

Today, one of the biggest users of sheet glass is the car industry, where ordinary glass would be highly dangerous in the event of an accident. As early as 1905, the French chemist Edouard Benedictus made safety glass by sandwiching a layer of celluloid between two sheets: if broken, such 'Triplex' glass cracked rather than splintered. In the 1930s, celluloid, which slowly turned yellow, began to be replaced with a plastic sheet consisting of polyvinyl acetate. Today, most safety glass is made by a form of heat treatment which creates some built-in stresses, which result in the glass crumbling, rather than splintering, if it is shattered.

One of the most interesting of recent developments in glass technology is fibre optics, to which we shall come later in the context of telecommunications. Briefly, this depends on sending signals not as electrical pulses along a conducting cable but as laser-generated light pulses within a very fine glass fibre. Fibre optic devices are also used for exploring normally inaccessible regions, such as the chambers of the heart.

Another important post-war development in glass has been the greatly increased use of borosilicate, low-expansion glass for high-temperature use, as in ovenware and sealed-beam headlights for cars.

Plastics

Plastics, we have seen, have a history running back into the 19th century and were well established by the start of the Second World War. Since then, however, growth has been phenomenal and in the modern world plastics are the dominant new material. Today, about 80 per cent of the output of the world's organic chemical industry is in polymers – plastics and synthetic fibres. This is not the result of the invention of many new ones – for most of those used now had appeared by 1939, even if they had not made much public impact – but of vastly larger quantities being used and, no less important, the development of techniques to mass-produce plastic articles far bigger than was attainable even in the 1950s. By the 1970s, it was the American chemical

development of metallurgy 71–81, 183–190 **early pottery** 38–41, 122 **porcelain** 150–152 **telecommunications** 314–315, 342

industry that dominated the world and by then it was making some 3 million tonnes of polythene a year, 1.7 million tonnes of polystyrene, 1.5 million tonnes of PVC, and half a million tonnes of polyurethane, mostly in the form of foam.

The use of plastics is now so all-pervasive that it is very difficult to think of any sphere of human activity in which plastics are not merely involved but have become essential. If, as it were, we disinvented plastics, the results would be catastrophic. The electrical industry is based on the use of metals as conductors and plastics (very largely) as insulators – though on high-voltage cables, spark-plugs working at high temperatures and for a number of other specialist applications ceramics or glass may be used. Virtually all electric cable is insulated with plastic, usually PVC. The telephone system would cease, not only through lack of insulated wire, but because virtually every receiving set in the world is now housed in moulded plastic. In the home a vast array of everyday equipment, from waste-pipes and buckets to baths and water tanks, from bowls to colanders, from television cabinets to flowerpots and garden pools, are all made of plastic. Huge quantities of polymers, such as PVC and polyurethane, are used in the paint industry, especially for the cheap emulsion paints. Through the packaging industry they have invaded the retail market – blister packs for every kind of small item; plastic wrap for food; shopping bags in which to carry goods away; sacks for potatoes and coal. Even sport depends very much on plastics, as in small dinghys, surf- and sail-boards, ropes for mountaineering, and snorkels for underwater swimming.

During the First World War Germany developed the manufacture of fibrous glass as a substitute for asbestos. Today, this has become a major industry. Glass fibre bonded with plastic is an important structural material – as in the hulls of small boats – and it is used also for acoustic and thermal insulation and for making fireproof fabric.

Fibres

So far we have, by implication, been considering what might be called 'solid' plastics. But many plastics have also made big inroads, in fibre form, into the traditional textile industry: nylon is something of an exception in that it is used in both forms, though the fibre form accounts for much the greater part of its market.

Although nylon made its debut in the USA in 1938, for the rest of the world it was for all practical purposes a post-war development. By that time, it had been joined by a rival, Terylene (Dacron), invented in Britain in 1941 by John Rex Whinfield and J. T. Dickson, then working with the Calico Printers Association. Chemically they are different – nylon is technically a polyamide and Terylene a polyester. A distinction important from the manufacturing point of view is that to complete the fibre-making process Terylene must be cold-drawn. As textile fibres they are not dissimilar, but Terylene

The first commercially important man-made fibre was viscose rayon, a form of reconstituted plant cellulose. The solubilized cellulose is extruded through spinnerets into a coagulating bath and then wound on to bobbins.

cannot be moulded, though it does make an excellent transparent film. Although a number of other synthetic fibres have been developed – notably the acrylic fibres – the polyamides and polyesters dominate the market.

The synthetic fibres have much to recommend them. They are resistant to shrinkage in washing and to the attacks of clothes moths and similar pests, they can be set with permanent creases and they can be quickly drip-dried. But against all this, they soften and melt at quite low temperatures – which is why they can be permanently creased. Additionally, they cannot be dyed in the same way as natural fibres, and new processes have had to be devised.

As the effect of drugs can be considerably varied according to the formulation in which they are offered to the patient, so synthetic fabrics can be offered in a variety of guises. Over the years, they have changed enormously as both the chemical and textile industries began to sense the subtleties of garment making: for example, the importance of such intangible qualities as drape and texture.

The enthusiasm of the chemical industry for these new products was not immediately shared by the notoriously conservative textile industry, which was understandably wary of this new rival to the age-old fibres, to which all of its machinery and marketing was geared. Equally, the world of fashion had to be persuaded that the new fibres were acceptable on their own merit – in the past so-called 'artificial silk' based on cellulose had made a poor impression – and a great deal of money had to be spent on promotion. A first requirement for the textile industry was that the new fibres were presented in a form suitable for existing machinery: manufacturers had no incentive to invest large sums for developing an untried product which in the end might not be acceptable. As part of a general strategy, the new fibres were presented not as rivals to traditional ones but as complements to them. If they were blended with cotton, wool or linen, rather than used alone, it was argued, the whole market would expand. In fact, this strategy paid off very well, for in the event many fibre mixtures did in fact prove very successful in their own right, the resulting fabrics combining the best qualities of both.

It is not only in the garment trade that synthetic fibres have found a place alongside natural ones, for they have invaded the textile trade generally. They are, for example, now widely used in the carpet industry, though here there are technical problems. Although the fibres are elastic, they are slow to recover: consequently depressions made by the feet of furniture or heavy objects may take a long time to disappear. Some synthetic fibres used for carpets can accumulate static electricity which can give a painful shock if a metal object resting on it – such as a filing cabinet – is touched. To overcome this, various antistatic sprays have been devised. Another important outlet is in the motor trade where cotton was first largely replaced by rayon and then by nylon in the walls of tyres.

In this highly competitive market, improvements in the basic manufacturing process were of great economic importance. The original process for the manufacture of ethylene involved the use of high pressures and temperatures, and it came to be accepted that this was inescapable. In 1953, however, the Italian chemist Karl Ziegler transformed the situation with the discovery of a catalytic process for polythene that could be worked at ordinary temperatures and pressures, a far simpler and cheaper operation.

This work was developed in Italy by Montecantini and in the following year one of their consultants, Giulio Natta, director of the Milan Institute of Industrial Chemistry, discovered that a related gas, propylene, could be polymerized in the same way. Polypropylene soon became a major new plastic and one which lent itself – as polythene does less readily – to fabrication in fibre form. Polypropylene has found wide use in home furnishings and in a variety of industrial applications, including the manufacture of rope, fishing nets and binder twine.

From the earliest days of the textile industry fibres have been turned into fabric by some sort of interlocking system, whether it be weaving or knitting. The one important exception is the making of felt, in which short fibres are matted together and then compacted by heat and pressure: the process is, indeed, very like that used to make paper. The result is a flexible product, very suitable for hats, boots and similar purposes, but far too stiff for garments. Generally speaking, only wool and hair are suitable for felting: this is because the fibres have a scaly surface – like the awns of barley – which, as it were, hooks them together.

For this reason, smooth synthetic fibres cannot be felted, but non-woven fabrics have been made from them in another way. The lightly compressed fibre pad is very carefully heated in such a way that the fibres will just fuse together where they intersect: on cooling, a firm union is made. As it lies somewhere between melting and welding, the process is known as melding.

While plastics have undeniably brought many social benefits, they have also created a considerable social problem. Some of the very qualities that make them so useful – resistance to water and to the normal processes of decay – also make them exceedingly difficult to dispose of. This is a problem with all plastic products but especially with those that start their lives as short-term disposables – bags, sacks, wrappings and containers of all kinds. When these are discarded as litter they accumulate year after year unless removed – and in the country can be hazards to farm animals. Even if they find their way to organized refuse tips, they are still not easily destroyed: if incinerated, for example, they may give off noxious fumes which must themselves be contained. There has, therefore, been much interest in biodegradable products – that is, plastics whose chemical structure has been modified so that natural destructive micro-organisms can attack them – and others which will spontaneously disintegrate after a period of time. To achieve these goals at all, let alone economically, is not easy and the problem of plastic waste disposal remains a serious one, despite recycling efforts.

After the Second World War a major use for plastics was in the manufacture of film for the wrapping and packaging industry. Pictured here is transparent polypropylene film.

CHAPTER THIRTY-THREE

Computers and Information Technology

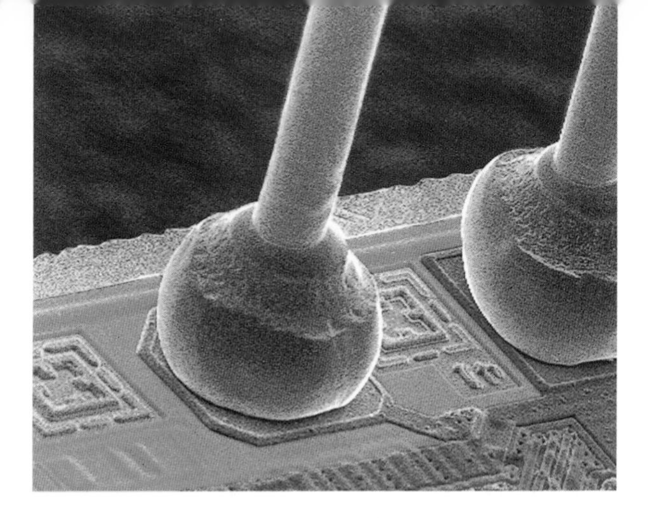

In just half a century since their inception, electronic computers have become ubiquitous. Expanding far beyond their initial intended purpose as calculating machines, they are now part of everyday life – serving as vital components in products from coffee-makers to automobiles.

If we consider information transfer in its broadest sense, the information industry, with all its complex ramifications, now employs more people than any other. To the old established methods of conveying information through the written word, in books, newspapers, magazines and by direct speech in traditional teaching establishments and at public meetings, we must now add the many new ones already mentioned. These include the transfer of information by telephone, fax, radio and television; photographs and films; records, cassettes and compact discs; and last – but certainly very far from least – by extraction from the vast store of knowledge electronically accumulated on the Internet. This extension of the realm of information, and of the means of conveying it, gives a great deal of indirect employment in the manufacturing of paper and photographic film; printing and processing machinery; a great range of complex electrical, electronic and reprographic equipment; teaching aids and so on.

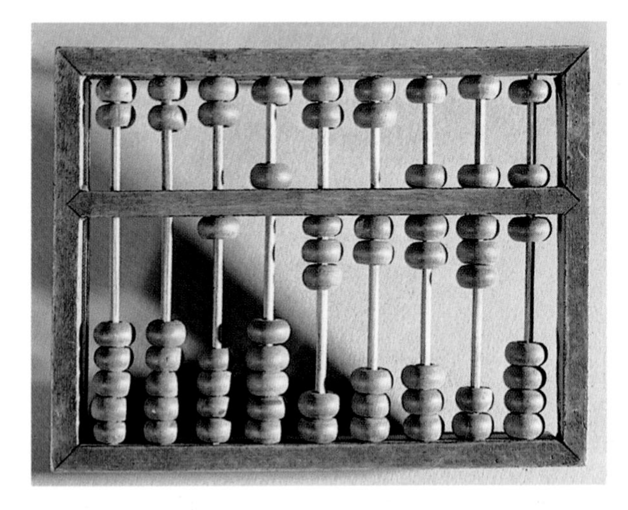

Above: The abacus, or counting board, is a calculating device of great antiquity, used by the Egyptians, Greeks, Romans and later the Chinese. This Chinese example was made in the mid-19th century: similar devices are still widely used.

Top right: A coloured scanning electron micrograph of two micro-wire traces on a silicon chip shown at approximately 240 x actual size.

Today, information exchange via computers has come to be regarded as a normal feature of daily life in the Western world. Communicating via computers, travel agents can, within minutes, check the availability of seats for a particular flight on the other side of the world; banks can check the credit worthiness of customers before dispensing cash via a machine hundreds of miles away; and investors can have an instant résumé of the financial status of companies in which they are interested.

By implication, computing is essentially a numerical operation but computers have been able to expand into the information field because images, numbers and alphabetic characters can be expressed similarly, a fact which cryptographers have long utilized. Thus, in the decimal system the alphabet could be rendered 01, 02, 03 ... etc. On this simple basis the word 'computer' would be rendered: 0315131621200518. In everyday life we use a decimal system of numbers acquired from the Arabs, who themselves derived it from Sanskrit, but computers use a binary system expressed as on/off code signals in their electrical circuits. Simple binary codes can indicate the basic mathematical operations so that all the operations necessary for recording and displaying information can be converted to logical equations that the computer can digest. Before embarking on this broader use of the computer, however, we can usefully consider its history as a calculating machine.

Calculating machines

In the world of classical antiquity the abacus was widely used and the Japanese and Chinese had a similar device known as a soroban. These, however, were intended mainly for everyday use to assist the simple transactions of shopkeepers, tax-collectors and so on. By the 17th century, however, more extended and complex calculations were becoming necessary for the preparation of tables of logarithms and trigonometrical functions and various aids to navigation. Logarithms, invented by John Napier in 1614, were of great importance because they enabled multiplication and division to be effected by the simpler operations of addition and subtraction. This led to the appearance of the linear slide-rule, invented by William Oughtred in 1621, by juxtaposing two logarithm scales devised by Edmund Gunther. Thereafter, a range of mechanical adding machines appeared, beginning with that of Blaise Pascal in 1642. In this a series of dials connected by gear trains recorded digits, tens, hundreds, etc: one complete revolution of the 1–10 digit dial moved the next dial on one-tenth of a turn and so on. In the 1670s Gottfried Leibniz improved its design and introduced the concept of an information store, or memory, to facilitate repetitive calculations, but the mechanics of his day could not satisfactorily translate his ideas into practice. In the event, such machines did not become available until about 1820: all of them depended on the input and output figures being displayed on dials.

A much more ambitious scheme was that of Charles Babbage. He conceived an 'analytical engine' which would not only make calculations but would also then print the result on paper. When he demonstrated the feasibility of this in 1822 using a partially constructed Difference Engine, the British government was greatly interested, for they were being subjected to considerable public criticism through errors in the *Nautical Almanac*, a set of important navigational tables published since 1727. This project never reached completion during his lifetime, for many reasons, not least of which was the departure of Joseph Clement, the engineer in charge of construction; it was finally built in the 20th century and can now be seen at the Science Museum in London.

Inspired by the Jacquard loom, Babbage's calculator was programmed by means of punched cards which, when suitably aligned, enabled a triggering rod to be passed through a given set of perforations. This idea was taken up by Herman Hollerith in the USA, who designed a punched card machine to speed up the processing of the US census returns of 1890. The census data was recorded as a set of perforations in cards measuring 7.5 × 12.5 centimeters (3 × 5 inches): for example, a perforation in a specific position might indicate age, another profession, another place of residence and so on. If a metal rod could be passed through a series of aligned perforations, it completed an electrical circuit. In this way it was feasible, for example, to discover quickly how many men within a particular age group worked as meat packers in Chicago. Such machines were widely used for information retrieval in the first half of the 20th century. The National Cash Register Company, in the 60 years after its foundation in 1884, sold some four million registers worldwide. Hollerith set up in business on his own in 1896, founding the Tabulating Machine Company, parent in 1924 of International Business Machines (IBM).

Early mechanical calculators – such as this Hollerith machine used to process the American census returns of 1890 – made much use of perforated cards punched out with a tabulator (below).

Konrad Zuse was a pioneer whose initial work was driven solely by his own intellectual interest. He started work on his first mechanical computer in 1934 and two years later switched to an approach using electromagnetic relays such as those used in telephone exchanges. In 1940 the war interrupted the development of his third computer, but after a year he was released from military service to complete Z3, a program-controlled, universal computer relying entirely on electric relays. The German Aeronautical Institute (DVL) supported his work, and the Z4 was used in the design of the V-2 rockets deployed in the attacks on London.

First generation computers

In their day these machines were important innovations but, as with television, the future lay not with electromechanical devices but with purely electronic ones which had no moving parts and were thus potentially enormously faster. A sort of half-way house was set up by the Automatic Sequence Controlled Calculator (ASCC) built jointly in 1944 by IBM and Howard H. Aiken of Harvard University. This was something of a computer dinosaur: programmed by perforated tape, it weighed 5 tonnes and contained 800 kilometers (500 miles) of wiring. Based on this use of thermionic valves, its power consumption was huge.

Arab mathematics 94–95 **databases** 310 **magnetic tape** 242 **origin of punched cards** 150 **thermionic valve** 239

The Automatic Sequence Controlled Calculator (ASCC), built during the Second World War, was a hybrid – part mechanical, being programmed with perforated tape, and part electronic, using hundreds of thermionic valves. By modern standards it was extraordinarily cumbersome and its consumption of electrical power was enormous.

A stronger claim to be the first of the modern generation of all-electronic computers was the ENIAC (Electronic Numerical Integrator and Calculator) designed by John Presper Eckert and John William Mauchly of the University of Pennsylvania. Its intended purpose was to calculate gunnery tables for the US government and was begun in 1943. It was even bigger than ASCC, containing 18,000 valves, occupying 150 square metres (1600 square feet) of floor space and consuming 100 kilowatts of electricity: the dissipation of the heat was a major problem. Nevertheless, it represented an important milestone: it brought computers into an area where normal methods of calculation were feasible in theory, though still unattainable in practice because they were so time-consuming.

These early machines were based on decimal notation, but in the early 1940s John von Neumann of Princeton University and, later, the University of Pennsylvania reverted to an early suggestion by Liebniz that a binary notation would be more appropriate: in this numbers are represented by only two digits, 0 and 1, instead of 10 (0–9).

Von Neumann also formalized the crucial concept of the stored program – also foreshadowed by Leibniz – which is the basis of the logical design of all modern computers. The program (the American spelling is universal in the computer world) is the sequence of instructions by which the computer is controlled. The earliest computers were programmed by changing circuits by means of leads and plugboards, as in manual telephone exchanges. In von Neumann's design the program became a sequence of numbers stored in the computer's memory, like the data on which it worked. This enormously increased the speed with which a program could be fed into the computer and revised if necessary.

Public awareness of the capability of computers was greatly stimulated when UNIVAC I correctly predicted that Eisenhower would win the 1952 presidential election. Here J. Presper Eckert (centre), its co-designer with John W. Mauchly, is seen discussing some of the early results with Walter Cronkite of CBS News (right).

Although von Neumann, a Hungarian by birth, worked in the United States, the first prototype machine to incorporate his principles, 'Baby', was built at the University of Manchester, in England, in 1948.

However, EDSAC – Electronic Delay Storage Automatic Calculator, built by Maurice Wilkes at Cambridge may validly be considered the first fully operational and practical stored-program computer as, from the time it ran its first program in May 1949, it was put to immediate use as a scientific tool. The world's first scientific paper to be published using computer calculations was a genetics paper written by R. A. Fisher using EDSAC for computed data. Both machines used storage methods inspired by experiences with radar during the Second World War. Interestingly, both relied on metaphors of time delay. EDSAC used the memory device proposed by von Neumann, Eckert and Mauchly: a mercury delay line to store its memory. Using principles of echoing acoustic waves, a crystal was vibrated in short binary pulses. The acoustic signals travelled through 1-metre (3-foot) tubes filled with mercury in which they were amplified and bounced back. In this way, signals were kept bouncing back and forth in a time delay

Baby used a cathode ray tube storage method developed by Frederic Calland Williams, and known thereafter as a Williams storage tube. A short pulse of negatively charged electrons focused to a small spot on a screen will knock off extra electrons and cause a brief positive charge to appear on the screen. If the beam is moved to one side, a row of bright spots can thus be 'written' on the screen. These spots of varying charge strengths could then be 'read' by the same beam, amplified and returned to the computer.

The advent of the 'chip' in the 1960s, by which complete microcircuits could be etched on a tiny slice of silicon, completely altered the scale of electronic devices, and effected a degree of miniaturization previously unthinkable; a chip is shown here on a bed of sugar crystals (right). In particular, it opened up completely new possibilities in the design of computers.

An important new use for computers is in design, as in the aerodynamic modelling of automobile bodywork (below right). The computer can ring the changes on a single basic design unit until a desired specification is met.

In 1950, the first computer intended for business use, the Universal Automatic Computer (UNIVAC I), came on the market. Like the Hollerith machine, it was designed to speed the US census but it became famous in quite a different context. In 1952 it was used to forecast the results of the presidential election and its correct identification of Dwight D. Eisenhower as the winner was widely acclaimed.

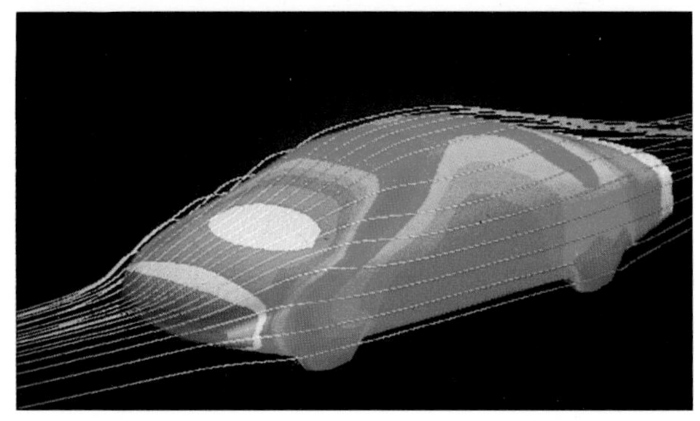

Hardware, software and the microchip

By this time, the terms hardware and software were coming into use. The hardware is all the physical equipment, while software encompasses all the programming material that directs its operation. Software evolved to keep pace with the hardware and a great deal of attention was paid to the development of programming 'languages'. The first stored-program machines were programmed directly by means of strings of numbers, a task that only highly skilled users with a mathematical turn of mind could perform, and even then many errors were inevitable. A programming language used terms that bore a resemblance to everyday English and could be used relatively easily by non-specialists. Additional software was required to translate a program written in such a language into the 'machine code' that was directly employed by the computer.

By the 1950s, the means of inputting and outputting information had got out of step with the computer's rapidly advancing speed and power: calculations could be made faster than the punched cards could supply the information. UNIVAC marked an important advance in that these methods were replaced by programs and data recorded on magnetic tape, which was already widely used, as we have noted, for sound recording.

The late 1950s saw two other extremely important advances. The expensive, power-consuming thermionic valve was replaced by the far cheaper, more compact and more reliable transistor. Although the transistor was invented in 1947, it was originally made with germanium, and it was not until 1960 that most used silicon.

In 1958, Jack St Clair Kilby of Texas Instruments assembled a few transistors and capacitors on a single support. The transistor and the integrated circuit ushered in a period of phenomenally rapid development, the age of the silicon chip. A small computer may contain several thousand circuits and a large one a hundred or more times as many. With modern technology, hundreds of thousands of electronic components can be constructed on a single silicon chip no more than 10 millimetres (¼ inch) square: these can then be combined to form large units.

In 1971, the Intel Corporation introduced the first commercial microprocessor, in which an entire central processing unit – the heart of a computer – was located on a single chip. Just a year before, Gilbert Hyatt had registered for a US patent detailing an integrated circuit containing all the necessary elements for a computer. He is therefore considered the inventor of the microprocessor. Today, the chip-based industry exceeds in gross annual turnover the steel industry which dominated the 19th century.

data storage 310 **early calculators** 306–307 **facsimile transmission** 236 **fifth generation computers** 312 **rise of television** 274–275

The first microcomputers, beginning with the Micral in France, were released in the early 1970s. The late 1970s and early 1980s saw their adoption by small businesses and in the 1980s and 1990s, due to lower prices and increased availability, families began to purchase them for personal use. In less than half a century, electronic computers had changed from being immense, specialized calculators to small household commodities. Processing speeds have doubled every two years since the introduction of the first electronic calculators: the latest personal computer is 60 million times faster than the 1951 UNIVAC I.

As processors improved, so did the means of storing information. Early electronic computers had moved quickly through several types of storage systems before settling on magnetic methods. Jay Forrester, recognizing that electrical polarity could be used to represent binary signals, had developed ferrite core memory in 1953. In one form or another, this storage method has remained the most prevalent. Floppy disks (introduced in 1970), hard disks and magnetic tape all encode data as a series of varying charges.

Although the audio compact disc (CD) appeared in 1982 (ten years after the Philips laserdisc), it was another two years before optical storage for computer data was introduced in the CD-ROM (Read-Only Memory). Compact discs use lasers to read and write data stored on a plastic surface as a series of tiny pits. As costs have fallen, optical storage has begun to supercede magnetic for many archiving needs, storing over 5¼ billion bits (a 1 or a 0) on a single CD.

Mainstream storage media have remained fairly stable for the lifespan of the microcomputers simply because the average computer user can ill afford to change frequently. This does not prevent the research and development of new methods. The latest optical method attempts to store greater amounts of data by using holographic techniques to place images in a three-dimensional matrix on crystal or plastic. By such means, 840 million bits (100 megabytes) have been stored on a 2⅕-centimetre (1-inch) crystal cube. One-tenth of that amount has been stored on a read-only plastic film which is ⅒ millimetre thick. Other researchers are endeavouring to store data on the atoms themselves. In the late 1990s Michael Noel and Carlos Stroud of the University of Rochester, New York used quantum lasers to store charges in electrons, storing up to 900 bits in a single atom.

While printers were first used with computers in 1953, it was not until the mid-1980s that the first laser printers were introduced for use with microcomputers. Coupled with Postscript – a page description language allowing users to print directly to a page – and software for integrating text with illustrations, the era of desktop publishing began. In addition to changing the printing industry forever, desktop publishing accelerated development in peripheral hardware – scanners, removable storage, input/output devices, digital photography and film output.

These developments have had far-reaching consequences. Calculators for everyday use have achieved a degree of miniaturization and cheapness that would have been unbelievable a generation ago: calculators are incorporated in everything from light switches to wristwatches. Tiny computers can be programmed to control a vast range of technical operations, from the operation of an oil refinery to the performance of an automobile engine or the cutting of a machine tool. Wholesalers and shopkeepers can use computers to keep track of stock, including details of how long particular items have been held. Information can also be displayed on a Visual Display Unit (VDU) – the monitor – and then printed out as a permanent record.

The VDU became an essential part of information exchange. In the 1970s specially adapted television receivers began to be available which could, on command, display a wide range of written information on the weather, sports results, traffic conditions, stock exchange quotations, radio and television programmes, and so on. In the British Teletext system, selection could be made from about 150 classified pages. The Prestel service, introduced in 1979, was far more ambitious for it enabled paying subscribers to key in, via their normal telephone sets, to a database containing tens of thousands of pages of information supplied by scores of independent contractors. When the World Wide Web (known as the Web or WWW) gained prominence in the mid-1990s, this information, and the way in which it was presented, expanded greatly.

The Internet and the World Wide Web

The Web emerged from a strategy to access research information stored on a global network of computers called the Internet. Vannevar Bush presaged the Web in 1945 with a prescient article in the *Atlantic Monthly* called 'As We May Think', in which he outlined a procedure for organizing and accessing data by 'associative indexing', where two or more pieces of information can be 'tagged', or tied together. It was to be many decades before these thoughts became reality.

The Internet had humble beginnings in 1969 as a networking research project called ARPAnet, sponsored by the Advanced Research Projects Agency (ARPA) within the US Department of Defense. The project linked several university and governmental agency computers in an attempt to share information and resources effectively and increase collaboration. This led to a series of evolving projects; continuous improvements were made as more scientists and researchers began to work together.

By 1971, a program for sending electronic mail (email) across a distributed network was developed. University College of London and the Royal Radar Establishment of Norway made the first international connection in 1973. The following year, satellite links across two oceans were tested and the first public data service, Telenet, was released; Queen

Elizabeth II sent her first email in 1976. The University of Essex created the first Multi-User Dimension (MUD) in 1979, enabling several people to interact simultaneously in a text-based space. In 1982, the word 'internet' was coined to refer to the connected set of networks. In 1990, the World became the first commercial provider of Internet dial-up access, and the following year Tim-Berners-Lee of the European Laboratory for Particle Physics (CERN) added a critical dimension of accessibility and functionality to the Internet by realizing Bush's dream with the Hypertext Transfer Protocol (HTTP) which 'tagged' information so that people could access it without knowing exactly where it was located in the vast web of data. In 1993, the National Center for Supercomputing Applications' (NCAS) release of a visual 'browser' interface, Mosaic, using the HTTP, set off an explosion. With an easily understood tagging system and a visual means to present information, large numbers of people from around the world began to get involved with the WWW, which proliferated for the next year or two at an astonishing 341,000 per cent annual growth of service traffic.

Applications of computers

An important use of computers besides those already mentioned is in design and simulation. Ivan Sutherland is considered by many to be the father of computer graphics. It was he who demonstrated, in the early 1960s, the potential of computers for design in his doctoral thesis at MIT. 'Sketchpad – A Man-Machine Graphical Communication System' was the first computer graphics software, and led to his development of some of the first input devices (other than a keyboard), including the first lightpen and the first computer-driven head-mounted display (HMD). He later co-founded Evans & Sutherland, a leading supplier of high-end simulation systems. As Sutherland said in 1965, 'A display connected to a digital computer gives us a chance to gain familiarity with concepts not realizable in the physical world.' Today, the computer serves not only as a design tool, but also as a medium for creative exploration and as a platform for communication. For instance, new engineering concepts can be generated rapidly within a computer space, visualized and tested entirely for structural and aesthetic integrity before an object's physical creation. New systems of production developed throughout the 1980s and 1990s enable components to be fabricated directly from the digital file by computer-driven equipment. So-called rapid-prototyping facilities create a finished product by using the three-dimensional data to drive milling machines or lasers which cut away hard material or build up molten plastic in successive layers.

Virtual reality interfaces offer new means of visualizing complex datasets. A scientist from the Human Interface Technology (HIT) lab in Seattle demonstrates how air traffic controllers may operate in the near future.

Internet 342, 345 **rapid prototyping** 343 **satellites** 315, 332–333, 334, 336, 340–341

Although simulation systems were developed initially after the Second World War by the military for training pilots, it was the entertainment industry which provided the world with one of the first visions of 'immersion' in a simulated space. Mort Heilig patented Sensorama in 1962 – an 'experiential cinema' which combined 3D movies, stereo sound, mechanical vibrations, fan-blown air and aromas into a multisensory experience driven from a motorcycle seat and viewed through Viewmaster-type goggles. Like Baird's television, it provided a mechanical vision which would inspire the development of electronic systems.

In 1981, after years of work, the US Air Force's Super-Cockpit project was completed at Wright-Patterson Air Force Base, Ohio. This system comprised a mock aircraft cockpit which integrated computers and HMDs to simulate a three-dimensional graphic space through which pilots could learn to fly and fight. A more affordable system, the Virtual Interface Environment Workstation, was subsequently developed by the NASA Ames Research Center in California for planning space missions and combined computer graphics, video imaging, three-dimensional sound, voice recognition and synthesis with immersive input devices.

It was these devices, an HMD based on miniature television sets bought at a local consumer electronics shop and a data glove created by Tom Zimmerman and Jaron Lanier that helped spark a new market in Virtual Reality (VR) hardware and software. They realized that such systems for interacting with simulated environments in real time could be created inexpensively, and formed the first VR system supply company, VPL Research, in the late 1980s. Such systems were quickly appropriated by universities, businesses and enthusiasts, and are now used for such varied purposes as modelling and testing aircraft, supporting battle simulations, visualizing and interacting with molecular models or statistical data, and engaging in games with people on the other side of the world.

Sequential developments in the speed and miniaturization of computer software and hardware continues to change rapidly their role in society. In addition to these incremental changes, however, progress continues laterally as well. In 1982, the Japanese launched a 10-year programme for the development of 'fifth generation' computers. By 1991 the Japanese Institute for New Generation Computer Technology had developed the Parallel Inference Machine (PIM) which handles words and images, not through numeric representation, but by logical inference alone. However, the Ministry of Trade and Industry subsequently abandoned the fifth generation computers and focused on sixth generation computing based on neural networks able to simulate the mental processes of inference, association, emotion and learning.

In addition to the miniaturization of existing components, entirely new systems of computing are being devised. In 1990, a prototype of an optical computer was introduced by Alan Huang, of AT&T, suggesting the possibility for calculation speeds a thousand times greater than existing machines by relying on photons of light, rather than electronics to transport data. However, the technical feasibility of effectively developing such computers has yet to be proven.

In 1999, a biological computer utilizing neurons from leeches was created by a team of researchers from Georgia Institute of Technology and Emory University. The prototype 'leech-ulator' could only perform simple sums, but the objective is to devise a new generation of fast and flexible computers which can grow and develop their own pathways to a solution rather than produce pre-programmed solutions defined by humans. Although microelectrodes are used to stimulate the neurons, it is the neurons' ability to form new connections and 'think for themselves' which provides the possible advantages to a biological approach to computing. Such a computer is particularly suited to pattern recognition tasks such as face, voice or handwriting recognition.

Pictorial communication

The communication of information by pictures received a considerable stimulus with the development of a variety of

Holography, invented in 1947, was a new departure in image formation, giving a remarkable three-dimensional effect. It was much improved after the laser became available 13 years later.

new processes in the 19th century for the mass reproduction of pictures in books, newspapers and magazines; the advent of the cinema, television and, later, visual browsers to the Web further enhanced the progress. Today, the public impact of sound and pictures through the various media is enormous; probably more information is now conveyed in this way than by the printed word.

Most of these pictorial techniques we have already discussed, but we have not so far mentioned holography, which is entirely a post-war development. It differs from photography in that while this records an image, a hologram records a wave pattern induced by an object. It was invented in 1947 by Dennis Gabor, a Hungarian electrical engineer working with the British firm BTH, whose original interest was to improve the image created by an electron microscope. The process involves two beams of monochromatic light (or electrons), one of which is reflected on to a photographic film from the object to be depicted. The other is a reference beam going directly to the film within which the two means 'interfere' with each other. If the transparent film is then examined by transmission of the same monochromastic light, a three-dimensional picture is obtained.

Originally, holograms were of very poor quality, but with the invention in 1960 of the laser, which emits an intense beam of monochromatic light, the situation was transformed. Today, holography is the basis of a multi-million dollar industry, used not only in scientific research but in advertising, as a security device on credit cards and even as an acknowledged art form. In 1997, Tung H. Jeong demonstrated a system he developed at Lake Forest College, Illinois, with Hans Bjelkhagen and others, in which multiple beams of coloured light are used to mass-produce full-colour holograms; released from the monochromatic limitations, this opened up new avenues for their use as accurate three-dimensional representations. In recent years a new form of holography, also proposed by Gabor, has been developed, based on the use of sound waves in place of light waves. This system of acoustical imaging has been used for a variety of purposes in circumstances where the medium surrounding an object is opaque to light rays – as, for example, in medical diagnosis or the observation of submerged objects.

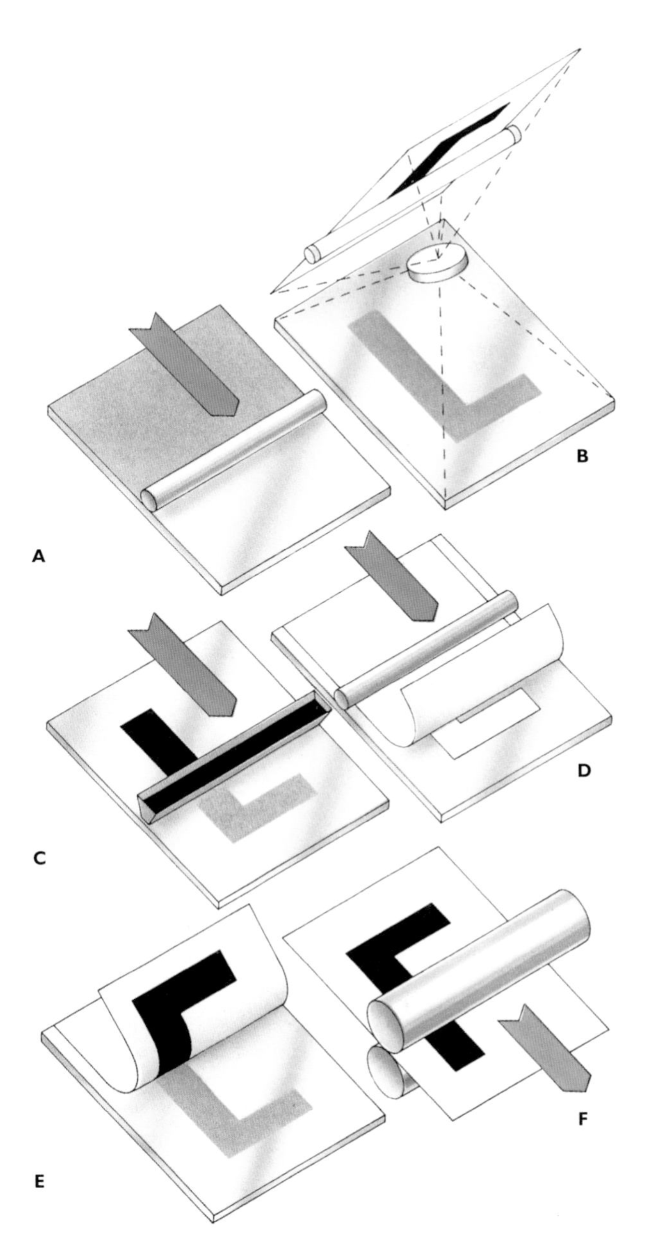

In photocopying, a metal plate is given an electric charge (A). An image of the document is then projected on to the plate (B), which loses its charge where light falls on it. Powder is then dusted over the plate (C) and adheres to the charged area (the dark part of the image). A piece of paper is then pressed against the plate (D), and the powder transfers to the paper (E). The copy is made permanent by heating the paper (F).

Electrophotography

Another photographic process which is for practical purposes an entirely post-war development – though its American inventor, Chester Carlson, patented it in 1937 – is that known as xerography, so called because it is completely dry (Greek *xeros*, dry): it was developed in the Batelle Institute, Ohio, by Roland M. Schaffert.

This has literally revolutionized the copying of documents in the office – with very serious implications for security and protection of copyright – and for short print runs. Although copying machines vary greatly in their design and performance, the basic principle – as set out by Carlson in his patent – is very simple. A metal sheet has a thin photoconductive layer of selenium deposited on it. This film can be electrically charged in the dark, but exposure to light will dissipate the charge. Consequently, a black-and-white picture can be delineated as an equivalent charged/uncharged pattern. If the plate is then exposed to an oppositely charged black powder, the particles will adhere to the charged area of the plate and can be transferred to a sheet of paper. Today's laser printers rely on much the same principles to generate images of computer files.

computer simulations 327 **printing processes** 233 **robotics** 343

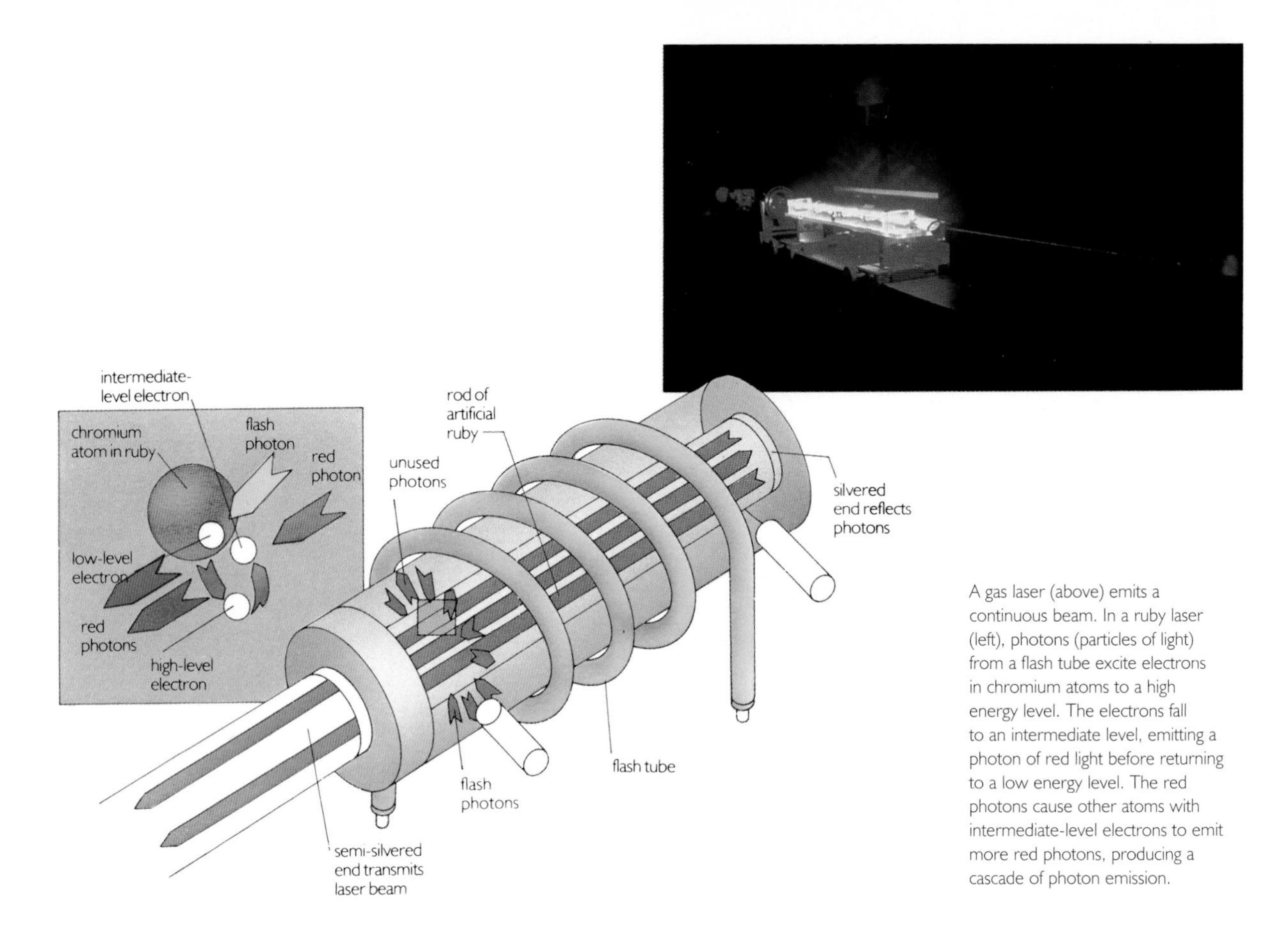

A gas laser (above) emits a continuous beam. In a ruby laser (left), photons (particles of light) from a flash tube excite electrons in chromium atoms to a high energy level. The electrons fall to an intermediate level, emitting a photon of red light before returning to a low energy level. The red photons cause other atoms with intermediate-level electrons to emit more red photons, producing a cascade of photon emission.

Lasers

The use of the laser in fibre-optic telephony and holography has already been noted, but this novel device has many other important applications. It depends on the fact that atoms can exist only at certain specific energy levels. If the atom goes from a high energy level to a low one, a burst of radiation will be emitted. In the early 1950s it was suggested that this effect could be used to amplify radiation in the short-wave region, such as is used for radar. Charles H. Townes in the USA and N. G. Basov and A. M. Prokhorov in the former Soviet Union developed this into the maser – named from the initials of *m*icrowave *a*mplification by *s*timulated *e*mission of *r*adiation – using first ammonia gas and later a ruby crystal as the irradiated material. The maser principle is used in low-noise amplifiers in applications such as radio telescopy.

Meanwhile, Townes, with Arthur Schawlow, had suggested the laser – from *l*ight *a*mplification by *s*timulated *e*mission of *r*adiation – the prototype of which was built by Theodore H. Maiman in 1960. The great importance of this was, as its name implies, that it amplified visible light, and this light has unusual properties. It can be several million times more intense than that of sunlight; it is produced as an extremely narrow pencil; and it is monochromatic, i.e. it is of a single wavelength. This first laser was a ruby laser but almost at once a gas laser was developed in the USA in the Bell Telephone Laboratories. This is less powerful but provides a continuous beam, compared with the ruby's rapid series of pulses.

Laser beams find many uses. In engineering, their narrowness and intense energy is used to cut and drill metals; in medicine they can be used to tack back a detached retina or make a bloodless incision. Because the beam does not spread out even over great distances, lasers can locate small targets at great distances and are used in military range-finders. During the 1970s the earth–moon distance was measured with unprecedented accuracy by means of laser beams reflected from mirrors placed on the moon's surface by American astronauts. At a more earthly level, they are used in stores as bar-code scanners and in compact disc players.

Such distance-measuring techniques demand extremely accurate chronometers. Modern instruments – presaged by W. A. Alvin's quartz crystal clock of 1929 – depend on the rate of vibration of molecules or atoms. An atomic clock, built in 1969 by the US Naval Research Laboratory and based on the vibrations of ammonia molecules, is accurate to one second in 1.7 million years.

Telecommunications

In recent years, telecommunications systems have expanded enormously and there have been some important technical changes. For the user, however, these changes have often not been very apparent. In telephony, for example, widespread use is now made of glass fibre 'cable' using laser-generated light impulses, instead of a modulated electric current flowing in a copper wire, to codify a sound signal. In Britain alone, there were 50,000 kilometres (31,000 miles) of such cable by 1985, yet the ordinary telephone subscriber would notice no difference in the nature of his service nor even have to change his instrument. The gain was in the efficiency of the service: for a given degree of traffic, a lighter, cheaper and more compact cable could be used. Equally, the introduction of electronic exchanges, initially by Bell Telephone in Illinois in 1960, meant that equipment became far more compact and connections were made much more quickly: in addition, customers' calls and charges could be recorded automatically. Similarly, long-distance telephony was improved by using satellite relay stations, the first being Telstar, launched in 1962.

However, these technical advances could be introduced only slowly; meanwhile, increasing demand had to be met by expanding conventional systems. For example, eight metallic submarine telephone cables were laid across the Atlantic between 1956 and 1976, having an overall capacity for about 4000 simultaneous calls. Not until 1988 was the first transatlantic glass fibre cable laid.

Radio and television developed rather similarly, with the customers' existing equipment slowly needing replacement. By the mid-1950s, transistors were being made at the rate of many millions a year and the valve is now obsolete, except for a few special purposes. One important development did, however, require a modification in receiving set circuitry: this was the introduction of frequency modulation (FM). All early radio systems operated by means of amplitude modulations: that is to say, the signal was transmitted as a variation in the strength of the wave. In 1933, Edwin H. Armstrong in the USA devised the alternative system of frequency modulation, primarily as a means of reducing the crackling background noise known as static. In this, as its name implies, it is the frequency of the wave that is varied to convey the signal. Frequency modulation began to be used in commercial broadcasting after the Second World War. However, as it requires the use of a relatively wide waveband, FM is suitable only for very high frequency (VHF) transmissions; it is particularly effective for high quality transmissions over limited areas.

The transistor radically reduced the size and power demand of radio sets, for the first time making available truly pocket-size sets. Its application to television receivers in the 1960s also reduced the size of sets, but this reduction was limited by the shape of the cathode ray tube. In 1979, however, the Japanese firm Matsushita introduced a flat tube in which the picture is displayed by making use of what are called liquid crystals. Briefly, these are complex chemicals in which molecular rearrangement occurs when an electrical voltage is applied to them. This rearrangement can be translated into a visible signal by means of polarized light: that is to say, light in which all the vibrations are in one plane instead of at random. Such liquid crystal display (LCD) devices demand very little power input, and they are widely used in laptop computers, digital watches, pocket calculators and similar small instruments.

There has been a great improvement in the quality of the picture. In Britain, for example, pre-war receivers displayed a 405-line picture and this was continued into the 1960s. Today most countries transmit a much sharper 625-line picture using a Phased Alternate Line (PAL) signal, though the USA and Japan still favour a 525-line system using a National Television Standards Commission (NTSC) signal. It is likely, however, that the future will see much sharper definition widely adopted, with lines in excess of 1200. The greatest limitations are a result of social and economic, rather than technical, constraints. Systems for High Definition Television (HDTV) have been available since the early 1990s, but as long as broadcasting stations continue to send lower resolution signals, the value of such systems will be under-utilized. Even after broadcast standards have been established, the economic stakes are still high, with an existing market in the USA alone of some $200 billion. Several countries are now broadcasting digital television signals, but the plan for obsolescence of the analogue signal and equipment that depends on it has yet to be defined.

Above: Satellites in geostationary orbit at a height of 35,900 kilometres (22,307 miles) take exactly 24 hours to orbit the earth, thereby remaining permanently above the same spot on the surface. From this position, they can effect communications over a large area.

Right: The Sony TR-55 – the world's first fully transistorized radio, introduced in August 1955.

fibre optics 303 **radio transmission** 236–240, 241 **satellites** 332–333, 334, 336, 340–341 **telecommunications** 342

CHAPTER THIRTY-FOUR

New Military Technologies

War, as we have previously remarked, is a great but uneven stimulus to technology because it creates an abnormal milieu for innovation. The economic, political and social constraints that generally regulate development are temporarily relaxed. In a fight for survival, great resources can be marshalled for special projects: time is telescoped, in that some programmes which would normally take years, supposing they were embarked upon at all in peacetime, get completed in months. Other programmes may be hindered by warfare, and prevented from achieving their potential for two reasons. The first is that expenditures are restricted to those projects deemed vital to national security. The second is that the enforced secrecy limits the collaborations necessary to foster innovation. Frequently, there is a duplication of effort from within the same governmental bodies as different divisions work on the same problem without knowledge of their compatriot's simultaneous endeavours. Most innovations continue to find a place in the reduced military needs of post-war years and many, suitably adapted, are subsequently the source of important developments in civilian life. Thus the air forces built up at great speed during the First World War formed the basis of a worldwide network of civilian airlines, providing a completely new mode of public transport. Civil aviation profited also from the military developments of the Second World War, notably with the advent of the jet engine and radar. Rocketry, to which Germany vainly pinned its hopes of final victory, spawned both a new generation of powerful intercontinental missiles and man's first successful venture into space, leading to the Apollo landings on the moon.

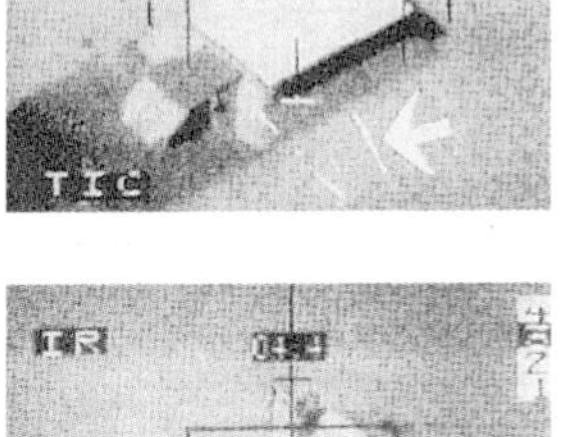

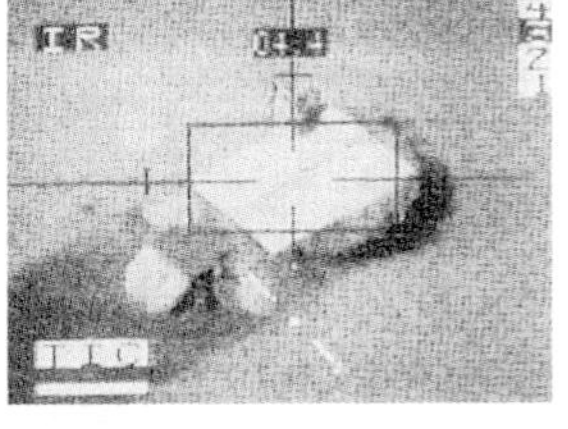

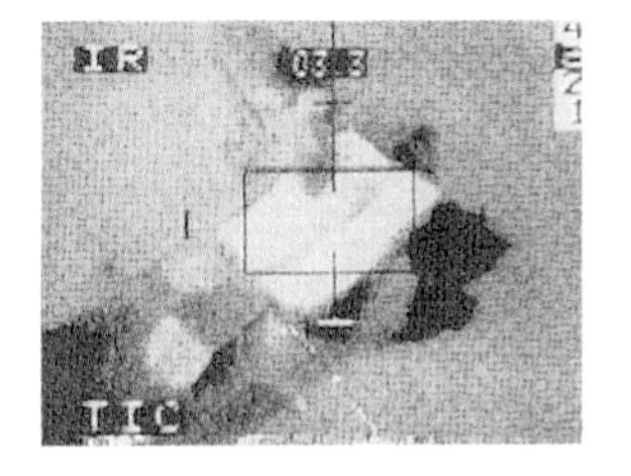

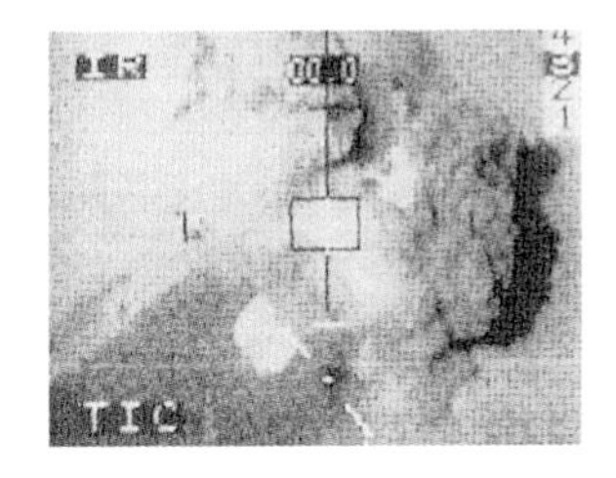

Important though they were, however, these military developments were overshadowed by one other, that of the atomic bomb. Too late to affect the course of events in Europe, it stopped the war against Japan in its tracks. Since the two attacks in 1945, no atomic bombs have been exploded as an act of war, but their existence has profoundly affected the military strategy of the great powers ever since and the political and economic repercussions have been enormous. The bomb has also led to an important new source of energy to complement the fossil fuels so rapidly being depleted. Because, for better or for worse, it is the most significant technological development in human history, atomic energy demands our attention first.

Atomic energy

The concept of atoms is as old as serious thought about the nature of the world about us. Thus Lucretius, in his *De rerum natura* of the first century BC, identified differences in the properties of substances with those in the ultimate particles, or atoms, of which they

Above: A precision weapon strikes a target using visual pattern recognition. A bomb's-eye view is provided by a camera in its nose.

Left: The dropping of the first atomic bomb on Hiroshima on 6 August 1945 opened an entirely new chapter not only in the history of warfare but of the human race. One consequence was that uranium, hitherto little more than a laboratory curiosity, suddenly became a very important industrial metal prospected for worldwide. Unlike the traditional metals, uranium is radioactive, and to the normal equipment of the prospector was added the Geiger counter, capable of detecting and measuring radioactivity in the field.

were composed. In the absence of experimental evidence, these were essentially philosophical speculations, based on two apparent alternatives: is matter a continuum or is there a point at which further subdivision is impossible?

The first to ascribe a real physical identity to atoms was the Manchester chemist John Dalton, who in 1803 showed that a great variety of chemical reactions could be logically explained on the assumption that all matter consisted of atoms, and that the atoms of an element were all alike and distinguished from the atoms of other elements only by differences in their weights. This was no mere speculation: although he had no means of determining the absolute weight of atoms, he was able to calculate the relative atomic weights of some elements.

This concept revolutionized chemical thought, but for nearly a century it was accepted that atoms were indestructible and immutable: whatever vicissitudes they might be subjected to, atoms of hydrogen or nitrogen, for example, would never lose their identity. Then, in 1896, Henri Becquerel discovered the phenomenon of radioactivity, showing that certain atoms spontaneously disintegrate with the emission of radiation: uranium, for example, breaks down to form radium and ultimately lead. This clear evidence that atoms were not, after all, immutable suggested that those of non-radioactive elements might be artificially disintegrated if they were struck by high-speed particles of atomic dimensions – just as a coconut can be shattered by a direct hit with a wooden ball. In 1919, Ernest Rutherford, at Manchester, succeeded in expelling a hydrogen atom from one of nitrogen by bombarding it with helium atoms. In 1932, at Cambridge, John Cockcroft carried out the famous atom-splitting experiment with E. T. S. Walton: in this atoms of lithium were not merely chipped, as it were, but shattered into two roughly equal parts, forming, with the particle that triggered the splitting, two helium atoms.

Meanwhile, other properties of the atom were becoming clear. Dalton's simple concept of all the atoms of a particular element having the same weight was no longer tenable. In 1913, Frederick Soddy advanced the idea that atoms could be chemically identical yet have different weights: thus magnesium atoms can have weights (relative to hydrogen) of 24, 25 or 26. For such atoms he coined the term isotope. What is important is the size of the charge on the relatively small central nucleus (+12 in the case of all isotopes of magnesium) balanced by the negative charge of outer clouds of electrons, each carrying a single negative charge.

The charge in the nucleus resides on small particles known as protons, each carrying a single positive charge. Additionally, the nucleus contains neutral particles, known as neutrons, roughly equal in weight to protons. Modern research has shown the atom to be vastly more complex than this, but it was on such a simple concept that the vital early work was done.

The first atomic bomb was exploded on 16 July 1945 in the desert of New Mexico. It was code-named TRINITY, after the site (top) at which the experimental development work was done. It was contained in a huge steel vessel, Jumbo (above), 8 metres (26 feet) long and weighing some 200 tonnes.

This was a period of rapid change in physics and two other basic tenets were being overturned. It had long been an article of faith, well attested by experiment, that both mass and energy are indestructible. Whatever changes, such as a chemical reaction, occur in a closed system the total mass remains unchanged, even though its nature may be different. Equally, as we have seen, if one form of energy disappears, an exactly equivalent amount of a different form takes its place. By and large, these laws of the conservation of mass and of energy are still true so far as all everyday operations are concerned. By the turn of the century, however, theoretical physicists were reaching the conclusion that in certain very special circumstances these laws were no longer valid and that matter and energy were not independent entities but themselves interconvertible. In 1905, Einstein put forward his famous equation: $E = mc^2$, relating to energy (E), mass (m) and the speed of light (c). The latter is very large indeed, 300,000 kilometres per second (186,300 miles per second) – fast enough to circle the earth in one-seventh of a second: the square of this constant is correspondingly far greater. If we go back to the atom-splitting experiment of Cockcroft and Walton, the law of conservation of mass would require that the weight of the two helium atoms formed should exactly equal that of the original lithium atom plus the incident

First World War 215–222 **medieval warfare** 133–140 **radioactivity** 212, 284 **rocketry** 322–324 **warfare in antiquity** 82–88

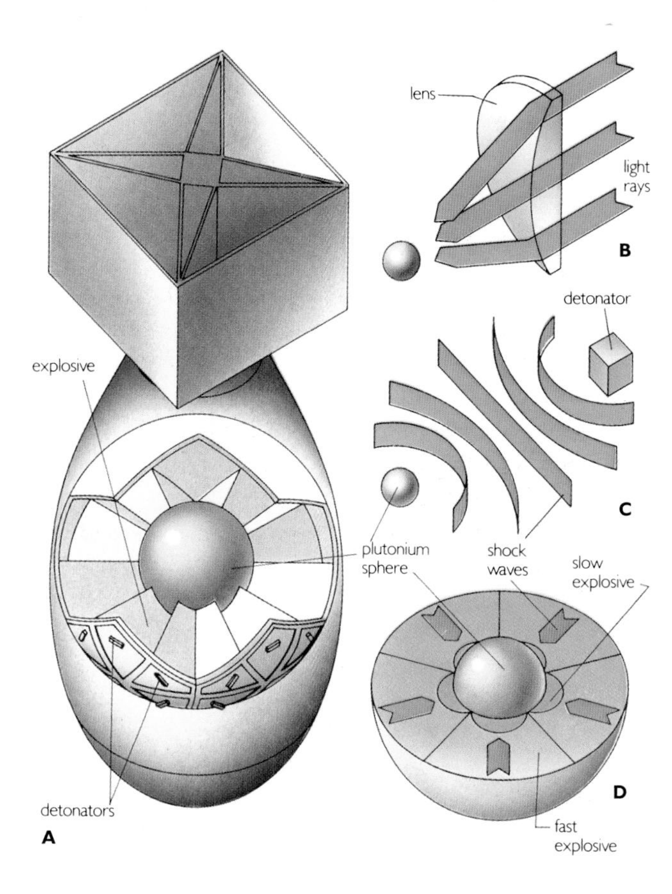

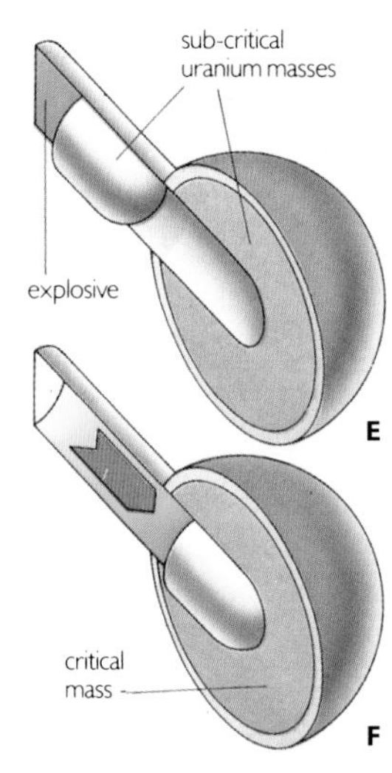

Inside the Fat Man atomic bomb (A) that destroyed Nagasaki was a hollow sphere of plutonium surrounded by layers of fast and slow explosive (D). Detonators ignited the explosive, which burned to produce a symmetrical shock wave that crushed the sphere, resulting in a nuclear explosion. By arranging for the explosive to burn at different rate, the shock waves were focused at the centre (C) in much the same way as a lens focuses light rays (B). The Little Boy bomb that destroyed Hiroshima contained two sub-critical masses of uranium (E). One was fired into the other to produce almost instantaneously a critical mass (F) that exploded.

particle that caused the reaction. In fact, experiment showed that not only did some mass disappear but also that an amount of energy was released that was in exact agreement with Einstein's equation. Thus a very profound new truth had been established – mass and energy are themselves interconvertible according to a fixed scale.

At the time, this momentous discovery seemed to have no practical relevance – these observations had been made by observing what happened when single atoms collided and single atoms are minute. A few imaginative scientists, such as Soddy, thought it portended the availability of energy on a hitherto unheard-of scale, but Rutherford – who had done more to unravel the secrets of the atom than any man alive – regarded the idea of atomic power as so much 'moonshine'. Rutherford died in 1937; within less than a decade moonshine had been translated into grim reality.

How this came about is too complex a story to relate in any detail here but the starting point was a series of experiments by Enrico Fermi, in Rome, in which he bombarded uranium with neutrons – the uncharged atomic particles present in the nucleus – as a result of which the uranium displayed increased radioactivity. At first it was not clear what had happened, but early in 1939 Lise Meitner and her nephew Otto Frisch, both refugees from Nazi Germany, correctly concluded that the uranium atom had been split into two roughly equal fragments, with a huge release of energy. In April of that year in Paris, Frédéric Joliot and his wife Irène, née Curie, showed that in the fission process itself not only were neutrons released but they appeared in greater numbers than in the original bombarding beam. These secondary neutrons could disintegrate more uranium atoms, and thus potentially set off a self-perpetuating, and accelerating, chain reaction (cascade reaction), spreading through the whole mass of uranium. Familiar counterparts are the domino trains beloved by the Japanese, in which the fall of a single domino triggers the sequential collapse of hundreds or thousands more. Finally, Niels Bohr and J. A. Wheeler, in Copenhagen, showed that a rare isotope of uranium, ^{235}U, present to the extent of only 0.7 per cent in natural uranium, is much more easily fissionable than the common isotope ^{238}U. By a strange irony, this crucially important result was published on 1 September 1939, the very day Germany invaded Poland and precipitated the Second World War which, six years later, was to be abruptly terminated by the two atomic bombs. Thereafter, research on atomic energy was conducted in great secrecy by a number of countries. When war broke out all the combatants stood possessed of the same basic knowledge, though in the event it was only the USA which made effective use of it.

Why this should be is at first sight puzzling, for the possibility of a new explosive vastly more powerful than any in existence was apparent. On the available information it was clear that above a certain critical mass – perhaps no more than 5 kilograms (11 pounds) – the number of neutrons produced by ^{235}U by the fission process would exceed those lost by escape or by absorption. When that happened, a chain reaction would be initiated which would in a fraction of a second release energy equivalent to the explosion of many thousands of tonnes of TNT. In practice the technological problems were enormous and probably at that time only the USA had the resources necessary to overcome them.

Although the full magnitude of the difficulties only slowly became apparent, it was clear from the outset that it would take a matter of years to develop a practical atomic weapon. Such a lengthy project did not encourage a major effort by Germany whose strategy was based on a successful *Blitzkrieg* bringing victory within a few weeks. By the time it was apparent that this had failed, Germany was under heavy aerial attack, essential materials were not available and many of the scientists essential for the success of such a project had been expelled because they were Jewish. Russia at that time was, for other reasons, also short of appropriate scientific manpower and after the German attack of 1941 was in no position to embark on a technological project of this magnitude and complexity. It is clear, however, that the Russians were well aware of the possibilities. In the autumn of 1941, their leading atomic physicist, Pyotr Kapitsa – who in 1934 had been barred from returning to Cambridge to resume research with Rutherford – stated categorically that an atomic bomb could destroy a large city. The Russians seem at that time to have been doing some basic research and it subsequently transpired that valuable information was sent to them from America by the spy Klaus Fuchs. The fall of France in the summer of 1940 ended tentative plans there to produce electricity from nuclear power.

In the event the atomic bomb began as an Anglo-American project. At the instance of two refugee scientists, O. R. Frisch and R. E. Peierls, the British government in April 1940 set up a high-level scientific committee, known under the code name MAUD, to consider the feasibility of making a bomb based on ^{235}U. A year later, the MAUD committee completed a report expressing the view that a uranium bomb requiring about 10 kilograms (22 pounds) of ^{235}U was feasible and might be decisive in the war. This report was made available to the US government, which already had a dual, but not very clearly defined, interest in atomic energy: they were interested in both bombs and power generation, especially for ship propulsion. However, after receiving the MAUD report and after the Japanese attack on Pearl Harbor in December 1941, which brought them into the war, they became firmly committed to the bomb. In June 1942, the great Manhattan Project was formally launched as a collaborative Anglo-American venture. However, for political reasons Britain was totally excluded by the USA at the end of 1942 and was only readmitted, and then only as a very junior partner, after the Quebec Agreement of August 1943. The British effort continued in a minor key under the title Tube Alloys.

The Manhattan Project

The Manhattan Project bristled with difficulties which were managerial as well as scientific and technological: not only had great quantities of complex instruments and equipment to be assembled but accommodation and facilities also had to be provided at short notice for tens of thousands of workers on remote green-field sites.

The major task was to separate a sufficient quantity of the ^{235}U isotope from the ineffective ^{238}U isotope in which it is present to the extent of only 0.7 per cent. As isotopes have, by definition, identical chemical properties, only physical methods could be employed. Of several possibilities, that most largely used was gaseous diffusion, which depends on the fact that light molecules will pass through a thin porous membrane faster than heavier ones. This posed the problem of finding a suitable volatile compound of uranium – to say nothing of a suitable membrane – and from a very limited choice uranium hexafluoride or HEX (UF6) was selected. Apart from the fact that this is a very difficult substance to work with, the masses of 235-HEX and 238-HEX are in the very close ratio 349:352 (i.e. 1:1.0086), so that a cascade of successive diffusions was necessary to effect a significant separation. Construction of this site was begun in May 1943, and it is indicative of the political and military pressure that even at that late stage no satisfactory membrane had been devised: the first useful product was obtained on 12 March 1945. Meanwhile, alternative separation processes – thermal and electro-magnetic – were being developed at other sites.

By the beginning of 1945, $2 billion had been spent on a project which at several points rested on very convincing, but nevertheless untested, theories: there was no absolute guarantee that when all was made ready success would be attained. Up to a late date there were, for example, doubts about the method of detonation: during manufacture and assembly it was vital, of course, not to exceed the critical mass, which had to be created when the target was reached. At that point, the critical mass had to be achieved in a tiny fraction of a second in order to produce a nuclear explosion.

Meanwhile, an alternative approach was being followed. If uranium is bombarded with neutrons under controlled conditions, in what is known as an atomic pile, the end product is plutonium. This is of great scientific interest, because it does not occur in nature at all. For the first time, man had created a new element: since that time about 16 new elements have been created, some of them very radioactive and with very short lives. In the context of the Manhattan Project, the interesting property of plutonium was that, like uranium, it could undergo an explosive chain reaction.

The first atomic bomb, code-named TRINITY, was successfully exploded in the desert 100 kilometres (60 miles) south of Los Alamos, the secret research centre established for the design of the weapon, on 16 July 1945. The energy released was equivalent to that of 20,000 tonnes of TNT. A similar bomb was detonated over Hiroshima on 6 August. Known as Little Boy, it weighed just over 4 tonnes and was 3 metres (10 feet) long. Three days later a plutonium bomb, of roughly similar size, known as Fat Man, was dropped on Nagasaki. On 14 August Japan surrendered, and the Second World War was finally ended after almost exactly six years. Some 100,000 people were killed outright in these two attacks, and subsequently thousands more died, or suffered severely, from the effects of the intense heat and radioactivity simultaneously released. As early as 1942, the possibility of a third type of atomic bomb had been suggested by the US

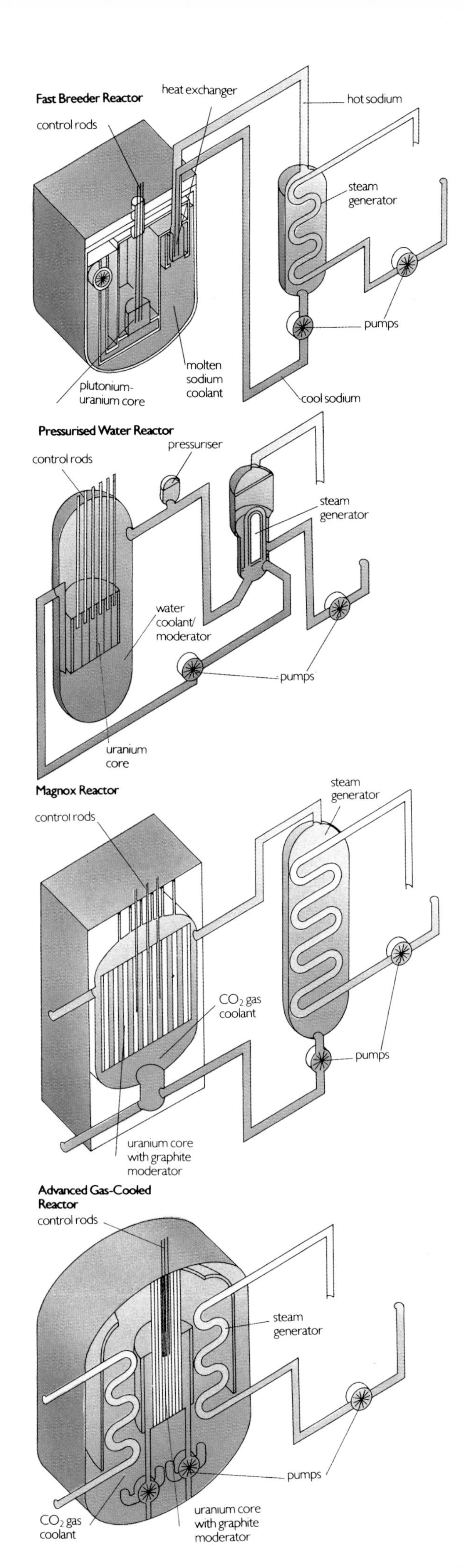

Above: The post-war arms race led to a new generation of atomic bombs vastly more powerful than those used against Japan. The first hydrogen bomb, IVY-MIKE, equivalent to 10 million tonnes of TNT, was exploded on Eniwetok Atoll on 1 November 1952.

Right: Four principal kinds of nuclear reactors. They all burn fissile nuclear fuel to produce heat and raise steam for electricity generation. However, the fast breeder reactor also converts non-fissile uranium around the core into fissile plutonium.

physicist Edward Teller. This depended on the release of energy not by fission but by fusing atoms of the heavy isotope of hydrogen, deuterium. Weight for weight, this could be three times as powerful as an explosion based on ^{235}U; additionally, deuterium was far more readily available than fission fuel. This opened up the fearful prospect of bombs a thousand times more destructive than that detonated in operation TRINITY. The technical problems were enormous, for the fusion would occur only at temperatures (100 million°C/ 180 million°F) comparable with that of the sun. In 1950, the necessary research programme was authorized and on 1 November 1952 the first hydrogen bomb, IVY-MIKE, was exploded by the USA on Eniwetok Atoll.

The post-war arms race

In the immediate post-war years many Americans cherished the belief that they could not be rivalled in the atomic field provided the knowledge and experience they had gained with so much trouble and expense were withheld from rivals. In fact, this belief had no solid foundation: not only had the scientific principles involved been published before the war, but rivals also had the certainty – which those managing the Manhattan Project had not – that ultimate success was assured. It was, therefore, a great shock to American public opinion when the Russians detonated an atomic bomb in 1949 and Britain, who had enjoyed only the half-hearted confidence of the USA, in 1952. A year later, the Russians detonated a hydrogen bomb and Britain another in 1956. Today, the principal nuclear powers are the USA, Russia, Britain, France and China. But a number of other countries have the potential to develop nuclear weapons, or have already done so, such as India and Pakistan. In 1977, the USA devised a neutron bomb, the blast of which would not damage substantial buildings but would release vast amounts of lethal neutrons.

By the early 1950s, hydrogen bombs individually equivalent to 50 million tonnes of TNT had appeared and even those who saw the proliferation of atomic weapons as the most effective deterrent to a third world war were becoming gravely concerned about both the economic consequences and the dangers of radioactive fallout from tests. In 1963 the USA, Britain, the then Soviet Union and later 49 other states signed a Treaty Banning Nuclear Weapon Tests in the Atmosphere, in Outer Space and Underwater. This effectively restricted tests to underground sites where the deadly products of explosion could be contained. Since then, political discussions have continued to try to limit and regulate the deployment of atomic weapons, but with very limited success globally. With the break-up of the Soviet Union, the Cold War was over, but left in its wake were multiple new countries with starving economies, and a readily available source of income residing in the stockpiles of nuclear weapons and uranium ore which had been building for decades. The threat of a global thermonuclear war begun by the superpowers no longer seems likely, but a potentially deadlier threat now faces the world, as the weaponry has been dispersed to less developed countries with potentially more to be gained by nuclear threat.

Atomic power

By an accident of fate the crucial discoveries that pointed the way to the release of atomic energy were made in the months preceding the outbreak of war in September 1939. In the circumstances, it was inevitable that thereafter the great concentration of effort was on military applications, culminating in the two atomic bombs – based on uranium and plutonium respectively – exploded in August 1945. There, the goal was the biggest release of energy in the shortest possible time, but if the release of energy similarly generated could be controlled, then it might be used – just like the heat generated by burning fossil fuels – to raise steam to generate electricity. To this alternative it is now appropriate to turn, for its origin is the same as that of the bomb and it ran in parallel with it. Moreover, power-generating reactors are a source of plutonium for bombs.

The first 'atomic pile' – which we now call a nuclear reactor – was activated by the Italian-born physicist Enrico Fermi at 3.20 PM on 2 December 1942 in a squash court at the University of Chicago: at that moment the Atomic Age was born. A preliminary stage in the Manhattan Project, which we have just described, it was designed to show that the kind of chain reaction on which the bomb would depend was in fact attainable. Such a reaction cannot be sustained in natural uranium, because most of the neutrons on which the fission of ^{235}U depends are absorbed by the 97.3 per cent of non-fissionable ^{238}U. However, the probability of this absorption taking place can be decreased if the neutrons are slowed down by passing them through what is called a moderator: the ^{238}U could then be transferred into a new 'synthetic' element, plutonium, which – as we have seen – is fissionable. The best moderator is 'heavy' water – water in which ordinary hydrogen atoms are replaced by doubly heavy atoms of its isotope deuterium – but this was not then available in sufficient quantity, although when France was invaded French scientists had escaped with all that was available there. As the next best thing, very pure graphite was used instead. The prototype Chicago reactor consisted of a pile of carefully machined graphite blocks, interspersed with ordinary uranium. Through this ran channels in which cadmium rods were placed. The small content of ^{235}U produced sufficient neutrons spontaneously to start a potential chain reaction; the cadmium, which strongly absorbs neutrons, could be used to restrain it. As the cadmium rods were slowly removed, a point was reached at which instruments showed that the neutron intensity was increasing, indicating that the critical point for energy production had in fact been reached. Immediately, the cadmium rods were thrust back into the pile before the reaction became uncontrollable. Among the spectators was Arthur Compton, one of the leaders of the American atomic energy project. He promptly telephoned his colleague James Conant, president of Harvard, with the cryptic and now historic message: 'The Italian navigator has landed.'

Waste disposal teams place low-level nuclear waste in a compactor at the Oak Ridge National Laboratory, Tennessee. Following new legislation, such waste is categorized before it is sealed in steel cannisters and stored in concrete vaults. The vaults are either buried in underground silos or in concrete bunkers.

Enrico Fermi 318 **first atomic bombs** 319–320 **fusion reactor** 275–276, 340 **future of nuclear power** 340 **nuclear submarines** 324

It was 12 years before the first reactor designed specifically to generate electricity came into operation, but this time it was the Russians who took the lead. In June 1954, after five years' development work, this reactor, located at Obninsk some 80 kilometres (50 miles) from Moscow, began to generate electricity at the rate of 5 megawatts. It used graphite as a moderator and enriched uranium as a fuel. The heat generated in its core was extracted by water under high pressure in a closed circuit system. This heat was circulated through a heat exchange system in which steam was generated to drive turbines, which in turn drove dynamos.

In Britain, a different system was developed. The Calder Hall reactor, which began to supply electricity to the national grid system in 1956, was also graphite moderated but cooled with gas under pressure. At about the same time, the USA – which in the immediate post-war years had shown rather little interest in the peaceful uses of atomic energy – developed yet another kind of reactor. In this, the fuel was enriched uranium but water under high pressures served as both moderator and coolant. As in the early Russian reactors, the heat was then transferred to a steam-raising system. The first such reactor was commissioned at Shippingport, Pennsylvania, in 1957. The Canadian CANDU reactor, developed in the 1950s, used pressurized heavy water as both coolant and moderator. The fuel elements were made of unenriched uranium and because heavy water does not absorb neutrons strongly, a particularly high burn-up can be achieved before the fuel elements need to be replaced.

The merits and otherwise of these cooling systems have been the subject of much controversy. The pressurized water reactor is relatively simple and cheap to build. Against this, it is argued that it is inherently dangerous. If an emergency occurs and too much heat is generated, the cooling water will be turned to steam and the reactor core will be left without means of cooling it. Gas, in contrast, does not change its state however high the temperature rises and thus retains its cooling power: it is, therefore, potentially safer. A further advantage of gas cooling is that gas-cooled reactors can be operated at a higher working temperature and are thus potentially more efficient. In the event, the USA proceeded very successfully with pressurized water reactors, while Britain and France favoured gas-cooled reactors.

In the 1950s, however, the possibilities of a new type of reactor – the fast breeder reactor – began to be explored. If a reactor is fuelled with a mixture of ^{239}plutonium and ^{238}uranium, neutrons are generated by fission of the plutonium and these interact with the uranium to produce more plutonium. Such a reactor can, as it were, 'breed' more fuel than it utilizes, thus extracting many times more energy from a given weight of uranium than is possible with the older type of reactor. The first breeder reactor was commissioned at Dounreay, Scotland, in 1959 and was shortly followed by Phénix at Marcoule, France.

Whatever the basis of their operation, nuclear reactors eventually create serious waste disposal problems. There is a gradual accumulation of ^{239}plutonium and also a variety of miscellaneous chemical isotopes, many of them intensely radioactive. Unspent uranium and newly generated plutonium are separated for re-use – some of the latter destined for nuclear weapons as well as for use in breeder reactors – but the safe disposal of radioactive waste is still a difficult and contentious problem. The material is exceedingly dangerous and will remain so for a great many years, some fractions of it perhaps for centuries. For such long periods, ordinary methods of storage are not suitable. Various solutions have been adopted, such as dumping in the deepest parts of the ocean and – most recently – fusing it with glassy, almost indestructible materials and storing the product in unused mines.

The handling of radioactive waste, and indeed all the basic operations of the nuclear industry, are made peculiarly difficult by the need to protect everybody concerned from the deadly effects of the radiation. It involves not only heavy screening with lead, a powerful absorber of radiation, but the carrying out of all operations – some extremely delicate and others involving the movement of considerable loads, by remote control. This has been a major field of application for the new techniques of robotics and virtual reality simulations.

The triumph of rocketry

The two bombs dropped on Japan in the summer of 1945 were carried by conventional B-29 bombers and detonated by remote radio control a short distance above the ground. Even then, new methods of delivery were being developed which could facilitate both the long-distance delivery of these fearful new weapons – by means of intercontinental ballistic missiles – and the relatively short-range tactical use of these weapons and of conventional high explosives.

Although this is a development which effectively dates from the last years of the Second World War, its history really begins at the turn of the century. From 1903 Konstantin Tsiolkovsky, in Russia, experimented with rockets at Kaluga University with the object of making scientific observations in space. He conceived the important principle of multi-stage rockets, in which successive fuel units are discarded as they become exhausted. Later, Robert H. Goddard, at Clark University in the USA, also took up rocketry as a means of carrying meteorological instruments aloft. By 1935, using liquid fuels, he had achieved a height of over 2 kilometres (1¼ miles).

It was in Germany that the most serious study was given to the development of rockets. There, Hermann Oberth had been active since the 1920s and the German Rocket Society was founded in 1927. In the early 1930s military interest was aroused, and under the technical directorship of Wernher von Braun an experimental rocket research station was

While the USA developed the atom bomb in great secrecy, Germany, in equal secrecy, pinned its hopes of victory on attacks with unmanned flying bombs. Here an A-4 (later V-2) rocket is launched from the German base at Peenemünde about 1942/3.

established at Peenemünde on the Baltic. There, by 1942, the liquid-fuelled A-4 rocket had been developed with a range of nearly 200 kilometres (125 miles) and a speed of about 6000 kilometres per hour (3600 miles per hour). This speed was far in excess of that of the fastest fighters, but the proximity fuse, developed by the Americans to detonate anti-aircraft shells at the moment when they would inflict most damage, did achieve some success against them.

Perhaps remembering the Allies' error in prematurely deploying tanks in the First World War, the Germans did not release their 'secret weapon' as the V-2 (Vengeance) weapon – with a 1-tonne warhead – until they had accumulated nearly 2000 of them. This was not achieved until September 1944. A few were launched against Paris but the main attacks were on London and Antwerp, which together suffered nearly 3000 rocket attacks. They were hard to combat at source as they were launched from either mobile bases or deep well-protected bunkers and the attacks ceased only when the launch areas were over-run in the Allied advance. After the war rocket development followed two separate, but inter-related, lines. For the moment we will continue to follow military applications; the other, the exploration of space, we will pursue in the next chapter.

As a generalization, military rockets can be categorized as tactical, for short range use in the field, and strategic, for long range bombardment: both can carry nuclear or high-explosive warheads. We shall consider tactical weapons first. By 1947, the USA was developing Firebird, an air-to-air missile (AAM) originally guided to its target by radio but later homing in automatically with the aid of radar. In the war against Egypt in 1956, the Israelis effectively used French SS-10 missiles against tanks. Ten years later the first surface-to-air missiles (SAM) appeared in the form of the British guided missile, Rapier. Russian versions of this were used successfully by the Egyptians against the Israelis in the Yom Kippur War of 1973. Among ship-to-ship missiles (SSM), one of the most successful was the French Exocet, introduced in 1971: flying just above the water, it readily evades radar detection. Although the American Tomahawk cruise missile programme had been in development since the 1970s, they were not used in active service until the Persian Gulf War of 1991. The Tomahawks utilize several advanced systems to strike their target successfully: a Tercom (*ter*rain *co*ntour *m*atching) system enables them to cruise under radar only a metre (3 feet) above the ground, then a Digital Scene Matching Area Correlator (DSMAC) uses visual pattern recognition software to home in on a precise target.

Unless a future war escalates to nuclear dimensions it is such weapons, with high-explosive warheads, that are most likely to decide the outcome. Immediately after the defeat of Germany, the USA began testing a series of high-altitude rockets, beginning with captured German V-2s. Drawing on the vast knowledge and experience of von Braun, the Americans confidently assumed that they possessed the lead in both peaceful and military rocketry. But the first giant inter-continental ballistic missile (ICBM) was the Russian SS-6, which made a dramatic impact, shaking US public opinion as much as when the Russians detonated their atom bomb in 1949, launched the first Sputnik satellite in 1957 and sent Yuri Gagarin into space on 12 April 1961. By the mid-1980s, the SS-18 had a range of 12,000 kilometres (7,500 miles) and could carry a multiplicity of megaton nuclear warheads. As Britain had ceased to be an island after Blériot's first cross-Channel flight of 1909, so the USA was no longer militarily insulated from Europe.

The first American response, the Thor missile of 1958, was relatively unimpressive, for it had a range of no more than 3000 kilometres (1800 miles), so that it could not usefully be fired against Russia from US home bases. Ten years later, however, they had developed an immensely powerful Saturn which in 1969 launched the first spacecraft to carry men to the moon. It had a launch velocity of 40,000 kilometres per hour (25,000 miles per hour).

In the 19th century, the Royal Navy was reluctant to adopt steam because it might be obliged to patrol waters remote from coaling stations at short notice. In the 20th century this sort of consideration remained a problem for naval strategists: ships could remain operational only as long as their fuel lasted. For this reason atomic power made a strong appeal for naval vessels, for a small mass of nuclear fuel could suffice for very long cruises without the need for replenishment. In particular, it was appealing for submarines, which

Chinese gunpowder rockets 137–138 **first atomic bombs** 319–321 **Louis Blériot** 220, 221 **Robert Goddard** 273 **tanks in First World War** 216–217

could thus avoid the necessity to surface in order to recharge their batteries: this problem had not been wholly solved by the introduction of the Schnorkel air intake system in the 1940s. It was such considerations that led to the building of the first nuclear-powered submarine, the *Nautilus*, for the US Navy in 1955. In the first two years of its commission it travelled 100,000 kilometres (62,000 miles) – equivalent to nearly three times round the earth – without refuelling. By the 1980s, the USA and Russia each had more than 100 nuclear-powered submarines and Britain and France about a quarter of this number. Some were conventionally armed with torpedoes; others with nuclear missiles, like Polaris and Trident, with a range of about 4000 kilometres (2500 miles), which can be fired from the obscurity of the ocean depths. In 1958, the Russians commissioned a nuclear-powered ice-breaker, *Lenin*.

Sonar and radar

Organization for gaining intelligence about the enemy's movements on land and sea has always been a major factor in military strategy. It was no less important, but more difficult to achieve, in the new three-dimensional world of submarines and aircraft.

In the submarine warfare of the First World War, various forms of hydrophones were developed to detect the noise of engines. This could give the bearing of the target, but not its distance – though an experienced operator might hazard a fair guess at this from the strength and nature of the signal. If two or more ships could obtain cross-bearings the target could be fairly precisely located and attacked with depth-charges set to explode at the appropriate depth. Equally, of course, submarines could use hydrophones to locate their targets or take evasive action. Such systems depend on the detection of sound emanating from the target: if the latter is totally silent – difficult in a submarine because of the need to maintain trim – no location is possible. This led, in the 1920s, to the development of various sonar systems in which a pulsed ultrasonic signal is emitted by the hunter: if the beam hits a target the pulse is received back like an echo, from the nature of which its position can be computed. Such systems can be used also for mapping the seabed and for locating wrecks or shoals of fish. Various physical factors, such as changes in the salinity or temperature of the sea, can give false readings and submariners are trained to take advantage of these. The most recent naval sonar devices, towed behind ships or submarines at variable depth, are passive: that is, they do not themselves emit pulses but detect and analyse sound waves emitted by hostile vessels.

The detection and location of aircraft is more difficult because of the greater speeds and distances involved and because aircraft are nimbler in their movements than submarines or ships. Neither of the latter can change speed or course rapidly and in principle it is not too difficult to make allowance for movement of the target during the time between firing a missile and its arrival. Relatively speaking, it is easier for aircraft to avoid the bursts of anti-aircraft shells, especially when these were simply set to burst at a predetermined height. The situation became very different with the introduction of the proximity fuse and of guided missiles which automatically home in on their target.

In many respects, the development of radar (*ra*dio *d*etection *a*nd *r*anging) parallels that of atomic energy. Before the Second World War all the major powers were equally aware of the scientific principles but they varied greatly in the speed and efficiency with which they applied them. In this case it was Britain that decisively took the lead, not so much as a result of exceptional success in devising detection units but through incorporating these in an operational system that could be very effectively used by an air force.

The possibility of radar was implicit in the earliest experiments of Heinrich Hertz in the 1880s, when he noted that radio waves could be reflected like light. As early as 1904, Christian Hülsmeyer patented the use of electro-magnetic radiation for range finding and in the 1920s Marconi was experimenting with radiolocation at sea. In the 1930s, radio operators near aerodromes reported re-radiation of signals from landing aircraft. In 1934 the French had plans to install a radiolocating device in the bows of the *Normandie*, but for technical reasons this did not materialize. It was also in 1934 that Henri Gutton developed the magnetron, later to become a critical component of radar systems. Maurice Ponte, director of Gutton's company, and married to an Englishwoman, was able to pass the invention to the English covertly during the war. Thereafter, the military possibilities and the growing threat of war led to further work being done in great secrecy.

In the 1930s, there was much speculation about the possibility of 'death rays' and in 1935 Robert Watson-Watt in Britain was asked by the Air Ministry whether a directed radio beam might come into this category. His conclusion was that it would not, but he went on to investigate the possibility of locating aircraft by means of the reflection of radio beams from ground stations. In this context light was excluded, because it is readily absorbed, especially by cloud, as was sound because of its relatively very slow rate of travel (331 metres per second/1086 feet per second). In a now classic memorandum of 12 February 1935 he spelled out with great clarity the basic principles involved in the design and use of a radiolocation system. Only five weeks later a research programme was authorized and on 11 June an aeroplane was detected at a range of nearly 50 kilometres (30 miles). Before the end of the year construction began of a chain of radar stations to protect London against approaching bombers – an event then expected within hours of the outbreak of war, though one that did not in fact occur until the summer of 1940. By September 1939, Britain had an early warning system that could locate approaching aircraft at a range of

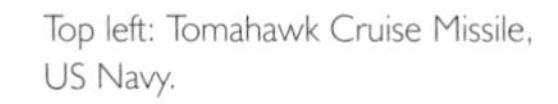

Top left: Tomahawk Cruise Missile, US Navy.

The 1960s were a period of intense rivalry in rocketry between Russia and America. This picture (above right) shows a range of American ballistic rockets at Cape Canaveral. On the left is the giant Saturn rocket that carried the first men to the moon in 1969.

Left: The French Exocet missile, introduced in 1971 for use against shipping, has a range of about 75 kilometres (47 miles). With a sophisticated control system it skims the water to elude radar detection.

Cruise missiles are the modern successors of the crude German 'doodle-bugs' of the Second World War. They are now highly sophisticated alternatives to ballistic missiles. This example (below left) is fitted with a W80 warhead.

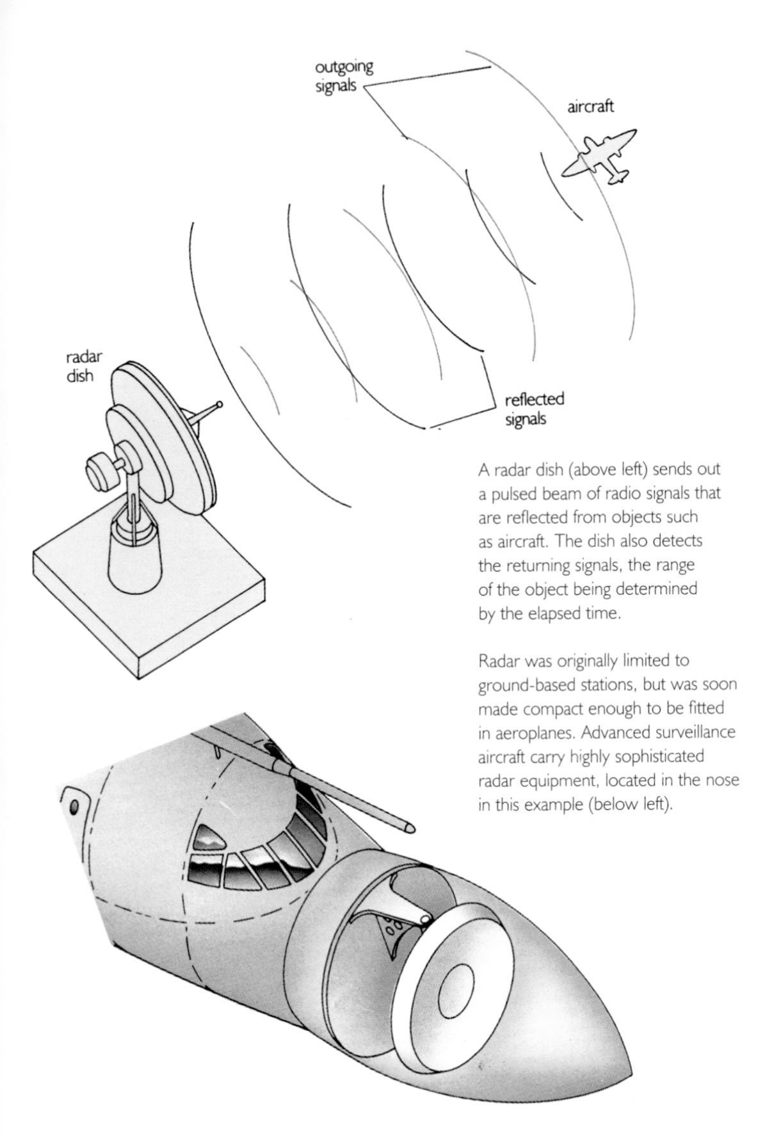

A radar dish (above left) sends out a pulsed beam of radio signals that are reflected from objects such as aircraft. The dish also detects the returning signals, the range of the object being determined by the elapsed time.

Radar was originally limited to ground-based stations, but was soon made compact enough to be fitted in aeroplanes. Advanced surveillance aircraft carry highly sophisticated radar equipment, located in the nose in this example (below left).

160 kilometres (100 miles). The existence of this was a crucial factor in the defeat of the German *Luftwaffe* in the Battle of Britain.

The principle was very simple. The threatened airspace was systematically scanned with pulsed radio waves and the orientation of a target was indicated by the direction of the aerial when the return signal was received. The time between signal and echo was a measure of the distance of the target. Initially, the relevant information was crudely displayed as a 'blip' on a cathode ray tube, but much more sophisticated detectors soon appeared. These displayed a map of the area scanned, with the positions of aircraft shown at their correct ranges and bearings. With the advent of the cavity magnetron in 1942 these were soon of a size that could be carried in aircraft and not confined to ground stations or ships, so that a bomber pilot could, in effect, see a plan of the ground beneath him as he approached his target. Equally, fighter pilots could pinpoint the position of hostile aircraft. At sea enemy ships could be monitored and their direction and range relayed to gunners, and submarines on the surface could be detected long before the spotter aircraft was in sight.

In 1939, as was appropriate for the needs of the day, the British radar network gave advance warning of aircraft approaching at around 500 kilometres per hour (310 miles per hour) and a height up to 10,000 metres (32,000 feet), at a distance of 150 kilometres (93 miles) – about 20 minutes' travelling time. Today, the situation is totally different but the basic principles remain the same. Early warning missile systems have to extend their vigilance to potential weapons approaching through space at many times the speed of sound at distances of around 3000 kilometres (1800 miles).

The cavity magnetron appeared because it was devised specifically to meet a need in the radar field for a very powerful source of radiation at centimetre (½-inch) wavelengths. After the war the principle served a totally different use. If such radiation is passed through food, the molecular vibrations set up generate local heat, and this is the basis of the microwave oven.

Precision warfare

With weapons of mass destruction sufficient to destroy the entire world already attained many times over, and increased reliance upon computers, modern military technologies have focused on waging precision strikes while minimizing the casualties on both sides. In addition, there has been a movement towards developing increasingly sophisticated simulation systems, and building weapons of information warfare. The Gulf War and the Kosovan conflicts ushered in a new era of military tactics waged by allied superpowers primarily through the use of remote, precision-guided air attacks orchestrated through computer-driven systems.

Developing visual pattern recognition systems, together with advanced guidance and tracking systems, have begun to increase the accuracy of missile and bomb strikes. The impact of these 'smarter' systems has consequences to infrastructure as well as personnel. Weapons may be guided to targets from increasingly greater distances by remote means, enabling bombers such as the American F-117 Stealth to fly at night and at high altitudes, out of range of enemy detection, thereby minimizing allied casualties. Guidance systems relying upon global positioning satellites (GPS) allow bombs such as the Joint Direct Attack Munition (JDAM) to be dropped in any weather. Precision weaponry may integrate visual pattern recognition software in their guidance systems, enabling them to strike precisely a particular point on a target. Consequently, these strikes are more effective – and humane – when focused upon destroying war machinery rather than human adversaries.

Smarter systems have been utilized in critical care as well as weaponry. Several ongoing research projects in the USA are developing the 'digital battlefield' of the future in which the status and information of each individual combatant and piece of equipment is available at any moment. 'Smart' uniforms are one component of the Personnel Status Monitor (PSM) system designed to provide information on the life state, condition and location of combatants. The clothing utilizes a screen-printed conductive mesh with elastic sensors and a tiny electronic processor to begin data acquisition if the uniform is penetrated, determine distance and wound location, measure respiration and heartbeat and send the data to a central location. Such a system will minimize casualties by directing medical attention to those who need it most, reduce resources risked on those who could no longer benefit (20 per cent of casualties occur while trying to help others) and reduce fratricide (which accounts for about 20–25 per cent of American combat deaths).

Although military simulation systems began shortly after the Second World War, it has been the extraordinary advances in computer processing which have permitted the successes of modern battle simulation. The first simulators were dedicated to training pilots, and focused on using real cockpits with simulated views and conditions. Today's simulations, however, support all branches of the military and have the capability to provide fully immersive virtual reality experiences integrating real soldiers, often separated by hundreds of miles, with simulated humans who can respond autonomously. They incorporate advanced technologies such as magnetic, infra-red and optical body trackers and bio-feedback sensors capable of controlling the simulation by responding to eye, head or limb movements or by the galvanic signals given off by muscular or brain activity. Such systems are supported by distributed computers connected through fibre-optic and satellite communications. They rely upon advances in information technology as well as upon those in hardware and software. The battlefield itself is becoming 'wired' with data collected from numerous sources and integrated into large databases. In 1991, a tank battle waged at night in the Gulf War became the first 'digital battlefield' to be documented in such full detail, and has been used to create a fully synthetic recreation in which a soldier may be trained by experiencing the action first-hand and reacting from the vantage point of any of the participants of the conflict. Data taken from sources such as radar, radio, infra-red, thermal imaging sensors and others located on each allied tank, global positioning satellites (GPS) and satellite imagery was used to model and create this digital training ground.

Television had begun changing the course of conflicts after the Second World War, when the horrifying results of warfare were brought into civilian homes. New strategies were developed for controlling media coverage and political wars at home were waged simultaneously with the conflicts abroad. The 1990s saw another change in media coverage as the world witnessed, for the first time, a 'missile-eye' view sent from a camera in the nose of a missile, guided by visual recognition systems, as it struck Iraqi bunkers during the Gulf War. NATO used footage from their bomb attacks during the Kosovan War as evidence to dispute or acknowledge the damage caused to civilian targets. Elements of these conflicts may prove to be a precursor of warfare to come.

Top: The Lockheed Stealth F-117 relies on a radical shape and advanced materials to avoid radar detection.

Above: This 'virtual battlefield' demonstrates how an officer may direct troops while receiving up-to-date views of aircraft and ground forces generated by real-time reconnaissance information.

In the 1990s military strategists began discussing a 'new wave' of military history which would move beyond weapons of mass destruction and embrace weapons of information warfare, in which conflicts are intended to minimize casualties but maximize critical data deletion. Information warfare has been defined as actions taken to affect adversary information, information systems and processes while defending one's own in an effort to achieve information superiority.

In both the Gulf and the Kosovan wars, the USA, in particular, used information technology to minimize casualties and make weaponry 'smarter'. While they did not utilize weapons such as computer viruses, electronic jamming or EMP bombs – which use electromagnetic pulses to destroy computerized systems – to delete their opponent's data, they did limit media coverage and direct strikes against the television and media broadcasting centres of their opponents, as well as direct physical assaults on military information strongholds. In turn, the Iraqi and Serbian media provided their citizens with a markedly different account of events than the Western press, while obfuscating information 'released' to the West. A modern age of information warfare was ushered in, as all parties involved sought to control public perception by editing or distorting events, and the Western forces used their information technology to minimize civilian and military casualties while directing attacks on the military infrastructure.

CHAPTER THIRTY-FIVE

The Shrinking Earth

Although the history of mankind is measured in millions of years, the organized societies of the earliest civilizations appeared only about 8000 years ago. For some 98 per cent of this time people travelled, whatever their means of conveyance, at no more than 20 kilometres (12 miles) per hour: for the most part they travelled by foot, at around 5 kilometres per hour (3 miles per hour). Distances were commonly measured in terms of time taken rather than actual length. This generalization is not invalidated by special situations: in the late 18th century a racehorse, for example, could average around 64 kilometres per hour (40 miles per hour) for the 2.5 kilometres (1½ miles) of the Derby course; an experienced skier could comfortably exceed this for short distances downhill. But at the end of the 18th century, the 652-kilometre (405-mile) journey from London to Edinburgh took four days by coach, with 42½ hours actually spent on the move.

Then, in the last 2 per cent of this period, the situation changed dramatically. Before the middle of the 19th century, the ordinary fare-paying citizen could travel by rail at an average speed of 100 kilometres per hour (62 miles per hour) and just after the turn of the century at 160 kilometres per hour (100 miles per hour). With the development of civil aviation the normal limits were pushed much higher, until Concorde – a joint Anglo-French aircraft which first flew in 1969 and was brought into service in 1976 – could carry passengers across the Atlantic at 2500 kilometres per hour (1550 miles per hour). By that time such speeds were common enough for a new nomenclature to have been introduced. High speeds were expressed in terms of the velocity of sound in air, which under standard conditions is about 1200 kilometres per hour (750 miles per hour), and referred to a scale named after an early aerodynamicist, Ernst Mach. Thus an aircraft or projectile moving at a speed of Mach 2.2, like Concorde, would be travelling at 2.2 times the speed of sound. The first aircraft to break this 'sound barrier' was the American Bell X-1, essentially a manned rocket, in 1947. Much higher speeds are now commonplace among military aircraft, but for big civil airlines – typified today by the 600-seater Boeing 777s introduced in the 1990s – top speeds are now around the 1000 kilometres per hour (620 miles per hour) level.

These increases in speed and capacity have been matched by big increases in traffic. In 1947, the total number of passengers carried by commercial airlines throughout the world was no more than 20 million: between 1971 and 1983 the number of domestic and international passengers carried by US airlines alone increased from 178 million annually to 318 million, and freight from 3.4 billion ton-miles to 6.1 billion ton-miles. In a surprisingly short space of time the world had, in effect, become a very much smaller place.

Evolution of the modern airlines

Our last references to the evolution of aircraft mentioned the introduction of the pressurized cabin for high-altitude flight; the development of the immensely powerful Wright Cyclone piston engines, such as were used in the four-engine B-29 bombers; and the advent of the jet engine as an alternative to the propeller for propulsion.

These developments were the basis of post-war civil aviation, which was dominated by the USA. There were two main reasons for the commanding position of the American aircraft

The development of Concorde (below) involved very complex aerodynamic research, including extensive wind-tunnel tests (right).

industry. First and foremost, much of European industry was in ruins, and transition to civilian needs was slow even in Britain. Secondly, the USA formed a very shrewd judgment of the future course of events – especially with regard to the carriage of freight – and planned their production accordingly. The possibilities for air freight were strikingly demonstrated in 1948–9 when a gigantic Allied airlift carried 2½ million tonnes of supplies into a beleaguered Berlin in 15 months. Immediately post-war, they could offer the world's airlines the Lockheed Constellation, the Douglas DC-4 and the Boeing Stratocruiser, developed from the B-29 bomber. New aircraft became increasingly expensive to design, and leading manufacturers such as Douglas and Lockheed introduced the concept of 'stretching' a basic airframe – enlarging it as necessary in successive versions without the expense and delay of going right back to the drawing-board.

During the war, jet propulsion had been successfully developed by both sides for fast fighter aircraft, but these were expensive both to make and to fly: only the over-riding need for military superiority had made their rapid evolution possible. With the cessation of hostilities, it became apparent that the future of aviation generally lay with jet propulsion in either pure jets or turbo-props in which propellers were driven by a gas turbine – but the high cost of development prevented their early appearance on the scene. The first generation of post-war transport aeroplanes were all powered by piston engines and some are still flying. In the event, the first jet-propelled airliner was the de Havilland Comet which first flew in 1949. This had a chequered history: soon after being put into service to South Africa in 1952 two unexplained mid-air disasters led to its abrupt withdrawal. The cause was eventually found to be metal fatigue, and after this design fault was corrected it was recommissioned in 1958. In the hands of the British Overseas Airways Corporation (BOAC), it then began the first commercial transatlantic service, beating the introduction of TWA's Boeing 707 airliner by a few days. At the same time, the Douglas DC-8 came into service.

These developments made the Comet obsolescent – yet another example of the hazard of being first in a new field. The Boeing 707 and the DC-8 had the advantage of bigger capacity, higher speed and longer range. A disadvantage was that they required longer runways and bigger airport facilities. In the mistaken belief that airports would be reluctant to make these improvements, BOAC responded with the smaller Vickers 10 but this, like the Comet, eventually appealed largely to smaller operators working short hauls.

While there is a continuing demand for relatively small airliners for medium-haul flights, and a considerable number have been introduced, it was clear by the early 1960s that the economics of airline operation favoured large, wide-bodied aircraft carrying up to 500 passengers. It was this that gave rise to the now familiar jumbo jet, which in one version or

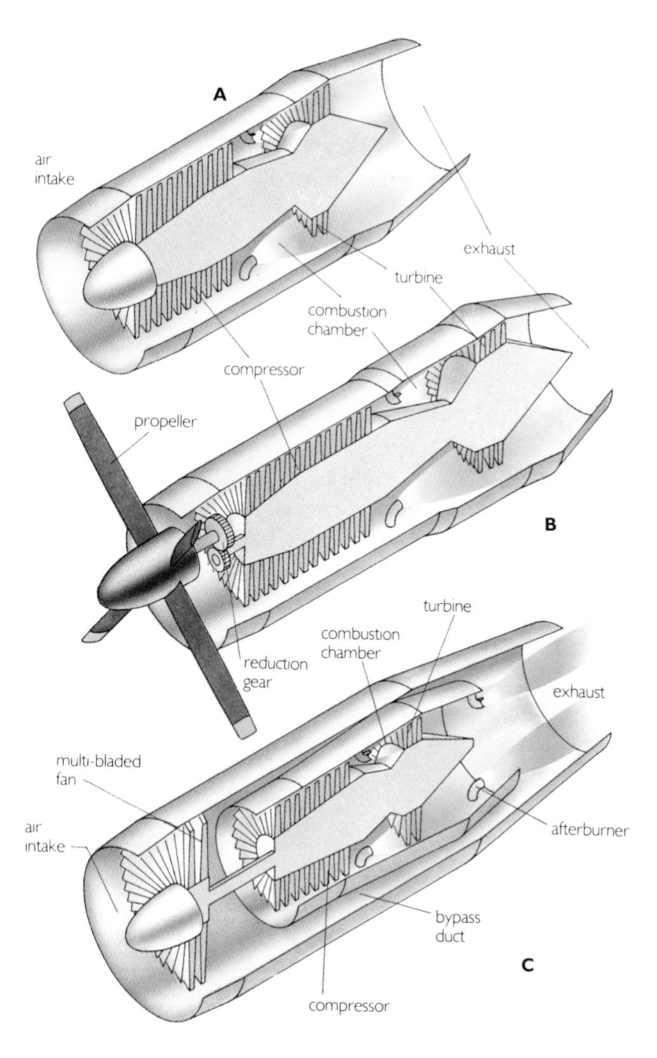

Top: With the arrival of large passenger jets, the aircraft assembly industry required space and equipment comparable with those of shipyards.

Above: A turbojet engine (A), turboprop engine (B) and a turbofan engine with afterburner (C).

aeronautics 256–257 **first aerial warfare** 219–222 **first jet aircraft** 257, 274 **engines** 166–169, 251 **invention of aircraft** 219–222

another now carries most of the world's long-distance travellers. The jumbo jet was made possible by the development of the turbo-fan engine, a kind of hybrid of the jet and the turboprop, which is relatively quiet and can generate powerful thrust at the low speeds of take-off and landing. In the face of rising oil prices it is economical in fuel and it also causes relatively less pollution, an important factor with the growing strength of the environmentalist lobby. Apart from the long runways required, its disadvantage from the airport point of view is that handling facilities of all kinds have to be increased when so many passengers arrive simultaneously.

Released in the mid-1990s, the Boeing 777 was created using lightweight materials and components and employed improved engines to create a smaller, more efficient jumbo jet capable of travelling greater distances. As a result of streamlined production processes and the use of modern composite materials such as titanium alloys, it provided a greater percentage of interior space while maintaining smaller external dimensions than the 747. It was the first of the jumbo jets light enough to require only two engines rather than four. It was also the first aircraft to utilize an integrated, three-dimensional database of design information in which every component was designed and tested on computer and the latest revisions were immediately accessible by every member of the development team, including external vendors.

Helicopters are slow and expensive to run, but can be uniquely valuable in special circumstances, such as ferrying men and supplies to offshore oil-rigs.

Helicopters

All forms of land and sea transport have one feature in common: they can of intent, or in an emergency, remain stationary. In contrast, an aeroplane supports itself in the air by its forward movement: if its speed falls below a certain level, it stalls. For the same reason, a conventional aeroplane must have a runway to gain flying speed before take-off. The only major exceptions are military Harrier jump-jets. It is, therefore, not surprising that a good deal of attention was given to the development of flying machines which could take off vertically and hover at will. The answer to this has been found in the helicopter, in which horizontal rotors provide a vertical lift: forward propulsion is achieved by altering the pitch of the blades. Although some experiments were made even before the First World War, it was not until the late 1930s that real success was achieved. In 1936, Heinrich Focke produced a prototype machine with two counter-rotating propellers, introduced to prevent the fuselage itself rotating when in flight. This was followed in 1940 by a larger six-seater machine. Meanwhile, in the USA, the Russian-born aeronautical engineer Igor Sikorsky – who had been interested in helicopters since 1909 – produced a prototype single-rotor machine, the VS 300. In this, the torque of the rotor on the fuselage was balanced by a small tail rotor. This design was subsequently widely adopted: from about the 1950s jet engines began to be used to power larger machines.

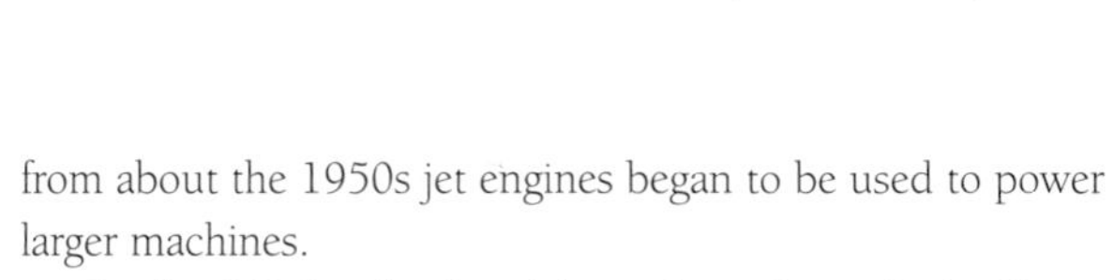

The main rotor of a helicopter rotates to produce lift and thrust. The tail rotor produces a force to counterbalance the torque of the main rotor in straight flying. Varying the speed of the tail rotor produces sufficient torque to turn the helicopter.

In the 1930s, the Spanish engineer Juan de la Cierva introduced the autogyro, something of a hybrid between an aeroplane and a helicopter. In this, a rotor turned by the airstream gave additional lift and a means of safe landing in the event of engine failure. It had a short vogue, but several hundred were built.

Apart from its relatively low speed, a very serious defect of the helicopter is its heavy fuel consumption, which makes it an exceptionally expensive form of transport. Nevertheless, its manoeuvrability makes it uniquely suitable for a variety of important purposes in the modern world. It has a strong military appeal for transport, requiring no specially prepared landing strips for troops or equipment, especially for the fast and comfortable evacuation of wounded. In recent years it has been adapted for offensive use as an armoured 'gunship'.

In the civilian field, helicopters find many uses. These include rescue operations from otherwise inaccessible positions, the supply of off-shore oil-rigs, providing relief services in major disasters such as floods and earthquakes, and various kinds of aerial survey.

Hovercraft and hydrofoils

Hovercraft – or air cushion vehicles – are something of an anomaly in the transport world. Although confined to the surface of the sea or land, they are regarded as aircraft in the sense that the officer in charge must normally hold a pilot's licence.

The principle of the hovercraft is very simple – friction between the vehicle and the surface is minimized by making it float on a cushion of air. The invention was patented in 1955 by Christopher Cockerell, who began manufacture on the Isle of Wight. His 7-tonne prototype, the N-1, was demonstrated in 1959 and was shortly improved by the addition of a rubberized 'skirt' to contain the air cushion created by compressors. The performance of this prototype was impressive. It was highly manoeuvrable and capable of a speed of over 100 kilometres per hour (60 miles per hour). On 25 July 1959, the 50th anniversary of Blériot's historic flight across the English Channel, the first hovercraft crossing was achieved. Today, cross-Channel hovercraft carry over 400 passengers and 50 cars on a frequent service. Elsewhere, similar vehicles have found uses as ferries and also as transport over flat terrain – such as mud-flats and desert – where no roads exist.

The air-cushion principle has found some application in quite different fields. A familiar one is the domestic lawnmower. Heavy industrial equipment can be designed for easy movement over factory floors by making provision for compressed air to be introduced underneath. In the 1970s, a number or railway systems experimented with tracked hovercraft but these have not so far been adopted for service.

Somewhat akin to the hovercraft is the hydrofoil, in which the hull of a boat is supported out of the water by planes immersed in it. This provides low frictional resistance, too, and thus the possibility of high speeds for an acceptable power input. The first commercial hydrofoil service was operated on the Rhine in 1937.

The conquest of space

The idea of escaping from the earth and exploring the heavens has long been a theme of imaginative writers, from the *True Histories* of Lucian of Samosata in the second century AD to John Wilkins in the 17th, Jules Verne in the 19th and H. G. Wells in the 20th. Their usual objective was the moon, as by far the most accessible of celestial bodies: in 1609

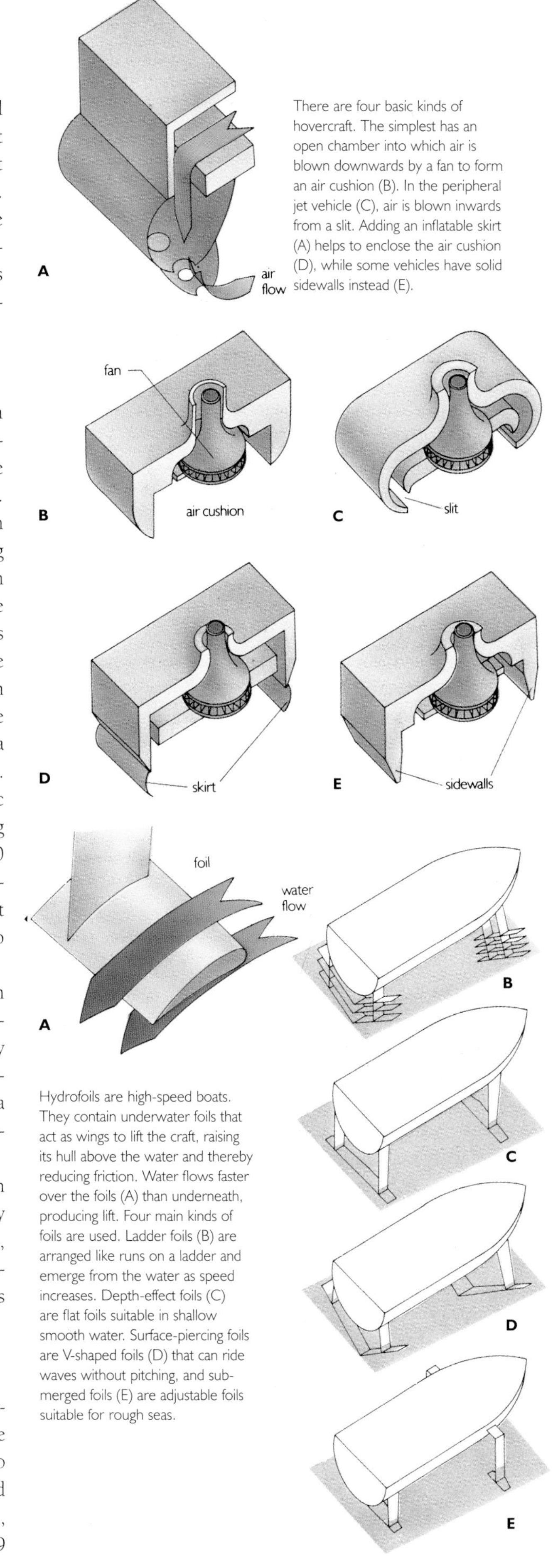

There are four basic kinds of hovercraft. The simplest has an open chamber into which air is blown downwards by a fan to form an air cushion (B). In the peripheral jet vehicle (C), air is blown inwards from a slit. Adding an inflatable skirt (A) helps to enclose the air cushion (D), while some vehicles have solid sidewalls instead (E).

Hydrofoils are high-speed boats. They contain underwater foils that act as wings to lift the craft, raising its hull above the water and thereby reducing friction. Water flows faster over the foils (A) than underneath, producing lift. Four main kinds of foils are used. Ladder foils (B) are arranged like runs on a ladder and emerge from the water as speed increases. Depth-effect foils (C) are flat foils suitable in shallow smooth water. Surface-piercing foils are V-shaped foils (D) that can ride waves without pitching, and submerged foils (E) are adjustable foils suitable for rough seas.

air transport industry 257, 273–274 **balloons and airships** 171–172, 256 **Blériot's cross-Channel flight** 220, 221

Almost unbelievably, less than 25 years separate the single-orbit space flight of Yuri Gagarin, strapped alone in the tiny capsule of Vostok 1, and the launch of the giant Russian space station MIR-1 in 1986. MIR-1 can provide accommodation for up to 20 people, and may be used for the start of a manned journey to Mars.

Galileo's telescope had explored its mountains and revealed it as a body whose topography was not dissimilar to that of the earth. In later years, very detailed maps of the moon were compiled on the basis of telescopic observation: a moon-map of 1960, 3 metres (10 feet) in diameter, detailed no less than 90,000 separate features.

The barrier was not simply one of distance. The insuperable obstacle was the force of gravity. To escape permanently from the earth's gravitational field, a projectile must attain a critical velocity of around 40,000 kilometres per hour (25,000 miles per hour) and until very recently there was no available means by which this could be achieved.

The seeds of modern success lay in the German rocket research programme conducted during the Second World War at Peenemünde. When the first V-2 rocket was successfully launched there on 3 October 1942, the Commanding Officer, W. R. Dornberger, is recorded as having said: 'We have proved rocket propulsion practicable for space travel.' It was, indeed, a continuation of the military programme that, after immense expenditure of money and effort, finally took man into space.

During the uneasy Cold War that followed the cessation of hostilities in 1945, the USA and the former Soviet Union – each with immense resources – settled down to the development of increasingly powerful rockets, with the ultimate object of being able to use them to land nuclear bombs from long range on each other's territory. Inevitably, this eventually reached a stage where the rocket that propelled the ICBM was powerful enough to project some sort of spacecraft beyond the gravitational pull of the earth. To the surprise and dismay of the American public, the first to achieve this goal were the Russians. On 4 October 1957, Sputnik 1 was launched into orbit: it was visible – and indeed audible, through the constant bleep of its radio transmitter – demonstration of the formidable strength of Russian technology. Only a month later the famous dog Laika survived the launching and made several orbits: in August 1960, two dogs were launched into space and successfully retrieved unharmed. While dogs do not exactly equate with man, their anatomies and physiologies are not dissimilar, so there were good grounds for believing that a man, too, would survive the severe stress of the launching process, in which he is subject to a force many times that of gravity and, equally, the state of weightlessness which he must endure in flight.

The crucial test came on 12 April 1961 when the Russian cosmonaut Yuri Gagarin successfully orbited the earth in Vostok 1 and returned safely – the final fall of the capsule in which he was enclosed being controlled by parachute. The blow to American self-esteem and sense of national security was enormous and led to the inauguration of the Apollo programme designed to land men on the moon. This was finally achieved on 21 July 1969 when Neil Armstrong and Buzz Aldrin stepped out of Apollo 11 on to the moon's surface, watched by television viewers throughout the world: the cost of the programme to that point is estimated to have been $24 billion. There were five more manned Apollo missions to the moon: the last landing was in 1972, and in all several hundred man-hours were spent on the surface.

These spectacular and much publicized landings were the culmination of a series of unmanned exploratory flights which, in the aggregate, have been the most productive of new scientific information about the nature of the moon. In all, there were 73 lunar missions – manned and unmanned, successful and unsuccessful – between 1958 and 1975, of which 40 were American. In this field the Russians – who, after major disasters in 1967 and 1971, about which little is known, abandoned plans for manned landings – made some very notable contributions. From 1958, the USA established a leading position in the successful launching of earth-orbiting satellites, but their first attempt at a moon shot – Pioneer 1, launched on 11 October 1958 – was a total failure. By contrast, the Russian series of Luna flights, beginning in 1959, initially went well. Luna 1 flew past the moon at a height of 7500 kilometres (4660 miles) and then went into orbit round the sun, becoming the first artificial planet. Luna 3 was a spectacular success. Launched on 4 October 1959 – the second anniversary of Sputnik 1 – it secured excellent photographs of the far side of the moon. This had never previously been seen by human eyes, because the moon keeps one side permanently turned towards the earth.

Both Russia and the USA then had a lean time. Subsequent Luna flights failed, as did the first American Ranger flights, designed to survey the moon's surface in detail at short range as a preliminary to a manned landing. In July 1964, however, Rangers 7, 8 and 9 sent back thousands of clear television pictures as they plunged towards possible future landing sites. Those from Ranger 9 appeared internationally on television, dramatically billed as 'live from the moon'.

By then, however, the Russians were back in business. In January 1966, Luna 9 made a soft landing on the moon and relayed back a series of very informative television pictures. Among other things, these gave the reassuring news that – at least in the Ocean of Storms, where it landed – there was no deep dust layer. In the same year, the USA made the first of a series of Surveyor landings, which up to February 1968 sent back tens of thousands of photographs to guide those responsible for planning the final stages of the critical Apollo flight which was to land the first men on the moon.

Apollo achieved well deserved success in July 1969, but the Russians continued to demonstrate the remarkable results that could be achieved by indirect methods. On 20 September 1970, Luna 16 landed safely in the Sea of Fertility. There, remotely controlled from earth, samples of rock and dust were taken from the surface and drawn into the spacecraft. This was then relaunched and returned safely to Russia with its precious samples. This was followed by even more remarkable achievements. In November 1970, Luna 17 delivered an eight-wheeled moon-car, powered by solar cells, which could be driven and guided by remote control from the Russian space centre on earth. In exploratory journeys, totalling over 10 kilometres (6 miles), it relayed photographs of its surroundings, collected and analysed rock samples, and measured cosmic radiation. A year later, a similar vehicle travelled nearly four times the distance, and sent back nearly 100,000 pictures before it failed. The Americans, on the Apollo 15 mission, used a similar moon buggy, but manned, to extend the range of their astronauts.

The planets

From the space travel point of view the moon can be regarded as providing for landings and probes the equivalent of nursery slopes for skiers. In cosmic terms the distance is relatively short – a mere 400,000 kilometres (240,000 miles): Mars, by contrast, is never less than 35 million kilometres (22 million miles) from the earth. For journeys to the planets we must think in terms of voyages lasting for many months, even for years. There is no present possibility of constructing a manned spacecraft with such endurance: apart from ordinary victualling needs, the crew must be uninterruptably supplied with oxygen in the vacuum of space. Additionally, there are the still undefined hazards of long exposure to weightlessness and intense cosmic radiation. No doubt such voyages will come, but even optimistic estimates do not expect them before 2010.

Unmanned probes are a very different matter, however, and both the USA and Russia have already achieved some astonishing results. In June 1963, the American Mariner 2 passed within 35,000 kilometres (22,000 miles) of Venus and sent back pictures of a seething mass of enveloping cloud; on reaching Mars, however, it detected from a distance of 10,000 kilometres (6,200 miles) an apparently arid landscape, pitted with craters, much like that of the moon. Then in November 1970, Mariner 9 was put into orbit round Mars for nearly a year, sending back more than 7000 photographs. These included many of Olympus, a vast volcano three times the height of Everest.

Such surveys paved the way for a Mars landing by pinpointing suitable sites. On 20 July 1976 – the seventh anniversary of the first manned landing on the moon – Viking 1, after a 12-month journey, went into orbit and released a landing capsule which immediately began to transmit clear pictures back to earth. The limited part of the Martian landscape within the range of the cameras could be seen as clearly as it might have been by a human observer standing on the spot. Three months later, Viking 2 landed a similar unit at a different site. From both sites an arid, red, stony landscape was revealed. From both landings, much scientific information was gained, including many soil analyses. Neither, however, gave any conclusive answers to the intriguing question of whether life exists on Mars, nor did the Sojourner rover, which landed on 4 July 1997 as part of the Pathfinder mission. However, the rover did bring back evidence that water had existed on the planet between 3 and 4½ billion years ago. The rover was a six-wheeled vehicle remotely operated from earth. It remained operational for 12 times its designed lifetime of seven days, and transmitted a huge amount of data, including 550 images from its camera.

If travellers could survive the long intervening voyage, a manned landing on Mars is perfectly feasible. The situation with regard to Venus is quite different, however. There surface conditions of atmospheric pressure 100 times that of the earth and temperatures above the boiling point of water are so severe as to make it out of the question. However, after a

The advent of spacecraft opened up new possibilities for observational astronomy. This close view of Jupiter was taken from Pioneer 10 in 1973.

MIR-1 space station 335, 344 **satellites** 315, 334, 336, 340–341 **space shuttle** 335, 340, 344 **V-2 rocket** 323

series of failures the Russians succeeded, in October 1975, in landing two large and widely separated units on the surface and each of these transmitted good pictures for nearly an hour. Each revealed a barren surface strewn with rocks. More than a year earlier, the American Mariner 10 had passed close enough to Venus to photograph its cloud formations, before being deflected by the planet's gravitational field and redirected to Mercury, the planet nearest the sun; from there thousands of pictures were sent back from a height of about 300 kilometres (186 miles). Again, the landscape appears stony and barren, resembling that of the moon. After a launch in 1989, Magellan reached Venus in 1990 and produced an extremely detailed map of the surface using radar.

Jupiter, too, has been investigated at close quarters by the American spacecraft Pioneer 10 and Pioneer 11 in 1973 and 1974 respectively. The trajectory of Pioneer 10 was such that it was accelerated by the gravitational field of Jupiter and went on to be the first man-made object to leave the solar system. Pioneer 11 was programmed to go on and survey Saturn. In March and July 1979 two Voyager spacecraft, launched in 1977, flew past Jupiter and observed directly the four moons first described by Galileo through his new telescope at the beginning of the 17th century; it also revealed three new satellites. In all, the Voyager missions discovered 22 new satellites in our solar system, including the three orbiting Jupiter, three orbiting Saturn, ten orbiting Uranus and six orbiting Neptune. Coupled with these discoveries was the revelation that Io, a moon of Jupiter, was actively volcanic. This knowledge prompted scientists to begin scrutinizing the planetary satellites for signs of life as closely as the planets themselves.

The earth itself is among the planets regularly surveyed from orbiting satellites. Apart from military surveillance and weather observation, some carry equipment for remote sensing. From computer-enhanced photographs of the surface much can be learnt not only about the pattern of vegetation but also about the distribution of mineral deposits.

Research in space has not been limited to observation of the sun and planets. In 1985 five probes – one British, two Russian and two Japanese – loaded with scientific equipment were dispatched to observe Halley's Comet, returning after an absence of 76 years.

Laboratories in space

The weightlessness of space and freedom from the radiation filtering effect of the earth's atmosphere provide conditions for research, and perhaps for certain very special manufactures such as growing crystals, that cannot be created in earthbound laboratories. For this reason, both America and Russia developed plans for establishing permanent laboratories in space – launched either in their entirety or by assembling modules after they had gone into orbit – in which scientists could work for long periods in reasonable comfort. The research teams could be relieved from time to time by small spacecraft serving as ferries. After grave troubles at the outset, Russia in 1971 established the first of their Salyut space laboratories. In 1978, two Russian astronauts completed a then record 139-day stint in space.

Meanwhile, in May 1973 the USA launched their huge 75-tonne Skylab, costing $2½ billion. The launch was successful, but as it reached orbit there was a major setback. The heat-shield disintegrated and one of the solar panels – the main source of power – ripped off and the other failed to open. As the satellite was unmanned, there was time for thought, and eventually a two-man crew was dispatched who were able to make improvised repairs and restore the power supply. In the event, Skylab was manned for a total of 172 days. It eventually entered the atmosphere and disintegrated in 1979. Then the repair episode was sensational: today such operations, though still difficult and hazardous, are accepted as a normal feature of space travel. For work outside a spacecraft astronauts normally remain in communication by a long umbilical cord, but in 1984 Bruce McCandless made history by stepping unattached into space, controlling his movements with small gas jets. In February 1986 Salyut and Skylab were superseded by the giant Russian MIR-1.

MIR (meaning 'peace') was put into operation for the purposes of analysing changes in the human body during long periods of weightlessness, for observing celestial bodies, and for conducting experiments in weightlessness. In 1987, Yuri V. Romanenko returned to earth from MIR after spending a record 326 days in space. Although it was originally planned for a seven-year mission, it was not completed until 1996 (six years behind schedule) because of delays in the completion of the other modules. It did, however, serve as a platform for growing co-operative endeavours with the United States which would lay the foundation for an international space station. From February 1994 to June 1998, the Shuttle flew a total of 11 missions to the MIR space station in an effort to build joint space experience and begin collaborative scientific research.

In addition to manned laboratories, numerous pieces of scientific equipment have been positioned in geosynchronous orbit, outside the atmospheric obstruction of various electromagnetic emissions. On 25 April 1990, the first optical telescope in space, the Hubble Space Telescope, was launched. The objectives of Hubble were to provide a long-term space-based research facility for optical astronomy. In particular, it would be used to study the evolution of the universe by examining the visible, ultra-violet and near infra-red spectrum of deep space and investigate the characteristics of celestial bodies. Unfortunately, a flaw in the main reflective mirror was discovered almost immediately. The first servicing by the Shuttle in December 1993 fully resolved this problem by installing a corrective optics package and upgrading other instruments.

The 'Hubble' became the first optical telescope in space when it was launched in 1990. High above the obscuring atmosphere, it has a clear view of deep space. Although a critical flaw was discovered in its optical equipment shortly after launch, a servicing by the Shuttle provided a corrective optics package.

The Shuttle

From a few figures quoted in earlier pages it is clear that technological success in space has been achieved only at enormous cost. Much of the latter must necessarily be written off as part of the complex development programme, but every launch is in itself expensive partly through the loss of successive booster rockets and partly through the elaborate organization needed when the spacecraft is to be recovered on landing.

In August 1977, a new era in space travel opened with the advent of the Shuttle, which is, as it were, a recyclable spacecraft intended to reduce the expense of space flights. As before, this is launched by rocket, but although the large liquid-fuel tank is expended the boosters are parachuted into the Atlantic and can be recovered for re-use. The Shuttle itself, after completing its mission in space, can then glide back to its base and touch down like a conventional aircraft at around 300 kilometres per hour (186 miles per hour). Within a couple of weeks it can be serviced and ready for relaunch: it was expected that each craft would make about 100 flights. The first trial flights of the Shuttle were made by carrying it piggy-back on a specially adapted Boeing 747 but, on 12 April 1981, *Columbia* lifted off unaided on its maiden flight from Cape Canaveral, Florida, and landed 54 hours later at Edwards Air Force Base in California. The Soviets initiated their own programme in 1988 with the space shuttle *Bourane*, but several factors resulted in this being the only flight of the 20th century. Initial technical setbacks, compounded with the financial and political instability following the dissolution of the Soviet Bloc put the programme under review. As relations improved with the US and successes were recorded with the MIR space station, enthusiasm for a Russian shuttle programme waned.

Barely 20 years separate Gagarin's single orbit in 1961 from the fully operational Shuttle, but they seem to – and indeed, do – belong to different ages. Whereas Gagarin was almost immobile in his tiny capsule, the Shuttle has accommodation for up to 10 people – though the normal crew is seven – and a huge cargo bay nearly 20 metres (66 feet) long. Fully laden it weighs 115 tonnes, of which 39 tonnes can represent cargo – a remarkable contrast to the total of 6 tonnes of Vostok 1.

The development costs of the Shuttle over some 10 years are reported to have been about $7 billion. Although it can scarcely be regarded as a viable commercial proposition, it has earned very large fees in return for launching into space a variety of large and expensive communication and meteorological satellites. Its customers have included the US government and private companies – including foreign ones. Apart from launching new satellites and retrieving exhausted ones, it has the capability of carrying out emergency repairs in space. After a promising start, the Shuttle programme received a more severe setback in January 1986, when one of the four initial craft exploded – as a result of design flaws in the O-rings separating the solid-fuel boosters – shortly after take-off, with the tragic death of all seven astronauts on board. This disaster led to a suspension of all flights until its cause had been established. On 29 September 1988, space shuttle flights resumed with the launch of the redesigned *Discovery*. Since that time, the S huttle programme has been vital in supporting the construction and maintenance of satellites and space stations.

International co-operation

Space technology is an exceedingly expensive business, within the scope of only the great powers: in practice, there have been only two major participants, Russia and the USA, and it is on these that we have necessarily concentrated our attention so far. Other countries have participated in a minor way either collectively, as through the European Space Agency (ESA), or individually, as in Japan. They have concerned themselves particularly with the development of rockets, both solid- and liquid-fuelled, to put satellites into orbit round the earth.

By 1984, some 3000 satellites had been launched into space, about 96 per cent of them by Russia and the USA. About four-fifths of these were for scientific research and the

international space collaboration 334–345 **MIR-1 space station** 332, 344 **SETI** 344 **space shuttle** 340, 344

The maiden flight of the American Shuttle craft took place on 12 April 1981. Carried into space by a semi-recoverable rocket assembly, it can glide back to earth when its mission is completed and land under pilot control.

remainder, more recently, for the practical purposes such as telecommunications or weather observation. The demand for reliable rocket vehicles capable of putting large and expensive satellites in orbit is such that it has become, as we have seen, very competitive.

The Japanese came late into the field, launching their first artificial satellite OSUMI – for scientific research – in 1970, but by 1977 their N1 rocket launched the first Japanese geostationary satellite. Such satellites orbit the earth in a day, so that they appear to remain stationary at a fixed point above it. In 1981, the Japanese started development of their bigger H1 rocket scheduled for 1986. This was competitive with their ESA rival Ariane 1 which, after an unhappy start, had acquired an acceptable risk-rating by the mid-1980s.

To get a satellite into the desired orbit is not sufficient in itself. Once there, its attitude must be accurately controlled: for example, its radio antennae and observational equipment must always point in the desired direction. To achieve this, two systems have been used; the older of the two stabilizes the satellite by rotating it on its axis like a top. The other, three-axial attitude control, was first used in 1974 in the US geostationary satellite ATS-6. In this stability is obtained either by using three flywheels to absorb any disturbing force, or else a single large gyro-wheel inside the satellite.

An international space station, complete with a Trans-Hab living accommodation for six, is due to be completed in 2004. The station requires at least 45 rocket or shuttle launches to place the 100 constituent components into orbit. When completed, it will measure some 110 × 88 metres (361 × 290 feet). The TransHab living quarters will be spacious by existing standards; formed by inflating Kevlar sections with air, they will measure almost 12 × 8 metres (40 × 27 feet) and provide a galley and meeting room, exercise and recreation area, as well as sleeping and work areas. Even before completion, the first crew of three, comprising two Russian cosmonauts and one American astronaut, will begin habitating the station in March 2000.

Radio astronomy

Space probes and satellite laboratories are only two of several new techniques that have been developed to increase our knowledge of the universe. Some of these involve considerable technological achievements.

In 1931 Karl Jansky, of the Bell Telephone Laboratories, detected radio waves emanating from outer space and thus opened up a completely new field of astronomy, largely developed after the Second World War. Many modern radio telescopes are massive engineering works, such as the great steerable dish, 80 metres (262 feet) in diameter, built at Jodrell Bank in England in 1957: unexpectedly one of its first major tasks was to track early spacecraft. It was dwarfed by later radio telescopes.

At Arecibo, in Puerto Rico, the bowl of an extinct volcano has been adapted to hold the 350-metre (1150-foot) dish of an enormous radio telescope. It is not steerable, but scans the sky as the earth rotates. Radio telescopes have even been used to generate radio signals to be bounced off the moon and some planets. The highest resolution imagery today is gained by utilizing series of antennae, such as the very long baseline array (VLBA), brought on-line in 1993. This is comprised of ten 25-metre (82-foot) parabolic dishes distributed across some 8000 kilometres (5000 miles) from Hawaii to the Virgin Islands. The VLBA can achieve resolutions 1000 times better than the finest optical telescopes.

The British Telecom Earth Station at Goonhilly Downs, South Cornwall.

CHAPTER THIRTY-SIX

Into the Future:

Convergence, Communication and Collaboration

Throughout this book, we have examined trajectories of technological invention across food, architecture, clothing, health care, materials, transport, communication and warfare covering some 2600 millennia from early hominid tools to the first man-made replication of a mammal. Yet, these divisions which we have established are no longer as meaningful when food is constructed, transport involves the transfer of information rather than matter, and the distinction between what is natural and what is manufactured becomes a subject of heated debate. The next millennium is poised to be one of increased communication, collaboration and convergence. A global network of interlinked computers and satellites, supported by warming international relations, has opened new avenues for communication and education; increasingly, exchanges take place in real time over thousands of miles through text, audio and video channels. Already the collaborations which have occurred as a result of this capability are radically altering our cultures with new financial and research unions forming where political ones have failed – international banking and space stations, distributed computing and simulations, remote education, health care and entertainment. Many of the technologies which are fostering this expansion have been in existence for decades, but it is the increased convergence – the cross-pollination of disciplinary endeavours – which has led to its manifestation. Food, medicine and materials are being manipulated at the molecular level, our machines can sense and act autonomously, and we grow organs and animals as once we grew plants.

Forecasting events in any field is a notoriously hazardous occupation, which is why soothsayers wisely took refuge in generalities or ambiguities. Technology is no exception. If we look back along the road we have travelled in this book there are very few points – even within the comparative stability of the Roman Empire or Victorian Britain – at which predictions were not liable to be quickly confounded by an unforeseeable turn of events. In 1000 BC, there was no means of telling that copper and bronze, which must then have seemed to provide all that could be expected of a metal, were very soon to be largely superseded by iron. Nor in AD 1300 could military leaders have foreseen that within 50 years the whole technique of war was to be revolutionized by the advent of gunpowder. If we stood on the threshold of this century, there were none who could have told us that not only would transport be revolutionized on land and on the sea but that people in their millions would take to the air as a normal way of travel. Nor could it have been predicted that people could sit in their own homes and watch events as they occurred not only in the remotest parts of the world but even on the moon; that some mortal diseases, long accepted as inescapable, could be virtually eradicated by advances in medicine; that machines could take over many of the functions of the human brain.

Reflecting on such matters indicates that the task of prediction is becoming harder rather than easier, for the tempo of change is rapidly increasing. The penetration of iron into the Western world was measured in centuries and the process of assimilation – except where primitive people suddenly encountered iron weapons for the first time – was correspondingly painless. Today, however, the situation is totally different. The first electronic computers appeared in the 1940s; by the 1970s they had begun to pervade every aspect of everyday life. Man made his first single orbit of the earth in 1961, yet only eight years later he walked on the moon.

As the complete genetic make-up of humans is catalogued, moral as well as scientific decisions must be faced as the creation of 'designer babies' becomes increasingly possible.

Latter-day seers face an even shorter run than those of the past before events catch up with them to measure their success.

Moreover, the volume as well as the pace of development is daunting. We can fairly liken the history of technology to a river which started thousands of years ago as a small stream but now, fed over the centuries by endless tributaries, flows as a great and still widening estuary. We can carry this metaphor further. It is tempting, for one's peace of mind, to suppose that man sails these waters in a craft whose speed and direction he can control: sadly, a truer comparison is that we are adrift on some great raft which will with luck keep us afloat, but which we can at best hope to steer well enough to avoid the worst hazards that appear in our path. Our final destination is far from certain.

Only with such considerable provisos can one begin with reasonable responsibility to speculate on what kind of new technological developments we can expect to see as we enter the 21st century and what their consequences may be. Before doing so, however, we should consider some of the overall constraints that must apply.

First, we can see the course of events being regulated by economic factors: by the 1980s, after the euphoria of the immediate post-war years, it was becoming increasingly apparent that what we could achieve technologically and what we could afford to do for the general good were two quite different things. For example, if all the advances in medicine were made available to all who might profit from them, no nation could afford the cost: many of the newer techniques, such as body scanners, involve very expensive equipment and highly skilled staff to operate them, interpret the results and carry out such treatment as may be diagnosed as necessary.

The very success of medicine to date has created its own problems in the Western world: there it has helped to increase an aging population – prone to all the diseases of old age – with a diminishing proportion of young and active people to support them with the minimum comfort society deems appropriate. Increasingly, as we have noted, governments are having to turn to the promotion of health – by, for example, warning the public against the dangers of smoking, drinking and excessive fat – rather than the combating of overt disease.

A second general constraint is an increasing awareness of the environmental consequences of certain technological developments. In this there is nothing new in principle: in the early 19th century the French government began to introduce legislation to control the operation of offensive industries generally and half-a-century later Britain began to regulate the working of specific processes, beginning with those of the alkali industry.

It is only since the Second World War, with growing concern about the cumulative toxic effects of agrochemicals, radioactive waste and the acid rain generated by burning fossil fuels, that preservation of the environment has become a major public issue. The environmental lobby has become a political force strong enough to have secured the abandonment of some important technological projects, such as a large hydroelectric scheme in Tasmania. How far such exercise of power without responsibility will prove beneficial remains to be seen.

The dramatic effects of acid rain are plainly visible in this once thriving forest in Siberia, where a nickel smelting plant (in the background) emits pollutants which form acid in the atmosphere. Acid rain affects our water supplies too, destroying the natural balance of aquatic life.

A third factor is the Third World: not as something new, for it has always been with us, but as something that the prosperous nations can no longer ignore on the grounds of either conscience or expedience. This awareness owes much to new methods of mass communication, which can project vivid pictures of human misery into every comfortable home. Equally, swift and substantial relief is possible only because of modern means of transport, especially air-lifts. It is generally recognized that while major catastrophes can be alleviated only by immediate outright gifts of urgently needed food and equipment, the only effective long-term aid is to provide the means by which stricken communities can assist themselves to be self-sufficient, if only at a subsistence level. If this is accepted politically on a scale commensurate with the need, the implication must surely be that the economic and technological growth of the wealthy nations must be retarded by the amount – not merely in terms of finance but of skilled teachers and advisers – that they set aside for the benefit of the Third World. In this context we are not, of course, talking about advanced technology but such basics as wells and irrigation systems, simple farm machinery and plants and animals selected for semi-arid conditions.

When talking of under-development in the above sense, we are considering countries so poor that their own resources are altogether too small to enable them even to begin to compete with the advanced Western nations. But we must not ignore the fact that since the Second World War the whole pattern of world industry has changed and many unexpected bonuses arise. Leadership in the heavy industries on which the fortunes of Europe and North America were based has been challenged by newly industrializing countries, particularly in the Middle and Far East. For example, the cost of carrying iron ore over long distances has been greatly reduced by the advent of giant ore-carriers, and it is feasible to turn this into sponge iron – suitable for further refining in electric arc furnaces – in the oil-producing countries of the Middle East, where vast quantities of unwanted natural gas – otherwise simply flared away to get rid of it – provide a virtually free fuel. Countries such as Korea, Singapore, Taiwan and Hong Kong can manufacture a wide range of goods – from textiles to stainless steel cutlery, electronic equipment to cameras and printed material – in fierce and effective competition with the West. China, with its long tradition of fine craftsmanship, is now committed to rapid industrialization in the Western style. Some Third World countries have found a degree of economic salvation in their possession of minerals urgently needed by industry – chromium in the Great Dyke of Zimbabwe and oil in Nigeria, for example. How the West, with its reluctance to cut its coat according to its cloth, will fare in this free-for-all is a matter for speculation. The developing world owes nobody a living.

The positions of Japan and of Russia and the former Eastern Bloc countries are, in different ways, anomalous. Japan, once dismissed as a mere imitator, was the only country in the post-war world to adopt a considered technological strategy and as a result has become a world leader in many aspects of modern high technology – such as microchips, robotics and optical instruments – as well as competing strongly with the West in old-established industries such as shipbuilding and automobile and motorcycle manufacture. It has still further to develop other fields, however, such as space technology, life sciences and new-material engineering. Japan has also to contend with such problems as a very dense population, so that the area of land for cultivation is severely limited, and an almost total dependence on imported oil.

Russia's formidable technological capacity has long been clear to the whole world, but it is also clear that it is seriously imbalanced. Its space technology is highly advanced, but by the standards of the Western world its car industry is backward and consumer goods generally are unsophisticated, scarce and expensive. A succession of Russian leaders have made no secret of the fact that its agriculture, of great importance in the economy, causes much concern despite investment that has been exceptionally heavy by the standard of any other country. In 1970, both the USA and Russia were exporters of grain: in 1981 the USA exported 115 million tonnes, whereas Russia had to import 45 million tonnes. Production of potatoes, vegetables, milk and meat has also declined since the late 1970s. How far this is due to excessive bureaucracy, rather than technical backwardness, is debatable, but the reality is there.

Invention can flower only in an atmosphere not only conducive to it but also able to provide the technical resources and finance necessary to transform a brilliant idea into a workable product or process. The day of the versatile individual inventor, such as Eli Whitney or Thomas Edison, is largely past: the future lies mainly with those who are not just highly trained but are also perceptive of the needs of the day, working within teams in large organizations, whether private, national or international, and herein lies a paradox. Large organizations must, by their very nature, be conformist or they are unmanageable; yet if they are to innovate they must be able to accommodate imaginative non-conformists not readily amenable to formal discipline.

However, the fact that there are many incalculable factors determining where the main centres of innovation are likely to be in future need not deter us from speculating on their nature. In this, we must keep two separate possibilities in mind. First, we can with reasonable confidence predict lines of development which evolve directly from an existing situation. Thus the automobile in substantially its present form is bound to be with us for some considerable time to come, simply because the sheer size and inertia of the road transport system throughout the world makes rapid change impossible. This does not imply, of course, that there may not be important peripheral changes such as the substantial replacement of the reciprocating piston engine by, say, the rotary Wankel engine or electric propulsion. Secondly, and with far more difficulty, we must try to make allowances for inventions that, as it were, come in from the side, rather than evolving from what has gone before. Such, in the fairly recent past, were the unexpected discoveries of electromagnetic radiation outside the visible region of the spectrum, radioactivity and the structure of the atom, heavier-than-air flight and antibiotics.

The biggest imponderable of all, of course, is whether the threat of a nuclear conflict can be averted: if not, predictions about the future of mankind, technological or otherwise, are pointless. Here we see, but on a vaster and more terrifying and dangerous scale, a repetition of the age-old political strategy of being technologically superior to the potential enemy in the military field.

agrochemicals 291–295 **automobiles** 248–250 **preventative medicine** 289–290 **Wankel engine** 250–251

Energy

The most vital factor in modern industry is a sufficient supply of energy. World energy consumption increased by 50 per cent in 1956–66, and a further 75 per cent in 1966–76. The OPEC crisis of 1973 demonstrated the importance of energy very forcibly. This was, of course, specifically an oil crisis and the severity of its impact on Western countries was a measure of their excessive economic dependence on it. No immediate response was possible, but over the next decade several means of adjustment were found. Among the most important of these were successful policies aimed at reducing energy requirements in all possible ways – from the heat-insulation of buildings to low-temperature detergents for washing machines – and the more intensive exploitation of existing and new oilfields, as in the North Sea and Alaska. Additionally, much attention was devoted to the development of new or improved methods of utilizing renewable energy sources, such as those of the tides and the sun.

In the circumstances of the time, this kind of reaction was entirely predictable but it is not easy to see now what the future holds. The Western world did, in fact, cut its oil requirements and by the mid-1980s the price of oil collapsed and there was a glut rather than a shortage. High moral talk about a duty to conserve fossil fuels for future generations has tended to be forgotten and with the date of exhaustion of reserves – always problematical – pushed comfortably far into the future, the world is tending to resume its old dependence on oil. There has, as a result, been correspondingly less urgency about the development of alternative sources of energy.

The exception to this is nuclear energy, which has been steadily though unevenly developed: France, for example, was the first country to produce more than half its electrical energy from nuclear sources. This is, in any case, not strictly a renewable source of energy like solar energy, in that it demands substantial amounts of fuel, such as uranium, and the world resources of this, though enormous, are strictly limited, like those of oil and coal. Workable world resources are currently estimated at about 2 million tonnes which, bearing in mind the enormous amount of energy released per unit weight, is sufficient for a very long time indeed. Additionally, it is possible that thorium – a metal far more abundant than uranium – may also be usable as a nuclear fuel. On taking up a neutron and losing two electrons, ^{232}Th is converted to ^{233}U, which is fissile.

The building of nuclear power stations is necessarily very slow. It is, therefore, not too difficult to make reasonable predictions of future output up to at least the year 2000. In 1985, nuclear power production within the 19 countries that belong to the Organization for Economic Co-operation and Development (OECD) – which include most of the European countries, together with the USA, Canada, Japan and Australia – was around 300,000 megawatts. This was expected to rise to about 500,000 megawatts by 1990 and one million megawatts by 2000. Thereafter, a new factor may be introduced. These figures all refer to fission-based reactors of one sort or another, but much systematic research is being done worldwide to develop fusion reactors, theoretically several times more efficient than those based on fission and requiring only light elements such as hydrogen (or one of its isotopes) and lithium. These are abundantly available and safe to handle. However, the technical problems are enormous and the most optimistic predictions see the second decade of the 21st century as the earliest possible date for an operational fusion reactor.

The generation of electricity thus seems likely to continue to depend largely on fossil fuels or nuclear fission with fusion as a long-term possibility. However, the Chernobyl disaster shook public confidence and gave new strength to the antinuclear power lobby, making the future course of events even harder to predict. In the storage of electricity, a major development seems probable rather than possible. The state of development work suggests that a sodium-sulphur storage battery may displace the long established lead-cell accumulator. This would have something like double the power, giving it a power-to-weight ratio which might make it a suitable power source for automobiles. This would be favoured by environmentalists, though it must be remembered that it would only transfer the pollution problem from the roads back to the power station.

Space

The conquest of space has occurred so rapidly that it is hard to believe that the headlong pace of the past four decades will not be maintained. On the whole, however, it seems that we are more likely to see a period of consolidation, on two fronts. First, scientific research under the unique conditions pertaining in space will continue with the aid of long-distance probes dispatched to the furthermost parts of the solar system and even beyond, of equipment lodged in unmanned satellites launched by rockets or craft like the Shuttle and in permanent space laboratories, many of them the results of international cooperation and manned by changing teams of scientists.

Secondly, the use of communication satellites of all kinds will continue to be more extensive and better organized. In particular, stationary satellites, to which we have already referred, are likely to have greater significance. Already, their utility is indispensible to modern society, enabling daily global commerce to proceed by supporting media communications, medicine, transportation, military and scientific projects by the swift relaying of information. Interestingly, their orbits depend on a law of planetary motion published by Johannes Kepler more than three centuries ago. This states that the square of the periodic time of a satellite is proportional to the cube of its distance from the earth. If this periodic time is exactly equal to that of the rotation of the earth – that is, 24 hours – the satellite will appear stationary, hovering like a hawk, in a so-called geostationary orbit, above a fixed point on the earth's surface. This condition is fulfilled by satellites placed in orbit at a height of roughly 36,000 kilometres (22,000 miles). In the international INTERSAT system, used for telephoning and distribution of television programmes, three such satellites can cover the globe.

Geostationary satellites can also be used to provide local television services and have the advantages of providing better reception through the more direct penetration of the radiation through the atmosphere and the elimination of the need for booster stations on the ground. An additional advantage is that receiving sets are not screened by topographical features on the earth's surface such as tall buildings or high ground: in mountainous country, good television reception is still often difficult to achieve. Plans are well advanced for a European service of this kind. As it is relatively cheap, it is a system with a particular appeal to countries such as Australia, whose television systems are still not fully developed: there would not, therefore, be the economic deterrent of writing off a great deal of expensive equipment already installed.

The ocean depths

Although we are unlikely to see any attempt at direct human exploration of a planet, specifically Mars, until at least well into the 21st century, the reality of space exploration has been accepted remarkably readily. It is salutary to recall, however, that a large part – indeed the greater part – of our own planet still remains unexplored. Apart from the 15 million square kilometres (6 million square miles) of the inhospitable Antarctic continent – which is still only sketchily surveyed – there remain the depths of the oceans, totalling around 400 million square kilometres (150 million square miles). Although coastal regions and continental shelf areas, where rocks and banks are dangers to shipping, have long been accurately charted, it is nevertheless true that we are, literally, abysmally ignorant of the topography and nature of the floors of the deep oceans and of the life that exists there. In a general way, this ignorance is likely to persist, for the hazards and difficulties of deep-sea diving ensure that it is pursued only when there is a very valid reason for doing so. Nevertheless, such reasons do exist and, as they are increasing in number and urgency, there will certainly be major developments in specialized diving and submarine technology over the next few decades

One reason is that increasing amounts of oil and gas are now being won from underwater wells – and many of these are located in deep water. This involves much complicated and expensive underwater structural work, pipelines, valves and so on, and it is essential to be able to inspect and service these as necessary. There are, moreover, enormous mineral deposits on some parts of the deep ocean floor and there is much interest in evaluating these with a view ultimately to mining them commercially. Of particular interest are the manganese-containing nodules, of which immense quantities lie in the Pacific at depths of 4000–6000 metres (13,000–20,000 feet). In the Red Sea, mineral deposits of manganese, iron, zinc and copper have been detected at depths of around 2000 metres (6500 feet). Not all of these are rich enough to warrant exploitation, but others are clearly commercially viable if they could be brought to the surface economically.

Vast regions of our planet have yet to be explored. The ocean's depths are being investigated by manned and robotic submersible devices such as this Deep-Star 4000.

Increasingly, highly specialized submarines are being developed for these purposes, as the limit of conventional deep-sea diving in the most modern suits is only around 500 metres (1640 feet) – though 700 metres (2300 feet) has been attained under simulated pressure conditions. A pioneer in this field was Auguste Piccard, famous also for his balloon ascents far into the stratosphere. In 1960, in his bathyscaph, he reached a record depth of 11,000 metres (36,000 feet), but this craft had poor mobility and limited powers of observation, and was not designed as a working vessel.

In this field, France and the USA have been the pioneers. The French *Cyana* and the American *Albion* reached depths of 3000 and 4000 metres (10,000 and 13,000 feet) respectively in the 1970s. In the 1990s, a new era of robotic submersible exploration began: 1993 witnessed significant advances in deep-sea exploration with the launch of two new compact 2-metre (6-foot) submersibles. An inexpensive, tetherless robot called *Odyssey*, designed by James G. Bellingham and others, was launched by scientists at MIT in Antarctic waters. With a depth of 5000 metres (16,000 feet) and a cost a thousandth of the human-piloted *Alvin*, it holds great potential as a prototype for inexpensive, safe exploration in deep, hazardous territory. The Autonomous Benthic Explorer, designed by a team led by Albert M. Bradley, was also launched in 1993 in Bermudan waters by a team from Woods Hole Oceanographic Insititute. It is capable of attaining depths of 6000 metres (19,000 feet) and remaining submerged for up to a year. Such submersibles are highly sophisticated craft capable of performing a variety of tasks at great depth, beyond those of scientific exploration, such as

energy 64–70, 223–231, 275–278 **fusion reactor** 275–276 **OPEC oil crisis** 255, 264, 275, 278, 297 **space** 331–336 **submarines** 218–219

attaching cables to wreckage ships and aircraft for recovery after accidents or maintaining off-shore industrial platforms.

Communication

In an age where telecommunications is an overused term, it is easy to forget that the word means more than television or data exchange. Communication, after all, implies that we hold something in common, deriving from the Latin root *communis*. Telecommunication – communicating over distance – does not mean simply sending data back and forth between two computers; it is a meaningful exchange based on an understanding that we have something in common, such as language, beliefs or values. The more individuals and cultures learn about each other, the more possibilities there are for finding common ground and improving their ability to communicate. Modern communications technologies are not simply providing faster, more accessible means to speak and interact with each other, they are changing the meaning of communication itself.

It has been said that Gutenberg saw his printing press as a means to spread Catholicism by mass-producing the Bible. Instead, by placing the word of God in the hands of the people, they had the potential to interpret the Bible for themselves, which led to the rejection of the Pope as intermediary by an increasing number of Protestants. Likewise, the growing global network of computers and the expanding digital interactive television industry has the potential to change educational systems and the role of educators in a similar radical fashion.

As access to the Internet becomes available to a wider audience, more people will have the ability to educate themselves in any subject; the world's knowledge will be brought to their fingertips as never before – in minutes, interpreted as scientifically or as simplistically as desired and frequently current to the day. Educators will still have a role, but increasingly this may be as a mentor or as one who teaches how to learn and communicate rather than one concerned with imparting collective knowledge.

This expansion of communication has brought with it a concomittent confusion of identity, where the sense of body, gender, culture, heritage, nationality and worth is no longer stable, but is continuously questioned and mediated by our use of technology.

Convergence

Transport no longer simply refers to moving matter, but instead to moving information. Nowhere is this phenomenon more evident than in health care, a field which has been limited principally by constraints of time and space. For example, many accident victims and military casualties could be saved simply by the rapid diagnosis of a trained physician who is located miles away. In 1996, the US Army launched an advanced telemedicine network providing medical care to US forces deployed in Bosnia. Operation Primetime III utilized satellite and Internet links to connect field hospitals and the 212th Combat Surgical Hospital in Tuzla, Bosnia with specialists around the world located in Tazar, Hungary; Landstuhl, Germany; and San Antonio, Texas, and Bethesda, Maryland, in the USA. High bandwidth signals carried diagnostic-quality video running at 30 frames per second to transfer data such as patient records, filmless radiology, CAT scans and colour ultrasound. In addition to trauma support, Primetime III also provided personnel with teledentistry and telepsychology services. A dentist in Landstuhl could see into a patient's mouth in Tuzla and direct a field dentist, while video-conferencing capabilities provided a platform for psychological consultations. This communications bridge kept doctors and patients safe by transporting data rather than people, while providing a quicker response and the best possible health care.

Likewise, if we examine everyday products as packets of solidified design information, we can see that much of our transportation costs are a result of moving matter from its manufacturing point to consumers. The Industrial Revolution centralized and homogenized our production processes to lower the costs of producing goods and achieve a consistency of quality. In so doing, it separated the processes of design, production and sales into discrete procedures. Rather than products being customized, produced and sold in one shop, they became produced in a large manufacturing facility and then shipped to a wholesaler who distributed

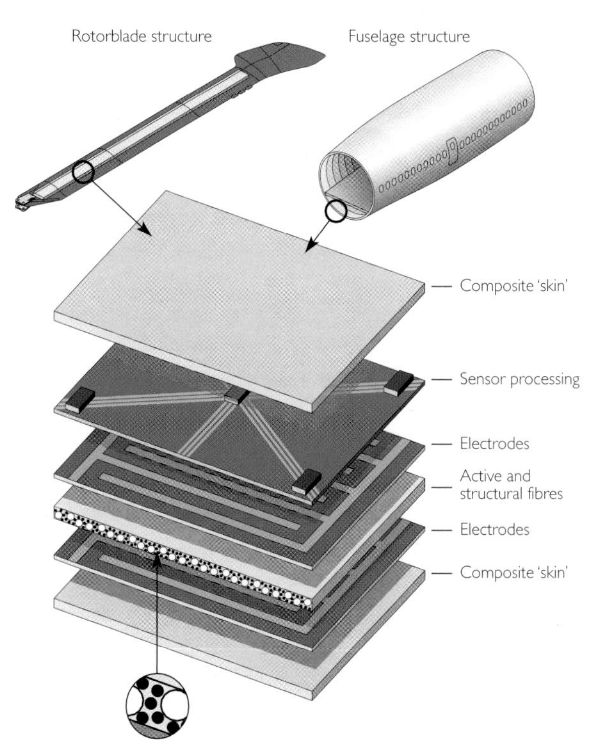

'Smart' materials enable products to respond to their environment. Fibre optics sense pressure changes. A sensor processing layer then sends a signal through electrodes to fibres which change the structure's shape.

them to various sales agents. Development of new computer-driven manufacturing techniques, combined with a desire for increased customization are beginning to reverse this trend, and once again, bring design, production and sales into one facility. Rapid prototyping facilities will soon permit low-volume, customized products to be produced inexpensively. This is exemplified by Levis stores, where a person is measured for a new pair of trousers, a pattern is customized, laser-cut and sewn all in one shop within a few hours.

Our objects are becoming 'smarter' – capable of sensing and responding. Computing technologies, combined with composite materials and electrical engineering knowledge are advancing the development of 'smart' materials which can sense variations in pressure, light or electrical flow and respond accordingly without human direction. In addition to the shape memory alloys – which have been used for such purposes as anti-scalding valves on water faucets – plastic electrolytic gels, piezoelectric plastics and ceramics, and fibre optics are being imbedded in manufactured goods as diverse as clothing, buildings, aircraft and artificial organs to provide greater safety and responsiveness to their users. Buildings that sense earthquakes by imperceptible shock waves are constructed with materials containing imbedded fibre-optic sensors. Aircraft wings have been formed with a thin piezoelectric film on one side which senses pressure changes and an electro-responsive material on the other which can change shape to respond accordingly.

Convergences between the health care and food industries, supported by developments in genetic engineering, smart materials, printing and computing have created a surge of miniscule bacterial detection devices which facilitate the quick detection of dangerous infection or tainted foods. *E. coli* test kits were sold to the meat-packing industry by Meridian in Ohio from 1997. Similar so-called integrated biological sensors continue to be developed in smaller packages, responding in improved and faster times to an increasing number of bacterial strains.

Nanotechnology, in which matter is manipulated and constructed at the atomic level, draws on the technologies and knowledge from engineering, chemistry and materials science, among others. Microscopic structures, such as engines, rotors and sensors, can be created measuring no more than 10 or 20 nanometres (1 or 2/100ths of a millimetre). In 1997, scientists in Tokyo and Yokohama used an enzyme (H+-ATP) as a motor – capturing the record at the time for the world's smallest – by attaching a fluorescent protein arm onto the 'spindle' and activating it with fluorescent light. Nanogen, in San Diego, California, integrates microelectronics with molecular biology to create DNA-testing devices which are scarcely bigger than the gene strand itself. Using the natural charge of biological molecules, they cause pieces of DNA to be attracted and chemically bound to specific locations on a microscopic chip.

Boundaries

Such convergences bring about a subsequent collapsing of boundaries and a re-evaluation of identities. Although many of the boundaries which are collapsing as a result of technological advances are those of professional trade or national identity, the greatest perhaps is that defining the boundary between synthetic and natural, between life and manufactured artefacts. One of the first boundaries which children learn is that of self, distinguishing what is part of themselves and what is their environment. Interestingly, it is this boundary which we are increasingly disintegrating. Fashion and societal beliefs have driven the practice of tattooing and body piercing for centuries, while medicine initiated implantation and prosthetics. In the late 20th century, fashion began to subsume certain forms of implantation and created a separate branch of surgery focusing on cosmetic alterations. Through the convergence of technologies in medicine, chemistry, robotics and computing, however, our capabilities and choice of implants and prosthetics grows annually.

We now have the ability to supplement our bodies with synthetic creations which surpass our natural abilities. Campbell Aird received the first fully mobile bionic arm in 1998 after some ten years' development by the Prosthetics Research and Development team at the Princess Margaret Rose Orthopaedic Hospital in Edinburgh. Also in 1998, a team of researchers at Warwick University, England released information on a bionic nose they had created which can detect the bacteria *staphylococcus aureus* with 100 per cent accuracy. Although neither of these inventions fully realized the potential for augmenting our bodies with superior systems, they represent key steps in that progression. The arm represents the successful convergence of robotics and medicine, the 'nose' – a fully artificial device not attached to a person – represents the advances in integrating miniaturized computing with chemistry.

We also have the ability to choose implants which enable us to communicate better with our products. In 1999, Kevin Warwick wore an infra-red implant for several weeks which enabled computer systems to track and respond to his presence, activating light switches and his business computer. Similar small-scale implants are also being explored by the military to monitor health and relay positional or health care information to centralized command centres. In 1998, researchers at Emory University, in Atlanta, Georgia provided severely paralysed volunteers with brain implants enabling them to operate a computer by thought power alone. Tiny electrodes were imbedded in the brain's motor cortex and, after several months of nerve growth, neurons grew onto the implants. When the person thought of an action, such as moving an arm, the electrical signal which would activate the muscular system of a healthy person was sent by wireless FM transmitter to a computer which responded by moving the cursor around a screen. Pointing at various icons allowed

Gutenberg Bible 102–104 **Internet** 310–311 **medicine** 279–290 **telecommunications** 314–315

them to communicate by synthesized speech generated by the computer.

Perhaps the greatest modifications possible are those brought about by genetic engineering, which manipulate our natural structures. Gene therapy has been used for several years to treat a range of disorders such as Parkinson's Disease or Down Syndrome. In this treatment, the gene which causes the disorder is replaced with a healthy gene in order to stimulate the growth of healthy cells and eradicate those that are diseased. This is called somatic therapy and has the potential to affect only the individual. Far more controversial is human gene line therapy in which manipulation of eggs, sperm or embryos has the potential to control the evolutionary development of generations to come, where traits are passed on to offspring.

The first rulings in cases in the USA may not have legalized patenting humans, but they did provide the possibility for patenting all human components. Some 35 years after James Watson, together with Francis Crick, identified the double-helix molecule of DNA as the source of genetic coding, he became the first director of the Human Genome Project, an ambitious government-funded effort to identify and catalogue every one of the three billion base sequences that comprise human DNA by 2003. What began as a government programme quickly expanded into an international collaboration, Human Genome Organization (HUGO), with the findings to be shared with the world. But as HUGO gained speed, so did commercial competition. In the late 1990s, as adult mammals were cloned for the first time, Craig Venter founded Celera Genomics with the intention to complete the project ahead of schedule and, in the process, to patent as many genes as he could.

Collaboration

Evidence of collaboration has indicated evidence of civilization to modern historians: the chamber tombs of Ireland, Stonehenge and the Great Pyramid of Cheops all suggest, through the scale and magnitude of their structures, that communities must have formed to undertake the construction. Humans must have collaborated, not simply to build, but also to sustain each other – a division of labour and food gathering. Our divisions are no longer necessarily those of labour and food gathering, but collaborations continue to sustain us and permit us to achieve more collectively than individually. Communication systems first arose to co-ordinate collaborative endeavours, but today's communication systems are creating new forms of collaboration. Three projects launched at the end of the 20th century exemplify new directions and possible potentials in global collaborative efforts, distributing wealth, knowledge, enthusiasm and computational power.

Scientists working on a Search for Extra-Terrestrial Intelligence (SETI) project devised a unique plan to help them cope with limited and inadequate computing resources. They realized that most of the time a computer is in operation, the processor itself is idle. If they could tap into the potential of millions of personal computers used daily by ordinary citizens, it would give an enormous boost to their research. In March 1999, SETI@home was launched as a free screensaver which could be downloaded off the Web by those interested in helping the research. The screensaver is a piece of data analysis software which allows a personal computer, during idle moments, to work on one small portion of a bigger equation. This portion is sent back to SETI where it can be processed. In the first two weeks, people from over 203 countries contributed almost 3000 years of computer processing time to the project. In many respects, this represents the world's largest supercomputer, a giant step in distributed computing and may suggest a new means of tackling some of the most challenging problems facing humanity through global, collective endeavours.

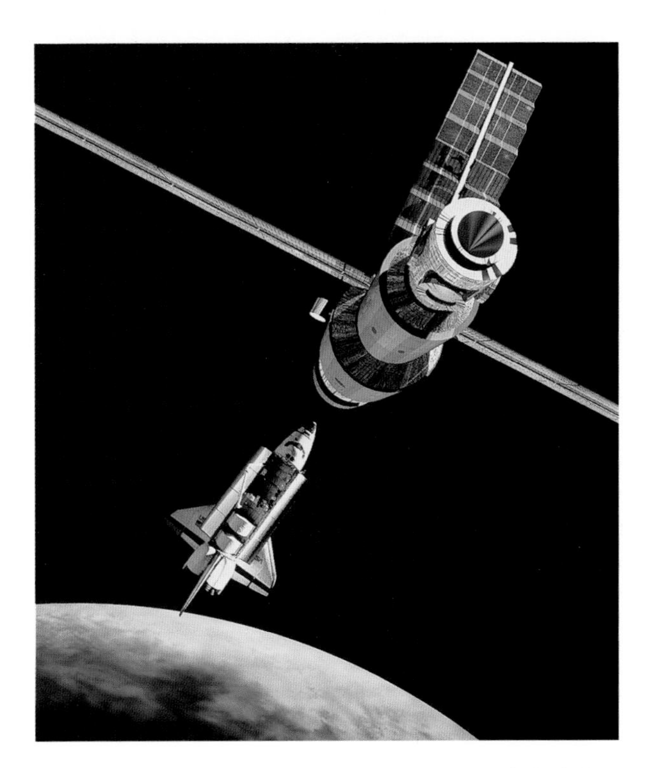

Once fierce competitors in the space race, now the US and Russia, along with some European nations, have become collaborators in developing space stations. The Shuttle is shown here prior to docking with MIR.

The International Space Station slated for completion in 2004 is the largest scientific cooperative programme of any kind in history, drawing on the resources and scientific expertise of 16 nations, including Brazil, Belgium, Canada, Denmark, France, Germany, Italy, Japan, the Netherlands, Norway, Russia, Spain, Sweden, Switzerland, the United Kingdom and the United States. Although the age of space travel was initially characterized by nationalistic and militaristic competition, recent developments have begun to demonstrate the advantages of international collaboration. Although the MIR station had been planned solely as a Russian venture, the improved relations with America throughout the late 1990s enabled the Shuttle to begin enhancing and supporting their efforts. The international station opens up additional possibilities for nations which, to date, have not had the financial capabilities to fund their own orbiting stations. Furthermore, a project to observe changes to earth, as well as to space, will no doubt be better supported by involving researchers from many nations.

On 10 April 1995 an email appeal was broadcast across the Internet for medical assistance in diagnosing a severe, unknown disease. Zhu Lingling, a young student at Beijing University, had lapsed into a coma and her friends used modern technologies in a desperate chance to save her life. The news quickly spread through numerous on-line newsgroups, and medical experts and well-wishers from 18 countries responded. Radiology images and other patient data were shared between medical practitioners who quickly diagnosed correctly that she had suffered thallium poisoning. After appropriate treatment, she began recovering, although full recovery in these cases frequently takes years. Perhaps it is collaborations such as this, above all, which will preserve our species in an increasingly fragile ecosystem. In a period of changing identities, collaboration may strengthen our own sense of identity. It is frequently through dialogue with others that we better understand our own position.

The quality of life

We began this history with the generalization that the basic needs of mankind are for food, clothing and shelter. The continuing importance of these needs is apparent from the amount of space devoted to developments over the centuries in agriculture and food processing, the making of textiles, building materials and construction methods. Many other technologies are closely related to these fundamental needs. Transport is a vital element in the distribution of goods of all kinds, including food from increasingly distant markets. The pottery industry provides both vessels for the kitchen and dining room and bricks for building. Survival depends still on these complex ramifications of technology functioning collectively, as is painfully apparent as soon as the balance is disturbed. A bad harvest, a widespread power failure, a prolonged strike or the interruption of supplies of important raw materials through political intervention can all have far-reaching repercussions.

The consequences of these perturbations have grown with changing ideas about what constitutes need. In Third World countries, it still means exactly what it did in Biblical times: a report issued by the Food and Agriculture Organization of the United Nations in 1985 predicted that, in the absence of very substantial agricultural development, nearly 500 million people in Africa – more than half the continent's population – would be below the starvation level by the year 2000. In the rich countries of the Western world, the necessities of life are quite differently defined. Here, it extends far beyond basic needs to all the manifestations of advanced technology that we have considered in the later chapters of this book – fast, comfortable travel; easy verbal communication by telephone; a constant flood of entertainment by radio and television in the home; a rich variety of foods imported and exported from all parts of the world; domestic appliances to remove the drudgery of housework; health services on demand.

How the means to satisfy these new needs have arisen and consideration of their consequences for the general pattern of life have been the substance of this history. This leaves unresolved, however, a question which has long troubled those concerned to identify the sources of invention: is it a response to a manifest need or does the very existence of an invention lead to uses being found for it? This problem is commonly debated as though these were incompatible alternatives, assuming that there is a single answer. In fact the lesson of history, as we have explored it, is surely that these are two sides of the same coin. What we have been observing is a constant interaction between technology and society, circumstances sometimes stimulating invention and sometimes favouring one already existing. To seek a single casual relationship is to seek a uniformity that simply does not exist within a very complex relationship.

That relationship has also delimited our choices; societal beliefs and values have influenced and constrained the boundaries of invention through laws, morals, religion, money and other resources. This book has highlighted many of the advances in technologies by tracing a series of inventions through changes in societies, but after millions of years of invention, we now stand questioning the meaning of invention itself, pondering the boundaries between natural and synthetic. These may prove to be the most difficult to resolve, because at the root of the issue lie some of humanity's most imponderable questions: What is life? What is sentience? What defines us – nature or nurture? Genetic engineering methods give us greater choices than ever before, but are we ready to make those choices, and on what basis will we make them? Life has been patented. It is inevitable that we will soon modify the attributes of our offspring by choice, but with which constraints will we limit our choices?

Finally, this brings into question the motivation of inventors themselves. Some have had financial gain as a major target, as evidenced not only by the care they have taken to secure their interests either by patents or by secrecy, but also by their own admission. Others have been fascinated by the challenge of a difficult technical problem and have devoted themselves unswervingly to its solution with little or no concern for personal advantage. These different attitudes still apply when inventors largely work not alone, but within large organizations: the motivation seems as diverse as human nature itself. If this is so, one prediction we can safely make is that our inventive genius will be as fruitful in the future as it has been from the dawn of civilization.

computers 306–312 **gene patenting** 297 **genetic engineering** 297–299 **international space collaboration** 335–336

Biographical Dictionary

An asterisk before a name indicates a separate biography.

Agricola Georg Bauer (1494–1555). German pioneer of mining and industrial chemistry. Trained as a physician in Italy, he settled in the Erzgebirge ('ore mountain') region of Saxony and Bohemia. He became proficient in the mining techniques of his day and is remembered for his massive, finely illustrated *De re metallica* (1555): this includes a section on chemical technology, mainly concerned with metallurgical processes. It owes something to the *Pirotechnia* of *Biringuccio (1540) and was standard work for more than a century.

Ampère André Marie (1775–1836). French mathematician and physicist, pioneer of electro-dynamics. A child prodigy in mathematics, he had a distinguished academic career. Ampère's law gave mathematical expression to *Oersted's discovery of the relationship between electricity and magnetism (1820). His name is remembered in the basic unit of electric current.

Appert Nicolas (François) (c. 1750–1841). French confectioner and inventor of food preservation by heat sterilization in glass vessels. He later moved into canning (1822; *Donkin).

Archimedes (c. 287–212 BC). Greek mathematician; pioneer of statics and hydrostatics. Generally regarded as the greatest mathematical genius of antiquity, he made important contribution to the geometry of plane and spherical surfaces, including an accurate calculation of π. He developed the theory of pulleys and levers and the concept of specific gravity, though the famous 'Eureka' experiment is probably apocryphal. He devised many mathematical devices including a planetarium and various engines of war, but whether he actually invented the Archimedean screw for raising water is doubtful.

Aristotle (384–322 BC). Greek philosopher who took as his province the whole realm of human knowledge. He studied under Plato in Athens before being called to Macedon to tutor the young Alexander the Great. About 335 Aristotle founded his own school in Athens, the Lyceum, but on Alexander's death (323) had to leave under a political cloud. Probably his greatest contributions lay in the fields of zoology and anatomy. In cosmology he conceived a finite spherical universe with the earth at its centre. He believed all matter to consist of four elements – fire, earth, air and water: the celestial bodies alone consisted of a fifth element, the aether. Aristotle's thinking dominated Western thought until the 16th century.

Arkwright Sir Richard (1732–92). English industrialist and inventor of the first successful cotton-spinning machine. Arkwright, who began life as a wig-maker, was a pioneer of the factory system.

Avicenna (Ibn Sina) (980–1037). Persian court physician, philosopher and polymath. His monumental *Canon of Medicine* (Latin translation, 12th century) embodied much Greek learning. Avicenna's interest in diseases of the eye led him to study optics and vision he concluded that light was corpuscular in nature. Like *Geber, whose alchemical writings he knew, he believed that all metals derived from two idealized elements, sulphur and mercury: however, he did not share Geber's belief in the possibility of transmutation of metals. He anticipated *Newton's first law of motion, but was not familiar with the concept of acceleration.

Babbage Charles (1792–1871). English mathematician and statistician. His 'analytical engine', programmed by punched cards, was designed to compute and print mathematical and astronomical tables. A forerunner of modern digital computers, it was not completed in his lifetime, owing, in part, to lack of finance and the limitations of contemporary technology.

Bacon Francis, 1st Baron Verulam (1561–1626). English philosopher and statesman (Lord Chancellor, 1618–21). Bacon reviewed the whole field of learning with the ultimate objective of systematizing it and turning it to the practical benefit of mankind. This was to be included in a 'grand instauration' (renewal) of human knowledge. His grandiose scheme, based on inductive reasoning, was never achieved but major work like the *Advancement of Learning* (1605) and the *Novum organum* (1620) would have been incorporated in it.

Bacon Roger (c. 1219/20–92). English Franciscan friar and philosopher, an early advocate of experiment as the confirmation of theory. His writings reveal his familiarity with Arab Aristotelianism, his deep knowledge of alchemy, and the breadth and power of his mind. He described, but did not invent, gunpowder.

Baekeland Leo Hendrik (1863–1944). Belgian-born US chemist and inventor. About 1891 he introduced the first photographic paper ('Velox') capable of giving satisfactory prints from negatives with the aid of artificial light. In 1898 he sold his business to *Eastman for $1 million. His invention (1909) of Bakelite – a thermo-setting phenol-formaldehyde resin – founded the modern plastics industry.

Baeyer Johann Friedrich Adolf von (1835–1917). Versatile German experimental organic chemist and teacher (he succeeded *Liebig at Munich), noted for his synthesis of indigo (1880). He developed the structure theory of organic molecules, and received the Nobel Prize for Chemistry in 1905.

Baird John Logie (1888–1946). Scottish pioneer of television. Using his photo-mechanical scanning apparatus he gave his first demonstration – transmitting a small, 30-line picture – in a London flat on 26 January 1926. Transmission by telephone line followed (London–Glasgow, 1927; London–New York, 1928) and in September 1929 Baird's newly formed company made the world's first television broadcast for the BBC. His system (improved to 240 lines) continued in use until 1935 when the BBC adopted an all-electronic 405-line system. Baird's work was not in the main line of progress but greatly stimulated public interest in television.

Bakewell Robert (1725–95). Pioneer of scientific agriculture. He improved the quality of his horses, cattle and sheep by systematic breeding: apparently by outbreeding to introduce desirable qualities and inbreeding to stabilize them. He kept his methods a close secret and made his profit by letting his best animals for stud, both at home and abroad.

Banting Sir Frederick Grant (1891–1941). Canadian physician: pioneer of insulin treatment for diabetes. In 1920 he conceived the idea of seeking a hormone in the islets of Langerhans – pancreatic cells known to be related to diabetes. Together with J. J. Macleod, *C. H. Best, and J. B. Collip he isolated insulin in 1922. The following year, he and Macleod received the Nobel Price for Physiology or Medicine.

Becquerel Antoine Henri (1852–1908). French physicist whose research on phosphorescent and fluorescent materials led to his discovery of radioactivity, in uranium (1896). He shared the 1903 Nobel Prize for Physics with *M. and P. Curie and did important research on the polarization of light and the earth's magnetic field.

Bell Alexander Graham (1847–1922). Scottish-born US inventor of the telephone (patent 1876, narrowly anticipating *Gray) and pioneer of sound recording with his wax-cylinder, spiral track 'gramophone'. Bell also developed deaf aids, among his numerous inventions.

Benz Karl (1844–1929). German engineer who pioneered the adaptation of the internal combustion engine to road vehicles. In the 1880s he designed a car powered by a petrol-fuelled modification of the *Otto four-stroke gas engine.

Berger Hans (1873–1941). German physiologist and psychiatrist, founder of encephalography. He observed the alpha rhythms of the brain and (1929) published the first of a series of papers on human encephalograms. The diagnostic value of the technique was rapidly appreciated.

Bergius Friedrich (1884–1949). German industrial chemist who developed a process for the manufacture of petrol from coal by hydrogenation at high pressure: this was of considerable strategic importance to Germany during the Second World War. Bergius shared the 1931 Nobel Prize for Chemistry with *Bosch, noted also for his association with *Haber in the development of the Haber-Bosch ammonia process.

Berliner Emile (1851–1929). German inventor of the disc record (1888). His 'gramophone' replaced the hill-and-dale groove of Edison's 1877 cylinder phonograph with the lateral helical grooved disc, since standard. His invention of the matrix record, from which multiple replicas could be pressed, pioneered the mass record industry.

Bessemer Sir Henry (1813–98). English engineer and inventor of the Bessemer process (patent, 1855) for the decarburization of molten iron by blowing compressed air through it. Combined with Bessemer's tilting converter, this became the standard process until rivalled by the *Siemens-Martin open hearth process. Bessemer developed it primarily for the manufacture of gun barrels: his Sheffield factory also produced steel rails. *Kelly independently developed a similar process in the USA.

Best Charles Herbert (1899–1978). Canadian physiologist who collaborated with *Banting, his professor at Toronto University, in the discovery (1922) of insulin and demonstrating its effectiveness for the control of diabetes: they shared the 1923 Nobel Prize for Physiology or Medicine.

Bevan Edward John (1856–1921). English chemist, a co-patentee (1892) with *Cross of the viscose rayon process in which cellulose was chemically dissolved and regenerated to yield spinnable fibres.

Biringuccio Vanuccio (1480–1538/9). Italian technologist, director of the papal arsenal, author of *Pirotechnia* (1540). It deals with mining and the extraction of metal from ores, the manufacture of cannon and gunpowder, and many other contemporary industrial processes.

Birkeland Kristian Olaf Bernhard (1867–1917). Norwegian physicist, university professor at Christiania (now Oslo) and co-inventor with *Eyde of an electric-arc process for the commercial fixation of atmospheric nitrogen.

Blériot Louis (1872–1936). French aviator, inventor and manufacturer, noted for the first aeroplane crossing of the English Channel on 25 July 1909. He had made his fortune in car accessories and from that time used it in the development and manufacture of aircraft.

Bohr Niels (1885–1962). Danish physicist who revolutionized atomic theory by applying quantum mechanics. In 1911 *Rutherford had established that atoms consisted of a heavy positively charged nucleus surrounded by a diffuse cloud of negatively charged electrons. In 1912 Bohr made the daring assumption that electrons could occupy only certain specific 'quantized' orbits, and that radiation was emitted when they jumped from one to another. This explained the line structure of the spectra of hydrogen and other elements. In the 1930s he made important contributions to knowledge of the compound nature of the nucleus. He was awarded the Nobel Prize for Physics in 1922.

Bosch Carl (1874–1940). German industrial chemist who designed (1913) with *Haber the massive high-pressure converters for the manufacture of synthetic ammonia by the Haber process. He shared the 1931 Nobel Prize for Chemistry with *Bergius.

Boulton Matthew (1728–1809). English manufacturer, noted for the silver and Sheffield plate ware of his Birmingham Soho factory. In 1775 he went into partnership with *Watt to manufacture the latter's improved steam engine. The partnership powerfully influenced the course of the Industrial Revolution. Boulton was a leading light in the Lunar Society of Birmingham, an association of industrialists and scientists.

Boyle The Honourable Robert (1627–91). Anglo-Irish philosopher, widely acclaimed as 'the father of chemistry'. He refuted *Aristotle's belief in air, earth, fire and water as elements and in his *Sceptical Chymist* (1661) taught that elements are simple bodies which combine to form compounds. His writings greatly strengthened a belief in atomism among European philosophers. His name is remembered in Boyle's Law, relating the volume of a gas to its pressure.

Braun Karl Ferdinand (1850–1918). German physicist who shared the 1909 Nobel Prize with *Marconi for improvements in wireless transmission and reception. In 1897 he invented the cathode ray oscilloscope, the principle of which – the deflection of an electron beam by a variable voltage – was subsequently incorporated in television receivers.

Braun Wernher von (1912–77). German rocket expert. Inspired by the work of Oberth, he joined the German Society for Space Travel in 1930. His postgraduate research at Berlin University was on rocket engines and in 1937 he became technical director of the Peenemünde Research Establishment set up to develop military rockets: the supersonic V-2 rocket was operational in 1944. After the war his team was taken to the US to work on ballistic weapons programmes and he led the Explorer satellite (1958) and the Saturn rocket projects.

Brindley James (1716–72). English canal builder. Following his design and construction of the Manchester–Worsley Canal (1759–61) for the Duke of Bridgwater, he built more than 563 km (350 miles) of canals. Never more than semi-literate, his approach was purely empirical.

Brunel Isambard Kingdom (1806–59). English engineer and surveyor, son of Sir Marc *Brunel. His father's assistant on the Thames Tunnel, he became chief engineer to the Great Western Railway (1833). He designed many outstanding bridges, docks and tunnels, introduced a wide-gauge 214-cm (7-foot) rail track and designed railways in Italy, Austria and India. He pioneered commercial transatlantic steam navigation with the SS *Great Western* (1838), the iron-hulled, screw-driven *Great Britain* (1845) and the *Great Eastern* (1858), the largest vessel of its time – a commercial disaster but a technological triumph.

Brunel Sir Marc Isambard (1769–1849). French-born British inventor and engineer, patentee of the first tunnelling shield. Fleeing the French Revolution, he became (1793) New York's city engineer. In England, from 1799, he patented mechanized mass-production machinery to make ships' pulley blocks and produced innumerable other inventions – notably improved sawmill machines. He was engineer of the Thames Tunnel, 1825–43.

Bunsen Robert Wilhelm (1811–99). German experimental chemist, professor at Heidelberg University, 1852–89. His application of spectroscopy to chemistry had far-reaching consequences, including knowledge of the composition of the sun and stars, the discovery of two new elements (caesium and rubidium), and the influence of light on chemical reactions. He is remembered for the Bunsen burner.

Burdin Claude (1790–1873). French engineer, professor at the St Etienne School of Mines. His paper describing a new type of power unit (which he called a 'turbine') exploiting the flow of water over a rotor, inspired the inventor *Fourneyron.

Campbell-Swinton Alan Archibald (1863–1930). Scottish engineer who, in 1908, outlined the principle utilized in modern television, namely the use of cathode ray tubes for transmission as well as reception. He elaborated these views again in 1911 and 1920, but not until the mid-1930s did electronic television finally oust *Baird's photo-mechanical system.

Carnot Nicolas Léonard Sadi (1769–1832). French physicist, pioneer of thermodynamics. Recognizing that Britain's industrial lead relied largely on steam technology, Carnot turned to the theoretical study of heat engines. In his *Reflections on the Motive Power of Fire* (1824) he formulated Carnot's Theorem, according to which all reversible heat engines working between the same temperature limits must have the same efficiency, the maximum possible for the temperatures in question. However, his work attracted little attention until rediscovered by *W. Thomson in connection with the formulation of the Second Law of Thermodynamics.

Caro Heinrich (1834–1910). German industrial chemist, a pioneer of synthetic dyes. After working at dyeworks in Manchester (1859–66), he joined BASF of Munich in 1868. He synthesized numerous dyes and devised industrial processes for their manufacture.

Carothers Wallace Hume (1896–1937). US industrial chemist who discovered nylon, the first commercial synthetic polymer fibre. As head of organic research for the Du Pont company, Wilmington, Delaware, Carothers inaugurated a systematic research programme into polymeric compounds of high molecular weight. The results included polyamides which could be moulded, melt-spun into fibres and made into transparent film. 'Nylon 66' was put into production in 1938, the year after Carothers, a depressive, had committed suicide. His programme also included polyesters, but the possibilities of these were independently developed as 'Terelyne' by *Whinfield.

Cartwright Edmund (1743–1823). English clergyman, manufacturer and versatile inventor, pioneer of the power loom (1785–87), in the development of which he was stimulated by *Arkwright. He built his own mill but retired to work on agricultural research for the Duke of Bedford at Woburn. Other inventions included a wool-combing and rope-making machine.

Castner Hamilton Young (1858–98). US pioneer of industrial chemistry with his method (1879) for making sodium – primarily for aluminium manufacture by reducing aluminium chloride with carbon. In 1897 he pioneered the Castner-Kellner process for manufacturing caustic soda by electrolysis of brine.

Cayley Sir George (1773–1857). English country gentleman and pioneer of the theory and practice of heavier-than-air flight. Cayley abandoned ideas of beating-wing machines and in 1853 built a fixed-wing glider which flew 500 metres (1640 feet). He had correctly calculated the power necessary but realized that contemporary technology offered no suitable engine: this situation did not change until the advent of the petrol engine.

Chain Sir Ernst Boris (1906–79). German-born biochemist who collaborated with *Florey at Oxford in the isolation and purification of penicillin. In 1945 he shared the Nobel Prize for Medicine or Physiology with Florey and *Fleming. Subsequently he held appointments in Rome and at Imperial College, London, during which time he made important contributions to the development of semi-synthetic penicillins.

Chappe Claude (1763–1805). French engineer, designer of a mechanical telegraph (1793–4). It comprised towers equipped with telescopes and semaphore arms, built in line of sight between Paris and Lille. Disputes over his primacy of invention led him to suicide.

Cochrane Archibald, 9th Earl of Dundonald (1749–1831). Scottish chemical industrialist whose ventures included a soda works using the *Leblanc process and experiments with the gasification of coal.

Cockcroft Sir John Douglas (1897–1967). English physicist who, with Walton, first split the atomic nucleus (1932). They received the 1951 Nobel Prize for Physics. Cockcroft played a leading role in radar development, was Jacksonian Professor at Cambridge (1939–46), and after the war became Director of Britain's Atomic Energy Research Establishment at Harwell.

Colt Samuel (1814–62). US inventor who pioneered mass production by making machine components so accurately that they were interchangeable. Colt's earliest venture was in the manufacture of the revolving breech pistol named after him: initially a failure, commercial success came with a government order during the Mexican War, 1846–8.

Cook James (1728–79). English circumnavigator and cartographer who transformed knowledge of the Pacific and Oceania, effectively dispelling belief in the existence of a great southern continent. In *Endeavour*, 1768–71, he observed the transit of Venus and made pioneering charts of the coasts of New Zealand and eastern Australia. He made two further expeditions, on the latter of which he was killed in an attack by natives in the Sandwich Islands (Hawaii). He revolutionized health at sea with dietary and hygiene regimens which largely eliminated scurvy.

Cooke Sir William Fothergill (1806–79). English scientist and inventor, co-patentee (1837) with *Wheatstone of an electric telegraph, perfected in 1845.

Cort Henry (1740–1800). English ironmaster, 'the Great Finer'. With capital acquired as a navy agent he opened his ironworks in the 1770s. He patented a grooved-roller machine for the production of iron bars and in 1784 introduced his 'puddling' process for decarburizing iron.

Crompton Samuel (1753–1827). English mill operative and inventor (1779) of the spinning 'mule', so called because it combined features from *Hargreaves's 'jenny' and *Arkwright's frame. It was an important advance and widely adopted. Too poor to patent his invention, Crompton eventually received a Parliamentary grant of £5000.

Crookes Sir William (1832–1919). Versatile British scientist, remembered particularly for his research in spectroscopy and radiation resulting from electrical discharges in gases. He discovered thallium in 1861 and collaborated with *Ramsay in the identification of the rare gases of the atmosphere (helium, neon, etc). His warning (1898) that the world faced starvation unless new sources of nitrogenous fertilizers could be developed had far-reaching consequences (*Birkeland, *Haber).

Cross Charles Frederick (1855–1935). English chemical technologist. With *Bevan he patented (1892) the viscose rayon process for making 'artificial silk' from cellulose and pioneered cellulose acetate.

Cugnot Nicholas Joseph (1728–1804). French military engineer. Commissioned by the army to devise a steam-propelled gun carriage, in 1770 he demonstrated a 5-kph (3-mph), three-wheeled locomotive.

Culpeper Nicholas (1616–54). English herbalist who contributed to the development of modern botany and pharmacology. His *Complete Herbal* (1653) was widely influential, though incorporating some erroneous ideas – such as the Doctrine of Signatures, which held that the appearance of a plant indicated its medicinal virtue.

Curie Marie (1867–1934) and Pierre (1859–1906), discoverers of radium and the nature of radioactivity. The Polish-born Marie Sklodowska came to Paris in 1891 to study physics and in 1895 married the brilliant physics professor Pierre Curie. Marie followed up *Becquerel's discovery of radioactivity in uranium by searching for it in other elements. In 1898 she and her husband discovered polonium and radium. They shared the 1903 Nobel Prize for Physics with Becquerel. In 1904 Pierre was created a professor at the Sorbonne; Marie succeeded him on his tragic death in a street accident. She won the 1911 Nobel Prize for Chemistry. She continued research into radium and radioactivity at the Curie Institute for Radium, Paris, and died of leukaemia, a then unrecognized hazard of excessive exposure to high-energy radiation.

Curtis Charles Gordon (1860–1953). US engineer, patent lawyer, and inventor of an improved steam turbine. Patented in 1896, it was widely adopted both for generating stations and marine power units.

Daguerre Jacques Louis Mandé (1787–1851). With *Niepce, French pioneer of photography. A scene painter, Daguerre first made a name with illusionistically painted and illuminated landscape 'Dioramas'. To prepare them he used a camera obscura and had long tried to 'fix' its images. From 1829 he partnered Niepce in research to reduce the eight-hour exposure time of the latter's 'heliographic' technique. After Niepce's death, Daguerre discovered that with mercury vapour he could develop latent images on silver iodide-coated copper plate in 30 minutes and fix such images with sodium thiosulphate (hypo). Bought by the French government for the public benefit, the process was demonstrated at the Institut de France on 19 August 1839.

Daimler Gottlieb (1834–1900). German inventor, engineer and automobile manufacturer. In 1885 he constructed a three-wheeled, petrol-powered vehicle. For many years he collaborated with Wilhelm Maybach in perfecting the *Otto engine. His 1883 patent involved a self-firing ignition system. In the 1890s he established his own motor works.

Dalton John (1766–1844). English chemist and pioneer of modern atomic theory, formulating the first table of atomic weights (1803). A teacher at the Manchester Academy, he also investigated colour blindness (Daltonism) and made important contributions to the study of gases, arising from his interest in meteorology.

Darby Abraham (1678–1717). English ironmaster who, at Coalbrookdale on the River Severn, in 1709, first successfully smelted iron with coal, in the form of coke. This was already used in the smelting of copper and lead but its sulphur content made it unsuitable for iron for which charcoal was the traditional, and increasingly scarce and expensive, fuel. The Coalbrookdale works expanded under Darby's son Abraham II (1711–63), casting great cylinders for *Newcomen steam engines. In 1779 Abraham Darby III (1750–91) built the world's first iron bridge, at Coalbrookdale.

Davy Sir Humphry (1778–1829). English chemist and physicist of manifold achievements, notably in electrochemistry. His isolation of potassium and sodium by electrolysis in 1807 was followed by that of calcium, magnesium, barium and strontium. An outstanding talent for experiment was matched by brilliance in exposition, and he was a fashionable lecturer at London's Royal Institution. He invented an effective miner's safety lamp and investigated the properties of nitrous oxide ('laughing gas'), subsequently an important anaesthetic.

De Forest Lee (1873–1961). US inventor of the triode thermionic valve (the 'Audion'). *J. A. Fleming successfully contested his 1907 patent on the grounds that it was implicit in his own diode (1900): it was little comfort to De Forest that this ruling was overturned 40 years later.

De Laval Carl Gustav Patrik (1845–1913). Prolific Swedish inventor (he employed 100 engineers to develop his ideas, which ranged from electrometallurgy to aerodynamics) and turbine technologist. His turbine-operated centrifugal cream separator (1878) was adopted worldwide in the dairy industry. Unlike *Parson's steam turbine, however, it was not suitable for high-power units.

Dewar Sir James (1842–1923). Scottish scientist noted for work on vacuum techniques and low temperature physics, liquefying hydrogen, supposedly a permanent gas, in 1895. As a laboratory tool, Dewar also invented the 'Dewar flask', a model for the domestic thermos flask.

Diesel Rudolf (1858–1913). German engineer and inventor. On theoretical principles he conceived, about 1890, the idea of an improved type of oil engine in which the fuel would be ignited simply by the heat generated by very high compression. After encountering serious practical difficulties, the new type of engine was widely adopted from the beginning of the 20th century. Burning relatively cheap oil, it was very economical.

Domagk Gerhard (1895–1954). German pathologist and pioneer of chemotherapy whose work led to the development of the sulphonamide drugs. From 1927 director of research in experimental pathology and bacteriology at IG Farbenindustrie, he conducted an extensive research programme for anti-bacterial agents. In 1932 he discovered that a dye (Prontosil) could control streptococcal infections in mice. When this was made public in 1935 it was quickly established that the dye itself was not the active agent but a part of it forming a sulphonamide group. This led to revolutionary developments in the treatment of a wide range of infections. He was awarded the Nobel Prize for Physiology or Medicine in 1939, but the Nazis refused to allow him to accept it.

Dondi Giovanni de' (1318–89). Italian physician and astronomer, renowned for the elaborate astronomical clock he designed and built in the 1350s.

Donkin Bryan (1768–1855). English engineer and entrepreneur, pioneer of commercial food canning. In 1813 his Bermondsey factory began supplying food preserved by *Appert's process but packaged in hermetically sealed tins rather than glass bottles. He also effected improvements in papermaking and printing.

Drake Edwin Laurentine (1819–80). American oilman whose strike at a depth of 21 metres (70 feet) at Oil Creek, Titusville, Pennsylvania on 27 August 1859 inaugurated America's petroleum industry.

Drebbel Cornelius Jacobszoon (1572–1633). Dutch-born engineer and inventor who, in 1621, demonstrated a submersible 'diving boat' on the River Thames. It was sealed with a leather cover, propelled by oars and supplied with air from tubes. He also introduced the use of tin compounds as mordants to fix dyes to fabrics.

Dunlop John Boyd (1840–1921). Scottish veterinarian who first patented (1888) a pneumatic tyre, developed for his son's bicycle. He lived to see this invention become the basis of a worldwide industry supplying the needs of the burgeoning car industry.

Eastman George (1854–19322). US photographic innovator and industrialist. His 'Kodak' camera (1888) was the first incorporating roll film. His company's mass production of the simple box camera and mass processing of the nitro-cellulose roll film established photography as a popular hobby.

Edison Thomas Alva (1847–1931). US inventor whose more than 1000 patents included the phonograph (1877), the incandescent filaments lamp (*Swan) and major contributions to telegraph technology. In 1876 he founded his 'invention factory' at Menlo Park, which anticipated the systematic rigour of modern industrial laboratories. In 1888 he opened the first central power station at Pearl Street, New York.

Ehrlich Paul (1854–1915). German bacteriologist, who pioneered scientific immunology and haematology. The term chemotherapy was born from his research programme for a chemical substance lethal to pathogenic bacteria but harmless to the human body: in 1911 his 606th specimen, salvarsan, proved successful for the treatment of syphilis. Ehrlich's many other major contributions included research on the use of synthetic dyes in microscopy.

Eiffel Alexandre Gustave (1832–1923). French engineer, renowned for the tower he built for the 1889 Paris Exhibition. It was almost the last major edifice to be built of iron: subsequently steel – increasingly cheap and abundant as a result of the researches of *Bessemer and others – was almost universally used.

Einthoven Willem (1860–1927). Dutch physiologist (professor at Leyden from 1886), inventor of clinical electrocardiography, a very important diagnostic technique: he won the 1924 Nobel Prize for Physiology or Medicine.

Elkington George Richards (1801–65). English industrialist, pioneer of electroplating. He patented a process for the application of electricity in gilding (1838), used cyanide plating baths (from 1840), and in 1841 installed a large plating plant at his Birmingham works.

Eratosthenes (*c.* 275–*c.* 195 BC). Greek astronomer and polymath; from *c.* 249 BC head of the great library at Alexandria. His interest in geography led him to calculate the circumference of the earth: his value of about 46,000 km (29,000 miles) is in fair accord with the true value of almost 40,000 km (25,000 miles).

Ericsson John (1803–89). Swedish-born engineer and inventor. In England (1829) he built the locomotive Novelty and patented a type of screw propeller, before going to America (1836). There he built screw-driven ships for the Navy (USS *Princeton*, 1844) and made many contributions to marine engineering. During the Civil War his heavily armoured *Monitor* successfully engaged the Confederate ship *Virginia* (1862) in the world's first battle between ironclads.

Evans Oliver (1755–1819). US inventor and pioneer of high-pressure steam technology: contemporary of *Trevithick. From 1807 he built about 50 high-pressure engines, some for marine use.

Eyde Samuel (1866–1940). Norwegian electrochemical engineer and industrialist, collaborator (1901) with *Birkeland in the Birkeland-Eyde process for the fixation of atmospheric nitrogen by an electric arc process.

Faraday Michael (1791–1867). English scientist who (1831) demonstrated electromagnetic induction, the principle of the electric motor and dynamo. He joined the Royal Institution (1812) as assistant to *Davy, becoming director of the laboratory in 1825 – when he discovered benzene and the first compounds of carbon and chlorine – and professor of chemistry in 1833. It is said that Faraday was Davy's greatest discovery. He postulated that electricity corresponded to vibrations at the molecular level. He formulated the laws of electrolysis and the Field Theory of Magnetism, thus inspiring the theoretical concepts of *Maxwell.

Fermi Enrico (1901–54). Italian nuclear physicist who, in 1942, achieved the first self-sustaining nuclear chain reaction. In the inter-war years his publications on the neutrono, ß-decay and quantum theory established his international reputation. Awarded the 1938 Nobel Prize in Stockholm, he did not return to Fascist Italy but, with his Jewish wife, went on to America. He taught first at Columbia University, New York, and then at Chicago where, on 2 December 1942, the pile he had designed went critical. In 1943 he joined the Los Alamos atom bomb project.

Ferranti Sebastian Ziani de (1864–1930). English electrical engineer and inventor: pioneer of high-voltage generating and transmission systems. In 1887, as chief electrician of the London Electric Supply Corporation, Ferranti headed a scheme to supply North London from a single high-power generating station at Deptford, designing and building generators and cables originally to deliver electricity at 10,000 volts. The high voltage was reduced through transformers for local supply.

Fessenden Reginald Aubrey (1866–1932). Canadian-born American scientist, electrical engineer and radio pioneer. After working with the Edison laboratories and as an academic, he joined the US Weather Bureau. In December 1906 he made a spoken broadcast from Brant Rock, Mass., which was heard in Norfolk, Va. Fessenden pioneered amplitude modulation and devised the heterodyne circuit (1912), among many other important inventions.

Fischer Emil (1852–1919). German organic chemist, professor at Berlin from 1892, and winner of the 1902 Nobel Prize for Chemistry. He is noted for research on natural products, including carbohydrates, sugars, and proteins. He discovered the drug barbital (veronal) in 1902.

Fitch John (1743–98). US steamboat pioneer, making his first successful launch in 1787, and establishing a regular service on the Delaware in 1790. The failure of later vessels discouraged financial backers and despair led him to suicide.

Flamsteed John (1646–1719). English astronomer and first Astronomer Royal at Greenwich Observatory (1675). His precise observations were the basis of two important star catalogues (published posthumously, 1725).

Fleming Sir Alexander (1881–1955). Scottish bacteriologist, the discoverer of penicillin (1928). Working at St Mary's Hospital, London, he had previously discovered (1922) the bacteriolytic enzyme lysozyme. Fleming shared the 1945 Nobel Prize for Physiology or Medicine with *Chain and *Florey for research on penicillin.

Fleming Sir John Ambrose (1849–1945). English electrical engineer and inventor (1904) of the diode thermionic valve, a major contribution to radio technology; this was followed (1906) by *De Forest's triode. Despite bitter litigation between the two men, the thermionic valve was an essential part of all radio transmitters and receivers until the advent of the transistor in 1948.

Florey Howard Walter, Baron (1898–1968). Australian physiologist and pathologist and 1945 Nobel Prize winner with *Fleming and *Chain for his work on the chemotherapeutic qualities of penicillin. As professor of pathology at Oxford (1934–61) after research on the antibacterial enzyme lysozyme (*Fleming), he began a general study of antibiosis with Chain and N. G. Heatley in 1939. In 1940, controlled experiments with infected mice indicated that penicillin was a uniquely powerful chemotherapeutic agent, but large-scale manufacture was impossible in wartime Britain. Florey succeeded (1941) in convincing the US pharmaceutical industry of the importance of the new drug and a crash programme of large-scale manufacture ensured sufficient material to treat all Allied casualties from the Normandy landings (1944) onward.

Ford Henry (1863–1947). US industrialist who, by adapting the conveyor belt and assembly line to automobile manufacture, established modern methods of mass production. The Ford Motor Company (founded 1903) introduced the Model T in 1908–9: by 1928 15 million had been sold and motoring was firmly established as a popular means of transport.

Fourneyron Benoît (1802–67). French designer of the first effective water turbine. Inspired by a paper by *Burdln he built a 50-hp machine which won a prize in 1833. Taken up worldwide, Fourneyron outward-flow turbines were installed at the great Niagara Falls hydroelectric station in 1895.

Francis James Bicheno (1815–92). English-born US hydraulic engineer, inventor of the Francis mixed-flow turbine.

Franklin Benjamin (1706–90). American statesman, a founding father of the Constitution, printer and writer, scientist and inventor. His scientific work included the demonstration that lightning is an electrical phenomenon: his inventions include the lightning conductor and bifocal spectacles. His 'one-fluid' theory of electricity unified a number of electrical phenomena: he is said to have 'found electricity a curiosity and left it a science'.

Freysinnet Eugène (1879–1962). French civil engineer who pioneered the use of pre-stressed concrete. His huge airship hangars at Orly (1916) were revolutionary in their day, and he built many outstanding bridges.

Frisch Otto Robert (1904–79) Austrian-born physicist, nephew of *Meitner and co-discoverer with her of nuclear fission in uranium. A refugee from Nazi Germany, Frisch was then working with *Bohr in Copenhagen.

Frontinus (1st century AD). Roman military governor of Britain and later Rome's water commissioner. He wrote a comprehensive historical and technical account of the city's water supply and aqueducts, *De Aquis*, and *Strategematicon*, a military treatise.

Fuchs Leonhart (1501–66). German botanist and physician, professor of medicine at Tübingen from 1535. His herbal *De historia stirpium* (1542) is noted for its accuracy and the beauty of its woodcuts. The plant genus *Fuchsia* is named after him.

Fulton Robert (1765–1815). Pioneer of steamboats. In 1807 he launched America's first commercially successful steamboat. Earlier, in 1801, he had demonstrated his submarine *Nautilus* – the name given to America's first nuclear-powered submarine in 1955 – in the Seine.

Gabor Dennis (1900–79). Hungarian-born physicist and inventor of holography with monochromatic light in 1947 while working with the British Thomson-Houston Company at Rugby. The advent of the laser in 1960 made possible much brighter and sharper images and led to the foundation of a major new industry. He received the 1971 Nobel Prize for Physics.

Galen (AD 129–c. 199). Greek physician, anatomist and physiologist whose writings – through Latin translations of Arabic versions – governed medical theory and practice until the 16th century.

Galileo (1564–1642). Italian mathematician, physicist and astronomer, one of the architects of modern science. Professor of mathematics at Pisa and then Padua, from 1610 he lived in Florence. He discovered, about 1602, the isochronism of the pendulum, subsequently applied to regulate the movement of clocks. In 1609, using a recent invention, the telescope, he examined the surface of the moon and discovered many heavenly bodies, including Jupiter's moons. His support of the heliocentric theory of the universe, postulated by Copernicus, brought him into serious conflict with the Catholic Church, leading to permanent house arrest.

Geber (Jabir ibn Hayyan) (721/22–815). Arab alchemist and physician. A prolific writer, familiar with a wide range of simple chemicals and chemical processes, Geber believed in the transmutation of metals, believing them all to be derived from sulphur and mercury. These were not, however, the familiar elements but some mystically sublime form of them.

Gerard John (1545–1612). English herbalist, a pioneer of botany and pharmacology as sciences. His profusely illustrated *Herbal* appeared in 1597, and described nearly 2000 plants.

Gilbert Sir Joseph Henry (1817–1901). English agricultural chemist and co-worker with *Lawes at the Rothamsted Experimental Station. Like Lawes, he had been a student of *Liebig at Giessen.

Gilchrist Percy Carlyle (1851–1935). Collaborated with his cousin S. G. Thomas in developing (1877/8) a process making it possible to smelt phosphoric iron ores. This opened up huge new deposits, notably those of Alsace-Lorraine and Luxembourg.

Goddard Robert Hutchings (1882–1945). US physicist and rocket pioneer. A professor at Clark University, in 1919 Goddard published his classic *A Method of Reaching High Altitudes*. His interest was primarily in meteorological observation. In 1926 he launched the first liquid-fuel rocket and, from 1930 at Roswell, New Mexico, pioneered many other aspects of rocket design.

Gramme Zénobe Théophile (1826–1901). Belgian electrical engineer and patentee (1869) of a d.c. (direct current) dynamo. He began manufacture in 1871 and in 1873 demonstrated its reversibility as a motor.

Gray Elisha (1835–1901). US inventor whose telephone patent was filed in 1876 only hours after *Bell's. Bell eventually won the lengthy legal battle against Western Union Telegraph which had acquired Gray's patent. He made many other inventions in this field, including the telautograph for transmitting writing and line drawing.

Guericke Otto von (1602–86). German lawyer and scientist noted for work on pneumatics and invention of the vacuum pump. Numerous experiments to demonstrate the elasticity of air and the properties of the vacuum included the famous 'Magdeburg hemispheres' of May 1654 when two eight-horse teams failed to separate the two halves of an evacuated metal sphere. In 1672 he published 'New Experiments Concerning Empty Space'.

Gutenberg Johann (*c.* 1400–1468). German goldsmith and inventor, pioneer of movable type printing in Europe. In Strasbourg (1403?–44?) Gutenberg began experiments which culminated, on his return to his native Mainz, in his first printed material *c.* 1450. A poor businessman, Gutenberg was backed (from 1452) by Johann Fust and also employed Peter Schoeffer; after a lawsuit Gutenberg was dispossessed and these two set up as successful commercial printers. Adopting a metal adhesive printing ink and a technique of manufacturing letter dies of standard shank length, Gutenberg combined these with printer's forme and mechanical press to create a revolutionary technology new in Europe.

Haber Fritz (1868–1934). German chemist, inventor (1908) of the Haber process for the synthesis of ammonia from hydrogen and nitrogen under pressure (*Bosch developed the industrial plant). Haber taught at Karlsruhe (1894–1911); headed Germany's chemical warfare research (1914–18); and in the 1920s made Berlin's Kaiser Wilhelm Institute for Physical Chemistry and Electrochemistry an internationally recognized research centre. In 1918 he received the Nobel Prize for Chemistry, but in 1933 he left Germany because of anti-semitic legislation and went to the Cavendish Laboratory in Cambridge.

Hall Charles Martin (1863–1914). US inventor (independently of *Héroult) of a high-temperature electrolytic process for the manufacture of aluminium, using cryolite (sodium aluminium fluoride) as a raw material. Success in 1886 led to commercial development in 1889 and the founding of the American Aluminum Company.

Hancock Thomas (1786–1865). English founder of the rubber industry. He made his first 'vulcanized' rubber in 1843 by treating it with sulphur under heat.

Hargreaves James (*c.* 1720–78). English hand-loom weaver and inventor of the spinning 'jenny'. Capable of spinning eight threads at once, it increased production but made a coarse yarn. He established his own cotton mill but, unable for technical reasons to patent his invention, died comparatively poor.

Harrison John (1693–1776). English horologist, father of the marine chronometer. Competing for a government-sponsored prize of £20,000 for a successful method of determining longitude at sea, Harrison built his first chronometer in 1735. His fourth (1763), little larger than a pocket watch, was awarded £5000 after rigorous proving on two transatlantic voyages. He received the whole award only in 1773, after the building of a duplicate by an independent maker at the intervention of King George III. Harrison's other inventions included a grid-iron pendulum (1726) to compensate for the effect of changing ambient temperature.

Hennebique François (1842–1921). French architect and engineer. In the 1880s he introduced the use of reinforced concrete for floors and pioneered the use of steel in place of iron. His spinning factory (1895) at Tourcoing, France, made uncompromising use of concrete frame and glass panels.

Hero of Alexandria (1st century AD). Mathematician and mechanician. His *Metrica*, on the mensuration of simple plane and solid figures, and *Mechanica*, on levers and gearing and other devices by which a small force can lift a large weight, were of major importance. He described such mechanisms as a double-screw press and the pawl and ratchet and a theodolite-type instrument depending on a refined screw-cutting technique. His *Pneumatica* – derived largely from Ctesibius – described various automata and mechanical contrivances (e.g. siphon, fountain, pump) operated by atmospheric pressure or heated gases, including the famous aeolipyle in which a globe rotated at high speed through the ejection of steam from side-arms.

Héroult Paul Louis Toussaint (1863–1914). French patentee (1886: independently of *Hall) of an electrolytic process (based on cryolite) for the manufacture of aluminium. The rival inventors reached a settlement only after 15 years of litigation. Héroult also designed electric furnaces for steel production.

Hertz Heinrich Rudolph (1857–94). German discoverer of the electric waves predicted by *Maxwell and thus the founder of radio communications. In the late 1880s he demonstrated the existence of these 'Hertzian' waves, showing them to have the speed of light and, like light, to be capable of refraction and reflection. he also studied electric discharges in rarefied gases.

Hoe Richard March (1812–86). US printer, inventor of the rotary printing press. Introduced in 1847 for the *Philadelphia Public Ledger*, his press had the type mounted on a central cylinder with the paper sheets fed to it by smaller cylinders. By 1870 Hoe had adopted the use of paper in continuous rolls rather than sheets.

Holland John Philip (1840–1914). Irish-born US inventor, noted for his development of the submarine. The *Holland*, accepted by the US Navy in 1900, was 16 metres (53 feet) long, cruised at 6 knots (11 kph/7 mph) with a crew of six and was powered by a petrol engine (surface) and battery-driven electric motors (submerged).

Hollerith Hermann (1860–1929). US engineer and inventor of a punched card system to record and analyse statistical information. Employed in the 1880 US census, he developed his system for use in the 1890 census. His Tabulating Machine Company (1896) became, through subsequent mergers, a constituent of IBM (International Business Machines Corporation), now a leading manufacturer of computers.

Hopkins Sir Frederick Gowland (1861–1947). English biochemist, a pioneer in the discovery of vitamins. He shared the 1929 Nobel Prize for Physiology or Medicine with C. Ejkman, another pioneer in this field, particularly noted for his discovery of the cause of beri-beri. He was professor of biochemistry at Cambridge (1914–43).

Huntsman Benjamin (1704–76). English inventor and steel manufacturer. From 1740 his Sheffield works was producing high quality steel by a secret crucible process, discovered by rivals about 1750. Local cutlers objected that it was too hard to work, so Huntsman exported to France, whence finished cutlery was sent back to England and proved so popular that the Sheffield cutlers had to follow suit.

Huygens Christiaan (1629–95). Dutch physicist, astronomer and mathematician, second only to Newton in his day. He refined the mathematical theory of the pendulum (*Horologium oscillatorium*, 1673), built the first pendulum clock, improved the telescope, discovered the rings of Saturn (1655/56) and propounded the first wave theory of light.

Jacquard Joseph Marie (1752–1834). French silk weaver and inventor. In 1801 he perfected a loom for pattern-weaving in which punched cards were used to control the threads, formerly manipulated by apprentices. Despite weavers' opposition, Jacquard looms were in general use by the 1810s.

Jenner Edward (1749–1823). English physician, discover of vaccination. In the late 1790s he demonstrated that controlled infection with cowpox gave immunity to smallpox. Jenner received Parliamentary grants totalling £30,000 and many European honours.

Joliot Jean-Frédéric (1900–58). French scientist who, with his wife *I. Joliot-Curie, demonstrated artificial radioactivity and the possibility of nuclear chain reactions (1939).

Joliot-Curie Irène (1897–1956). French scientist, daughter of *M. Curie who, with *J-F. Joliot, received the 1935 Nobel Prize for Chemistry for their discovery of artificial radioactivity. Their experiments with the radiation emitted from polonium paved the way for the identification of the neutron by J. Chadwick in 1932.

Joule James Prescott (1818–89). English scientist who determined the mechanical equivalent of heat, publishing his definitive results in 1849. Early in the 1850s he began collaboration with *W. Thomson (Lord Kelvin) in work of seminal importance for the newly emerging sciences of thermodynamics, from which evolved the law of conservation of energy. This, otherwise known as the First Law of Thermodynamics, is one of the most fundamental of all scientific generalizations. His name is remembered in the international unit of energy.

Kapitsa Pyotr Leonidovich (1894–1984). Soviet physicist, noted for his work in magnetism and low temperature physics. From 1921–34 he worked with *Rutherford at Cambridge, but was then prevented from returning while on a visit to Russia, where much of the equipment he used in Cambridge was sent. In the Second World War, as head of the Institute for Physical Problems, evacuated to Kazan, he designed apparatus for the mass-production of liquid oxygen. From this he was dismissed in 1946 on the ground of non-cooperation. Subsequently he was reinstated and did important research on gas liquefaction, ball lightning and the properties of plasma. Belatedly, he was awarded the Nobel Prize for Physics in 1978, when he was 84 years of age.

Kay John (1704–c. 1778). English inventor (1733) of the flying shuttle. This automatic return shuttle speeded hand-loom weaving and enabled a single weaver to produce broadcloth. Ruined in patent litigation, Kay died a pauper.

Kelly William (1811–88). US ironmaster who developed (independently of *Bessemer) an 'air boiling' process for the manufacture of steel. Kelly won the US patent rights (1857), but eventually had to come to terms with Bessemer.

Kelvin 1st Baron *see* W. Thomson.

Kipping Frederic Stanley (1863–1949). English organic chemist whose work on the organic compounds of silicon laid the foundations of the silicone industry. He was professor at Nottingham University (1897–1936).

Krupp Alfred (1812–87). German steelmaster and industrialist who introduced new methods for the mass production of cast steel. He introduced the *Bessemer process into Germany (1862) and established Krupps as the largest Continental steel producer, notably the manufacture of armaments.

Lanchester Frederick William (1868–1946). English engineer and pioneer of the British car industry. He improved carburettor design and, as early as 1902, experimented with disc brakes. His *Aerial Flight* (1907–8) was influential in early aircraft design.

Lavoisier Antoine Laurent (1743–94). French scientist, the founder of modern chemistry. He identified the true nature of oxygen, and demonstrated its crucial role in both combustion and respiration, thus invalidating the traditional phlogiston theory – though not immediately. This rationalized much of the accumulated chemical knowledge of the day: in particular, it identified water as a compound of hydrogen and oxygen. He was a pioneer of chemical nomenclature and in his *Traité élémentaire de chimie* (1789) – which identified 33 elements – laid out the lines of future developments of chemistry. A holder of numerous government posts, he was guillotined during the Reign of Terror.

Lawes Sir John Bennet (1814–99). English agricultural scientist and pioneer of the synthetic fertilizer industry. Founder of the now world-famous Rothamsted Experimental Station (on his family estate, inherited 1834), Lawes introduced rigorous experimental procedures: systematic field trials, for example, determined the mineral requirements of various important crops. In 1842 he patented a process for the manufacture of superphosphate fertilizers from mineral phosphates (rather than bones, the raw material used until then) and opened his factory in 1843. In the same year he was joined by *Gilbert.

Leblanc Nicolas (1742–1806). French inventor of a process for the manufacture of alkali from salt and sulphuric acid, a landmark in industrial chemistry. Prompted perhaps by the Duke of Orleans, whose physician he was, Leblanc competed for and won (1783) a national prize offered for a process, urgently needed in France for strategic reasons, to manufacture soda from a non-vegetable source. However, the prize was never paid. Orleans was executed and Leblanc's 1791 patent confiscated during the Revolution. Impoverished and embittered, he committed suicide. His process remained in large-scale use into the 20th century, when it was wholly superseded by that of *Solvay.

Lebon Philippe (1767–1804). French pioneer of gas lighting. He patented a process for the manufacture of gas from sawdust (1799), designed and exhibited a gas-burning lamp and in 1801, patented a forward-looking design for a coal-gas engine.

Leclanché Georges (1839–82). French inventor whose easily manufactured, compact electric cell (1866) was forerunner of the dry-cell battery. It comprised zinc and carbon electrodes in an ammonium chloride electrolyte and was a rapid commercial success for intermittent use, as for electric bells.

Leibniz Gottfried Wilhelm, Baron von (1646–1716). German philosopher, polymath and mathematician. He was a prolific writer, his interests ranging from astronomy and law to geology and genealogy, but he is most noted for his original work – independent of *Newton – on the differential and integral calculus and mathematical logic. He improved the mechanical calculators of *Pascal by introducing an information store (memory) for iterated operations, an essential feature of all modern computers.

Lenoir Jean Joseph Étienne (1822–1900). Luxembourg-born French engineer, inventor and manufacturer of the first industrially successful gas engine (patented 1860). Despite some initial success and an improved, four-stroke model (1883), Lenoir could not compete with the *Otto Motor Company.

Leonardo da Vinci (1452–1519). Italian painter and universal genius. In 1466 he joined the Florentine studio of the painter Verrochio and thereafter worked as painter, consulting architect and engineer, both civil and military, in Milan, Florence and Rome. In 1516, invited by Francis I, he retired to France. His famous notebooks (published long after his death) record scientific instruments, military drawings and marvellously beautiful and accurate anatomical drawings which reveal his genius as a dissector. Apart from all this and his masterly paintings, he made contributions to geology, botany, hydraulics and mechanics. However, many of his ideas were far ahead of his time, being quite incapable of realization by the available technology of his day.

Leupold Jacob (1674–1727). German engineer whose nine-volume *Theatrum machinarium generale* (1723–39) has been called the first systematic treatment of mechanical engineering. Its 472 engravings include designs for a high-pressure, non-condensing steam engine, but this was never built, being beyond the technological resources of the day.

Liebig Justus von, Baron (1803–73). Distinguished German chemist, a pioneer of agricultural chemistry and head of influential teaching laboratories at Giessen (1825–52) and Munich (1852–73). An advocate of artificial manures, he emphasized the role of chemicals, as opposed to 'humus', in plant growth, but underrated the role of nitrates. He made major contributions to organic chemistry, especially in analysis. He remained a believer in vitalism when it was being generally discounted.

Lilienthal Otto (1848–96). German pioneer of flight. Studying storks in flight, he recognized the importance of upward air currents. Building his first fixed-wing glider in 1881, he made more than 2000 flights before dying in an accident. He helped to establish gliding as a sport in Germany in the 1890s.

Linde Carl von (1842–1934). German engineer, founder of the liquid air industry. Leaving a teaching post in Munich, he worked in refrigeration during the 1880s, and in 1895 devised a large-scale plant for making liquid air utilizing the Joule-Thomson effect.

Lister Joseph, 1st Baron Lister (1827–1912). English surgeon, founder of antiseptic surgery. As a student he witnessed Robert Liston's first operation with ether anaesthesia in 1846, but was impressed by the fact that although the scope of the surgeon had been greatly extended his success was often negated by the onset of sepsis. In 1865, while professor of surgery in Glasgow and inspired by *Pasteur's work, Lister surmised that post-operative mortality (up to 50 per cent in many hospitals) was caused by micro-organisms and, after experiments, found carbolic acid an effective antiseptic. Despite initial scepticism, his procedures became general in the 1870s. Lister was subsequently professor of surgery at King's College, London (1877–92).

Lodge Sir Oliver Joseph (1851–1940). English physicist: demonstrated in 1894 that electromagnetic waves (*Hertz) could be used for telegraphic signalling in *Morse code (patent 1897).

Lucretius (*c.* 95–55 BC). Latin poet and philosopher whose six-book *De rerum natura* ('On the nature of things') presents a philosophical account of the universe based on the atomic theories of Democritus (*d.c.* 370 BC) and Epicurus (*d.* 270 BC). According to this, the physical world and also the human soul have resulted from the evolutionary combination of atoms and their differing shapes account for sensory phenomena.

Lumière the brothers Auguste (1862–1954) and Louis (1864–1948). French pioneers of cine- and colour photography. In December 1895 they gave the world's first public film show in Paris. Their *cinématographe* with intermittent claw movement and perforated 35 mm film established modern practice. Their Autochrome process, the first commercial colour film, appeared in 1907.

Macintosh Charles (1766–1843). Scottish inventor of waterproof fabrics. His patent process (1823) involved the impregnation of cloth with naphtha-rubber solution. He opened a factory in partnership with *Hancock (1834).

Magellan Ferdinand (*c.* 1480–1521). Portuguese seaman who discovered the Magellan Strait (October 1520) and whose heroic navigation of the uncharted South Pacific Ocean effected the first circumnavigation of the globe, though he died *en route*.

Marconi Guglielmo, Marchese (1874–1937). Italian inventor and pioneer of the development of radio. Inspired by the work of *Hertz on electromagnetic waves, and developing a practical antenna and receiver (coherer), in 1895 Marconi transmitted Morse code messages by wireless telegraphy over a distance of 1.6 km (1 mile) near Bologna. He patented his system in England in 1896; formed a company for its commercial exploitation in 1897; and transmitted signals across the English Channel in 1899, and across the Atlantic in 1901. Later he worked on short-wave transmissions. He shared the 1909 Nobel Prize for Physics with *K. Braun.

Mauchly John William (1907–80). US research engineer and physicist, with *Eckert pioneer of the electronic digital computer.

Maudslay Henry (1771–1831). English engineer and machine-tool builder. He designed a compact power unit, or 'table engine' (1807), built *M. Brunel's block-making machinery at Portsmouth, and made important improvements to lathe design. He trained such distinguished engineers as *Nasymth and Whitworth.

Maxwell James Clerk (1831–79). Scottish physicist and mathematician, creator of the electromagnetic theory of light. In such powerful works as his *Treatise on Electricity and Magnetism* (1873), he interpreted mathematically *Faraday's concept of the electromagnetic field: he identified light as being a form of electromagnetic radiation and postulated the existence of the radio waves later discovered experimentally by *Hertz. Maxwell made major contributions to the study of heat and the kinetic theory of gases. As a creative and imaginative genius, he ranks with *Newton and Einstein.

McAdam John Loudon (1756–1836). Scottish road engineer. Adopting the principles of *Trésaguet, he devised the macadamized road in which a comparatively light pavement protects the underlying soil, the real load bearer. His system was rapidly adopted in Europe.

Mège Mouriés Hippolyte (1817–80). French chemist and manufacturer, inventor of margarine. His first tallow-based recipe (patent 1869) yielded an unappetizing product but won a French government prize for a butter substitute suitable for the army and the poorer members of the community: it was soon improved by the addition of emulsified skim milk.

Meitner Lise (1878–1968). Austrian physicist who, with *Frisch early in 1939, identified and described nuclear fission in uranium. They reached their momentous conclusions after studying a report of his bombardment of uranium atoms with neutrons by her former collaborator Otto Hahn.

Mendel Gregor Johann (1822–84). Austrian monk and botanist, the pioneer of modern genetics. His interest in hybridization in plants led to a series of brilliant experiments on the common edible pea. Based on controlled pollination and careful statistical analysis, they led him to postulate that an inherited characteristic is determined by two hereditary units (later termed genes), one from each parent. He published his results in 1866, but unfortunately only in an obscure local natural history journal, and it was only after their independent rediscovery in 1900 by *de Vries and others that their importance was recognized.

Mercator Gerardus (Gerhard Kremer) (1512–94). Flemish mathematical instrument maker, geographer and cartographer. He directed his own institute of geography in Louvain from 1534 to 1592, when he became professor at Duisburg. He produced numerous maps and, in 1569, his chart of the world, using the famous projection named after him. This gave true bearings between any two points.

Messel Rudolph (1848–1920). German industrial chemist and manufacturer. In 1875, he patented, with W. S. Squire, an improved (contact) process for the manufacture of sulphuric acid.

Montgolfier the brothers Joseph Michel (1740–1810) and Jacques Etienne (1745–99). French pioneers of flight. Paper manufacturers from Annonay, near Lyons, they sent up their first hot-air balloon in 1782. A small-scale model, it was followed in 1783 by one which carried three animals and, later that year, one carrying two passengers.

Morgan Thomas Hunt (1866–1945). US geneticist and embryologist, originator of the gene theory. He developed the chromosome theory of inheritance (1911), which satisfactorily accounted for the simple laws described by *Mendel in 1865.

Morse Samuel Finley Breese (1791–1872). US artist and inventor, pioneer in America of the electric telegraph and deviser of the Morse code. His telegraph had been anticipated in Europe, however, notably by the 1837 *Wheatstone-Cooke patent. Nevertheless, encouraged by a Congress commission (1844), Morse was to make a fortune from his US patents.

Murdock William (1754–1839). English engineer, pioneer of coal-gas lighting. An employee of *Boulton and *Watt, he illuminated with coal-gas a house in Redruth, Cornwall, in 1792. This led to Boulton and Watt taking a serious interest in gas lighting, and they recalled him to Birmingham. Unfortunately, Boulton and Watt abandoned their interest in 1814.

Napier John, Laird of Merchiston (1550–1617). Scottish mathematician, inventor of logarithms. His publications included *Mirifici logarithmorum* (1614) and *Rabdologia* (1615), on systems of arithmetical calculation with mechanical aids (Napier's Bones).

Nasmyth James (1808–90). Scottish engineer, inventor of the steam hammer (1839), which made it possible to forge much larger items than previously. Trained under *Maudslay, he set up business in Manchester (1834) as a manufacturer of steam locomotives and machine tools.

Natta Giulio (1903–79). Italian chemist who developed a catalytic process for the manufacture of polypropylene. A university teacher and from 1938 director of the Milan Institute of Industrial Chemistry, in 1954 he derived his high-strength plastic by polymerizing propylene. He shared the 1963 Nobel Prize for Physics with *Ziegler.

Newcomen Thomas (1663–1729). English inventor of the first effective steam engine. An ironsmith by trade, Newcomen developed *Savery's Miner's Friend into a beam-engine incorporating a piston and cylinder and operating at atmospheric pressures. The first model began working in 1712 and for 70 years Newcomen-type engines were widely used in mines throughout Europe.

Newton Sir Isaac (1642–1727). English physicist, natural philosopher and mathematician, renowned for his formulation of the laws of motion and the theory of gravitation, summarized in the *Principia* (1687). His researches in optics revealed Newton as a master of experiment; with the aid of a prism, he demonstrated white light to be a mixture of differently coloured rays, and advanced a corpuscular theory of light. His *Optics* (1704) was an exceptionally influential work. Newton developed the calculus of variations and formulated the binomial theorem. Newtonian physics dominated science for two centuries, and he ranks among the most original thinkers of all time.

Niepce Joseph Nicéphore (1765–1833). French pioneer of photography: he made his first experiments, while an army officer, in 1793. Niepce took the world's first surviving photograph (*héliographie*, 'sun drawing') in 1826, using a camera obscura and a pewter plate coated with a bitumen solution which hardened on exposure to light. Bitumen on the dark parts, remaining soluble, was washed away with a solvent. The exposure, however, was some eight hours even in strong sunlight and only a single picture could be made. In 1829 Niepce went into partnership with *Daguerre, who subsequently succeeded in developing a commercially viable process.

Nipkow Paul Gottlieb (1860–1940). German inventor (1884) of an optical scanning system employing his patent Nipkow disc and selenium (known, since 1873, to be photoconductive). Pierced with a spiral of holes, the disc scanned an image of one revolution, the light passing through it being focused by a lens on the selenium-coated plate which registered the varying light intensities. Nipkow's dream of 'seeing by telegraphy' was realized in 1926 by *Baird's photo-mechanical television which incorporated his disc.

Nobel Alfred Bernhard (1833–96). Swedish chemist, inventor and philanthropist. Son of an armaments manufacturer and inventor, he perfected dynamite (1866–7) in which kieselguhr was combined with a nitroglycerine-based explosive to make it safer in working. In 1875 he improved on dynamite with a blasting gelatine which combined nitroglycerine with guncotton. On this basis he established a highly profitable international business. Nobel also had large holdings in the oil industry. A man of wide general culture, and disturbed by the military potential of his inventions, he bequeathed his large fortune to establish annual prizes in physics, chemistry, physiology or medicine, literature and the promotion of international peace. The first were awarded in 1901.

Oersted Hans Christian (1777–1851). Danish physicist and chemist, pioneer of electro-magnetism. Professor at Copenhagen from 1806, in 1820 he discovered the magnetic effect of an electric current. He also investigated thermoelectric effects.

Ortelius Abraham (1527–98). Flemish geographer, friend of *Mercator, who inspired his principal work, the atlas *Theatrum orbis terrarum* (1570). Many geographers believed a southern continent must exist simply to balance the world's land masses: the word 'Australia' derives from Ortelius's term for it.

Otis Elisha Graves (1811–61). US engineer and inventor who built (1857) the first modern passenger lift. It incorporated his automatic safety device (1852) to prevent the cage falling in the event that the cable broke.

Otto Nikolaus August (1832–91). German engineer who introduced (1876) the four-stroke cycle basic to most modern internal combustion engines. Unfortunately, his patent for this was invalidated in 1886. In 1867 he and Eugen Langen had demonstrated the first successful gas engine.

Papin Denis (1647–*c.* 1712). French scientist and inventor whose 'atmospheric engine' of 1690 pioneered the principles of the steam engine. Assistant both to *Huygens and *Boyle, he was professor of mathematics at Marburg, 1687–96. Papin's 'digester' (1679) anticipated the modern domestic pressure cooker and the hospital steam sterilizer.

Pappus (late 3rd century AD). Greek mathematician of Alexandria whose *Synagoge* (Mathematical Collection) is the principal source for our knowledge of much ancient mathematics.

Parsons Sir Charles Algernon (1854–1931). British engineer and manufacturer, designer of the first successful industrial and marine steam turbines. The 1884 Parsons' turbine coupled to a Parsons-designed dynamo could generate 7.5 kW at 100 volts. His company was to install improved high-voltage turbo-generators in power stations in Britain and abroad and build sophisticated geared turbines. In 1897, his experimental vessel *Turbinia* – spectacularly demonstrated at the 1897 Spithead Review – achieved speeds of 34 knots and Parsons marine turbines were soon adopted by the Royal Navy and the Cunard steamship company. Interested in optics, he also developed numerous types of high-performance optical glass and built large reflecting telescopes.

Pascal Blaise (1623–62). French mathematician and physicist who, at the age of 19, built a calculating machine. As a mathematician Pascal was an infant prodigy and before he was 16 he made important contributions to the projection geometry of conics. He also contributed to the mathematical theory of probability. As an experimentalist he is remembered for his work on hydrostatics, forging a link between the mechanics of fluids and solids. He extended the work of *Torricelli on the barometer and demonstrated – by having a barometer carried up the Puy de Dôme – that the pressure of the atmosphere diminishes with height. Pascal's Principle states that in a fluid at rest pressure is transmitted equally in all directions.

Pasteur Louis (1822–95). French chemist, renowned as the pioneer of vaccines against infectious diseases, including anthrax and (1885) rabies. Studies in natural fermentations led to his demonstration that they are caused by micro-organisms (1857); later he showed that such organisms, or bacteria, are also the cause of certain diseases. His work led to the Pasteurization process of sterilization. His research discredited the traditional belief in spontaneous generation. He also did important work on the structure of organic molecules, demonstrating that certain crystals could exist in two forms, which were mirror images of each other (stereochemistry).

Perkin Sir William Henry (1838–1907). English chemist, pioneer of the modern dye industry, with his accidental discovery of mauve (1856) or 'aniline purple', the first aniline dye. He opened a factory for its manufacture, improved methods of silk dyeing and patented a process for the synthesis of alizarin, an important natural dye. In the case of the latter, he was anticipated by one day by three German chemists, but a friendly accommodation was reached. He also did important work on molecular structures of organic compounds. His son, William Henry Perkin (1860–1929), who studied chemistry with Baeyer at Munich, became influential as research chemist and teacher at Edinburgh, Manchester and, from 1912, Oxford.

Playfair Lyon, 1st Baron (1818–98). Scottish chemist and statesman of science. One of the organizers of the Great Exhibition (1851), he realized that despite its outstanding success this was the end of an era for Britain and urged reforms of Britain's technical education system as a means of retrieving the situation.

Pliny the Elder (Gaius Plinius Secundus) (*c.* AD 23–79). Roman naturalist and encyclopaedist. His *Historia naturalis* is a vast compilation from other authors (many of whose works are now lost) covering geography, anthropology, zoology, botany, mineralogy and the nature of the physical universe. It is a valuable record of classical learning in the natural sciences and includes various practical observations, e.g. on medicinal properties of plants, mining technology and the manufacture of pigments.

Polhem Christopher (1661–1751). Swedish engineer, pioneer, at his Stjärsund works, of advanced water-powered iron-working machinery. Polhem also directed an experimental mechanical workshop, and supervised docks and canal works: he has been called 'the father of Swedish mechanics'.

Polo Marco (*c.* 1254–*c.* 1324). Venetian traveller whose book of his journeys through central Asia, China and South East Asia (1271–95), was a prime Western source on these regions for centuries.

Poncelet Jean Victor (1788–1867). French mathematician and army engineer, a pioneer in projective geometry and the theory of turbines. Taken prisoner (1812–14) during Napoleon's Russian campaign, he subsequently taught at Metz (establishing a course in applied mechanics) and Paris, at the Faculty of Sciences and at the École Polytechnique.

Popov Alexander Stepanovich (1859–1906). Russian physician and electrical engineer, a pioneer of radio. Working at the Kronstad naval base – contemporaneously with *Hertz at Karlsruhe – he experimented with electromagnetic waves, building a primitive transmitter and receiver in 1896. Within two years he had developed a ship-to-shore radio communications system.

Ptolemy, Claudius Ptolemaius (2nd century AD). Alexandrian astronomer, mathematician and geographer. His 'Mathematical Treatise' (or the *Almagest*, Arabic 'Great Book') remained the standard work on astronomy until the 17th century. Brilliant mathematics enabled Ptolemy to interpret the motions of the heavenly bodies on a geocentric theory, postulating that they moved in perfect circles ('epicycles') whose centres moved round the earth, also in perfect circles ('deferents'). His *Geography*, virtually unquestioned until the 16th century, contained errors and inaccuracies resulting often from questionable data. He also wrote on optics, describing the phenomena of reflection, refraction and acoustics, and made important contributions in trigonometry.

Roebling John Augustus (1806–69). German-born engineer, who pioneered the manufacture of steel cable and designed many notable suspension bridges. His Niagara Falls bridge was completed in 1855. He died in an accident shortly after work began on his famous Brooklyn Bridge.

Roebuck John (1718–94). Scottish chemical engineer and inventor whose lead chamber process for the manufacture of sulphuric acid (1749) continued in use into the 20th century, until displaced by the contact process of *Messel.

Rømer Ole (1644–1710). Danish astronomer, the first man to measure the speed of light, on the basis of astronomical observations. He made his calculation while an assistant at the Royal Observatory, Paris (1672–9).

Röntgen Wilhelm Conrad (1843–1923). Outstanding German physicist noted for his discovery (1895) of X-rays, which were quickly adapted as the basis of one of the most valuable of all techniques of medical diagnosis, besides finding many other practical applications.

Rumford Benjamin Thompson, Count (1753–1814). American-British scientist and man of affairs. Experiments with the boring of cannon (he directed the Arsenal at Munich in the 1790s) led him to a theory relating the mechanical work done to the heat produced. He thus anticipated, on an essentially qualitative basis, the precise later work of *Joule on the mechanical equivalent of heat. In 1800 he founded the prestigious Royal Institution in London, but was unhappy there and settled in Paris, where he married the widow of *Lavoisier (1805).

Rutherford Ernest, 1st Baron (1871–1937). New Zealand nuclear physicist, winner of the 1908 Nobel Prize for Chemistry for research on radiation. Rutherford identified the alpha particle (positively charged helium nucleus) and beta particle (negatively charged electron), using the former (1919) to bombard nitrogen atoms to detach hydrogen atoms. He formulated the nuclear model of the atom with orbital electrons, which was developed by *Bohr, invoking Planck's quantum theory, to explain how such a system could remain stable (1912). Under Rutherford (1919–37) the Cavendish Laboratory at Cambridge led the world in nuclear physics.

Savery Thomas (*c.* 1650–1715). English military engineer who in 1698 patented a steam 'engine' – essentially a pump – for raising water 'by the impellent force of fire'. It was not a great success because the technology of the day made it difficult to contain the high-pressure steam needed. It was soon replaced by *Newcomen's beam engine, incorporating a piston working in a cylinder.

Siemens Sir Charles William (1823–83) German-born British engineer and co-inventor (1861) with Friedrich (Frederick) von Siemens of the open-hearth steel process, later improved in conjunction with Pierre Martin. By 1900 more steel was being made by this process than by that of *Bessemer.

Sikorsky Igor Ivanovich (1889–1972). Russian-born US pioneer of the helicopter and of aviation design. In 1919 he founded his own aircraft company in the USA and from the mid-1930s took a particular interest in helicopters. His first major success was the VS 300 (1939) which went into production in 1941: more than 400 were used by the Allied forces during the Second World War.

Singer Isaac Merritt (1811–75). US industrialist and pioneer of sewing-machine manufacture. His 1851 model was held to infringe Elias Howe's 1846 patent, and after litigation he had to come to terms with him. However, Singer patented numerous improvements, including the foot treadle, designed a standard model and went into large-scale and very profitable production.

Smeaton John (1724–92). English engineer. A successful instrument maker, he turned to engineering and toured the Low Countries (1755) to study harbours and drainage works. His famous Eddystone Lighthouse (1756–9) was followed by the Forth–Clyde Canal and many bridges. He made important improvements in the design of water- and windmills and of steam engines by conducting careful experiments with models – an important departure from the empirical approach of his predecessors.

Soddy Frederick (1877–1956). English chemist who first identified isotopes (coining the word), as atoms that could be chemically identical and yet have different weights. He received the 1921 Nobel Prize for Chemistry.

Solvay Ernest (1838–1922). Belgian industrial chemist, inventor (1861) of the Solvay process for the manufacture of soda from ammonia and brine. For 50 years chemical engineers, looking for an alternative to the *Leblanc process with its noxious spoil heaps of alkali waste, had been frustrated by the engineering problems caused by the volatility of ammonia. Assisted by his brother's business acumen, Solvay, who took several years to perfect his process, patented not only his crucial innovation, the carbonating tower, but every other stage and was soon granting licences worldwide. Nevertheless, despite its technical inferiority, the Leblanc process survived into the 20th century.

Spode Josiah (1754–1827). English pottery manufacturer who expanded the flourishing company founded by his father, also Josiah (1733–97), becoming Royal Potter in 1806. He continued the firm's underglaze blue transfer-printed ware and about 1800 introduced a porcelain based on feldspar and bone-ash.

Sprengel Hermann Johann Philipp (1834–1906). German-born British chemist and physicist. Holder of many patents on the new high explosives, he is also noted for the invention (1865) of a mercury air pump of much improved efficiency. This proved of great technological importance as it provided an effective and economic means of producing the high vacuum necessary for incandescent filament electric lamps, and, later, thermionic radio valves.

Stephenson George (1781–1848). English engineer and pioneer of the railway. He built his first locomotive as enginewright at Killingworth colliery (1814). Surveyor for the Stockton–Darlington Railway (1825), which used his *Locomotion* with patent steam-blast, he surveyed the Liverpool–Manchester Railway and, in 1829, won its Rainhill trials with his *Rocket*: the latter embodied all essential features of future steam locomotives, including a tubular boiler. In the 'railway mania' that followed, Stephenson was called in as engineer by other railway companies both in Britain and abroad.

Swan Sir Joseph Wilson (1828–1917). English chemist and inventor of the incandescent filament electric lamp. Working first for a photographic chemical supplier, he invented the carbon (Autotype) process for making permanent prints. He demonstrated his carbon filament vacuum lamp in Newcastle-upon-Tyne in 1878. It was the result of years of experiments, and its development was facilitated by the availability of *Sprengel's mercury vacuum pump and the development of dynamos as a source of electricity. In 1882, he successfully contested a charge of patent infringement by *Edison and in 1883 the Edison and Swan United Electric Light Company was formed.

Swedenborg Emanuel (1688–1772). Swedish engineer, scientist and mystic. Renowned as founder of the Swedenborgian religious sect, he was also an active man of affairs, being appointed assessor of the Board of Mines in 1715. His handsomely illustrated *Regnum subterraneum* (1734) is an authoritative review of the mining and smelting technology of his day.

Swinburne Sir James (1858–1958). Scottish electrical engineer and inventor, his many patents including improvements to dynamo design and electric lamps. He founded his own company in London in 1899. Beaten by one day by *Baekeland to his patent for a phenol-formaldehyde thermo-setting plastic, Swinburne did patent a highly profitable synthetic lacquer (Damard), and later became chairman of Bakelite Ltd, which absorbed his Damard Lacquer Co. in 1910.

Talbot Henry William Fox (1800–77). English landowner and scientist, pioneer of the negative–positive principle on which modern photography is based. His first negative picture, on silver chloride paper and fixed with common salt, dates from August 1835. Capable of yielding multiple copies from a single original – his *The Pencil of Nature* (1844) was the first book to be illustrated with photographs – it was a radical advance on the single-picture system of *Daguerre and *Niepce.

Telford Thomas (1757–1834). Scottish civil engineer. In 1793 he was appointed engineer of the Ellesmere Canal (on the English-Welsh borders), famous for its great Chirk and Pont-y-Cysylltau aqueducts. Following his 1802 survey of the Scottish Highlands, Telford built various harbours, some 1450 km (900 miles) of road, 1200 bridges (making much use of cast iron), and the Caledonian Canal. Other major achievements included the Gotha Canal, Sweden, and the Menai suspension bridge – crowning glory of his London–Holyhead road. Magnificently surveyed (maximum gradient 1:30), this road combined solid foundations with good drainage: as a road builder he ranks second only to *McAdam. Telford also founded Britain's Institute of Civil Engineers.

Tesla Nicola (1856–1943). Croatian-born US physicist and electrical engineer who, in the 1880s, devised the first a.c. induction motor. He settled in America in 1884, working first for Edison and then Westinghouse. His motor, and improvements to transmission systems, strongly influenced the electrical industry.

Theophrastus (c. 327–c. 285 BC). Greek philosopher and scientist, *Aristotle's successor as head of the Lyceum. His *Enquiry into Plants* and *Causes of Plants* made important contributions to classification and the study of plant distribution and ecology. He is rightly regarded as the father of scientific botany.

Thomson Elihu (1853–1937). English-born electrical engineer and inventor; head (1920–3) of the Massachusetts Institute of Technology. With E. J. Houston he founded the Thomson-Houston Electric Co. at Lynn, Mass., a forerunner of the giant General Electric Co.

Thomson William, 1st Baron Kelvin (1824–1907). Irish physicist, formulator, with *Joule, of the Second Law of Thermodynamics. Professor at Glasgow (1846–99), Thomson developed *Faraday's electromagnetic field theory and assisted materially with the design of the first transatlantic cable (1866). He made important contributions to low temperature physics: the Kelvin scale of absolute temperature is named after him.

Torricelli Evangelista (1608–47). Italian physicist and mathematician. An assistant to *Galileo and then professor of mathematics at Florence, Torricelli made important contributions in dynamics and hydrodynamics. His demonstration of atmospheric pressure with a mercury-filled tube (1643) led to the development of the mercury barometer. A skilled mechanic, he constructed microscopes and telescopes.

Trésaguet Pierre Marie Jérôme (1716–96). French road engineer. Appointed Inspector-General of Roads and Bridges from 1775, he introduced a lighter and cheaper form of road construction than any then in use. A foundation of large stones was covered by cambered layers of other stones of diminishing size, all firmly compacted.

Trevithick Richard (1771–1833). English engineer and inventor, pioneer of high-pressure steam engines. Son of a Cornish mine captain, he began with improvements to the Boulton and Watt pumping engines. He built the world's first steam carriage (1801). In 1804 he built for the Pen-y-Darran ironworks a locomotive incorporating an important modification to improve the draught in the fire-box by diverting waste steam up the chimney. He made and lost fortunes, as steam engineer and prospector, in South America (1816–27), but eventually returned home, to die in poverty.

Tsiolkovsky Konstantin Eduardovich (1857–1935). Russian pioneer of spaceflight and aerodynamics. His *Exploration of Cosmic Space by Means of Reaction Devices* (1903) outlined the mathematical principles of rocket travel. He presented a design for a multi-stage rocket in the 1920s. The Russian launch of the first artificial satellite, Sputnik 1, was planned for the centenary of his birth but was in fact three weeks late.

Tull Jethro (1674–1741). English agricultural innovator. He developed a seed-drill (1701) – possibly on the basis of an Italian version of a Chinese original – and in *The Horse-hoeing Husbandry* (1733) recommended pulverization of the soil, row culture and regular hoeing to keep down weeds.

Vauban Sebastien, Marquis de (1633–1707). Marshall of France. Military and civil engineer. He systematized siege warfare (successive parallel trenches were advanced towards the beleaguered fortress by zigzag connecting trenches), designed hundreds of fortifications calculated on geometric principles to maximize enfilading fire, and built many waterways.

Vavilov Nikolai Ivanovich (1887–1943). Russian botanist and geneticist, noted for his studies of cultivated plants, especially cereals. He achieved high academic distinction but his opposition to approved Communist Party doctrine on genetics, identified particularly with Lysenko, led to political victimization.

Vesalius Andraes (1514–64). Flemish anatomist, professor at Padua University (1540s) and subsequently physician to Emperor Charles V and Philip II of Spain. His superbly illustrated *De humani corporis fabrica* (1534; 'The Structure of the Human Body') revealed his skill as a dissector and acute observation and was the first book to contest *Galen. With his assertion that the interventricular septum is not pierced by even the minutest channels, he came close to discovering the circulation of the blood (Harvey, 1628).

Vitruvius (Marcus Vitruvius Pollio) (1st century BC–1st century AD). Roman architect, mechanical and civil engineer. His *De architectura* was an encyclopaedic survey of classical architecture and civil engineering. Its topics included the Greek orders of architecture; city planning; the construction and interior decoration of public and private buildings; the construction of machines, both civil and military; the design of water-wheels and filtration beds for water purification; and timepieces. It became a standard reference book for Renaissance architects and engineers.

Volta Alessandro (1745–1827). Italian physicist, pioneer in the study of electricity; professor of physics at Pavia University. He invented the voltaic cell (1800), in which plates of dissimilar metals are immersed in acid or salt solution. He also did important research on the effect of electricity on animal muscles.

Von Neumann John (1903–57). Hungarian-born US mathematician, a pioneer in the design of electronic computers. He greatly speeded operations by replacing the expression of computer words as a succession of pulses on a single wire by pulses on a group of wires in parallel, each wire carrying a 'bit' in the word. This development was further supported by the advent of the magnetic core store (memory) replacing the dielectric store.

Vries Hugo de (1848–1935). Dutch plant physiologist, noted for his work on heredity and genetics, professor at Amsterdam, 1888–1918. Systematic experiments led him to the rediscovery (1900) of *Mendel's laws of heredity and to his own theory of mutations as the cause of new species in the process of evolution: this had to be qualified with the discovery that some mutants are due not to gene mutation but to rearrangements of chromosomes.

Watson-Watt Sir Richard Alexander (1892–1973). Scottish pioneer of radar. During the First World War, he worked in meteorology and the radio detection of thunderstorms hazardous to aviators. In 1921 he became head of the government-sponsored radio research station at Slough. In 1935 he presented proposals for locating aircraft by radio-pulse technique. The idea itself was not new; his achievement was to develop a system of great accuracy operable by non-specialists under war conditions. Britain's radar defences, the world's first, were largely in place by the outbreak of the Second World War in 1939.

Watt James (1763–1819). Scottish engineer and inventor, pioneer of the improved steam engine. His experiments began when, as an instrument maker at Glasgow University, he was asked to repair a demonstration model of a *Newcomen engine and was impressed by its poor performance. In 1769 he patented his own design incorporating a separate condenser and in 1775 entered into partnership with *Boulton of Birmingham to manufacture it. Later Boulton and Watt improvements included the double-acting rotative engine; sun-and-planet gearing to convert reciprocal into rotary motion; and a centrifugal governor to maintain constant speed at varying loads. For a quarter of a century the Boulton and Watt partnership dominated the world of steam.

Wedgwood Josiah (1730–95). English potter and industrialist. His improved 'useful' wares, such as the cream (also Queen's) ware, ensured commercial success; his elegant ornamental wares, notably the porcelain-like Jasper stoneware, brought prestige. At his Burslem factory (Etruria), he built a model village for his work people.

Weizmann Chaim (1874–1952). Russian-born chemist, renowned Zionist leader and first President of Israel. Graduating in Berlin (1900), he later taught at Manchester University and was director of Britain's Admiralty laboratories (1916–19). In the First World War he invented a crucially important process for producing acetone (for explosives' manufacture) by a fermentation process. In later years he founded Israel's Weizmann Institute of Science.

Wells Horace (1815–45). US dentist, a pioneer of anaesthesia who successfully used nitrous oxide ('laughing gas') in 1844. But ether and chloroform were rapidly adopted as alternatives and Wells, depressed by failure to recognize his innovation, committed suicide.

Welsbach Carl Auer von (1858–1929). Austrian chemist who patented the incandescent gas mantle (1885), utilizing the fact that the oxides of certain rare metals (ceria, thoria) emit a bright light when strongly heated.

Westinghouse George (1846–1914). US engineering inventor, incorporating the Westinghouse Electric Company in 1886. His many important inventions included a rail airbrake (1868), various pneumatic devices and automatic signalling. He also pioneered the adoption in America of high-voltage a.c. electric transmission.

Wheatstone Sir Charles (1802–75). English physicist and inventor, co-patentee with *Cooke of the electric telegraph (1837), and professor of experimental physics at King's College, London (from 1834). He made improvements to the dynamo, developed the Wheatstone bridge for measuring electrical resistances and invented an automatic telegraph transmitter and an electric recording machine.

Whinfield John Rex (1901–66). English research chemist, inventor of Terylene. Trained under *Cross, inventor of rayon, in the 1930s, he was encouraged by *Carothers's discovery of nylon (polyamide) to concentrate on polyesters; in 1941, with J. T. Dickson, he derived Terylene (Dacron) from terephthalic acid and ethylene glycol. This established a large worldwide market after the Second World War.

Whitehead Robert (1823–1905). English engineer and inventor of the torpedo (1866). After working for his family's firm in Marseilles, he set up his own company in Fiume in 1856. From 1866 he modified and improved his torpedo until it was eventually capable of some 53 kph (29 knots/33 mph), with an effective range of 1000 metres (3280 feet). Its military importance was demonstrated in the defeat of the Russian fleet by the Japanese in 1904.

Whitney Eli (1765–1825). US inventor of the cotton gin and a pioneer of mass-production methods. A Boston-born lawyer, he developed a natural bent for invention on a visit to Georgia when, invited by his hosts, he devised a machine to clean green seed cotton – complete with revolving cylinder, brushes and a fan – in just ten days. His 1794 patent was widely infringed and cotton production in the South soared. In 1798, winning a large government contract for muskets, Eli produced precision-made interchangeable parts that could be assembled by unskilled labour.

Winsor (Winzer) Frederick Albert (1763–1830). German promoter of gas illumination. In London in 1804, he patented a system for manufacturing 'inflammable air' from coal and founded a company – eventually the London Gas Light and Coke Company – for gas illumination.

Wollaston William Hyde (1766–1828). English pioneer of powder metallurgy for producing malleable platinum, and discoverer of rhodium and palladium; Secretary of the Royal Society. After systematic private experiments, he devised a high-pressure sintering process for working the previously very intractable metal platinum, which found a ready market in the chemical industry, especially among sulphuric acid manufacturers.

Wright the brothers Wilbur (1867–1912) and Orville (1871–1948). US builders of the world's first successful aeroplane. Stirred by *Lilienthal's achievements, the brothers flew their first glider in 1900; wind tunnel tests followed. They designed a 12-hp petrol engine to power two rear-mounted airscrews in an aeroplane steered by ailerons and a vertical rudder. It first flew at Kitty Hawk on the morning of 17 December 1903. In 1905 they flew a 40-km (25-mile) circuit. An improved model was patented in 1906 and in 1908 Wilbur went to France, licensing a French syndicate to build it there. In 1909 the US Wright Company was founded on a military contract.

Young James (1811–83). Scottish chemist and entrepreneur who made his fortune as the first to market paraffin lighting oil (1856), derived from shale. This was the basis of a profitable new industry, though later eclipsed by the advent of petroleum.

Zeiss Carl (1818–88). German manufacturer of optical instruments, founder of the world-famous Zeiss Company of Jena (1846). From 1866 Ernest Abbe (1840–1905), professor of physics at Jena, was an associate, becoming owner of the company in 1888. He made important improvements in camera and microscope lenses, notably in eliminating chromatic aberration by using compound lenses containing different glasses.

Zeppelin Ferdinand, Graf von (1838–1917). German inventor who built the first rigid airships. An army officer, he had retired in 1891 to pursue research. In 1908 he founded his Zeppelin company at Friedrichshafen for the manufacture and exploitation of airships: until the outbreak of the First World War he carried 25,000 passengers without accident.

Ziegler Karl (1898–1973). German chemist who, in 1953, devised the catalytic process for the commercial production of polythene at ordinary temperatures and pressures from ethylene. Using Ziegler's catalyst *Natta developed a similar process for polypropylene. The two men shared the 1963 Nobel Prize for Chemistry.

Zworykin Vladimir Kosma (1889–1982). Russian-born US electronic engineer, pioneer of television technology. Emigrating to the US in 1919, he joined Westinghouse in the hope of developing his ideas on electronic television: when he found that their interest was mainly in radio, he resigned but rejoined them in 1923, with important rights in respect of his own patents. In 1923 he devised the iconoscope, the first practical television camera. Twelve years later, with G. A. Morton and L. Malter, he invented the multipler phototube, a highly efficient current amplifier than can be linked with a cathode ray tube.

Bibliography

The literature on the history of technology is enormous and constantly increasing This short bibliography is intended not for the specialist, but for the general reader who may need some preliminary guidance in order to study particular aspects in greater detail. To keep it within bounds, it is restricted to books which have been published during roughly the last 50 years in English – though in a few instances translated from foreign-language originals – and which are fairly readily accessible through libraries. Most of the works cited themselves contain extensive bibliographies of a more specialized nature.

A great deal of relevant information is to be found not in books but in a wide range of scholarly journals. Of these the most important – again restricting the choice to general works published in English – are the *Transactions of the Newcomen Society* (from 1920); *Technology and Culture* (from 1959); and *History of Technology* (from 1976).

General

Barnes, Harry Elmer, *The History of Western Civilisation*, Harcourt Brace, New York, 1935

Birdsall, Derek and Cipolla, Carlo M., *The Technology of Man: a Visual History*, Wildwood House, London, 1980

Cardwell, Donald S. L., *Technology, Science, and History*, Heinemann, London, 1972

Carter, E. F., *Dictionary of Inventions and Discoveries*, Philosophical Library, New York, 1967

Chinese Academy of Sciences, *Ancient China's Technology and Science*, Foreign Languages Press, Beijing, 1983

Daumas, Maurice (ed.), *Histoire Générale des Techniques* (5 vols.), Presses Universitaires de France, Paris, 1962–79 (vols. 1–3 in English translation, Crown, New York, 1969)

Derry, T. K. and Williams, Trevor I., *A Short History of Technology from the Earliest Times to AD 1900*, Clarendon Press, Oxford, 1960

Eco, Umberto and Zorzoli, G. B., *The Picture History of Inventions: from Plough to Polaris*, Weidenfeld and Nicholson, London, 1962

Forbes, R. J., *Studies in Ancient Technology* (6 vols.), Brill, Leiden, 1955-8

Forbes, R. J., *Man the Maker*, Constable, London, 1958

Forbes, R. J. and Dijksterhuis, E. J., *A History of Science and Technology*, Penguin Books, Baltimore, 1963

Giedion, S., *Mechanization Takes Command*, Oxford University Press, New York, 1948

Gillispie, C. C. (ed.), *A Diderot Pictorial Encyclopedia of Trade and Industry*, Dover Publications, New York, 1959

Habakkuk, H. H., *American and British Technology in the Nineteenth Century*, Cambridge University Press, London, 1962

Hill, Donald, *A History of Engineering in Classical and Medieval Times*, Croom Helm, London, 1984

Hodges, H., *Artifacts: an Introduction to Primitive Technology*, John Baker, London, 1964

Kirby, R. S., Withington, S., Darling, A. B. and Kilgour, F. G., *Engineering in History*, McGraw-Hill, New York, 1956

Klemm, F. A., *A History of Western Technology*, Allen and Unwin, London, 1959

Kranzberg, Melvin and Pursell, Carroll W., Jr (2 vols.), *Technology in Western Civilization*, Oxford University Press, New York, 1967

Landels, J. G. *Engineering in the Ancient World*, Chatto and Windus, London, 1978

Larsen, Egon, *A History of Invention*, Roy, New York, 1969

Li Guohao, Zhang Mengwen and Cai Tianqin (eds.), *Explorations in the History of Science and Technology in China*, Shanghai Chinese Classics Publishing House, Shanghai, 1982

Lilley, S., *Men, Machines, and History*, Cobbett, London, 1948

Mumford, L., *Technics and Civilization*, Harcourt, New York, 1934

Musson, A. E. and Robinson, Eric, *Science and Technology in the Industrial Revolution*, Manchester University Press, Manchester, 1969

Needham, J., *Science and Civilisation in China* (vols. 1–6), Cambridge University Press, Cambridge, from 1954

Neuberger, Albert, *The Technical Arts and Sciences of the Ancients*, Methuen, London, 1930 (reprinted 1969)

Oakley, Kenneth, *Man the Toolmaker*, British Museum, London, 1950

Oliver, John, W., *History of American Technology*, Ronald Press, New York, 1956

Piggott, Stuart (ed.), *The Dawn of Civilization: the First World Survey of Human Cultures in Early Times*, Thames and Hudson, London, 1961

Rochlin, Gene I. (ed.), 'Scientific Technology and Social Change', *Scientific American*, Freeman, Reading, 1974

Ronan, Colin A., *The Shorter Science and Civilisation in China* (3 vols.), Cambridge University Press, Cambridge, from 1978

Ronan, Colin A., *The Cambridge Illustrated History of the World's Science*, Cambridge University Press/Newnes Books, Cambridge, 1983

Semenov, S. A., *Prehistoric Technology*, Cory, Adams and Mackay, London, 1964

Singer, Charles, Holmyard, E. J., Hall, A. R. and Williams, Trevor I. (eds.), *A History of Technology* (8 vols.), Clarendon Press, Oxford, 1954–84

Sung Ying-hsing, *Chinese Technology in the Seventeenth Century*, Pennsylvania State University Press, 1966

Taylor, Gordon Rattray (ed.), *The Inventions that Changed the World*, Reader's Digest, London, 1982

Temple, Robert, K. G., *China: Land of Discovery*, Patrick Stephens, Wellingborough, 1986

Usher, A. P., *A History of Mechanical Inventions* (2nd edn), Harvard University Press, Cambridge, Mass., 1954

Wenke, Robert J., *Patterns in Prehistory: Mankind's First Three Million Years*, Oxford University Press, New York, 1980

Wolf, F., *History of Science, Technology and Philosophy in the 16th and 17th Century* (2nd edn), Allen and Unwin, London, 1950

Wolf, F., *History of Science, Technology and Philosophy in the 18th Century* (2nd edn), Allen & Unwin, London, 1952

White, Lynn Jr., *Medieval Technology and Social Change*, Clarendon Press, Oxford, 1962

Williams, Trevor I., *A Short History of 20th Century Technology: c. 1900–c. 1950*, Clarendon Press, Oxford, 1982

Medicine

Major, Ralph H., *A History of Medicine* (2 vols.), Thomas, Springfield, 1954

Parish, H. J., *A History of Immunization*, Livingstone, Edinburgh, 1965

Wain, Harry, *History of Preventive Medicine*, Thomas, Springfield, 1970

Mechanical Engineering

Armytage, W. H. G., *A Social History of Engineering*, Faber, London, 1961

Burstall, A. F., *A History of Mechanical Engineering*, Faber, London, 1963

Dickinson, H. W., *A Short History of the Steam-engine*, Cambridge University Press, London, 1939

Kilon, I., *The Evolution of the Heat Engine*, Longman, London, 1972

Parsons, R. H., *The Development of the Parsons Steam Turbine*, Constable, London, 1936

Rolf, L. T. C., *A Short History of Machine Tools*, MIT Press, Cambridge, Mass., 1965

Schlaifer, R. and Herm, S. D., *Development of Aircraft Engines and Fuels*, Harvard University Press, Boston, 1950

Steeds, W. A., *A History of Machine Tools 1700–1910*, Clarendon Press, Oxford, 1969

Woodbury, Robert S., *History of the Grinding Machine*, MIT Press, Cambridge, Mass., 1959

Woodbury, Robert S., *History of the Lathe to 1950*, MIT Press, Cambridge, Mass., 1960

Woodbury, Robert S., *History of the Milling Machine*, MIT Press, Cambridge, Mass., 1960

Woodbury, Robert S., *Studies in the History of Machine Tools*, MIT Press, Cambridge, Mass., 1972

Transport

Abell, Westcott, *The Shipwright's Trade*, Cambridge University Press, London, 1948

Allen, G. Freeman, *Railways: Past, Present and Future*, Orbis, London, 1982

Beaubois, H., *Airships: an Illustrated History*, Macdonald and Janes, London, 1973

Berghaus, Erwin, *The History of Railways*, Barrie and Rockcliff, London, 1964

Brooks, P. W., *The Modern Airliner: its Origins and Development*, Putnam, London, 1961

Chant, Christopher, *Aviation: an Illustrated History*, Orbis, London, 1983

Davies, R. E. G., *History of the World's Airlines*, Oxford University Press, London, 1964

Emme, Eugene M., *The History of Rocket Technology*, Wayne State University Press, Detroit, 1964

Georgano, G. N. (ed.), *A History of Transport*, Dent, London, 1972

Georgano, G. N. (ed.), *The Complete Encyclopaedia of Motorcars 1885 to date*, Ebury Press, London, 1973

Gibbs-Smith, C. H., *A History of Flying* Batsford, London, 1953

Gibbs-Smith, C. H. and Brooks, P. W., *Right Through the Ages*, Hart-Davis MacGibbon, London, 1974

Gibbs-Smith, C. H., *The Aeroplane: an Historical Survey of its Origins and Development*, HMSO, London, 1960

Gregory, J. W., *The Story of the Road, from the Beginning down to the Present Day*, A&C Black, London, 1938

Hadfield, Charles, *British Canals: an Illustrated History*, Phoenix House, London, 1950

Hindley, George, *A History of Roads*, Davies, London, 1971

Hornell, James, *Water Transport: Origins and Early Evolution*, Cambridge University Press, London, 1946

Johnstone, Paul, *The Archaeology of Ships*, Bodley Head, London, 1974

Kemp, Peter, *This History of Ships*, Orbis, London, 1978

Montagu of Beaulieu, Lord and Bird, Anthony, *Steam Cars 1770–1970*, Cassell, London, 1971

Nayler, J. L. and Ower, E., *Aviation: its Technical Development*, Owen, London, 1965

Nock, O. S. (ed.), *Encyclopedia of Railways*, A&C Black, London, 1977

Page, Robert Morris, *The Origin of Radar*, Doubleday, New York, 1962

Ridley, Anthony, *An Illustrated History of Transport*, Heinemann, London, 1969

Rowland, Keith T., *Steam at Sea: History of Steam Navigation*, David and Charles, Newton Abbot, 1970

Simonson, G. R., *The History of the American Aircraft Industry*, MIT Press, Cambridge, Mass., 1968

Smith, E. C., *A Short History of Naval and Marine Engineering*, Cambridge University Press, London, 1938

Tarr, Laszlo, *The History of the Carriage*, Vision, London, 1969

Taylor, E. G. R., *The Haven-Finding Art*, Hollis & Carter, London, 1956

Tooley, R. V., *Maps and Map-makers* (4th edn), Bonanza Books, New York, 1970

Power

Barty-King, Hugh, *New Flame*, Graphmitre, Tavistock, 1984

Hogben, Lancelot, *Men, Missiles, and Machines: the Wonderful World of Power*, Rathbone Books, London, 1957

Skilton, C. P., *British Windmills and Watermills*, Collins, London, 1947

Wailes, R., *The English Windmill*, Routledge and Kegan Paul, London, 1954

Williams, Trevor I., *A History of the British Gas Industry*, Oxford University Press, Oxford, 1981

Williamson, H. F. and Daum, A. R., *The American Petroleum Industry 1859–1899*, Northwestern University Press, Evanston, 1959

Williamson, H. F., Andreano, R. L., Daum, A. R. and Klose, G. C., *The American Petroleum Industry: the Age of Energy 1899–1959*, Northwestern University Press, Evanston, 1970

Communication and Record

Beeching, W. A., *Century of the Typewriter*, Heinemann, London, 1974

Butler, P., *The Origin of Printing in Europe*, Chicago University Press, Chicago, 1940

Fabre, Maurice, *A History of Communications*, Hawthorn, New York, 1964

Fielding, R. (ed.), *A Technological History of Motion Pictures and Television*, University of California Press, Berkeley, 1967

Gernsheim, H. and A., *The History of Photography*, Oxford University Press, London, 1955

Marland, E. A., *Early Electrical Communication*, Abelard-Schuman, London, 1964

Michaelis, A. R., *From Semaphore to Satellite*, ITU, Geneva, 1965

Moholy, Lucia, *A Hundred Years of Photography 1839–1939, London*, 1939

Quigley, M., *Magic Shadows: the Story of the Origin of Motion Pictures*, Georgetown University Press, Washington, 1948

Stenger, Erich, *The History of Photography its Relation to Civilization and Practice*, Easton, 1938

Agriculture: Food

Barker, Graeme, *Prehistoric Farming in Europe*, Cambridge University Press, Cambridge, 1985

Bitting, A. W., *Appertizing or the Art of Canning: its History and Development*, Trade Pressroom, San Francisco, 1937

Curwen, E. Cecil and Hatt, Gudmund, *Plough and Pasture: the Early History of Farming*, Schuman, New York, 1953

Francis, C. A., *A History of Food and its Preservation*, Princeton University Press, Princeton, 1937

Fussell, G. E., *The Farmer's Tools 1500–1900*, Melrose, London, 1952

Fussell, G. E., *Farming Technique from Prehistoric to Modern Times*, Pergamon Press, New York, 1966

Gras, N. S. B., *A History of Agriculture in Europe and America* (2nd edn), Crofts, New York, 1940

Hyams, Edward, *Soil and Civilisation*, Thames and Hudson, London, 1952

Hyams, Edward, *Plants in the Service of Man: 10,000 Years of Domestication*, Lippincott, Philadelphia, 1972

Jenkins, James Travis, *A History of the Whale Fisheries*, Willerby, London, 1921

Kerridge, Eric, *The Agricultural Revolution*, Allen and Unwin, London, 1967

McLachlan, T., *History of Food Processing*, Pergamon Press, Oxford, 1975

Mathias, P., *The Brewing Industry in England, 1700–1830*, Cambridge University Press, London, 1959

Stuyvenberg, J. H. van, *Margarine: an Economic, Social and Scientific History*, Liverpool University Press, Liverpool, 1969

Tannahill, Reay, *Food in History*, Eyre Methuen, London, 1973

Rasmussen, W. D. (ed.), *Readings in the History of American Agriculture*, University of Illinois Press, Urbana, 1960

Schlebecker, J. T., *Whereby We Thrive: a History of American Farming 1607–1972*, Iowa State University Press, Ames, 1975

Slicher van Bath, B. H., *The Agrarian History of Western Europe*, AD *500–1850*, Arnold, London, 1963

Storck, John and Teague, Walter Dorwin, *Flour for Man's Bread: a History of Milling*, University of Minnesota Press, Minneapolis, 1952

Trow-Smith, Robert, *English Husbandry, from the Earliest Times to the Present Day*, Faber, London, 1951

Zeuner, Frederick, *A History of Domesticated Animals*, Hutchinson, London, 1963

Chemicals

Furter, William F. (ed.), *History of Chemical Engineering*, American Chemical Society, Washington, 1980

Haber, L. F., *The Chemical Industry during the Nineteenth Century*, Oxford University Press, London, 1958

Haber, L. F., *The Chemical Industry 1900–1930*, Oxford University Press, London, 1971

Haynes W. Haynes, *The American Chemical Industry* (6 vols.), Van Nostrand, New York, 1945–54

Kaufman, M., *The First Century of Plastics – Celluloid and its Sequel*, Plastics Institute, London, 1962

Morgan, G. T. and Pratt, D. D., *British Chemical Industry: its Rise and Development*, Arnold, London, 1938

Multhauf, R. P., *The Origins of Chemistry*, Oldbourne, London, 1966

Singer, Charles, *The Earliest Chemical Industry*, Folio Society, London, 1948

Taylor F. S., *A History of Industrial Chemistry*, Heinemann, London, 1957

Warrington, C. J. S. and Nicholls, R. V., *A History of Chemistry in Canada*, Pitman (Canada), Toronto, 1949

Williams, Trevor I., *The Chemical Industry: Past and Present*, Penguin, Harmondsworth, 1953

Building: Civil Engineering

Beaver, Patrick, *A History of Lighthouses*, Davies, London, 1971

Briggs, M. S., *The Architect in History*, Clarendon Press, Oxford, 1927

Copplestone, Trewin (ed.), *World Architecture: An Illustrated History*, Hamlyn, London, 1963

Chinese Academy of Sciences, *History and Development of Ancient Chinese Architecture*, Science Press, Peking, 1986

Goodman, W. C., *A History of Woodworking Tools*, Bell, London, 1964

Jope, E. M. (ed.) *Studies in Building History*, Odhams Press, London, 1961

Knoop D. and Jones G. P., *The Medieval Mason*, Manchester University Press, Manchester, 1933

McKelvey, Blake, *The Urbanisation of America 1860–1951*, Rutgers University Press, New Brunswick, 1963

Pannell, J. P. M., *An Illustrated History of Civil Engineering*, Thames and Hudson, London, 1964

Robins, F. C. W., *The Story of Water Supply*, Oxford University Press, London, 1946

Sandström, Gosta E., *The History of Tunnelling: Underground Workings through the Ages*, Barrie and Rockcliffe, London, 1963

Shirley-Smith, H., *The World's Great Bridges* (2nd edn), Phoenix House, London, 1964

Straub, H., *A History of Civil Engineering*, Hill, London, 1952

Smith, N. A. F., *A History of Dams*, Peter Davies, London, 1971

Mining: Metallurgy

Aitchison, L., *A History of Metals* (2 vols.), Macdonald and Evans, London, 1960

Ashton, T. S., *Iron and Steel in the Industrial Revolution* (2nd edn), Manchester University Press, Manchester, 1951

Barraclough, K. C., *Steelmaking Before Bessemer* (2 vols.), The Metals Society, London, 1984

Carr, J. C. and Taplin, W., *History of the British Steel Industry*, Blackwell, Oxford, 1962

Davies, M. E., *The Story of Steel*, Burke Publishing, London, 1950

Dennis, W. H., *A Hundred Years of Metallurgy*, Duckworth, London, 1963

Schubert, H. R., *History of the British Iron and Steel Industry*, c. *450* BC–AD *1775*, Routledge and Kegan Paul, London, 1957

Warren, K., *The American Steel Industry 1850–1970: a Geographical Interpretation*, Clarendon Press, Oxford, 1973

Warren, K., *The British Iron and Steel Industry since 1840*, Bell, 1970

Instruments

Bruton, Eric, *The History of Clocks and Watches*, Orbis, London, 1979

Daumas, Maurice, *Scientific Instruments of the 17th and 18th Centuries*, Batsford, London, 1972

Gordon, G. F. C., *Clockmaking Past and Present*, Technical Press, London, 1949

Hartree, D. R., *Calculating Instruments and Machines*, Cambridge University Press, London, 1950

Lloyd, H. Alan, *Some Outstanding Clocks 1250–1950*, Leonard Hill, London, 1958

Ceramics and Glass

Angus-Butterworth, L. M., *The Manufacture of Glass*, Pitman, London, 1948

Charleston, R. J., *Roman Pottery*, Faber, London, 1955

Dickson, J. H., *Glass*, Hutchinson, London, 1951

Douglas, R. W. and Frank, S. A., *A History of Glass Making*, Foulis, Henley-on-Thames, 1972

Textiles and Allied Products

Bobart, Henry Hodgkinson, *Basketwork through the Ages*, Oxford University Press, London, 1936

Clapperton, R. H., *The Paper Machine: its Invention, Evolution and Development*, Pergamon Press, London, 1967

Hunter, David, *Papermaking: the History and Technique of an Ancient Craft* (2nd edn), Knopf, New York, 1957

Leggett, William F., *The Story of Wool*, Chemical Publishing Co., New York, 1947

Lipson, E., *The History of the Woollen and Worsted Industries*, Black, London, 1950

Reed, R., *Ancient Skins, Parchments, and Leathers*, Seminar Press, London and New York, 1973

Wadsworth, A. P. and Mann, Julia de L., *The Cotton Trade and Industrial Lancashire, 1600–1780*, Manchester University Press, Manchester, 1931

Waterer, J. W., *Leather and Craftsmanship*, Faber, London, 1950

Military Technology

Carman, W., *A History of Firearms from the Earliest Times to 1941*, Routledge and Kegan Paul, London, 1955

Haber, L. F., *The Poisonous Cloud: Chemical Warfare in the First World War*, Clarendon Press, Oxford, 1986

Humble, R. (ed.), *Naval Warfare: an Illustrated History*, Orbis, London, 1983

Hogg, Ian, *The History of Fortification*, Orbis, London, 1981

Toy, S., *A History of Fortification from 3000* BC *to* AD *1700*, Heinemann, London, 1955

Electrical Industry

Braun, Ernest and MacDonald, Stuart, *Revolution in Miniature: the History and Impact of Semiconductor Electronics*, Cambridge University Press, London, 1978

Dummer, G. W. A., *Electronic Inventions and Discoveries 1745–1976* (2nd edn), Pergamon Press, Oxford, 1978

Dunsheath, Percy, *A History of Electrical Engineering*, Faber, London, 1962

Gowing, Margaret, *Britain and Atomic Energy 1939–1945*, Macmillan, London, 1964

Gowing, Margaret, *Britain and Atomic Energy 1945–1952* (2 vols.), Macmillan, London, 1974

Hennessy, R. A. S., *The Electric Revolution*, Oriel Press, Newcastle-upon-Tyne, 1972

MacLaren, M., *The Rise of the Electrical Industry during the Nineteenth Century*, Princeton University Press, Princeton, 1943

Parsons, R. H., *The Early Days of the Power-Station Industry*, Cambridge University Press, London, 1940

Index

Figures in **bold** refer to the Biographical Dictionary; and figures in *italic* refer to the illustrations

C

D